Sugar Cane
(Saccharum Officinarum)

SUGARS
IN NUTRITION

THE NUTRITION FOUNDATION
A Monograph Series

SUGARS
IN NUTRITION

Edited by

HORACE L. SIPPLE
The Nutrition Foundation
New York, New York

KRISTEN W. McNUTT
The Nutrition Foundation
Washington, D.C.

1974

ACADEMIC PRESS   New York   San Francisco   London

A Subsidiary of Harcourt Brace Jovanovich, Publishers

ACADEMIC PRESS, INC.
111 Fifth Avenue, New York, New York 10003

United Kingdom Edition published by
ACADEMIC PRESS, INC. (LONDON) LTD.
24/28 Oval Road, London NW1

Library of Congress Cataloging in Publication Data

Main entry under title:

Sugars in nutrition.

Based on papers presented at an international con-
ference on sugars in nutrition held Nov. 8-11, 1972 at
Vanderbilt University, School of Medicine, Nashville,
Tenn.
Includes bibliographies.
1. Sugars in human nutrition. I. Sipple, Horace
Lawson, Date ed. II. McNutt, Kristen W., ed.
[DNLM: 1. Carbohydrates—Congresses. 2. Dietary
carbohydrates—Congresses. 3. Nutrition—Congresses.
QU75I59s 1972]
TX553.S8S93 641.1'3 73-18974
ISBN 0−12−646750−1

PRINTED IN THE UNITED STATES OF AMERICA

CONTENTS

SUGARS IN FOOD: OCCURRENCE AND USAGE

SUGARS IN FOOD: RECENT TECHNOLOGICAL DEVELOPMENTS

DIGESTION AND ABSORPTION OF SUGARS

METABOLISM OF SUGARS

DISORDERS RELATED TO SUGAR METABOLISM: METABOLIC ABNORMALITIES

DISORDERS RELATED TO SUGAR METABOLISM: GALACTOSE-CONTAINING SUGARS AND THE EYE

DISORDERS RELATED TO SUGAR METABOLISM: OBESITY, CARDIOVASCULAR DISEASE, AND HYPERTRIGLYCERIDEMIA

DISORDERS RELATED TO SUGAR METABOLISM: DIABETES

THERAPEUTIC USE OF SUGARS

SUGARS IN THE ORAL CAVITY

Chapter 38   *Carbohydrate Metabolism in Caries-Conducive,*
            Oral Streptococci
            ALBERT T. BROWN

LIST OF CONTRIBUTORS

Numbers in parentheses indicate the pages on which the authors' contributions begin.

CARL AMINOFF (135), Research and Development, Finnish Sugar Company Ltd., Helsinki, Finland

W. R. AYKROYD (3), Queen Anne House, Charlsbury, Oxford, England

S. BRILLER, Diabetic Unit, Hebrew University Hospital, Jerusalem, Israel

M. BRIN (591), Department of Biochemical Nutrition, Hoffman-La Roche, Inc., Nutley, New Jersey

ALBERT T. BROWN (689), Department of Oral Biology, The University of Connecticut Health Center, Farmington, Connecticut

ROBERT H. CAGAN (19), Monell Chemical Senses Center and Department of Biochemistry, University of Pennsylvania, and Veterans Administration Hospital, Philadelphia, Pennsylvania

SIDNEY M. CANTOR (111), Sidney M. Cantor Associates, Inc., Haverford, Pennsylvania

SAM CHAN (613), Vanderbilt University School of Medicine, Nashville, Tennessee

A. M. COHEN (483), Diabetic Unit, Hebrew University Hospital, Jerusalem, Israel

MARVIN CORNBLATH (451), Department of Pediatrics, University of Maryland, School of Medicine, Baltimore, Maryland

E. CRISTOFARO (313), Nestle Products Technical Assistance Co. Ltd., La Tour de Peilz, Switzerland

OSCAR B. CROFFORD (513), Departments of Medicine and Physiology, Vanderbilt University, School of Medicine, Nashville, Tennessee

ARNE DAHLQVIST (187), Department of Nutrition, University of Lund, Lund, Sweden

WILLIAM J. DARBY (11), The Nutrition Foundation, Inc., New York, New York

J. B. EDWARDS (567), Division of Clinical Chemistry, Institute of Medical and Veterinary Science, Adelaide, South Australia

R. G. EDWARDS (567), Division of Clinical Chemistry, Institute of Medical and Veterinary Science, Adelaide, South Australia

H. FÖRSTER (259), Institut für Vegetative Physiologie, Johann-Wolfgang Goethe Universität, Frankfurt, Germany

BERTA FRIEND (93), Consumer and Food Economics Institute, Agricultural Research Service, United States Department of Agriculture, Hyattsville, Maryland

E. R. FROESCH (241), Department für Innere Medizin der Universität, Medizinische Klinik, Stoffwechise Labteilung Kantonsspital, Zurich, Switzerland

KENNETH H. GABBY (469), Children's Hospital, Medical Center and the Department of Pediatrics, Harvard Medical School, Boston, Massachusetts

H. EARL GINN (607), Vanderbilt University Medical Center, Nephrology Division, Thayer Veterans Administration Hospital, Nashville, Tennessee

FRANCISCO GRANDE (401), Laboratory of Physiological Hygiene, University of Minnesota, Minneapolis, Minnesota

PAUL GYÖRGY (215), Department of Pediatrics, University of Pennsylvania, Philadelphia, Pennsylvania

R. G. HANSEN (281), Department of Chemistry and Biochemistry, Utah State University, Logan, Utah

ROBERT H. HERMAN (145), Metabolic Division, U.S. Army Medical Research and Nutrition Laboratory, Fitzsimons Army Medical Center, Denver, Colorado

H. G. HERS (337), Laboratorie de Chimie Physiologique, Universite Catholique de Louvain, Louvain, Belgium

L. HUE (357), Laboratoire de Chimie Physiologique, Universite Catholique de Louvain, Louvain, Belgium

A. JAKOB (241), Metabolic Unit, Department of Medicine, University of Zurich, Switzerland

SYNTHIA M. KING (613), Department of Biochemistry, Vanderbilt University School of Medicine, Nashville, Tennessee

JIN H. KINOSHITA (375), Laboratory of Vision Research, National Eye Institute, National Institutes of Health, U.S. Department of Health, Education and Welfare, Bethesda, Maryland

IAN MACDONALD (303), Department of Physiology, Guy's Hospital Medical School, London, England

KAUKO K. MÄKINEN (645), University of Turku, Institute of Dentistry, Turku, Finland

H. C. MENG (527), Department of Physiology, Vanderbilt University School of Medicine, Nashville, Tennessee

O. N. MILLER (591), Department of Biochemical Nutrition, Hoffmann-La Roche, Inc., Nutley, New Jersey

E. L. MITCHELL (127), California Canners and Growers, San Jose, California

HOWARD R. MOSKOWITZ (37), Pioneering Research Laboratory, U.S. Army Natick Laboratories, Natick, Massachusetts

F. MOTTU (313), Nestle Products Technical Assistance Co. Ltd., La Tour de Peilz, Switzerland

ESKO A. NIKKILÄ (439), Meilahti Hospital & Department of Medicine, University of Helsinki, Helsinki, Finland

LOUISE PAGE (93), Consumer and Food Economics Institute, Agricultural Research Service, U.S. Department of Agriculture, Hyattsville, Maryland

E. ROSENMANN, Diabetic Unit, Hebrew University Hospital, Jerusalem, Israel

NORTON S. ROSENSWEIG (173), Department of Medicine, College of Physicians and Surgeons of Columbia University, St. Luke's Hospital Center, New York, New York

A. L. RUSSELL (635), University of Michigan Schools of Dentistry and Public Health, Ann Arbor, Michigan

GOTTHARD SCHETTLER (389), Medizinische Universitätklinik, Heidelberg, West Germany

GUENTER SCHLIERF (389), Medizinische Universitätklinik, Heidelberg, West Germany

E. SHAFRIR, Diabetic Unit, Hebrew University Hospital, Jerusalem, Israel

R. S. SHALLENBERGER (67), Department of Food Science and Technology, New York State Agricultural Experiment Station, Geneva, New York

VORAVARN S. TANPHAICHITR (613), Departments of Biochemistry and Pediatrics, Vanderbilt University School of Medicine, Nashville, Tennessee

A. TEITELBAUM (483), The Diabetic Unit and the Department of Pathology, Ophthalmology, Clinical Biochemistry, Hadassah University Hospital, and Hebrew University Hadassah Medical School, Jerusalem, Israel

D. W. THOMAS (567), Division of Clinical Chemistry, Institute of Medical and Veterinary Science, Adelaide, South Australia

OSCAR TOUSTER (229), Departments of Molecular Biology and Biochemistry, Vanderbilt University, Nashville, Tennessee

J. VAN EYS (613), Departments of Biochemistry and Pediatrics, Vanderbilt University School of Medicine, Nashville, Tennessee

Y. M. WANG (613), Department of Biochemistry, Vanderbilt University School of Medicine, Nashville, Tennessee

ARVID WRETLIND (81), Nutrition Unit, Karolinska Institutet, Stockholm, Sweden

J. J. WUHRMANN (313), Nestle Products Technical Assistance Co. Ltd., La Tour de Peilz, Switzerland

L. YANKO, Diabetic Unit, Hebrew University Hospital, Jerusalem, Israel

PREFACE

The Nutrition Foundation Monograph Series was originated to provide systematic coverage of new nutritional knowledge in important areas, to define research needs, and to stimulate more effective application of scientific knowledge in the field of human nutrition. The subject for this initial volume of the series was selected because of the fundamental importance of sugars in all aspects of nutrition. Revolutionary technologic changes are altering the sources and forms of sugars used in foods. Remarkable advances are being made in knowledge concerning the metabolic and nutritional role of the many sugars or derivatives that are now components of the human diet.

The merit of looking broadly at this subject becomes evident as one reads the contributions which cover a variety of disciplines including nutrition, food technology, biochemistry, medicine, pediatrics, physiology, psychology, dentistry, economics, nephrology, clinical chemistry, and pharmacology. The most recent developments in major segments of the field of sugars in the nutrition of man from some thirty-five academic, industrial, and government institutions in Europe, Australia, and the United States are presented. The comprehensive nature of the material is evident from an examination of the contents. Each section cites examples of the effective use of new scientific knowledge in dealing with current human nutrition problems and defines the areas in which more research would be beneficial.

A section dealing with occurrence, usage and technology of sugars in food delineates usage trends. Major advances in sugar technology are making available a greater amount and variety of sugars, leading toward further change in the use of sugars in food. The section devoted to digestion and absorption includes chapters on adaptive effects of dietary sugars on disaccharidases, enzyme deficiencies and malabsorption, and the effect of carbohydrates on intestinal flora. Chapters on the metabolism of polyols; xylitol, sorbitol, fructose, and glucose; lactose and galactose; raffinose sugars; and maltose and higher saccharides bring together a comprehensive compilation of the most recent knowledge regarding metabolism of various sugars. The section on disorders related to sugar metabolism contains chapters on inborn errors of carbohydrate metabolism, the metabolism and toxic effects of fructose, cataractogenic effects of lactose and galactose, obesity, cardiovascular disease, diabetes, and metabolic effects of dietary fructose and sucrose in hyperlipidemia.

New and important observations are reported in the section devoted to the therapeutic use of sugars. The chapters deal with use of glucose, fructose, xylitol, and sorbitol in parenteral nutrition, the toxicity of parenteral xylitol, the safety of oral xylitol, uses of mannitol, and the efficacy of xylitol as a therapeutic agent in glucose-6-phosphate dehydrogenase deficiency. The section on sugars in the oral cavity combines data from both studies in the field of epidemiology and metabolism of oral microflora and provides a broader approach to the question of the etiology of dental caries.

The sweet taste is an important sensation but little is known about its underlying biochemical mechanism. Review of the information available in this field, presented in the section entitled Biochemistry and Psychology of Sweetness, emphasizes the need for extensive studies to elucidate the mechanisms by which chemical information is processed by living organisms.

No recent publication has brought together information of the type presented in this book. It is hoped that it will prove useful to food and nutrition scientists, educators, economists, and all others interested in understanding and improving the nutrition of man.

Contributions to this book are based on papers presented at an International Conference on Sugars in Nutrition held November 8–11, 1972 at Vanderbilt University School of Medicine, Nashville, Tennessee. In some instances authors have included additional material not presented at the conference.

HORACE L. SIPPLE
KRISTEN W. McNUTT

ACKNOWLEDGMENTS

The selection of topics and of authors of chapters in this monograph was made by a Planning Committee from the faculty of Vanderbilt University School of Medicine in Nashville, Tennessee. That Committee, comprised of John G. Coniglio (Chairman), Oscar Touster, Jan van Eys, and H. C. Meng, initially hosted the International Symposium on Sugars in Nutrition on which this publication is based. The Committee received generous assistance in the identification of topics of interest from a wide variety of colleagues at Vanderbilt University and elsewhere. To all of them the Committee is grateful.

The Nutrition Foundation, Inc., is greatly indebted to the Planning Committee for the selection of the well-considered topics, for the distinguished scientists who authored the papers, and for assembling the manuscripts. Especial thanks are due Dr. John G. Coniglio, Chairman, who carried the major burden of the correspondence relative to the development of these manuscripts.

WILLIAM J. DARBY, *President*
The Nutrition Foundation, Inc.

SUGARS
IN NUTRITION

State of the Art,
Past and Present

CHAPTER 1

Sugar in History

W. R. AYKROYD

A. The Role of Foods in History

Some foods have played an important part in human history, apart from providing calories and nutrients. The seeds of wheat, it has been said, were the seeds of civilization. The domestication and cultivation of wild wheats in western Asia, as long ago as the eighth millenium B.C., enabled man to break away from his dependence on hunting and food gathering and to found settled communities with an assured supply of food. These in turn became the bases of cities and early civilizations. Barley, a sort of poor relation of wheat, played a useful but secondary role. Rice, a crop which thrives in the humid tropics, has been the staple food of other early civilizations in south and east Asia. The original inhabitants of the Americas domesticated only one cereal, namely, maize or corn, but they also discovered and used a whole range of other foods. These include a number of legumes or pulses which are now grown all around the world; legumes, as we know, are valuable supplements to cereals because of their high protein content. The list of familiar modern foods which originated in the New World is surprisingly long.

Among them is the potato, *Solanum tuberosum*, which originally grew wild on the slopes of the Andes. Carried to Europe at the end of the seventeenth century, it proved to be especially suited to Ireland and became the staple food of the Irish peasant. Adam Smith, author of *The Wealth of Nations*, writing in 1776, remarked that the Irish nation pro-

duced "the strongest men and most beautiful women in the British do-
minions," and ascribĕd this to the "nourishing quality" of the potato.
In the autumn of 1845, potatoes in Ireland were attacked by the blight
Phytophera infestans which destroys first the leaves, then the stalk, and
finally the tubers themselves; thus, potato beds become a mass of decay
and the fields reek of corruption. The blight persisted for 5 years and
during that period almost two million people, out of a population of eight
million, died of famine and famine diseases. The famine left scars which
even yet have not entirely disappeared. The potato has certainly molded
the history of Ireland.

B. Sugar Cane

The sugar cane, like the cereals, belongs to the grass family—the
Gramineae. It is, in fact, like the bamboo, a gigantic grass. But with
the cane it is the stems and not the seeds which are eaten. Sugar, extracted in bulk from the cane and during the last two centuries also from
the sugar beet, is a relatively recent human food. For millennia after
the birth of agriculture, cereals and legumes were the main source of
calories for man. Occasional sweetness was provided by honey, too rare
and precious a food to be used except for ceremonial occasions, although
it was habitually used by the nobility. The craft of beekeeping or apicul-
ture is an ancient one. Early civilizations in western Asia had elaborate
codes of law governing hives and the sale of honey. The mountain of
Hymettus near Athens was famous for its honey in Ancient Greece.

Noël Deere's great work, *The History of Sugar* (1949, 1950) published
in two volumes, is the main source of information on sugar production
and consumption down the centuries until recent times. Deere, an
Anglican vicar's son born in 1874, spent his life working for the sugar
industry in both hemispheres. He was the author of various technical
books and papers. On retirement in 1937, he decided to write the history
of sugar, though without training or experience as a historian. This took
him nearly 10 years. His contemporaries justifiably described him as a
remarkable blend of a scholar and a man of affairs.

The cane was first cultivated in India, on alluvial soil in the Ganges
valley where the State of Bihar is located today. It is probably a native
of India, but no wild ancestor is known with certainty, and it may have
originated elsewhere in east and south Asia. Nearchus, an officer in Alex-
ander's invading army, wrote of sugar in the Punjab in the third century
B.C. The cane spread eastward and westward from India, reaching China
about the first century B.C. But it did not reach the Mediterranean until

much later. The Ancient Greeks and Romans went without sugar. There is no mention of sugar in the Bible; the Promised Land flowed with milk and honey, not with milk and sugar. When Pliny the Elder, the Roman scientist who was killed when observing the eruption of Vesuvius in 55 B.C., wrote of sugar, which he had heard of as a curiosity, he called it "a kind of honey made from reeds."

In the westerly direction the cane moved slowly from India along the shores of the Persian Gulf and in due course came to be cultivated in the delta south of Bagdad and later on the coast of the Caspian Sea. About the sixth century A.D. it reached the Mediterranean. One of the results of the Arab conquests and expansion inspired by Mohammed was the introduction of the cane into the conquered Mediterranean territories—Syria, Cyprus, Crete, Egypt, Morocco, Sicily, and Spain. The Nile valley is well suited for sugar cultivation. Egyptian sugar was considered the best in the world—"white as snow and hard as stone."

The introduction of the cane into lands which are now Israel, Lebanon, and Syria is of particular interest since it was there that western Europeans, represented by the Crusaders, encountered sugar for the first time. Like Pliny, they called it "honey from reeds." No doubt the Crusaders brought home small samples of sugar to be tasted by their wives in the cold castles of their native lands. Knowledge of sugar in central and western European countries dates from this time. A small trade between these countries and lands bordering the Mediterranean came into existence, with Venice as the principal trading center. Venice also engaged in sugar refining. From the eleventh until the fifteenth century sugar was a well-known but scarce and highly expensive commodity in Europe, one of its principal uses being to disguise the taste of the dreadful medicines of the period. As a delicacy it was eaten only in the great houses of the king and the nobility. In 1226, King Henry III of England personally asked the Mayor of Winchester to procure him three pounds of sugar at the Winchester Fair. Special jars of "rose" and "violet" sugars, flavored with aromatic substances and fantastically expensive, were made for sickly royal children. The delicate son of King Edward I, Henry, who suffered from perpetual colds, was given rose and violet sugars and sugar-sticks like modern barley sugar—unfortunately to no avail, since he died early in life.

So far the story of sugar has been tranquil enough—the story of the slow spread of an Indian plant, yielding a desirable sweet substance, eastward to China and westward to Europe. In the nineteenth and twentieth centuries the story hinges on production and consumption statistics as sugar becomes a cheap and common food for ordinary people. But during the intervening three centuries sugar was the prime mover in one of the grimmest chapters in human history, the transatlantic slave trade and the establishment of sugar plantations in the New World worked by slave labor—first in South America, then in the Caribbean, and later in North America. The cane reached the New World via Spain and Portugal and their colonies in the Atlantic and West Africa. It has been reckoned that during the centuries when sugar-slavery flourished, some ten million Africans were carried across the Atlantic, mainly to man the sugar plantations. Nearly all the sea-faring nations of Europe, as well as North America, were involved. Sugar did not create Negro slavery—at first Europeans were more interested in silver and gold and needed labor for mines—but without sugar the whole dark episode in human history, causing untold suffering and leaving consequences which curse the world today, would not have taken place (Aykroyd, 1967).

The busy and highly lucrative sugar industry based on slaves and the sailing ship made sugar a more abundant and cheaper commodity in Europe and North America, and created some enormous fortunes. Most of the sugar reached Europe in crude form and was refined in various centers specializing in the operation. Per caput consumption rose appreciably, the demand being stimulated by the growing popularity of two other tropical products, coffee and tea. But sugar remained a relatively expensive luxury, largely beyond the means of the poor. Europe's craving for sweetness was still far from satisfied.

C. Beet Sugar

Two new developments made sugar a cheap article of food, habitually eaten by all sections of the population in Europe and North America. One was the spread of the cane throughout much of the tropics, leading to sugar production on a large scale in countries not dependent on slave labor. The other was the discovery of the sugar beet as a source of sucrose. Beet sugar is one of the few important human foods whose production and processing have been based almost entirely on science and technology, beginning with its extraction and crystallization by Marggraff in 1740. Its earlier development owed much to the drive of Napoleon when the British Royal Navy sought to cut off supplies of cane sugar to France during the Napoleonic wars. Hearing of a particularly successful beet sugar factory, Napoleon rushed off to visit it, accompanied by his magnificent bodyguard. He was presented with two loaves of beet sugar and in return took from his own breast the medal of the Legion of Honour and decorated the owner of the factory with it. He then informed the English, through a speech to the Chamber of Commerce in

Paris, that as far as he was concerned they could throw their sugar from the West Indies into the River Thames. France had no need of it.

But after Waterloo the Napoleonic impulse faded and other countries in Europe took the lead in beet sugar production. In the second half of the nineteenth century the quantity of beet sugar produced in the world began to rival that of cane sugar. Current proportions (1971) are 60 cane sugar to 40 beet sugar. Beet sugar is a product of the temperate zone; the countries which make it in greatest amounts are the U.S.S.R. and the United States. There is, of course, no difference in the refined white substance, for practical purposes sucrose, which emerges from the processing of the cane and the beet.

D. World Production of Sugar

The world possesses resources for producing sugar on a large scale. World production of "centrifugal," i.e., factory-processed sugar, rose from 8 million tons in 1900 to 70 million tons in 1970 (see Fig. 1), involving a huge expansion both in cultivation and factory capacity, and the Food and Agriculture Organization of the United Nations (FAO) has predicted that about 93 million tons will be reached in 1980. No other human food has shown an increase in production of this order during the same period.

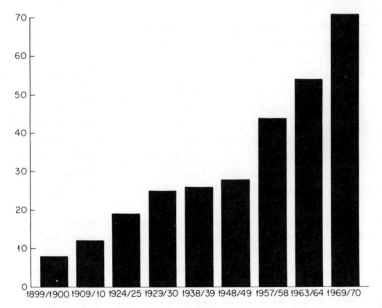

Fig. 1 World sugar production 1899–1900 to 1969–1970 (million metric tons).

In a series of commodity studies, FAO has presented the curious picture of world sugar consumption, which reflects human craving for sweetness. The countries with the largest consumption, namely, Australia, the United Kingdom and Switzerland, consume between 50 and 60 kg of sugar per head annually, which supplies about 18% of total calories—a very appreciable percentage. This seems to be the upper limit of consumption; no country can, or wants to, swallow more sugar than that. In the United Kingdom, consumption has remained almost stationary for 10 years. The United States comes a little behind, with just under 50 kg. In some affluent countries with special dietary patterns—in France, for example—the summit is a good deal lower, between 30 and 40 kg annually. There are plenty of countries well below this level, but mounting the ladder all the time. In most of these, urbanization is quickly extending, and sugar is a convenient food highly favored by town-dwellers. It is a cheap source of calories and fits into urban dietary practices. During recent decades, according to FAO, the greatest consumption growth has been in Latin America, the Near East, and Africa. At the outbreak of the Second World War the ten high-consuming countries of northwestern Europe, the United States, Australia, and New Zealand accounted for 52% of the world utilization of centrifugal sugar; in 1958 their share was only 39%. During the same period consumption also rose rapidly in southern and eastern Europe and in many parts of Asia. The rates of increase in some countries well down the consumption scale have been remarkable—from 100 to 200% over 10 years and even, in a few instances, up to 2000%. The number of countries with rapidly growing consumption has been increasing all the time (FAO, 1961).

This was the picture in 1958; more recent studies have shown that since that date similar trends have prevailed in accentuated form. Consumption continues to "falter" in a few affluent countries and to rise steeply in developing countries. Trends are so clear-cut and universal that FAO can confidently predict a world output of over 93 million tons in 1980. The rising curve of sugar production from 1900 to 1980 is much steeper than that of world population—8 to 93 million tons as compared with 26 to 35 hundred million people, according to a "medium" United Nations estimate (FAO, 1971; Viton, 1971).

The International Sugar Organisation, located at 28 Haymarket, London, which sponsors international sugar agreements from time to time, is more concerned with abundance than shortage of sugar and seeks to ensure that producing countries find markets for their sugar at reasonable prices. It is remarkable that the rare substance so prized by kings and queens in the Middle Ages should have become an ordinary everyday food for the man in the street.

To the nutritionist, sugar is not a satisfactory food; it compares unfavorably with the cereals in nutritive value. Eaten in certain forms it is bad for our teeth and some people think that it causes other diseases. One of the most striking changes in the human diet during the last century has been the large increase in sugar intake and when some new, or apparently new, disease makes its appearance it is likely to be ascribed to too much sugar. But, in the author's view, convincing evidence of such associations have not been forthcoming, in spite of a large literature on the subject.

References

Aykroyd, W. R. (1967). "Sweet Malefactor: Sugar, Slavery and Human Society." Heinemann, London.

Deere, N. (1949). "The History of Sugar," Vol. 1. Chapman & Hall, London.

Deere, N. (1950). "The History of Sugar," Vol. 2. Chapman & Hall, London.

Food and Agriculture Organization of the United Nations. (1961). "Trends and Forces of World Sugar Consumption," Commodity Bull. Ser. No. 32. FAO, Rome.

Food and Agriculture Organization of the United Nations. (1971). "The State of Food and Agriculture." FAO, Rome.

Viton, A. (1971). "The Prospects and Problems of Sugar," Int. Symp. Comité Européens des fabricants de sucre, Brussels.

CHAPTER 2

Present Knowledge
of Sugars in Nutrition—
An Overview

WILLIAM J. DARBY

To most persons "sugar" is equated with the common household food-stuff, crystalline sucrose. Historically, this naturally occurring sweetener is but one of many sugar substances long sought by man to satisfy his inborn appreciation for sweet taste. How man's quest for sugar has at times altered the course of history is delightfully set forth in Chapter 1, this volume, by Aykroyd.

A. Changes in Western Diet

Gradual alterations in Western man's diet have resulted in many un-recognized changes in the amount and kind of sugars consumed. Commonly quoted statistics on sugar consumption fail to take account of the variety and amount of sugars naturally present in foods of all mankind, of the numerous derivatives such as polyols that widely occur naturally in the vegetable kingdom, and of those sugars derived from digestion of complex carbohydrates or glycoproteins. Such statistics often fail to

11

distinguish between disappearance and consumption, and they do not take into account the nonfood, industrial uses of sugar and unconsumed food wastes.

The amounts and proportions of sugars in dietaries are changing. Sugars such as fructose will appear in the diet in greater amounts because of the increasingly wide adoption of new enzymatic and other production methods. The importance of reliable data showing the intake of individual sugars and their contribution to the nutrient composition of various foods is comparable to that for specific fatty or amino acids. Such information is needed in order to begin objective assessment of the significance of sugar in the diet to health. Indeed, the tabulated approximation of intake provided by Page and Friend (Chapter 7), and the limited data available on occurrence of various sugars in foods presented by Shallenberger (Chapter 5), MacDonald (Chapter 19), and Wuhrmann *et al.* (Chapter 20) exemplify a basic, though yet budding, body of knowledge which future research should greatly expand.

B. New Industrial Technology

Major innovations in the food industry's methods for producing sweeteners have resulted from the application to industrial technology of accumulated biochemical knowledge. These advances are especially well summarized in Chapter 9 by Mitchell and Chapter 10 by Aminoff. Cantor (Chapter 8) indicated how the changing chemistry and technology of sugars make readily available cheaper concentrates of glucose, fructose, lactose, the polyols and other derivatives with useful properties of taste and function. Such changes will require periodic review of world sugar production and usage like that presented in this volume by Wretlind (Chapter 6). The wide acceptability and the economic importance of sugars likewise make it essential to evaluate cost-benefit relationships, especially in meeting the caloric needs of those parts of the world population with the lowest income levels.

C. Nutritionally Important Questions

Vital questions remain about the role of sugars in nutrition: What are the desirable levels of sugars in the diet? What are the benefits of particular levels? What constitutes excessive intake? What are the risks of excessive intake?

Too often unwarranted conclusions which lack balanced considerations are summarily reached and passed on to the public as nutritional or medical facts; for example, when the cataractogenic effect in young rats of excessive feeding of lactose, galactose, and, later, xylose was discovered in the 1930's, a major consumer advisory agency warned against consumption of milk! This ill-advised overextension of experimental observations was made without recognition of the principle that an excessive amount of *any* nutrient—water, essential amino acids, fats, calories from any source, vitamins, macrominerals, trace elements, even oxygen— proves toxic. Thus, with sugars, we need to examine the *levels* that produce desired effects and those levels that are in fact excessive. Interesting and important experimental observations such as those reported by Cohen *et al.* (Chapter 29) on the effect of large intakes of certain carbohydrates on the development of diabetes in susceptible rats must be interpreted with recognition of this principle of the inevitable toxicity of excessive levels.

Much recent attention has been given to the influence of sugars, particularly fructose and sucrose, on the level of serum lipids in cardiovascular diseases and in diabetes. These issues are considered by a number of authorities in this monograph. Grande (Chapter 25) examined the matter of dietary influences on cardiovascular disease with special reference to evidence pertaining to the role of sugars; so also did Nikkila (Chapter 26). Different interpretations of the significance of the evidence for man can be expected to continue. There is emerging, however, an appreciation of the need to distinguish between diet-induced effects in normal individuals and those in the smaller number of persons who are metabolically prone to one or another type of hyperlipemia. The multiplicity of other aspects of lifestyle that influence metabolism is likewise gaining recognition. Failure to appreciate these considerations leads to imposition of unnecessary dietary restrictions on the majority of persons because of metabolic aberrations present in a few.

Grande (Chapter 25) was of the opinion that current evidence indicates that the practical significance of differences in dietary carbohydrate is minimal compared to that of dietary lipids. He noted that certain hyperlipemic patients respond to dietary manipulation, such as increasing the sugar intake, by an elevation of plasma triglycerides. However, a critical review of the relationship between dietary sugar and cardiovascular disease led him to conclude: "The weight of evidence seems to be against any direct association between high sucrose intake and development of coronary heart disease."

A related issue, the influence of dietary fructose and sucrose on serum triglycerides in hypertriglyceridemia and diabetes, was considered by Nikkila (Chapter 26). He concluded that insulin-dependent diabetics can use a moderate amount of dietary fructose without short-term deleterious effects on the metabolic parameters of diabetes. Such examinations are especially pertinent because of the increased potential for change in the consumption of specific dietary sugars resulting from technological developments.

A clearer understanding of the effects of sugars upon oral biology is emerging. The shift of emphasis from the earlier hypothesis of a key role of *Lactobacillus acidophilus* to that of the plaque-forming bacterium, *Streptococcus mutans,* is elucidated in the discussion of the metabolic biochemistry of the microflora by Brown (Chapter 38). Current knowledge of the role of the dental plaque, composed of salivary proteins and glucan, in the etiology of dental caries and gingival disease was summarized by Maekinen (Chapter 37). He also emphasizes the importance of practical preventive measures such as brushing and rigorous regular cleaning of the teeth. Russell (Chapter 36) noted that since *S. mutans* utilizes sources of energy other than sucrose and that human dental plaque can be formed from a variety of sugars including glucose and fructose, sucrose cannot be regarded as the only cariogenic carbohydrate. He stated: "It is clear that the principle agent in the etiology of dental caries is the cariogenic microflora. The role of carbohydrates in the disease process is that of supplying energy and the site of attack to the active organisms. . . . Dental caries cannot be induced in germ-free animals no matter how cariogenic the diet on which they subsist." He emphasized (1) that dental caries is spread via human carriers throughout the world and hence must be considered a communicable disease, and (2) that caries occurs in peoples who have never used processed foodstuffs or sugar and hence that caries should not be considered as resulting from a single specific etiologic agent dependent upon one single type of substrate.

Thus viewed, the potential effectiveness of antibiotic or other cariostatic agents, or other biologic methods including the dietary use of certain polyols, clearly emerge as fruitful areas where intensified search may lead to improved oral health.

D. Implications of Modern Biochemical Knowledge

Significant implications, including potential benefits to human health, can be seen in the growth of knowledge about the biochemistry of polyols. Elucidation of the sorbitol metabolic pathway has recently opened an entirely new approach to understanding some of the most distressing complications of insulin-controlled diabetes. The excessive formation of polyols and their limited cellular metabolism or transfer under certain conditions appear to be responsible for the cataractogenic effects of excesses of certain sugars. Because of this common biochemical pathway involving dehydrogenase enzymes in metabolic processes related to both sugar-induced cataract formation and the neurologic complications in diabetes, Gabbay (Chapter 28) suggested that the use of aldose reductase inhibitors ultimately may yield new means for prevention and treatment of these aspects of diabetes.

Another illustration of the unanticipated benefits of basic nutritional biochemical research is the potential for use of xylitol as a therapeutic agent in glucose-6-phosphate dehydrogenase (G6PD) deficiency, discussed by van Eys *et al.* (Chapter 35). This enzyme defect occurs in favism and other extreme drug-induced hemolyses, neonatal jaundice, and congenital nonspherocytic hemolytic anemia. Despite the enzymatic defect in glucose metabolism, xylitol is utilizable by the erythrocyte because of the presence in the cell of NADP-linked xylitol dehydrogenase.

Experiments on G6PD-deficient cells indicate that xylitol dehydrogenase activity is normal and the cells are better·protected by this polyol than by glucose. When xylitol administration is increased gradually, it protects drug-treated rabbits against excessive hemolysis. These observations provide considerable hope for developing a rational therapy for several diseases common to man.

Polyols are among the sugars which can be used intravenously. Striking applications of total parenteral nutrition during the past decade extend from the treatment of the premature infant to the long-term maintenance of surgical and medical patients. Search for the ideal intravenous infusion mixture continues, however. The present position relative to the comparative usefulness of a number of sugars as a caloric source and questions regarding their safety for both parenteral and oral administration are examined in detail by several contributors to this monograph.

Perspectives concerning clinical aspects of pediatric diabetes by Cornblath (Chapter 27) and the advantages of the use of experimental models by Cohen *et al.* (Chapter 29) relate advances in the application of fundamental biochemistry to the understanding of the basic etiology of this disease. Innovative methods of management of diabetic patients are described by Crofford (Chapter 30). Obesity is discussed by Schettler and Schlierf (Chapter 24).

Newer knowledge of the relationship of fructose metabolism to that of several polyols under varying conditions is discussed by Touster (Chapter 15), Froesch and Jacob (Chapter 16), Förster (Chapter 17), Hers (Chapter 21), and Hue (Chapter 22).

The design of therapeutic "bypasses" for inborn errors of metabolism discussed by Hansen (Chapter 18) and Hers (Chapter 21) promises to be but a portion of the beneficial application of detailed knowledge of biochemical pathways.

Recognition of enzyme deficiencies that affect digestion and absorption, such as those reviewed by Dahlqvist (Chapter 13) and the remarkable advances in knowledge concerning the physiology of absorption and digestion in health and disease discussed by Herman (Chapter 11), have greatly improved the management of different clinical syndromes of malabsorption.

E. The Focus—Human Nutrition

Not every aspect of sugars and their importance in nutrition can be covered in a single volume. The focus has been the nutrition of man. Important determinants of acceptability of foodstuffs by individuals, such as the psychology and physiology of the sweet taste and the importance of naturally occurring raffinoses and their effects upon gas formation in the intestinal tract have been included. However, the metabolism of sugars by intestinal microorganisms has been but touched upon by György (Chapter 14), and the complex subjects of ruminant metabolism and many of the laboratory derivatives of sugars could not be treated in this volume.

The subjects embraced in this monograph are the broad considerations that have ushered nutritionists again into an age of the sugars. Recent biochemical research affords the opportunity to markedly alter patterns of usage of sugars in the diet. Concomitantly, nutritionists now have new tools of knowledge by which to capitalize upon the potential benefits as well as to define the possible risks of such changes. This potential for change underscores the importance of the science of nutrition and its understanding throughout the fields of curative and preventive medicine.

Biochemistry and
Psychology of
Sweetness

CHAPTER 3

Biochemistry of Sweet Sensation

ROBERT H. CAGAN

Sweetness is an important sensation to humans and probably to many species of animal as well. Although the relative importance of taste or of calories in determining sucrose ingestion may be difficult to quantify, it is certainly clear that the enormous consumption of sucrose (Page and Friend, Chapter 7; Wretlind, Chapter 6) is largely because people like its sweet taste. Nevertheless, we understand very little about the basic mechanisms underlying sweet taste sensation.

Studies on the biochemistry of the taste system should take cognizance of results obtained at other levels such as electrophysiological recordings and particularly behavioral responses to taste stimuli. When the term "sweet" taste is used in a strict sense it applies only to studies on humans because the description of taste *quality* is a verbal response. By inference, when the stimulus compound is known to be sweet to man, positive behavioral responses in animals (i.e., preferences) are usually considered to result from the sweet taste. In some cases, however, an observed behavioral or electrophysiological response may result from other taste qualities. The comparative aspects will not be stressed further but should of course be considered in extending the term "sweet" to animal studies.

The biochemical events that are discussed are those that occur, or are suggested to occur, in taste receptor cells beginning at the instant the

19

taste stimulus molecule interacts with the cell until the membrane of the receptor cell is depolarized. These are peripheral events. Two major areas of peripheral sensory mechanisms in taste that are amenable to study by the biochemist are specificity and membrane events. These are areas that can be subjected to quantitative experimental treatment. The incursion of biochemical approaches into these areas has barely started. Neither the biological system best suited for these studies nor the ligands most appropriate for demonstrating specific interactions is entirely clear, although efforts to develop experimental systems are underway, and some of these will be discussed here. This report focuses on four general areas, each including a brief critique of past work and a discussion of work now in progress in the author's laboratory. The area of chemical structure–sweet taste was recently extensively reviewed (Shallenberger and Acree, 1971) and will not be discussed here.

A. Choice of Biological System

1. Human, Bovine, and Rat Tongue Epithelium

An experimental approach to study mechanisms of taste was proposed by Lawrence and Ferguson (1959) that involved an attempt to correlate sweet taste with chemical properties of the stimuli and to correlate sweet taste with adsorption of the stimuli onto different surfaces, including charcoal, hair, and two types of tongue tissue. No data were reported for the latter system because the results were not reproducible. No positive generalizations were found (Lawrence and Ferguson, 1959). Later work (Barnes and Ferguson, 1960) showed that the amino acid compositions of proteins extracted from epithelium of different regions of the human tongue are similar; these results produced little information about taste mechanisms.

A few years later, biochemical studies of the sweet taste were reported (Dastoli and Price, 1966) in which an attempt was made to show an interaction between sweet compounds and a protein-containing extract from bovine tongues. Epithelium from the tip of the bovine tongue was used (see also Price and Hogan, 1969) and a protein was later purified (Dastoli *et al.*, 1968) that was called a "sweet-sensitive protein." Small changes were measured in refractive index on mixing one of several sugars or saccharin with the protein preparation. It was assumed that the proteins in the extract underwent changes in conformation upon interaction with added sugars or saccharin and that these conformational changes would be reflected by changes in refractive index of the mixture. From such changes in refractive index of the proteins, binding constants of the sugars and of saccharin to the protein-containing fraction were calculated. Unfortunately, the conclusion that the protein purified is a specific receptor protein seems to have been premature because the following evidence was recently reported (Koyama and Kurihara, 1971). Proteins that were extracted from circumvallate and fungiform papillae (which contain taste buds) and from epithelium without taste buds were analyzed by disc gel electrophoresis. The "sweet-sensitive protein" was found in large amounts throughout the epithelium, whether or not taste buds were present, and it is therefore not unique to taste buds. This finding strongly suggests that what was originally concluded to be a "sweet-sensitive protein" is probably a widely distributed protein(s) without any special function in the taste sensing mechanism. The conclusion that the "sweet-sensitive protein" is a receptor protein was earlier criticized (Cagan, 1971b) because the reported specificity of "binding" does not correspond well with data on relative sweetness of the sugars and saccharin in humans, nor do their data correspond with the taste preference data for the bovine. These studies did, however, mark the beginning of an attempt to study the biochemistry of taste receptors. A later report (Price and Hogan, 1969) showed that the epithelial protein preparation contains glucose dehydrogenase activity with a high specificity for glucose. It is difficult to explain how this protein could function as the receptor site for all sugars.

In a recent study (Nofre and Sabadie, 1972) the bovine tongue epithelial protein was prepared and assayed for binding activity as described by Dastoli *et al.* (1968; Dastoli and Price, 1966). Using the refractometric assay, they showed that sucrose did affect the refractive index, but that the effect was not specific for sweet compounds. Nor was the effect specific to the lingual protein since they found the same effect using bovine γ-globulin. Nofre and Sabadie (1972) concluded that the "sweet-sensitive protein" is not a receptor protein for sweet compounds.

Studies similar to those on "sweet-sensitive protein" of bovine were recently reported utilizing rat tongue (Hiji *et al.*, 1971). An extract of rat tongue epithelium was used to prepare an ammonium sulfate fraction. The protein fraction prepared in this manner showed increases in difference spectra in the region of 280 nm upon adding various sugars. The correspondence of the extent of the spectral increases induced by various sugars paralleled extremely well the electrophysiological response magnitudes of the chorda tympani nerve, which innervates the anterior two-thirds of the tongue. These experiments are interesting and ought to be confirmed. Particular attention should be paid to control preparations without taste buds to check for the specificity of the effect. Experiments in the author's laboratory in which binding of ^{14}C-labeled sugars was measured to intact rat tongue epithelium or to homogenates of epithelium resulted in binding data that were not reproducible. This probably results from the small proportion of taste bud cells in the rat tongue tissue and from the relatively large amount of nonspecific binding that occurs to the rest of the tissue.

2. Bovine Circumvallate Papillae

Direct measurement of binding of taste stimuli to suspensions of bovine taste papillae was reported recently (Cagan, 1971b). A suspension of homogenized bovine papillae was mixed with ^{14}C-labeled sucrose. The mixture was rapidly centrifuged in a microcentrifuge and the pellet was washed once with unlabeled sucrose. The mixture was again centrifuged and the ^{14}C-sucrose in the sedimented pellet determined by liquid scintillation counting. The fungiform papillae at the anterior of the cow tongue contain only a small percentage of the taste buds; most of the taste buds in this animal are localized in the circumvallate papillae found at the posterior of the tongue (Kare, 1970). Although stimulation of the calf tongue with sugars gave poor electrophysiological responses from both of the two nerves that innervate the tongue, the posterior region of the tongue did appear to be somewhat more sensitive to sugars than the anterior area (Bernard, 1964; Bell and Kitchell, 1966). It was found (Fig. 1) that circumvallate papillae show the highest extent of binding of ^{14}C-sucrose. Fungiform papillae, from the anterior of the tongue, bind somewhat less sucrose, and filiform papillae, which lack taste buds, bind very little sucrose under these conditions. The binding to circumvallate papillae also appears to be of a different character than the binding to the control papillae since binding to circumvallate papillae could be inactivated by first heating the tissue preparation prior to the binding assay (Fig. 2). Of the four sugars studied (sucrose, glucose, fructose, and lactose), the first three are strongly preferred by the bovine in two-choice preference tests, and the fourth is less preferred, if at all (Kare and Ficken, 1963). The binding data (Fig. 3) are in accord with these observations. The circumvallate papilla of the cow tongue therefore seems to be a suitable experimental system for further biochemical studies on taste.

3. Isolated Taste Bud Cells

Although the tissue preparation of circumvallate papillae is enriched in taste receptor cells relative to the other tissues that have been used, it is nevertheless a heterogeneous cell population. There are, for example, other epithelial cells, connective tissue elements, and nerve fibers present

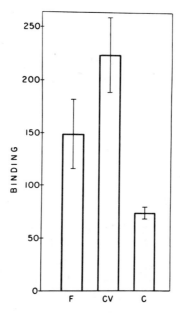

Fig. 1 Binding of ¹⁴C-sucrose to bovine papillae. Fungiform (F), circumvallate (CV), and filiform (C) (control) papillae were prepared and assayed using ¹⁴C-sucrose. Binding is expressed as cpm per 100 μg protein per 10^4 cpm per μmole of sucrose. Values are mean ±SEM. Figure taken from Cagan (1971b) with permission of the publisher.

in this preparation. We are therefore attempting to obtain a preparation that is enriched further with taste bud cells. Results of Dr. J. Brand in his laboratory show that starting with the bovine circumvallate papilla it is possible to prepare a fraction that contains largely taste bud cells. Further work is needed to define the cells and their biochemical properties.

4. Isolated Plasma Membranes

The likelihood seems good that the initial event in taste sensation is an interaction of the stimulus molecule with the plasma membrane ("gustatory microvilli") of the taste cell, as has been suggested in the past (Beidler, 1962, 1967; Kimura and Beidler, 1961; Cagan, 1971a). Beidler (1962) noted that several observations, including the rapidity of eliciting a taste response and the lack of receptor damage caused by stimuli that are toxic, suggested that taste stimuli act at the surface of the receptor cell. Recently, additional indirect evidence bearing on this question was obtained. A protein, monellin, which evokes an intense sweet taste was isolated from the tropical plant, *Dioscoreophyllum cumminsii* (Morris

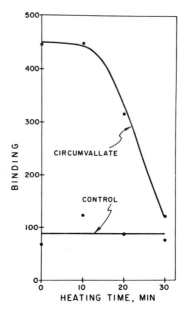

Fig. 2 Inactivation of binding of ^{14}C-sucrose to taste papillae by heating. Preparations of circumvallate and of filiform (control) papillae were prepared as noted
in Fig. 1 and heated in a boiling water bath for the times indicated. Assays for
binding activity used ^{14}C-sucrose, as noted with Fig. 1.

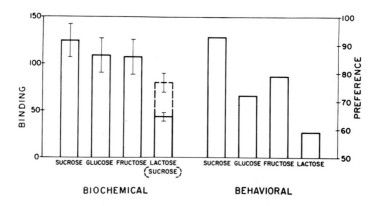

Fig. 3 Comparison of binding of sugars to taste papillae with behavioral taste
preferences in bovine. Papillae were prepared and assayed as noted with Fig.
1. Binding in each case is the difference between the sugar bound to circumvallate
and that bound to filiform (control) papillae. The dashed line for sucrose is an
additional control at the lower concentration used for the lactose assays because
of the lower solubility of lactose. Behavioral data are the results from two-choice
preference tests obtained by Kare and Ficken (1963). Figure taken from Cagan
(1971b) with permission of the publisher.

and Cagan, 1972). This unique sweet taste stimulus has been obtained in pure form and was chemically characterized in the author's laboratory and in collaboration with Martenson's laboratory (Morris *et al.*, 1973) (Table I). It is likely that monellin, because of its large size (molecular weight 10,700), interacts initially at the surface of the cell. (Alternatively, it is, of course, possible to postulate that the protein first passes through the cell membrane before exciting the receptor cell, but this seems less likely.) A "taste-modifier protein," miraculin, is known to cause acids to taste sweet after the tongue has been treated with the glycoprotein (Kurihara and Beidler, 1968; Brouwer *et al.*, 1968). It was postulated to act at the surface membrane (Kurihara and Beidler, 1969) by causing, in response to acid, a conformational change of the receptor cell membrane that has bound the miraculin (see also Section C).

Direct evidence for a role of the plasma membrane in taste sensation could be forthcoming by directly examining interactions of taste stimuli with isolated plasma membranes. Preliminary studies (R. H. Cagan, unpublished data) showed that a plasma membrane fraction can be readily prepared from bovine circumvallate papillae by adapting known preparative methods, but the membranes of course originate from a wide variety of cell types as noted in Section A,3. Attention has therefore been directed toward the preparation that contains isolated taste bud cells, which should provide a more desirable starting cell preparation for isolating plasma membranes.

5. Catfish Barbels

The catfish, *Ictalurus*, has a very large number of taste buds. These occur not only in the oral cavity but also are widely dispersed over the surface of the body including a dense concentration on the barbels (Her-

TABLE I

Properties of Monellin[a]

Molecular weight	10,700
Isoelectric point	9.3
Amino acid residues	91
1 Tryptophan, 1 methionine, 1 cysteine	
No histidine, no cystine	
Ultraviolet absorption: λ_{max} at 277 nm; $\epsilon = 1.47 \times 10^4\ M^{-1}\ cm^{-1}$	
Fluorescence emission: λ_{max} at 337 nm	
Intense and persistent sweet taste	

[a] Data from Morris *et al.* (1973).

rick, 1904; Bardach and Atema, 1971). Early electrophysiological recordings from the facial nerve complex (which innervates the barbels and lips) gave only weak responses to various chemicals (Hoagland, 1933). More recent work with isolated barbels of the catfish showed that they respond electrophysiologically to hydrochloric acid and various chloride salts (Tateda, 1964) and to certain amino acids (Bardach *et al.*, 1967a). The catfish also shows a positive feeding response to cysteine (Bardach *et al.*, 1967b) although additional studies of both behavioral and electrophysiological responses need to be made. Preliminary experiments by Mr. James Krueger in the laboratory show that the barbels bind certain radioactively labeled amino acids, and the catfish taste system is being studied further for its possible utility for biochemical studies of taste.

B. Choice of Ligand

As intimated above, sugars have been the most extensively used stimuli both in human and in animal studies. In the experiments using the bovine epithelial protein (Dastoli and Price, 1966), several sugars and saccharin were used as ligands. Some shortcomings of the experiments and conclusions were noted above (Section A,1).

^{14}C-labeled sugars have recently been used (Cagan, 1971b), enabling direct measurements of attachment of the sugar to the tissue preparation. Surprisingly, use of isotopic tracer techniques for studying binding of ligands to taste tissue had not previously been reported. In selecting sugars for these studies, there is an inherent difficulty in measuring the interaction between the ligand and the tissue preparation. This stems from the extremely weak binding of sugars to taste receptors (Beidler, 1962; Cagan, 1971b); for example, half-saturation values for the sugars in eliciting taste responses are of the order of several tenths molar in a variety of systems, and these high concentrations were also found to be necessary in binding studies. Additional confirmation of the weak binding was seen by the ease of washing out the bound ligand from the tissue pellets. As pointed out (Cagan, 1971b), the "binding" in these studies is an operational term. The weak binding is not only inherently difficult to measure but also leads to difficulties in interpretation because of the possibility of nonspecific effects as a result of the high concentration of ligand. The potential usefulness of the sugars as ligands in biochemical studies of this type seems limited. Unfortunately, most of the behavioral data available for bovines are for sugars (Kare and Ficken, 1963; Goatcher and Church, 1970a) although a few other types of stimuli have also recently been used (Goatcher and Church, 1970a,b).

Many other sweeteners might serve as ligands, but it would be desirable to first know if these are "sweet" taste stimuli to the species being studied; for example, saccharin is at least several hundred times more effective in evoking a sweet sensation in man than is sucrose (Schutz and Pilgrim, 1957; Stone and Oliver, 1969), and it might be expected to be a ligand with a higher affinity for the receptor. This appears not to be the case for cows, judging from behavioral studies which showed that they apparently do not taste saccharin (Kare and Ficken, 1963). Certain dihydrochalcones (Krbechek *et al.*, 1968; Horowitz and Gentili, 1969), the dipeptide ester L-Asp-L-Phe-Me (Mazur *et al.*, 1969), stevioside (Wood *et al.*, 1955; Mosettig and Nes, 1955), certain substituted *m*-nitroanilines (Lawrence and Ferguson, 1960), osladin (Jizba *et al.*, 1971), perillartine and other oximes (Acton *et al.*, 1970), and monellin (Morris and Cagan, 1972) are some of the potential candidates as ligands in this type of study. The potential usefulness of monellin seems especially high because its sweet taste is intense and persistent to man. In addition, because monellin is a protein (Table I) there may be many possibilities for labeling it in order to study its binding directly. It might also be possible to prepare an antibody to it and then localize the monellin that is bound to taste receptor cells.

Certain amino acids are likely to be good ligands using the catfish taste system; for example, electrophysiological recordings showed that cysteine, alanine, and phenylalanine are particularly good stimuli, while tyrosine and methionine are ineffective [apparently all L-isomers were used (Bardach *et al.*, 1967a)]. The latter two amino acids should therefore serve as good controls for nonspecific binding. The likelihood seems good that the binding constants of cysteine and alanine are favorable for biochemical studies.

C. Specific Activator and Inhibitor

A specific activator and a specific inhibitor of the sweet taste sensation are known and could be more fully exploited as tools to probe the biochemistry of taste. Miraculin is a glycoprotein of molecular weight close to 44,000 (Kurihara and Beidler, 1968; Brouwer *et al.*, 1968). When the tongue is treated with miraculin, subsequent exposure to acid causes the normally sour tasting acid to taste sweet. The ability of an acid to induce the sweet taste, after treatment of the tongue with miraculin, is closely related to the intensity of sourness of the acid (Kurihara and Beidler, 1969). The mechanism of action of the glycoprotein was suggested (Kurihara and Beidler, 1969) to involve a conformational change of the taste receptor membrane by the acid; this change would then allow a sweet taste to be stimulated at the sweet receptor site by the arabinose and xylose moieties of the bound miraculin molecule. It is not known if the sugar residues are essential for the biological activity of miraculin, but this could be tested by studying the activity of the molecule after removing and then replacing the sugar residues.

Extracts of the leaves of *Gymnema sylvestre* contain gymnemic acid. This natural product is actually a mixture of a few closely related compounds as it is usually prepared. They are known to be D-glucuronides of a hexahydroxytriterpene that contain several fatty acid residues (Stöcklin, 1969; Sinsheimer *et al.*, 1970), although the structures have not been completely elucidated. The active forms of gymnemic acid are able to specifically inhibit the ability by man (Warren and Pfaffmann, 1959; Diamant *et al.*, 1965) and other vertebrates (e.g., Andersson *et al.*, 1950; Faull and Halpern, 1971) to taste sweet materials, although extracts of the leaves were ineffective in altering the responses of two invertebrates to sugars (Larimer and Oakley, 1968). The effect appears to be widespread for virtually all sweet sensation because gymnemic acid inhibits the sweet taste of a variety of molecules (Warren and Pfaffmann, 1959; Kurihara, 1969) as well as inhibiting the change of sour to sweet induced by miraculin (Kurihara and Beidler, 1969).

In recent experiments in the author's laboratory, the effect of gymnemic acid on binding of sucrose to papillae was studied. The binding of ^{14}C-sucrose to circumvallate and filiform (control) papillae of the bovine was measured after treatment of the tissue preparations with ca. 10 mM and ca. 50 mM gymnemic acid (Fig. 4). The data show no evidence for inhibition of binding of ^{14}C-sucrose by gymnemic acid, but the results are equivocal because of an artifact induced in the assay system by the presence of gymnemic acid. A large increase in binding of ^{14}C-sucrose in the presence of gymnemic acid (Fig. 4, note right-hand ordinate), which is not dose-dependent, is observed in both circumvallate and control tissues, but the greater binding of sucrose by the circumvallate preparation is still seen. It was observed during these experiments that the tissue pellets recovered at the end of the binding assay, whether derived from circumvallate or control papillae, had become quite swollen and gelatinous and had entrapped a considerable amount of fluid (containing ^{14}C-sucrose). This led to the large increase in extent of binding by both preparations and was probably responsible for the greater variability as well. If, indeed, gymnemic acid inhibits binding of sucrose, it cannot be seen with this type of assay procedure. The question is still open, but the limited biochemical evidence suggests that gymnemic acid acts at a step subsequent to the initial binding.

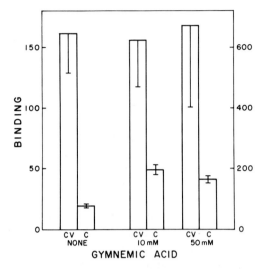

Fig. 4 Lack of effect of gymnemic acid on binding of ^{14}C-sucrose to bovine taste papillae. Circumvallate (CV) and filiform (control) (C) papillae were prepared and assayed for binding of ^{14}C-sucrose as noted with Fig. 1 (mean ±SEM, $N = 4$). The different scales on the ordinates are explained in the text. Prior to the binding assay, the tissue preparations were incubated for 15 min at 25°C in the presence of gymnemic acid in 0.01 M tris buffer, pH 8.4, or with the buffer alone. The gymnemic acid was a gift of H. van der Wel and had been prepared as described (Warren and Pfaffmann, 1959; Yackzan, 1966). It was converted to the potassium salt (Warren and Pfaffmann, 1959). The sample was analyzed in the author's laboratory using thin-layer chromatography on silica gel G (Analtech). Plates developed with solvent systems III, IV, or VII of Sinsheimer *et al.* (1970) and stained with either the Liebermann-Burchard reagent (Procedure No. 1 in Stahl, 1969) or the ceric sulfate-sulfuric acid reagent (Procedure No. 37 in Stahl, 1969) showed two spots. Using solvent system II there were four or five spots. The sample of gymnemic acid was biologically active as shown by the inhibition of the author's ability to taste sucrose.

This conclusion is in agreement with findings from a behavioral experiment on humans. The presence of sucrose or saccharin during treatment with gymnemic acid failed to alter the effect of the gymnemic acid (Warren *et al.*, 1969). This suggested to Warren *et al.* (1969) that there was no competition between gymnemic acid and sweet compounds for the same receptor site. The negative findings in both the biochemical and the behavioral studies are in agreement; neither set of data is, however, conclusive. The deficiency in the binding results is noted above. One uncertainty in the behavioral study stems from the weak binding of sweet compounds. If binding by gymnemic acid were much stronger, as seems likely from the persistence of its effect (Kurihara, 1969), then even in the presence of a sweet compound, gymnemic acid could still be effective because it would tend to remain bound. The evidence from both studies suggests, however, that the site of action of gymnemic acid is elsewhere than at the binding step of sweet molecules.

D. Monitoring Membrane Events

In addition to binding as a method to study early events in taste sensation, other parameters of cellular function that respond to taste stimuli could be examined. The concept of cyclic AMP as a "second messenger" has been well developed in the hormone field (Robison *et al.*, 1971), and the role of cyclic AMP in the nervous system is now receiving attention (e.g., Greengard and Costa, 1970). Evidence has been reported regarding its possible involvement in functioning of visual receptors (Bitensky *et al.*, 1971; Miller *et al.*, 1971; Brown and Makman, 1972). It was speculated (Riddiford, 1971) that a second messenger might act in insect chemoreceptors. The presence of adenyl cyclase in rabbit olfactory epithelium has been reported (Bitensky *et al.*, 1972), but the enzyme activity was unaffected by several olfactory stimuli.

A hormone may interact with a specific receptor in the plasma membrane of a responsive cell type. This leads to activation or inhibition of the membrane-localized enzyme adenyl cyclase by as yet little-understood mechanisms. Adenyl cyclase catalyzes the conversion shown in reaction (1):

$$ATP \rightarrow cyclic\ AMP + PP_i \tag{1}$$

Cyclic AMP then exerts effects on intracellular events, leading to biochemical expression of the action of the hormone that originally occurred at the cell surface.

Intracellular cyclic AMP levels are regulated not only by its rate of formation by adenyl cyclase but also by its rate of breakdown by reaction (2) which is catalyzed by a specific phosphodiesterase.

$$Cyclic\ AMP \rightarrow 5'\text{-}AMP \tag{2}$$

Two possible control points for altering the intracellular levels of cyclic AMP are therefore adenyl cyclase and phosphodiesterase. Hormones that are known to act via cyclic AMP as a second messenger do so at the level of adenyl cyclase. On the other hand, there is mounting evidence that a variety of drugs act at the phosphodiesterase step.

Recent findings (Table II) show that the enzyme adenyl cyclase is present in intact bovine circumvallate papillae as well as in fungiform papillae, both of which contain taste buds. It is present in a lower amount

TABLE II

Adenyl Cyclase Activity of Bovine Lingual Tissues

Tissue[a]	^{14}C-cyclic AMP formation[b] (cpm/mg protein)
Circumvallate papillae	65.3
Fungiform papillae	57.9
Filiform papillae	28.0
Epithelium	60.0

[a] Intact papillae or blocks of tongue epithelium were prelabeled by incubating with [8-^{14}C] adenine (e.g., Humes *et al.*, 1969) for 30 min at 37°C in a buffer of ionic composition similar to bovine saliva, pH 8.6. The papillae were assayed for adenyl cyclase by incubating them for 5 min at 37°C in the presence of 10 mM theophylline. Details of the experimental procedure will be published (Cagan, 1973). ^{14}C-cyclic AMP was isolated by the method of Krishna *et al.* (1968).

[b] The data are preliminary and are results obtained using a single bovine tongue. Samples contained the following amounts of tissue: circumvallate, 19 papillae, 72 mg wet weight, 11.8 mg protein; fungiform, 100 papillae, 188 mg wet weight, 31.8 mg protein; filiform, 30 papillae, 97 mg wet weight, 28.0 mg protein; and epithelium, 20 blocks, 138 mg wet weight, 26.9 mg protein. Only circumvallate and fungiform papillae contain taste buds.

in filiform papillae, which lack taste buds. The mere presence of the enzyme, however, does not indicate the presence of taste receptors as might be expected from the known wide distribution of the enzyme (Sutherland *et al.*, 1962), including its presence in mammalian skin (Mier and Urselmann, 1970). It is, in fact, present also in epithelium of the cow tongue at a level comparable to that in taste papillae (Table II). This is in contrast to a recent report of Kurihara and Koyama (1972) showing that there was a high level of the enzyme in a sedimentable fraction prepared from circumvallate papillae but very little in the preparation from nontaste epithelium.

One critical point for establishing a role for adenyl cyclase in taste sensation concerns its regulation by extracellular chemical stimuli and not merely its presence in the papillae. The adenyl cyclase in circumvallate papillae has been found, in preliminary experiments, to be activated by 0.2 M sucrose, a known taste stimulus in cows (Table III). Further enhancement of the incorporation of ^{14}C precursor into ^{14}C-cyclic AMP

TABLE III

Effects of Taste Stimuli on Adenyl Cyclase of Bovine Taste Papillae[a]

Additions		^{14}C-cyclic AMP formation (cpm/mg protein)
Exp. No. 1	None	157.2
	0.2 M sucrose	201.2
	10 mM theophylline	310.8
	0.2 M sucrose + 10 mM theophylline	427.1
Exp. No. 2	None	80.6
	10 mM theophylline	151.1
	3 mM caffeine	152.3
	2 mM quinine	130.5

[a] Experimental conditions as in Table II. The additions were made at the start of the 5-min assay period. The different levels of incorporation into cyclic AMP in the experiments in Tables II and III are because the preincubation fluids in these experiments contained different levels of radioactivity. Circumvallate papillae were used in all samples in Table III. These data are preliminary.

occurs upon adding 10 mM theophylline, a known inhibitor of phosphodiesterase. The results in Table III should be regarded as tentative. The author's observations to date are that the stimulation by sucrose is not always demonstrable but that the increased incorporation caused by theophylline is a consistent feature. A number of control experiments will be necessary to establish the specificity for different taste stimuli and for taste and nontaste tissues.

Theophylline tastes bitter to man, and the possible involvement of phosphodiesterase in mediating bitter taste sensations is also being explored. The preliminary findings (Table III) show that not only theophylline but also caffeine and quinine, which are bitter to man, give rise to an elevated incorporation of ^{14}C into cyclic AMP. Quinine is known to be an aversive taste stimulus in the bovine (Goatcher and Church, 1970b). Judging from findings in other systems, these effects most likely result from inhibition of phosphodiesterase rather than activation of the cyclase.

One alternative possibility to the role of cyclic AMP as a second messenger in the taste receptor cells should be considered. The changes in cyclic AMP could be a reflection of secondary changes occurring in the innervating nerves. The intact papillae do contain nerve fibers that synapse in the taste bud. Further research will be necessary to answer this question.

E. Research Needs

The sweet taste is an important sensation, but little is known about the underlying biochemical mechanisms. The bovine circumvallate papilla has been found to be a favorable tissue for studying biochemical mechanisms. The cow tongue is large and has large circumvallate papillae. Relative to most other tissues, and certainly relative to other regions of the cow tongue, the twenty circumvallate papillae contain the highest proportion of taste buds. The circumvallate papillae also show the greatest extent of binding of ^{14}C-sucrose. Although only preliminary attempts have been made, it appears possible to isolate taste bud cells from the circumvallate papillae of the cow, and these might provide the biochemist with a suitable experimental system for further study. The ability to prepare isolated taste bud cells would greatly enhance the experimental options open for studying the biochemistry of taste. In particular, such cells would provide a desirable cell population for isolation of receptor cell membranes. The catfish barbel appears to be a favorable tissue for taste biochemistry because of the dense concentration of taste receptors on this appendage. Preliminary studies on binding of radioactively labeled amino acids have been encouraging, and the catfish barbel is being studied further. The rat tongue has certain disadvantages for studying taste biochemistry. The papillae are small and only a small proportion of the tongue epithelium is taste receptor cells. This is unfortunate since a significant proportion of the behavioral and electrophysiological data on taste have been obtained with rats.

Although ^{14}C-labeled sugars have been used as ligands to demonstrate binding to bovine taste papillae, the binding is weak, and suitable ligands for binding studies need to be developed. The availability of ligands that are specific and that bind more tightly may be critical for isolation of receptor molecules. In addition to specific ligands, the specific activator miraculin and the specific inhibitor gymnemic acid could be used more extensively to probe the biochemistry of taste.

Monitoring membrane events may provide another approach to study the biochemistry of sweet and other taste sensations. Preliminary findings suggest that cyclic AMP formation by adenyl cyclase and its breakdown by phosphodiesterase may be involved in mediating taste sensations and should be studied further.

Sensory physiology has received extensive attention by biochemists only in the field of vision. Biochemical studies of chemoreceptors have barely started, but the mechanisms by which chemical information is processed by living organisms is an important and challenging problem.

Acknowledgment

The work from this laboratory was supported in part by USPHS Research Grant No. NS-08775 from the National Institute of Neurological Diseases and Stroke.

References

Acton, E. M., Leaffer, M. A., Oliver, S. M., and Stone, H. (1970). Structure-taste relationships in oximes related to perillartine. *J. Agr. Food Chem.* **18,** 1061–1068.

Andersson, B., Landgren, S., Olsson, L., and Zotterman, Y. (1950). The sweet taste fibres of the dog. *Acta Physiol. Scand.* **21,** 105–119.

Bardach, J. E., and Atema, J. (1971). The sense of taste in fishes. *In* "Handbook of Sensory Physiology" (L. M. Beidler, ed.), Vol. IV, Pt 2, pp. 293–336. Springer-Verlag, Berlin and New York.

Bardach, J., Fujiya, M., and Holl, A. (1967a). Investigations of external chemo-receptors of fishes. *Olfaction Taste 2, Proc. Int. Symp., 2nd, 1965* pp. 647–665.

Bardach, J. E., Todd, J. H., and Crickmer, R. (1967b). Orientation by taste in fish of the genus *Ictalurus*. *Science* **155,** 1276–1278.

Barnes, C. J., and Ferguson, L. N. (1960). Amino acid composition of proteins from surface tissue of the tongue. *Nature (London)* **186,** 617–619.

Beidler, L. M. (1962). Taste receptor stimulation. *Progr. Biophys. Biophys. Chem.* **12,** 107–151.

Beidler, L. M. (1967). Anion influences on taste receptor response. *Olfaction Taste 2, Proc. Int. Symp. 2nd, 1965* pp. 509–534.

Bell, F. R., and Kitchell, R. L. (1966). Taste reception in the goat, sheep and calf. *J. Physiol. (London)* **183,** 145–151.

Bernard, R. A. (1964). An electrophysiological study of taste reception in peripheral nerves of the calf. *Amer. J. Physiol.* **206,** 827–835.

Bitensky, M. W., Gorman, R. E., and Miller, W. H. (1971). Adenyl cyclase as a link between photon capture and changes in membrane permeability of frog photoreceptors. *Proc. Nat. Acad. Sci. U.S.* **68,** 561–562.

Bitensky, M. W., Miller, W. H., Gorman, R. E., Neufeld, A. H., and Robinson, R. (1972). The role of cyclic AMP in visual excitation. *Advan. Cyclic Nucleotide Res.* **1,** 317–335.

Brouwer, J. N., van der Wel, H., Francke, A., and Henning, G. J. (1968). Miraculin, the sweetness-inducing protein from miracle fruit. *Nature (London)* **220,** 373–374.

Brown, J. H., and Makman, M. H. (1972). Stimulation by dopamine of adenylate cyclase in retinal homogenates and of adenosine-3′:5′-cyclic monophosphate formation in intact retina. *Proc. Nat. Acad. Sci. U.S.* **69,** 539–543.

Cagan, R. H. (1971a). Counterparts of gustatory receptors. *In* "Gustation and Olfaction" (G. Ohloff and A. F. Thomas, eds.), pp. 243–244. Academic Press, New York.

Cagan, R. H. (1971b). Biochemical studies of taste sensation. I. Binding of ¹⁴C-labeled sugars to bovine taste papillae. *Biochim. Biophys. Acta* **252,** 199–206.

Cagan, R. H. (1973). In preparation.

Dastoli, F. R., and Price, S. (1966). Sweet-sensitive protein from bovine taste buds: Isolation and assay. *Science* **154,** 905–907.

Dastoli, F. R., Lopiekes, D. V., and Price, S. (1968). A sweet-sensitive protein from bovine taste buds. Purification and partial characterization. *Biochemistry* **7,** 1160–1164.

Diamant, H., Oakley, B., Ström, L., Wells, C., and Zotterman, Y. (1965). A comparison of neural and psychophysical responses to taste stimuli in man. *Acta Physiol. Scand.* **64,** 67–74.

Faull, J. R., and Halpern, B. P. (1971). Reduction of sucrose preference in the hamster by gymnemic acid. *Physiol. & Behav.* **7,** 903–907.

Goatcher, W. D., and Church, D. C. (1970a). Taste responses in ruminants. III. Reactions of pygmy goats, normal goats, sheep and cattle to sucrose and sodium chloride. *J. Anim. Sci.* **31,** 364–372.

Goatcher, W. D., and Church, D. C. (1970b). Taste responses in ruminants. IV. Reactions of pygmy goats, normal goats, sheep and cattle to acetic acid and quinine hydrochloride. *J. Anim. Sci.* **31,** 373–382.

Greengard, P., and Costa, E., eds. (1970). "Role of Cyclic AMP in Cell Function," Vol. 3. Raven Press, New York.

Herrick, C. J. (1904). The organ and sense of taste in fishes. *Bull. U.S. Fish. Bur.* **22,** 237–272.

Hiji, Y., Kobayashi, N., and Sato, M. (1971). Sweet-sensitive protein from the rat tongue: Its interaction with various sugars. *Comp. Biochem. Physiol. B* **39,** 367–375.

Hoagland, H. (1933). Specific nerve impulses from gustatory and tactile receptors in catfish. *J. Gen. Physiol.* **16,** 685–693.

Horowitz, R. M., and Gentili, B. (1969). Taste and structure in phenolic glycosides. *J. Agr. Food Chem.* **17,** 696–700.

Humes, J. L., Rounbehler, M., and Kuehl, F. A., Jr. (1969). A new assay for measuring adenyl cyclase activity in intact cells. *Anal. Biochem.* **32,** 210–217.

Jizba, J., Dolejš, L., Herout, V., and Šorm, F. (1971). The structure of osladin—the sweet principle of the rhizomes of *Polypodium vulgare* L. *Tetrahedron Lett.* **18,** 1329–1332.

Kare, M. R. (1970). Taste, smell and hearing. *In* "Dukes' Physiology of Domestic Animals" (M. J. Swenson, ed.), 8th ed., pp. 1160–1185. Cornell Univ. Press (Comstock), Ithaca, New York.

Kare, M. R., and Ficken, M. S. (1963). Comparative studies on the sense of taste. *Olfaction Taste, Proc. Int. Symp., 1962* pp. 285–297.

Kimura, K., and Beidler, L. M. (1961). Microelectrode study of taste receptors of rat and hamster. *J. Cell. Comp. Physiol.* **58,** 131–139.

Koyama, N., and Kurihara, K. (1971). Do unique proteins exist in taste buds? *J. Gen. Physiol.* **57,** 297–302.

Krbechek, L., Inglett, G., Holik, M., Dowling, B., Wagner, R., and Riter, R. (1968). Dihydrochalcones. Synthesis of potential sweetening agents. *J. Agr. Food Chem.* **16,** 108–112.

Krishna, G., Weiss, B., and Brodie, B. B. (1968). A simple, sensitive method for the assay of adenyl cyclase. *J. Pharmacol. Exp. Ther.* **163,** 379–385.

Kurihara, K., and Beidler, L. M. (1968). Taste-modifying protein from miracle fruit. *Science* **161,** 1241–1243.

Kurihara, K., and Beidler, L. M. (1969). Mechanism of the action of taste-modifying protein. *Nature (London)* **222,** 1176–1179.

Kurihara, K., and Koyama, N. (1972). High activity of adenyl cyclase in olfactory and gustatory organs. *Biochem. Biophys. Res. Commun.* **48,** 30–34.

Kurihara, Y. (1969). Antisweet activity of gymnemic acid A₁ and its derivatives. *Life Sci., Part I* **8,** 537–543.

Larimer, J. L., and Oakley, B. (1968). Failure of *Gymnema* extract to inhibit the sugar receptors of two invertebrates. *Comp. Biochem. Physiol.* **25,** 1091–1097.

Lawrence, A. R., and Ferguson, L. N. (1959). Exploratory physicochemical studies on the sense of taste. *Nature (London)* **183,** 1469–1471.

Lawrence, A. R., and Ferguson, L. N. (1960). Dissociation constants of some sweet and tasteless isomeric *m*-nitroanilines. *J. Org. Chem.* **25**, 1220–1224.

Mazur, R. H., Schlatter, J. M., and Goldkamp, A. H. (1969). Structure-taste relationships of some dipeptides. *J. Amer. Chem. Soc.* **91**, 2684–2691.

Mier, P. D., and Urselmann, E. (1970). The adenyl cyclase of skin. I. Measurement and properties. *Brit. J. Dermatol.* **83**, 359–363.

Miller, W. H., Gorman, R. E., and Bitensky, M. W. (1971). Cyclic adenosine monophosphate: Function in photoreceptors. *Science* **174**, 295–297.

Morris, J. A., and Cagan, R. H. (1972). Purification of monellin, the sweet principle of *Dioscoreophyllum cumminsii*. *Biochim. Biophys. Acta* **261**, 114–122.

Morris, J. A., Martenson, R., Deibler, G., and Cagan, R. H. (1973) Characterization of monellin, a protein that tastes sweet. *J. Biol. Chem.* **248**, 534–539.

Mosettig, E., and Nes, W. R. (1955). Stevioside. II. The structure of the aglucon. *J. Org. Chem.* **20**, 884–899.

Nofre, C., and Sabadie, J. (1972). A propos de la protéine linguale dite "sensible aux sucres." *C. R. Acad. Sci., Ser. D* **274**, 2913–2915.

Price, S., and Hogan, R. M. (1969). Glucose dehydrogenase activity of a "sweet-sensitive protein" from bovine tongues. *Olfaction Taste, Proc. Int. Symp., 3rd, 1967* pp. 397–403.

Riddiford, L. M. (1971). The insect antennae as a model olfactory system. *In* "Gustation and Olfaction" (G. Ohloff and A. F. Thomas, eds.), pp. 251–253. Academic Press, New York.

Robison, G. A., Butcher, R. W., and Sutherland, E. W. (1971). "Cyclic AMP." Academic Press, New York.

Schutz, H. G., and Pilgrim, F. J. (1957). Sweetness of various compounds and its measurement. *Food Res.* **22**, 206–213.

Shallenberger, R. S., and Acree, T. E. (1971). Chemical structure of compounds and their sweet and bitter taste. *In* "Handbook of Sensory Physiology" (L. M. Beidler, ed.), Vol. IV, Part 2, pp. 221–277. Springer-Verlag, Berlin and New York.

Sinsheimer, J. E., Rao, G. S., and McIlhenny, H. M. (1970). Constituents from *Gymnema sylvestre* leaves. V. Isolation and preliminary characterization of the gymnemic acids. *J. Pharm. Sci.* **59**, 622–628.

Stahl, E., ed. (1969). "Thin-Layer Chromatography," 2nd ed., pp. 855–861. Springer-Verlag, Berlin and New York.

Stöcklin, W. (1969). Chemistry and physiological properties of gymnemic acid, the antisaccharine principle of the leaves of *Gymnema sylvestre*. *Agr. Food Chem.* **17**, 704–708.

Stone, H., and Oliver, S. M. (1969). Measurement of the relative sweetness of selected sweeteners and sweetener mixtures. *J. Food Sci.* **34**, 215–222.

Sutherland, E. W., Rall, T. W., and Menon, T. (1962). Adenyl cyclase. I. Distribution, preparation, and properties. *J. Biol. Chem.* **237**, 1220–1227.

Tateda, H. (1964). The taste response of the isolated barbel of the catfish. *Comp. Biochem. Physiol.* **11**, 367–378.

Warren, R. M., and Pfaffmann, C. (1959). Suppression of sweet sensitivity by potassium gymnemate. *J. Appl. Physiol.* **14**, 40–42.

Warren, R. P., Warren, R. M., and Weninger, M. G. (1969). Inhibition of the sweet taste by *Gymnema sylvestre*. *Nature (London)* **223**, 94–95.

Wood, H. B., Jr., Allerton, R., Diehl, H. W., and Fletcher, H. G., Jr., (1955). Stevioside. I. The structure of the glucose moieties. *J. Org. Chem.* **20**, 875–883.

Yackzan, K. S. (1966). Biological effects of *Gymnema sylvestre* fractions. *Ala. J. Med. Sci.* **3**, 1–9.

CHAPTER 4

The Psychology of Sweetness

HOWARD R. MOSKOWITZ

There is scarcely any area of food habits today that does not in some way involve sweetness. Our use of sugar and sweeteners has increased, and with the diversity of natural and fabricated products available commercially, the use of sweet-tasting additives promises to increase even more. Psychological studies of man's "sweet tooth" thus enter at different levels of behavior. At the most molar level, studies of consumer behavior to foods are an appropriate focus for the investigation. At more molecular levels, sensory and perceptual studies of taste behavior, conducted in laboratories, and often with simple sugar solutions as stimuli, have been the rule. The present paper concerns the latter, or traditional focus of behavioral studies. The hope is that an elucidation of the previously reported results from the long history of sensory science can help clarify the behavior of individuals toward sweet-tasting foods.

We shall be concerned with three separate areas in the study of perceived sweetness. These three areas, the qualitative, the quantitative, and the hedonic (preference) aspects of taste, have received inputs from a diverse array of fields such as physics, chemistry, animal behavior, physiology, food science, pharmaceutical science, and psychology. Rather than dealing with the contribution of each field, we shall investigate the types of questions that have been asked concerning the sweet taste, and take our answers from the contributions of each field as they are appropriate.

A. The Quality of Sweetness

The history of sensory science has long sought to reduce the array of color experiences to the interaction of a limited set of color primaries, and by so doing explain the mechanism of color vision. Similar thinking has pervaded the chemical senses, taste and smell, so that during the past centuries different sets of taste primaries have been postulated. Figure 1 shows a list of these primaries as they have developed. Originally, the sets included sensory experiences that would today be classified as primarily odor-derived and not true taste sensations. Usually, two to nine primary tastes were assumed to suffice in the realm of taste experience, in contrast to three color primaries. The slight nuances of taste for a single primary (*viz.*, the sweet taste of sucrose differing from that of glucose) were ignored in these classification systems.

Human behavior is governed by social environment in which the taste classifier lives, so it is not surprising that many arrived at overlapping classifications. In order to establish that tastes such as "sweet" and "salty" are truly primaries, a set of anthropological studies were undertaken and reported more than a half-century ago. Various languages were studied. One of these, the Algonquin language, contained a word for sweet that derived from the description of the bipolar dimension "good–bad" (Chamberlain, 1903). The natives of the Torres Straits, located halfway around the world between New Guinea and Australia, were also studied in a separate anthropological investigation. They were reported to possess only one taste word "sweet," that referred to the sensation itself of taste. The rest of their words derived from objects; for

Taste primaries

Bravo (1592)	Linnaeus (1751)	Haller (1763)	Fick and Henning (1864) (1924)	Zenneck (1894)	Bekesy (1964)
Sweet	Sweet	Sweet Spiritous	Sweet	Sweet	Sweet-bitter
Sour	Acid Astringent	Acid	Sour		
Sharp Pungent Harsh	Sharp	Sharp			
	Viscous				
Fatty	Fatty				
Bitter	Bitter	Bitter	Bitter	Bitter	
Insipid	Insipid Aqueous				
Saline	Saline	Saline	Saline	Salty	Salty-sour
	Nauseous				

Fig. 1 Taste primaries of different periods [adapted from Boring (1942)].

example, "saltiness" was a derivative of their noun for seaweed (Myers, 1903). Finally, in the Indo-European language, the mother tongue of many European languages, the word for sweet, *swad*, also refers to a sensation, without concurrently typing the word as an object.

Although these studies suggest that sweetness is a taste primary, with perhaps the same status as the color primaries, it may eventually turn out that taste sensations are not describable by a single collection of discrete primaries. In contrast to color vision, taste experience may actually lie along a continuum, with some points of familiarity standing out in our experience. To these points we ascribe the word "primary," and today they are the sensations of "salty," "sour," "bitter," and "sweet." To a society whose cuisine revolves around sweet foods, the nodal point "sweet" on this continuum may actually comprise quite a number of smaller classes. A suggestion of this continuum was made back in 1828 by Greeves. Greeves, in his arrangement of tastes and smells and materials in the material medica of that era, suggested a host of different sweet sensations, ranging from the faint, the fruity, the insipid, to much stronger ones.

Studies of the nature of sweetness as a taste primary have not been limited to psychologists and anthropologists. With the advent of electrical recording from the taste nerve, the chorda tympani, scientists have been able to "tune in" to the neural traffic between the tongue and brain. When the tongue is stimulated by a stream of sugar solution, and the chorda tympani recorded from, there appears a discharge, representing electrical impulses traveling along the nerve. With refined techniques, single, small strands can be disengaged from the larger fiber mass, much as a single line can be unraveled from a tangled mass of communication cables. The messages conveyed by that single strand tend to represent more localized and specific taste responses, presumably from a more limited range of receptor cells at the tongue.

Although some neural fibers respond to sweet-tasting substances placed on the tongue, others do not. Indeed, the pattern of sensitivity is often a complicated one. Rarely can one find fibers that are specifically sensitive only to "sweet" or to "salty" substances. Rather, some fibers are sensitive to two, three, or even all four classes of taste chemicals. Others have an entirely different spectrum of sensitivities and may respond strongly to one sweetener and very weakly to another. Pfaffmann (1959a) has discussed how two fibers, one with one pattern of sensitivity to taste, and the other with a different pattern, can signal two different taste qualities, even though neither is specifically sensitive to either taste alone. Erickson (1963), working with an array of sensitive single neural units, each with its own unique pattern of responses to stimulus chemicals, has shown how the brain may conceivably reproduce our unique taste sensations. The mechanism does not rely upon single responses from one tuned neuron but rather from a complex pattern of responses obtained from an entire array of responding neurons. In spite of the complexity of the pattern, the taste may still be perceived as a coherent sensation.

On the perceptual side, the single unitary sensation can be decomposed, and the perception of sweetness can be analyzed into its nuances. When observers are presented with a collection of different taste materials that have been made equally sweet experimentally by subjective measurement, the observers attempt to detect differences among the substances and to expand their "unitary" sensation of sweet. Figure 2 summarizes the results of a study (Moskowitz, 1972b) in which observers compared the flavor difference of test sugars (*viz.*, fructose, maltose, lactose, etc.) to equally sweet levels of glucose and fructose. Four different levels of reference sugars were selected (0.25 *M*, 0.5 *M*, 1.0 *M*, and 2.0 *M* glucose) and each of the other sugars was compared to its equally sweet glucose partner and to an equally sweet fructose solution.

When the judgments of flavor difference were treated as distances between points in a "subjective space," the technique of multidimensional scaling (Shepard, 1962) provided an estimate of where those points lie in that space. Distances between points then corresponded to the subjective estimates of flavor difference originally collected by experimental procedures. Sugars as typical sweet stimuli cannot be located as points along a line. Instead, at least two different perceptual attributes or dimensions are taken account of by observers when they analyze the supposed unitary percept of sweetness. One component, viscosity vs. fluidity,

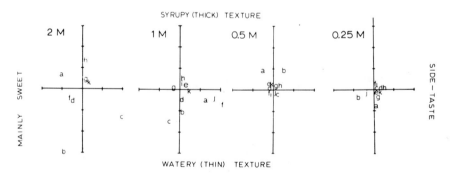

Fig. 2 Two-dimensional "taste space" for the flavor of sugars. The vertical dimension ranges from syrupy to watery and is primary textural. The horizontal dimension is the bipolar dimension "mainly sweet" to "side-taste." Abbreviations: a, glucose; b, fructose; c, xylose; d, arabinose; e, galactose; f, sorbose; g, sorbitol; h, sucrose; i, maltose; j, lactose; and k, glycerol.

is not a taste-related one, but applies more to the texture impressions. It can be discarded. The second appears to be the presence vs. absence of "side-tastes," not necessarily sour, salty, or bitter, that accompany sweetness. It may well turn out that with refined languages to describe taste impression these side-tastes can be named. However, they may also be slight variants in sweetness, possessed by these sugars. When the observer is instructed to concentrate on the nodal point "sweetness," he may find a significantly more complex sensation than in situations where sweet is opposed to, say, sour, salty, and bitter as a taste descriptor.

B. The Intensity of Sweetness

Historically, approaches to sweetness measurement have been found predominantly in the published scientific and technological literature of chemistry, food science, and psychology. Much of our present-day knowledge of sweetness intensity, both at the threshold level where taste begins and above the threshold level, derives from the application of psychological techniques. We shall look at these in detail, from the perspective of different types of information gained by the measurements of sweetness. It should be borne in mind that, historically, the interest in sweetness measurement runs parallel to interests in measuring the sensory capabilities of our other senses. In particular, however, refined measures of sweetness intensity grew out of the interest in replacing sugar with a cheaper, more potent and, in wartime, more available sugar substitute. The discovery of saccharin and the synthesis of various sugars prompted questions about the relative taste impressions provided by these substances.

There are four major types of measures that will be of concern here: (a) threshold measures or estimates of the physical level at which the sensation of sweetness begins, (b) equal-sweetness matches between sugar and other sweeteners, (c) category or rating scales (similar to the Fahrenheit-temperature scale), and (d) ratio scales (similar to the Kelvin or absolute scale). Each method has found its adherents and uses, and each possesses specific advantages and defects that indicate its use for one application, but contraindicate its use for another. Those applications will be noted where appropriate.

1. Threshold Measures

The "delicacy" of the sense of taste, as the threshold was labeled by Bailey and Nichols (1887), measures the lowest physical level at which taste is either detected (called the detection threshold) or recognized as to its quality (called the recognition threshold). Recognition thresholds are usually higher than their detection counterpart—more substance is needed in taste to provoke a response about "type" than to recognize the presence or absence of the chemical agent. Stimuli lower than threshold are usually not detected.

Thresholds are not "fixed points" along the scale of concentrations but may vary because of a host of factors. The method of presenting the taste material may introduce biases, the variety of threshold measuring techniques may introduce a constant error, and the fluctuations of attention may yield a threshold level higher than would be found were the observer to maintain a strict focus on the threshold detection task.

Some representative values for threshold compiled by Pfaffmann (1959b) are shown in Table I. There is a large variation of physical concentrations needed by different chemicals to reach threshold; thus, artificial sweeteners such as saccharin are far more potent than sugar. Even within a taste compound there is a considerable variation of published threshold values. The reader is referred to standard texts in the field that discuss the assessment of thresholds, both for taste and for the other senses (e.g., Woodworth and Schlosberg, 1954).

It has been tempting to some to use ratios of threshold values as a rapid assessment of the relative taste intensities of diverse compounds. In this system, dulcin, an artificial sweetener used in Germany during the First World War, would be quoted as being 70 to 350 times sweeter than sugar (Dermer, 1946), because its threshold is 1/70 to 1/350 that of sucrose. The ratio should be labeled something different to distinguish it from a true measure of the relative taste impressions produced by two substances at fixed concentrations. *Taste potency* may be more appro-

TABLE I

Sweet Thresholds in Man (in Molar Concentrations)[a]

Substance	Molecular weight	Median	Range
Sucrose	342.2	0.01[b]	0.005–0.016
		0.17[c]	0.012–0.037
Glucose	180.1	0.08	0.04–0.09
Saccharin (sodium)	241.1	0.000023	0.00002–0.00004
Beryllium chloride	80.0	0.0003	
Sodium hydroxide	40.1	0.008	0.002–0.012

[a] From Pfaffmann (1959b).
[b] Detection threshold.
[c] Recognition threshold.

priate, in order to distinguish the two. Relating threshold matches to each may be seen to be nothing more than allowing two chemicals to be matched on the basis of a sensory criterion (*viz.*, threshold). The match is of the class called "null balances." The only measurement that occurs on the part of the human subject is the balancing one. Numbers elicited from the procedure must derive from the different measures of physical concentration.

Thresholds are useful to the psychophysicist because they tell him the level of physical energy at which sensation begins. To the model builder in the chemical senses they provide the relative amounts of different chemicals that evoke precisely the same sensory effect, namely, threshold. Finally, to food scientists and technologists the threshold values are important because they tell him the physical concentration at which taste quality provided by the chemical becomes noticeable and participates in the overall impression of a flavor. It should always be recalled, however, that extrapolations from threshold to levels clearly perceived as sweet may bear little or no relation to relative sweetness values across different chemicals. Extrapolations at their best are too simple, and at their worst highly misleading.

2. Equal Sweetness Matches

The logic of threshold measurement, i.e., the balancing of two chemicals to achieve the same sensory response, may be extended above the threshold. With equal-sweetness matches one can do this by determining the concentration of different sweeteners that the observer finds equally intense. Usually one substance (often sucrose) is fixed as the reference, and the other varied to match it. Sometimes the matches are done in both directions.

A number of equal-sweetness values, in addition to representative threshold ones, are listed in Table II. They represent values taken from Dermer's considerably longer list, detailing published values for the four tastes as of 1946 (Dermer, 1946). They also show the distribution of values across different experimental conditions and laboratories. It is worthwhile noting that the range of concentrations needed to match sucrose varies from a minimal of 1/4100% (for the artificial sweetener 1-propoxy-2-amino-4-nitrobenzene discovered by the Dutch chemist Verkade) to the minimally sweet sugars lactose, quebrachitol, and raffinose.

A series of equal-sweetness matches between two substances, when they vary in concentration, trace out an "equal-sweetness contour." The function so described is far more valuable than the individual estimates of

TABLE II

Relative Sweetness of Various Substances[a]

Substance	Relative sweetness	Method
1-Propoxy-2-amino-4-nitrobenzene	5000	—
	4100	1% sucrose
	3300	1% sucrose
1-Allyloxy-2-amino-4-nitrobenzene	2000	1% sucrose
a-Antiperillaldoxime	2006	—
1-ethoxy-2-amino-4-nitrobenzene	1400	1% sucrose
	1000	—
	950	—
6-Iodo-3-nitroaniline	1250	1% sucrose
1-n-Butoxy-2-amino-4-nitrobenzene	1000	1% sucrose
6-Bromo-3-nitroaniline	800	1% sucrose
syn-5-Benzyl-2-furfuraldoxime	690	2% sucrose
Saccharin, as sodium salt	675	2% sucrose
	200–700	Varied sucrose
	190–675	Varied sucrose
1-Isopropoxy-2-amino-4-nitrobenzene	600	1% sucrose
n-Amylchloromalonamide	400	2% sucrose
6-Chloro-3-nitroaniline	400	—
6-Chlorosaccharin	ca. 340	—
4-Nitro-2-aminotoluene	330	1% sucrose
1-Methoxy-2-amino-4-nitrobenzene	330	1% sucrose
	300	—
	220	1% sucrose
1-Propoxy-2-amino-4-nitro-6-methylbenzene	310	1% sucrose
n-Hexylchloromalonamide	310	6% sucrose
N-Methyl-N-p-ethoxyphenylurea	ca. 265	—
p-Ethoxyphenylurea	265	2% sucrose
	70–350	Varied sucrose
Furylacrylonitrile	200	2% sucrose
2-Amino-4-nitrophenol	200	1% sucrose
"p-Methylsaccharin"	200	—
Sodium N-cyclohexylsulfamate	170	
	70	Threshold
2-Nitro-4-aminobenzoic acid	120	1% sucrose
Sodium aminotriazinesulfonate	100	
Anti-5-benzyl-2-furfuraldoxime	ca. 100	2% sucrose
Furonitrile	100	2% sucrose
Sodium 2-thiazolylsulfamate	55	—
Antiphenylacetaldoxime	50	—
m-Nitroaniline	40	1% sucrose
6-Flouro-3-nitroaniline	40	1% sucrose
Chloroform	40	—
Ammonium N-cyclohexylsulfamate	35	—

TABLE II (*Continued*)

Substance	Relative sweetness	Method
Sodium salicylate	28	1% sucrose
4-Nitro-2-aminobenzoic acid	25	1% sucrose
syn-Phenylacetaldoxime	25	—
p-Methoxyphenylurea	18	—
Sodium *N*-(2-methylcyclohexyl) sulfamate	17	—
Dichloromalonamide	9	6% sucrose
Ethylchloromalonamide	9	6% sucrose
n-Propylchloromalonamide	9	6% sucrose
Isopropylchloromalonamide	9	6% sucrose
n-Butylchloromalonamide	9	6% sucrose
Salicylic acid	4	1% sucrose
Chloromalonamide	3	6% sucrose
Methylchloromalonamide	3	6% sucrose
Furfuraldoxime	2.5	2% sucrose
DL-Erythritol	2.4	Threshold
Fructose	1.7	Threshold
	1.35	Threshold
	1.03	3% sucrose
	1.08	3% sucrose
	1.11–1.20	Varied sucrose
Ethylene glycol	1.3	Threshold
	0.49	3% sucrose
DL-Alanine	0.93–1.70	Varied sucrose
	0.92	3% sucrose
Pentaerythritol	1.1	Threshold
Glycerol	1.08	Threshold
	0.56–0.74	Varied sucrose
	0.48	3% sucrose
Sucrose	1.00	(The standard)
L-Arabitol	1.0	Threshold
Glycine	0.46–1.19	Varied sucrose
Glucose	0.62–1.00	Varied sucrose
	0.53–0.88	Varied sucrose
	0.53–0.80	Varied sucrose
	0.52	3% sucrose
	0.80	Threshold
	0.75	Threshold
	0.74	Threshold
i-Dulcitol	0.74	Threshold
	0.41	3% sucrose
D-Alanine	0.73	3% sucrose
Sarcosine	0.62	3% sucrose
D-Mannitol	0.57	Threshold
	0.45	3% sucrose

TABLE II (*Continued*)

Substance	Relative sweetness	Method
DL-Sorbitol	0.54	Threshold
	0.48	3% sucrose
Inositol	0.50	Threshold
D-Xylose	0.40	Threshold
Maltose	0.36–0.57	Varied sucrose
	0.45	Threshold
	0.32	Threshold
Rhamnose	0.32	Threshold
Galactose	0.32	Threshold
Lactose	0.31–0.37	Varied sucrose
	0.33–0.60	Varied sucrose
	0.27–0.28	3% sucrose
	0.31	Threshold
	0.16	Threshold
Quebrachitol	ca. 0.3–0.5	—
Raffinose	0.22	Threshold

[a] From Dermer (1946).

matching sweetness at isolated points of concentration. One immediate advantage is that the contour provides a handy means of computing any desired match of equal sweetness to the reference sugar, even one that has not been directly tested in experimentation. A second advantage, which becomes of great interest in the study of the sweetness mechanism, is the relative growth of sweetness provided by the contours. If a small percentage increment of sugar (sucrose) requires a substantially larger percentage increment of saccharin to maintain the sweetness matches, then one may infer that sucrose sweetness grows more rapidly with concentration than does saccharin sweetness.

Figure 3 shows a series of equal-sweetness contours derived from the experimental work of Cameron (1947), a Canadian researcher, who in the 1940's considerably advanced the science of sweetness measurement. The contours are arranged first in linear coordinates and then in double logarithmic (log–log) ones. In both parts of the figure sucrose has been assumed to vary in order to match the sweetness of selected levels of fructose, dulcin, glucose, etc. In linear coordinates the equal-sweetness contours are almost straight lines described by the equation $S = kC$ (where S is the matching level of sucrose, C the concentration of the criterion sugar, and k the constant of proportionality). When k exceeds 1.0, e.g., for dulcin, sucrose is the less potent sweetener. When the proportionality constant is less than 1.0, then the sugar in question is less potent,

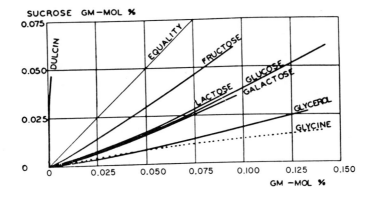

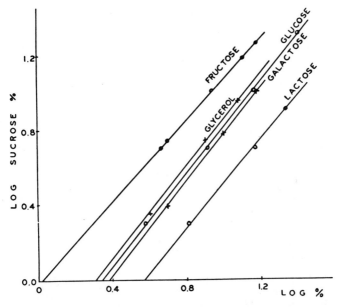

Fig. 3 Equal-sweetness matches between sucrose and other sweeteners. The top figure represents matches expressed in linear coordinates. The bottom figure represents similar matches but plotted in log–log coordinates (data from Cameron, 1947).

by the factor $1/k$. The conversion of the matches to log–log coordinates, as shown by the bottom of Fig. 3, linearizes the function, and in several cases removes severe distortion in the match contour. For log–log coordinates the straight line is written: $\log S = 1.0 \log C + \log k$. The slope of the line is unity, and the intercept corresponds to the logarithmic value of the proportionality constant.

Not all equal-sweetness contours are described by lines with slope equal to 1.0 in log–log coordinates. Figure 4 shows some previously published data for matches between sucrose and saccharin taken from tabulated values by the German investigator teams Taufel and Klemm (1925) and Magidson and Gorbatchow (1923). The slope is steeper than 1.0 when sucrose is the independent variable. In fact, the data are best expressed by the function $\log Sa = 1.7 \log Su + \log k$ (where Sa is the matching saccharin level corresponding to Su, fixed sucrose level). In linear coordinates the matching concentrations of saccharin and sucrose trace out a curved function, with severe convexity, or concave upward. The equation

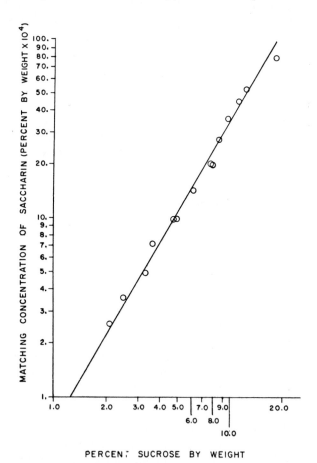

Fig. 4 Equal-sweetness matches between sucrose and saccharin, plotted in log–log coordinates (adapted from Moskowitz, 1970). Slope = 1.7. Data from Magidson and Gorbachow (1923) and Taufel and Klemm (1925).

would then be expressed as $Sa = k(Su)^{1.7}$. In a simple example, an increase in the level of sucrose from 1 concentration unit to 10 units requires more than a tenfold increase in saccharin to maintain the equal-sweetness match. The saccharin must be increased by the amount $(10)^{1.7}$, or approximately 50 times.

The importance of such equal-sweetness contours for a measurement of sweetening ability is that saccharin obtains varying estimates of sweetness depending upon the criterion level of sucrose. In addition, however, the suggestion made by the contour for saccharin is that a fixed (e.g., 10) percentage change in concentration of sucrose produces similar sweetness changes for other sugars, but a much lower sweetness change for matched saccharin. The conclusion, in terms of measuring the change of sweetness with concentration, is that sugar sweetness is far more affected by increments and decrements in concentration than saccharin is, and that all sugars appear to be equally affected.

From threshold studies, single estimates of matches in sweetness and equal-sweetness contours, one can obtain some indication of the relative capacity of different chemicals to sweeten foods. None of the techniques, however, truly provides an appropriate measure of subjective *sweetness* per se. Each method assumes that the numerical manipulations for relative intensity values remain within the domain of physical concentrations; thus, all ratios of intensities are taken with respect to matching concentration values. For sensory analysis to proceed further it is vital that appropriate methods of measurement be used to quantify the sensory responses and to measure the perceptions of sweetness, not simply the physical aspects of the sweetening agent.

3. Subjective Sweetness Scales

Measurement, according to one psychophysicist (Stevens, 1951) is primarily the technique of assigning numbers to things according to rules. Weights, temperatures, heights, and masses are provided with appropriate numbers according to the specifiable rules of physics. Sensory measurement is no different. The past three decades have witnessed the development of two different approaches to sensory scaling. Each approach comes equipped with its own set of guiding assumptions about the operations used by the human observer in making his subjective measurements, and about the potential uses and limitations of the measurements themselves.

The two methods, category and ratio scaling, can be best understood by comparing them to the standard centigrade and Kelvin (absolute scale) scales of temperature. Category or interval scales of intensity are akin to centigrade temperature measures. The scales do not possess a

fixed zero, denoting lack of strength. The zero can shift anywhere up
and down the scale. The only fixed piece of information that can be
obtained is the magnitude of difference between two scale values. In
the centigrade scale readings of 100 and 50, for example, do not mean
that the higher is twice the lower. Indeed, the zero point could be shifted
to +40, in which case the two values would then become $(100 - 40 = 60)$
and $(50 - 40 = 10)$. In that case, the ratio would be shifted to 60/10, or
6 to 1. Instead, the difference, 50 degrees on the scale, is the only invariant,
or fixed information that the scale provides. The ratio scale is more rigid
in the allowable transformation. Since the zero point is fixed, the ratio
of 100 and 50 on the absolute or Kelvin scale means 2/1.

Category or interval (centigrade-type) scales long have been popular
as tools for measurement. In their simplest form the observer is instructed
to assign to each taste stimulus its appropriate category value, much like
selecting a temperature from a set of permissable values. The category
scale may contain as few as two categories, or upward of 20 or more,
depending upon the measurement goal. Often, for the convenience of the
observer, the experimenter will label the categories as follows:
"tasteless," "weak," "moderate," "strong," "extremely intense," etc.

The outcome of one extensive series of studies with the category procedure is shown in Fig. 5. The curves, obtained by Schutz and Pilgrim
(1957), show how the ratings for sweetness increase with the logarithm
of concentration. As Fig. 5 shows, some curves (e.g., alanine) are concave
upward; thus, the category ratings increase faster than the logarithmic
value of concentration. Other curves may be concave downward or ap-

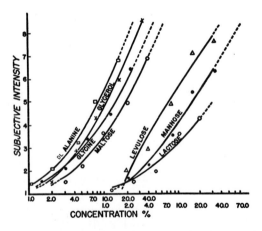

Fig. 5 Category scaling of sweeteners. Average rating is presented as a function
of log concentration. [Reprinted from Schutz and Pilgrim (1957). Copyright © by
Institute of Food Technologists.]

proximately linear. Steep curves indicate that the sweetness ratings grow faster with concentration than curves exhibiting less steep ones.

Category or interval scales are simple to use by the observer and are easily analyzed by the experimenter. Thus, the procedure of measurement is tempting to adopt as a reliable standard technique. It is appropriate for applications such as quality control, and as a rapid method for assaying relative sweetness for diverse materials. However, there are severe limitations to this method, just as in physics there are limitations to the Fahrenheit and centigrade scales. The lack of a meaningful zero point prevents the experimenter from gaining a true measure of percentage sweetness relative to a standard, and he must be content with differences in sweetness. In the category measurement, this shows up in statement of the form "Sugar A is 4 units sweeter than sugar B." No information is available as to whether A is twice as sweet as B, five times as sweet, etc.

Ratio scaling, akin to the Kelvin scale of temperature, is the strongest form of measurement known today and provides the greatest potential for scientific and practical applications. Its aim is to assign numbers to stimuli (here sweeteners, but often loudness or brightness) so that ratios of numbers reflect ratios of subjective intensities. In the ideal situation, a sweetener (e.g., 1 M glucose) assigned the value 10 would be 33% as sweet as another sweetener (e.g., 1 M sucrose) assigned a value 30.

To effect this measurement, psychologists have devised a method known as *magnitude estimation*. The observer samples the first stimulus (called the standard) and assigns it a convenient, positive number (called the "modulus" number). Both the standard and the modulus number may be randomly selected since only the ratio of numerical assignments conveys information. In some cases the experimenter may wish to designate a certain sweetener at a fixed concentration as the standard and request that the observer assign it a prearranged value. The observer then assigns to the subsequent stimuli numbers that reflect relative intensities, referring back to the standard. If the second taste is twice as strong (or sweet) as the first, he assigns it a number twice the first, whereas if it is one-ninth the first, he assigns it a number one-ninth as large.

Usually, but not always, the measurements fall along a curvilinear function when the geometric mean of the estimates is plotted against physical concentration of the sweetener. A logarithmic transformation serves to linearize the curve, and it has been suggested that the method of magnitude estimation yields a *power function* for judgments of sweetness against sweetener concentration (Moskowitz, 1970):

$$\text{Sweetness} = k(\text{concentration})^n \quad \text{or} \quad \log(\text{sweetness}) = n \log(\text{concentration}) + \log k$$

The exponent of the power function, n, measures how rapidly the estimated sweetness grows with concentration. If n is greater than 1.0, then changes in concentration are reflected in increasing returns on sweetness. If n equals 1.0, then sweetness and concentration grow commensurately, whereas if n is less than 1.0, changes in concentration are reflected in diminishing returns; thus, sweetness fails to grow as rapidly.

A long series of studies aimed at uncovering representative values for sweetness exponent n suggest that for sugars the value lies around 1.3 (Moskowitz, 1970, 1971). Table III shows a list of exponents for sugar sweetness obtained by various experimenters who have used the procedure of magnitude estimation (Meiselman *et al.*, 1972). The exponent value of 1.3 means that sweetness for this class of substances grows faster than concentration. A tenfold change in concentration for glucose, for example, produces a twentyfold change in the sweetness judgment. The exponent value seems to be approximately constant for most sugars so far tested

TABLE III

Exponents for Sucrose Sweetness in Power Functions of the Form[a] (sweetness) = k (concentration)^n

Sucrose[b]	Flow
2.93	Production of taste to match number
1.80(S)	
1.62	Geometric mean
1.60(S)	
1.47(S)	
1.40(S)	
1.40(t)	
1.30(t)	
1.30(S)	
1.10(S)	
0.98(t)	
0.93	Estimation of taste intensity
0.79(D)	
0.70(t)	
0.67(D)	
0.62(t)	
0.46(D)	

[a] From Meiselmann *et al.* (1972).
[b] S stands for estimates of sweetness; t stands for estimates of total taste intensity; and D stands for dorsal flow of liquid over the tongue.

with the possible exception of mannose. Mannose has a bitter side-taste that interferes with its sweet aspect.

Not all sweeteners are governed by the exponent 1.3, or similar values. Saccharin and cyclamate, for example, are probably lower (Moskowitz, 1970). Good estimates for saccharin lie around 0.6–0.8, and for cyclamate between 0.8 and 1.0. An increase of 10 times in the concentration of each of these sweeteners will produce, therefore, either the same, or, more likely, a lower increase in reported sweetness. The cause for this difference among sugars and artificial sweeteners is not known, although one possibility is the bitter taste that inheres in the overall perception of the taste of artificial sweeteners. At the same time that sweetness increases with concentration (potentially at the same rate as sugar sweetness grows), the off-taste, usually bitter, increases as well. The two may suppress each other partially, allowing a diminished sweetness and bitterness to emerge. At higher concentrations, the suppression may become increasingly severe.

Sensory measurement of sweetness does not stop after the intensity curves are determined for diverse sweeteners. These curves may be used to determine the sweetness ratios of different sugars. Figure 6 shows a tree of sugars. Each sugar is associated with two numbers in parentheses. The left-hand number gives the relative sweetness of 1 mole of the sugar relative to 1 mole of glucose; for example, sucrose at 1 mole is rated at 3.20, compared to glucose, which is rated at 1.0. In terms of the observer's behavior, a mole of sucrose would obtain a magnitude estimate approximately 3.2 times that given to a mole of glucose. On the right-hand side the same index numbers apply, except that the numbers are relative to a value of 1.0, assigned to 1% of glucose in solution; 1% sucrose would be rated 1.4 times higher than 1% glucose (Moskowitz, 1971).

Other uses of ratio scales are listed in Table IV. There are three different sets of results listed. One is a study of the effects of viscosity upon perceived taste intensity. A power function $(T = kV^n)$ predicts how taste intensity (T) diminishes with physical viscosity if a single concentration of sweetener is mixed with successively more viscous solvents. The exponent n is -0.25 if the viscosity-imparting agent is the compound sodium carboxymethyl cellulose. The power function means that it requires a 10,000-fold *increase* in viscosity to drop taste intensity to one-tenth of its original value.

The second set of studies concerns the additive nature of sweetness when two sweeteners are sampled together in a single sip. It appears possible to predict mixture sweetness by adding together the sweetness estimates (or the sweetness functions) of the components. The generalization

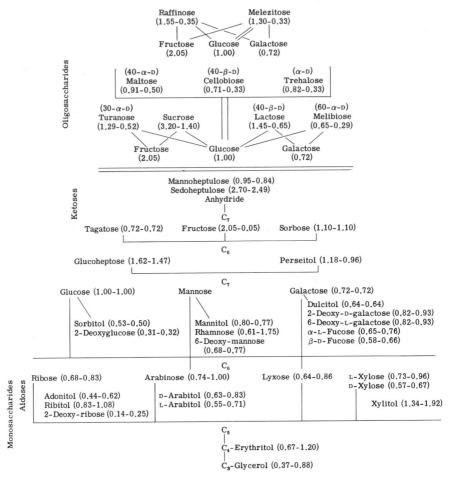

Fig. 6 "Tree" of relative sweetness. Numbers in parentheses on the left represent relative sweetness of the sugar compared to the sweetness of 1 *M* glucose (arbitrarily assigned a value of 1.0). Numbers on the right represent the same relative sweetness, but compared to 1% glucose (assigned a value 1.0). Figure from Moskowitz (1971).

of this additivity model may also apply to mixtures of acids with each other as well (Moskowitz, 1973).

Finally, the third set of studies concerns how tastes may suppress each other. If a sweetener (either glucose or fructose) is mixed with salt, citric acid (sour), or quinine sulfate (bitter), the mixture is less intense than the components. The degree of suppression can be assessed through the method of magnitude estimation and ratio scaling. It appears that the mixture strength is reduced to about 60% of its component strength, al-

TABLE IV

Extensions of Intensity Measurement of Sweetness

I. Effects of viscosity upon perceived sweetness (Arabie and Moskowitz, 1971)
 Usual effect: Reduction of sweetness
 Characteristic function: $S = kV^{-0.25a}$

II. Sweetness of glucose or fructose in a mixture with either NaCl, citric acid, or quinine sulfate (Moskowitz, 1972a)
 Usual effect: Reduction of sweetness as well as reduction of the other taste
 Characteristic function: $S_{mix} + O_{mix} = k(S_{sim} + O_{sim})^b$

III. Sweetness of mixtures of diverse sweeteners (Moskowitz, 1973)
 Usual effect: Sweetness adds by one of two different ways
 Model I (summation of apparent sweetness)—the gustatory system adds together the sweetnesses of the components
 Model II (summation of concentrations)—the gustatory system treats the two sweeteners as a higher concentration of a single sweetener, adds the concentrations, and then converts the sum to an impression of sweetness
 Characteristic functions: Model I $S_{a,b} = k_a C_a{}^m + k_b C_b{}^{n\ c}$
 Model II $S_{a,b} = k_a[C_a + (k_b C_b{}^n/k_a)^{1/m}]^{m\ c}$

---

[a] Here S stands for sweetness and V for viscosity in centipoises. Concentration fixed for glucose or saccharin.

[b] Here S stands for sweetness; O for intensity of saltiness, sourness, and bitterness; mix = intensity rated in a mixture; sim = intensity of same concentration rated in unmixed, simple solution. Values for k are approximately 0.59 for mixtures with NaCl, 0.49 for mixtures with citric acid, and 0.46 for mixtures with quinine sulfate.

[c] Here C_a stands for concentration of sweetener a, S_a for sweetness of sweetener a, and $S_{a,b}$ for sweetness of the mixture.

though the components themselves may not be reduced by the same degree (Moskowitz, 1972a).

In retrospect, the choice of an appropriate measuring tool for the intensity of subjective sweetness concerns two questions that the user must answer: (a) How easy is the scale to use? (b) How powerful are the measurements that are derived, what conclusions can be drawn, and how meaningful are the numerical operations that may be necessary to draw those conclusions? Ratio scales are the most versatile scale known today. On one hand, they provide the most powerful measurements and have the greatest potential for uncovering various perceptual aspects of sweetness; on the other hand, they may be, on occasion, difficult to construct. Where simple scales are needed, e.g., for quality control, ratio scales may not be required, and categories or intervals may suffice. With out ratio scales, however, useful and general equations for sweetness are difficult, if not impossible, to obtain.

C. The Pleasantness of Sweetness

Hedonics, or the attribute of pleasantness–unpleasantness, is inherent in our conception of taste qualities, and highly so for the sweet taste. Colors, sounds, and pressures may evoke memories and conjure up impressions, and by doing so produce a hedonic response of liking or disliking. The hedonic response is probably secondary, however, and does not inhere in the sensation or perception of the stimulus. On the other hand, tastes and smells appear to produce reactions of liking and disliking. Few people appear to enjoy the taste of bitter water when quinine sulfate is dissolved in the taste solution. However, when the stimulus is a bitter drink, the taste, formerly unpleasant, now becomes an acceptable and expected one.

Reactions that are labeled "hedonic" may occur in invertebrates as well as in higher animals. It is, of course, difficult to ask an animal without the capacity of language to indicate its degree of liking or disliking of a food, but a criterion response, *viz.*, extending the tongue to drink or withdrawing from the stimulus, may be used instead as an indicator. Some flies possess specific sense organs that react to sugars, and in the presence of the so-called pleasant-testing stimulus, the fly extends its proboscis to the sugar water (Minnich, 1926).

In humans, it is still not conclusively known whether neonates or even babies like sweets. They react as if they did, however. As the child grows he is continually surrounded by sweet foods, including candies. Anecdotal reports suggest that the child actively accepts and eats candies and sweetened foods that adults find overly sweet and distasteful. The novice drinker often is seen to prefer mixed, sweet drinks, although with age the behavior shifts, so that preferences for more bitter, and "straight" drinks becomes the norm. Some of this change undoubtedly results from social influences and expectations, but a part can be correlated with the possible loss of acuity to sweetness, and the growing experience of the individual with foods that possess nonsweet tastes.

Much of the published information on taste preferences deals with "model systems," usually sugar solutions, with an occasional flavor added to make the solution appear more like a food. Much valuable information on hedonics of taste has been published by Young (1966), but those studies have concerned animal preferences for sugar solutions, and for sugar solutions adulterated by unpleasant tastes such as that imparted by bitter quinine. Much of what we know, however, may pertain only to these artificial systems, and may fail in generalizability to the real world situations of complex foods, textures, and flavors.

One of the earliest questions that was asked in studies of human preferences toward taste stimuli concerned whether humans prefer sweet solutions to nonsweet ones, and whether increasing sweetness at the same time increase pleasantness. Engel (1928) performed a relatively comprehensive study to answer these two questions. He presented his observers with different concentrations of sugar, salt, acid, and quinine, and asked them to indicate, for each taste stimulus, whether they were indifferent, liked the taste, or disliked it. In general, for salty, bitter, and sour the increasing concentrations produced a lower proportion of tasters finding the taste pleasant. Figure 7 shows the percentage of respondents who found cane sugar in water solution acceptable with varying concentrations of sugar. At a level of 9% the majority of the population stabilized in their hedonic impression. With increasing concentration a few individuals actually began to dislike the taste and no longer classed the stimulus as a pleasant one.

Information obtained by the percentage statistic, or the rough classification of tastes into the categories pleasant, neutral, and unpleasant is a valuable first step in the study of hedonics. However, one cannot learn about the way in which *relative* pleasantness or acceptance of a sweet taste changes with concentration. Are sweeter solutions actually more pleasant than less sweet ones, even though both are pleasant to begin with? Within any category there is no gradation of degree of liking or disliking.

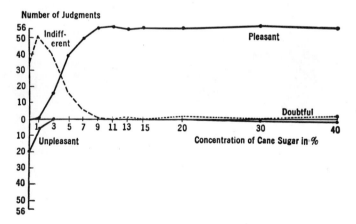

Fig. 7 Percentage of the population reporting "pleasant," "indifferent," and "unpleasant," to varying concentrations of sucrose (data from Engel, 1928).

More quantitative estimates of acceptability can be obtained by the procedure of category scales, which was discussed with respect to measurements of sweetness. The typical category scale for hedonic measurement is called the 9 point Hedonic Scale. Originally developed by the U.S. Government for the sensory assessment of foodstuffs (Peryam and Pilgrim, 1957), the Hedonic Scale has shown itself to be a robust measuring tool for human judgments of food and flavor acceptance. The categories 1–4 represent degrees of disliking (1 indicates most disliked), category 5 is reserved for the neutral or indifference judgment, and categories 6–9 represent increasing degrees of liking. Ratio scales of pleasantness, still in the development stage, provide even more powerful measurements of hedonic responses to sweetness. Percentage or relative acceptability compared to a standard is useful in comparing responses to different sweeteners (sugars vs. artificial sweeteners, one sugar vs. another), and the ratio scale can provide that percentage value conveniently.

Figure 8 shows a set of curves relating human judgments of both sweetness and pleasantness to concentration (Moskowitz, 1971). The curves,

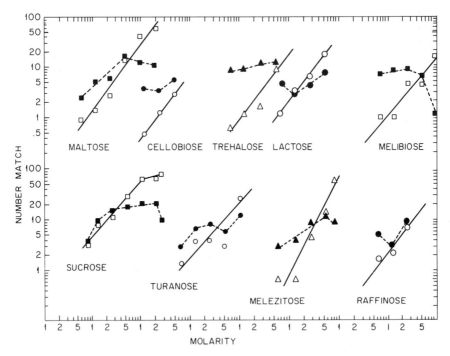

Fig. 8 Changes in (——) sweetness and (- - -) pleasantness (acceptability) as a function of sugar concentration. The coordinates are log–log (data from Moskowitz, 1971).

representing ratio scales, were obtained by the method of magnitude estimation. In one session a group of observers estimated the sweetness of various concentrations of different sugars. In another session, run one day later, another group of observers estimated the degree to which they found the sugar solutions pleasing to the taste.

In Fig. 8 one can compare how rapidly the judgments of both sweetness and pleasantness increase with increases in concentration. In virtually all cases sweetness increases more rapidly than pleasantness does, even though both judgments are presumably related to sugar concentration. Hence, the attribute of pleasantness is entirely distinct from that of sweetness, and the two subjective concepts conform to entirely different laws of growth. At high concentrations of sugar the pleasant taste may actually diminish, presumably because intensely sweet concentrations of sugar are no longer as palatable, even for simple tasting without swallowing.

Measurements of the hedonic or pleasantness functions, in contrast to asking only "like" or "dislike," are useful for still a further application. If the relative acceptabilities of different sugars are of interest, either at a common concentration or at a common sweetness level, then one needs only refer to the functions shown in Fig. 8. At fixed concentrations the relative acceptability may be read directly from the graph. At fixed sweetness levels, one need only locate the concentrations that correspond to the sweetness level and then compare the two estimates of acceptability or pleasantness.

Before leaving model systems and inquiring about actual food products, one final question is appropriate. Does the point at which pleasantness or acceptability drop off at the high end of the concentration scale correspond to either a fixed sugar concentration, a fixed sweetness level, or neither? Figure 9 presents one approach to an answer. The previous data, shown in Fig. 8, have been replotted with the estimate of pleasantness as the ordinate and the estimate of sweetness as the abscissa value, respectively. In addition, data from other sugars have also been added. Again the coordinates are log–log. The overall trend is for the relation to assume a linear form with a slope (or exponent of the power function) varying between 0.3 and 0.5. A break in the line occurs between 50 and 100 units of sweetness at which region the curve drops. Because the slope is relatively low, and considerably less than 1.0, a 10-fold increase in the sweetness of a sugar can be expected to produce no more than a 2.5 to 3.2-fold increase in the report of acceptability.

In the real world, food materials encountered are significantly more complex than the sugar solutions tested in the laboratory. Hedonics of food can be severely affected by taste, but the interplay of color, aroma,

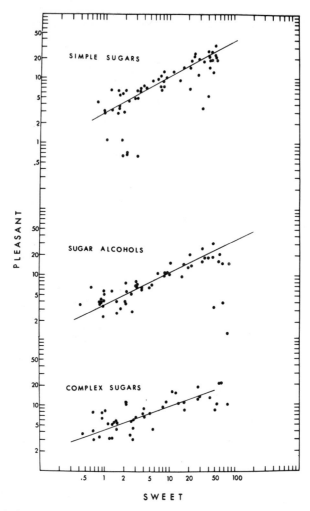

Fig. 9 Relation between pleasantness judgments and sweetness judgments for different sugars. Each point represents a different concentration of sugar, based upon the mean of 28–36 individual observation for sweetness and for pleasantness, respectively. The coordinates are log–log and the slope (up to 50 sweetness units) lies between 0.3 and 0.5. Data from Moskowitz (1971).

texture, and the experience of the observer can often overshadow the presence or absence of a desired taste in a food. In this respect, the addition of sweet tastes to foods is especially interesting. Often, the sweetness of a beverage, especially soda, may be counterbalanced by the realization that the sugar sweetening agent contains calories. To those who are conscious of their weight such considerations produce consummatory behavior not in accordance with one's expectation, were taste alone to be considered. Thus, in order to properly investigate the role of hedonics in sweetness of food, the psychologist must focus upon a potentially wider variety of behaviors, especially when sweetness connotes and contains aspects that are experiential, motivating, but not simply gustatory. And yet, paradoxically, in order to understand fully the taste triad of sweetness quality, intensity, and acceptability the psychologist must necessarily limit the range of his vision in order to manipulate his stimuli in a well-controlled manner. Only by exploring in depth the facets of the observer's behavior toward sweet foods can the investigator hope to arrive at significant statements of generality.

One appropriate extension of the laboratory analysis of taste hedonics into real world stimuli concerns the generality of the sweetness-pleasantness equation, previously shown to obey a power function $P = kS^n$ (n between 0.3 and 0.5). Are the equations for foods members of the power function family, and, if so, are they governed by the same exponent as that of sugar solutions in water?

Figure 10 presents the results of one study (H. R. Moskowitz, unpublished manuscript, 1972), in which three products—cherry-flavored beverage drink, vanilla pudding, and white cake—were prepared according to standard recipes, except that sucrose was replaced by four other sugars. The sugars—maltose, glucose, fructose, and sorbitol—were substituted by weight, but the amounts were selected so that in water solutions they each matched the sucrose in sweetness.

Since the sugars were originally rated as equally sweet, it is noteworthy

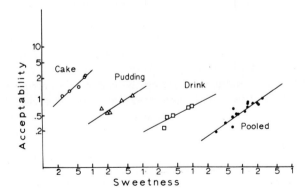

Fig. 10 Relation between acceptability (pleasantness) judgments and sweetness judgments for different foods. Each point represents the judgments of 20 observers for sweetness and another 20 observers for pleasantness. The points represent the judgments made to products in which sucrose was replaced by sorbitol, glucose, fructose, and maltose.

that they differed considerably in the finished food product. More un-usual, however, is the reappearance of the sweetness-pleasantness func-tion in a form similar to that previously seen for the taste of sugars in water. The relationship again appears to be fairly well described by a power function. The exponent is 0.64 for the beverage, 0.49 for the pud-ding, and 0.80 for the cake. In a practical sense, one may infer that small changes in the sweet taste does not always produce the same change in acceptability. Some foods, such as beverages and puddings, can tolerate more variability in sweetness and lose less of their acceptability than others such as cake.

As psychologists explore sweetness, and indeed the chemical senses, they are constantly required to emulate Janus—looking one way toward the behavior of model systems in the search for regularities and laws, but also to actual foods, where consumption occurs and where regularities give way to irregularities and laws of behavior to abundant exceptions.

D. Research Needs

A psychology of sweetness, or of flavor perception as a whole, grows and develops *de facto* from the work of numerous sensory scientists. Their investigations into the relationships between physical properties and taste perception continue to provide pieces of the developing picture.

Integrated studies on taste may well require the convergence of diverse disciplines, both for substantive input and more importantly for direction. Areas of potential application can be divided into (a) those investigating the behavior of the human organism, his nutritional state, etc., in which the focus is upon taste as a biological process; and (b) those investigating the properties of taste chemicals, with the human being serving as an "instrument," on par with other hardware equipment in the laboratory.

In the first group, studies concerning the changes in perception of taste and attractiveness of food with biological state have received some atten-tion (Cabanac, 1971; Jacobs and Sharma, 1969) and are continuing to attract investigators. Research investigations into the relations between physiological or nutritional state and taste perception and hedonics may show unexpected correlations between mental or experiential pro-cesses and sensory ones. In the second group are those studies that concern the measurement of sweetness, with special reference to the sweetening material rather than to the human organism. In this group are emerging systematic studies on effects of physical parameters upon sweet-ness, such as viscosity. Continuing research might disclose to food scien-tists and technologists ways of modifying sweetness perception in order to produce a more desirable product. With the increasing requirements of nutritional adequacy for the world's undernourished, knowledge of these modifications may prove ultimately to be the most significant step in fostering the acceptance of nutritious, fabricated foods.

References

Arabie, P., and Moskowitz, H. R. (1971). The effects of viscosity upon perceived sweetness. *Percept. & Psychophys.* **9,** 410–413.

Bailey, E. H. S., and Nichols, E. L. (1887). On the delicacy of the sense of taste. *Publ. Amer. Ass. Advan. Sci.* p. 138.

Beebe-Center, J. G. (1932). "The Psychology of Pleasantness and Unpleasantness," p. 176. Van Nostrand-Reinhold, Princeton, New Jersey.

Boring, E. G. (1942). "Sensation and Perception in the History of Experimental Psychology," p. 453. Appleton, New York.

Cabanac, M. (1971). Physiological role of pleasure. *Science* **173,** 1103–1107.

Cameron, A. T. (1947). The taste sense and the relative sweetness of sugar and other sweet substances. *Bull. Sugar Res. Found.* pp. 1–72.

Chamberlain, A. F. (1903). Primitive taste words. *Amer. J. Psychol.* **14,** 146–153.

Dermer, O. (1946). The science of taste. *Proc. Okla. Acad. Sci.* **27,** 9–20.

Engel, R. (1928). Experimentelle Untersuchungen uber die Abhangigkeit der Lust and Unlust von der Reizstarke beim Geschmacksinn. *Pfluegers Arch. Gesamte Psychol. Menschen Tiere* **64,** 1–36.

Erickson, R. P. (1963). Sensory neural patterns and gustation. *Olfaction Taste, Proc. Int. Symp., 1962* pp. 205–213.

Greeves, A. (1828). "An Essay on the Varieties and Distinctions of Tastes and Smells and on the Arrangement of the Materia Medica." Edinburgh.

Jacobs, H. L., and Sharma, K. N. (1969). Task vs. calories: Sensory and metabolic signals in the control of food intake. *Ann. N.Y. Acad. Sci.* **157,** 1084–1125.

Magidson, O. J., and Gorbachow, S. W. (1923). Zur Frager der Sussigkeit des Saccharins. *Ber. Deut. Chem. Ges. B* **56,** 1810–1817.

Meiselman, H. L., Bose, H., and Nykvist, W. (1972). Magnitude production and magnitude estimation of taste intensity. *Percep. & Psychophys.* **12,** 249–252.

Minnich, E. E. (1926). The chemical sensitivity of the tarsi of certain muscid flies (*Phormia regina, Phormia terrae nova* R.D. and *Lucilia sericata* Meigen). *Biol. Bull.* **51,** 166–178.

Moskowitz, H. R. (1970). Sweetness and intensity of artificial sweeteners. *Percept. & Psychophys.* **8,** 40–42.

Moskowitz, H. R. (1971). The sweetness and pleasantness of sugars. *Amer. J. Psychol.* **84,** 387–405.

Moskowitz, H. R. (1972a). Perceptual changes in taste mixtures. *Percept. & Psychophys.* **11,** 257–262.

Moskowitz, H. R. (1972b). Perceptual attributes of the tastes of sugars. *J. Food Sci.* **37,** 624–626.

Moskowitz, H. R. (1973). Models of sweetness additivity. *J. Exp. Psychol.* **99,** 88–98.

Myers, C. S. (1903). Taste. *In* "Reports of the Cambridge Anthropological Expedition to the Torres Straits," Vol. II, Part 2, pp. 186–188. Oxford Univ. Press, London and New York.

Peryam, D. R., and Pilgrim, F. J. (1957). Hedonic scale method of measuring food preferences. *Food Technol.* **11**, 9–14.

Pfaffmann, C. (1959a). The afferent code for sensory quality. *Amer. Psychol.* **14**, 225–232.

Pfaffmann, C. (1959b). The sense of taste. *In* "Handbook of Physiology" (Amer. Physiol. Soc., J. Field, ed.), Sect. 1, Vol. I, pp. 507–533. Williams & Wilkins, Baltimore, Maryland.

Schutz, H. G., and Pilgrim, F. J. (1957). Sweetness of various compounds and its measurement. *Food Res.* **22**, 206–213.

Shepard, R. N. (1962). The analysis of proximities: Multidimensional scaling with an unknown distance function. *Psychometrika* **27**, 219–246.

Stevens, S. S. (1951). Mathematics, measurement and psychophysics. *In* "Handbook of Experimental Psychology" (S. S. Stevens, ed.), p. 1. Wiley, New York.

Taufel, K., and Klemm, B. (1925). Untersuchungen uber naturliche und kunstliche Susstoffe. *Z. Unters. Lebensm.* **50**, 264–273.

von Békésy, G. (1964). Duplexity theory of taste. *Science* **145**, 834–835.

Woodworth, R. S., and Schlosberg, H. (1954). "Experimental Psychology." Holt, New York.

Young, P. T. (1966). Hedonic organization and regulation of behavior. *Psychol. Rev.* **73**, 579–586.

Sugars in Food:
Occurrence and Usage

CHAPTER 5

Occurrence of Various Sugars in Foods

R. S. SHALLENBERGER

The sugars which occur in foods are either natural constituents of the food, are generated during processing, or they are intentionally added. In either case, many different sugars may be present. Chemically, sugars are either monosaccharides or relatively low molecular weight polymers (oligosaccharides) of the monosaccharides. Monosaccharides are polyhydroxy aldehydes or ketones which usually exist in a hemiacetal ring structure (Pigman, 1957). They are alcohol and aldehyde or ketone derivatives of pentane and hexane. Thus, the aldohexoses are hexane-2,3,4,5,6-pentol-1-als. Oligosaccharides contain two to about ten monosaccharide units, but the dividing line between oligosaccharides and polysaccharides is arbitrary.

A. Naturally Occurring Sugars

1. Monosaccharides

Natural monosaccharides which occur in abundance in foods are D-glucose and D-fructose. D-Galactose, D-mannose, and certain pentoses such as D-xylose, L-arabinose, and D-ribose occasionally are found in trace

67

amounts. The cyclic hemiacetal structures for the D-aldohexose and the D and L series of aldopentoses in the C1 conformation are shown in Fig. 1.

The C1 conformations shown for these sugars are that of a left handed twofold screw structure, but the C1 conformation is *not* necessarily the favored conformation for each individual sugar. However, for the D series of aldohexoses, it is *usually* favored. The conformational structure shown for D-glucose is highly favored, and since it is "left-handed" lends one of the elements of dextrorotation to this compound. D-Glucose is known also by the name "dextrose."

With few exceptions, the naturally occurring sugars belong to the chiral family designated as D, which is a configurational assignment for the disposition of carbon atom substituents furthest removed from the primary functional aldehyde or ketone group. Free L-arabinose occurs naturally, but the L designation tends to be confusing unless the convention just described is applied. Structurally, L-arabinose is related to D-galactose, and since the favored conformation for both of these sugars is that shown in Fig. 1, both compounds are dextrorotatory.

The favored conformation of the cyclic hemiacetal structure for D-fructose is 1C. Since this conformation describes a right-handed twofold screw structure it is levorotatory and is known also as "levulose."

In solution, the monosaccharides exist in a state of mutarotational equilibrium, which is a dynamic balance between several structural forms which monosaccharides can assume. This situation prevails in foods.

a. Mutarotational Equilibrium

When crystalline sugars, which possess a free anomeric center, are dissolved in a solvent, an equilibrium is established between the various isomeric forms possible (Pigman and Isbell, 1968). The phenomenon can be monitored by the changing optical rotation of the system, and hence the term "mutarotation."

The various forms possible are six-membered ring (pyranose), five-membered ring (furanose), and acyclic (open chain) structures.

Glucose establishes a simple equilibrium between the α-D- and the β-D-pyranose forms, and the percentage of each isomer in water is 36 and 64%, respectively. Disaccharides with a terminal reducing glucose moiety such as lactose and maltose have about the same anomeric distribution in solution as glucose. D-Mannose also establishes a simple equilibrium between the α-D- and the β-D-pyranose forms, but the distribution of these forms is the reverse found for glucose.

Galactose yields a complex mixture of 32% α-D-galactopyranose, 64% β-D-galactopyranose, 3% β-D-galactofuranose, and 1% α-D-galacto-

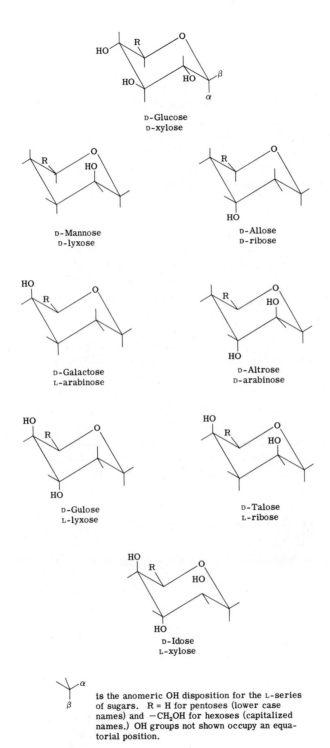

D-Glucose
D-xylose

D-Mannose
D-lyxose

D-Allose
D-ribose

D-Galactose
L-arabinose

D-Altrose
D-arabinose

D-Gulose
L-lyxose

D-Talose
L-ribose

D-Idose
L-xylose

is the anomeric OH disposition for the L-series of sugars. R = H for pentoses (lower case names) and —CH$_2$OH for hexoses (capitalized names.) OH groups not shown occupy an equatorial position.

Fig. 1 Pyranoid ring structure and C1 conformation of aldohexose and aldopentose sugars. Redrawn from Shallenberger and Acree (1971) by permission of the publisher.

furanose (Acree *et al.*, 1969). D-Fructose has, at equilibrium, 76% β-D-fructopyranose, 20% β-D-fructofuranose, and 4% of an unknown compound presumed to be α-D-fructopyranose (Shallenberger, 1973).

(1) SIGNIFICANCE OF ANOMERS IN FOODS.   The significance of the various anomers in foods is that they have varying properties and undergo different reactions. β-D-Glucopyranose is the only metabolizable form of glucose for the human, and β-D-fructofuranose is the only form of fructose fermented by yeast. Both α-D- and β-D-glucose are known as crystalline substances, and β-D-glucose is the more soluble of the pair. β-D-Fructopyranose is the only known crystalline form of fructose, and it is perhaps the sweetest of the naturally occurring sugars. The furanose form in solution seems to be void of sweet taste. The disaccharide lactose is also known in two crystalline forms. That of α-D-lactose monohydrate is difficult to dissolve in water, but the anhydrous β-D-anomer is readily soluble.

(2) FACTORS AFFECTING THE POSITION OF THE EQUILIBRIUM.   The position of a mutarotated sugar solution is affected by temperature, concentration, and the polarity of the solvent. An increase in either the temperature or the concentration of glucose in solution increases the concentration of the α-D-anomer slightly. As the polarity of the solvent is decreased, the position of the equilibrium approaches that of an equal concentration of each anomer.

Increasing the temperature and the concentration of fructose solutions markedly increases the concentration of the furanose form(s). This property of mutarotated fructose solutions has important bearings on the use of this sugar in various foods. Decreasing the polarity of the solvent also increases the concentration of furanose forms.

b. Epimerization and Transformation

Upon standing in solution for some time, especially at an alkaline pH, the equilibrium

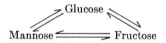

is established, with the glucose–fructose transformation being favored (MacLaurin and Green, 1968). The formation of mannose is an example of epimerization. Even at an initial pH of 6.6, the autoclaving of D-glucose leads to the formation of significant amounts (ca. 7%) of D-fructose

(Englis and Hanahan, 1945). The transformations are possible through the enolization of the open chain form of the sugars, followed by rearrangement of carbon atom substituents and ring closure. Progressive enolization of acyclic D-fructose, followed by ring closure leads to the occasional occurrence of D-psicose in foods. D-Psicose is the 3-epimer of D-fructose.

2. Oligosaccharides

The naturally occurring oligosaccharides are invariably polymers of D- glucose, fructose, and galactose, in varying combination. About forty oligosaccharides are known to occur naturally, but only several occur in abundance. These are known by their trivial names such as sucrose, lactose, maltose, raffinose, and stachyose. Nonreducing disaccharides are named as a glycosyl glycoside, and sucrose can be named as either β-D-fructofuranosyl α-D-glucopyranoside or α-D-glucopyranosyl β-D-fructofuranoside. A reducing disaccharide can be named as a glycosyl glycose. α-D-Lactose is, therefore, 4-*O*-β-D-galactopyranosyl-α-D-glucopyranose. The name indicates that the anomeric OH group of β-D-galactopyranose is linked to the OH group on carbon atom number 4 of α-D-glucopyranose with the removal of a proton. Reducing disaccharides can also be named according to the rules of nomenclature used for tri- and higher oligosaccharides (Rules of Carbohydrate Nomenclature, 1963). Beginning with the first nonreducing component, the first glycosyl portion with its configurational prefixes is delineated. This is followed by two numbers which indicate the respective positions involved in the glycosidic union. These numbers are separated by an arrow pointing from the glycosyl carbon atom to the number for the hydroxylic carbon atom involved, and are enclosed in parentheses inserted into the name by hyphens. The next dissacharide linkage is treated similarly. Using this system, α-D-lactose, becomes *O*-β-D-galactopyranosyl-$(1 \rightarrow 4)$-α-D-glucopyranose. This type of nomenclature is almost impossible to use in spoken language, but unambiguously specifies sugar structure, and it is more convenient to name in print than to draw the oligosaccharide structures.

a. Distribution of Sugars Occurring Naturally in Foods

This section will describe the occurrence of sugars, which occur naturally in foods in significant amounts, and sometimes in abundance. They are glucose, fructose, sucrose, raffinose, stachyose, and lactose. Lactose is the sugar of mammalian milk, found to the extent of about 7.5% in human milk and 4.5% in cow milk. In dairy products such as ice cream, yogurt, and buttermilk, about 4–5% lactose is present. On the other hand, lactose accounts for 52% of the total solids of dried skim milk, and 38% of the solids of dried whole milk (Hardinge *et al.*, 1965). This sugar is also frequently reported to be a minor component of many plants and plant food substituents; for example, it has been reported to be present in acacia and linden nectars, in apples and apple juice, and in onion and garlic. In no case, however, has its occurrence in plants been confirmed by isolation and crystallization. In all probability, it may be confused with the trace natural plant oligosaccharides melibiose and galactobiose, which are component parts of the plant oligosaccharide stachyose.

A number of "rare" oligosaccharides may be found in certain foods. α,α-Trehalose (α-D-glucopyranosyl-α-D-glucopyranoside) is found in young mushrooms. Occasionally, bees may collect the manna of various firs, or honey dew, and the resultant honey may contain the trisaccharide melezitose [O-α-D-glucopyranosyl-$(1 \rightarrow 3)$-O-β-D-fructofuranosyl-$(2 \rightarrow 1)$-α-4-glucopyranoside].

Many plants accumulate high concentrations of a galactosyl-sucrose series of oligosaccharides (French, 1954). The tetrasaccharide stachyose is typical. Stachyose is O-α-D-galacto-pyranosyl-$(1 \rightarrow 6)$-O-α-D-galactopyranosyl-$(1 \rightarrow 6)$-O-α-D-glucopyranosyl-$(1 \rightarrow 2)$-β-D-fructofuranoside. It may be considered to be made up of a number of other reducing and nonreducing oligosaccharides having the partial structures of the parent compound as follows:

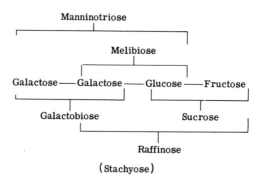

(Stachyose)

Stachyose in plants and in foods is usually accompanied with the non-reducing oligosaccharides raffinose and sucrose. Traces of the reducing oligosaccharides manninotriose, melibiose, and galactobiose may also be encountered, along with the corresponding trigalactosyl sucrose homolog verbascose.

The occurrence of stachyose in certain foods, particularly those of legume origin, presents an interesting nutritional question. Mammalian invertase is an α-D-glucosidase which will not hydrolyze sucrose if the glucose moiety is substituted. Because the human digestive tract does not possess an α-D-galactosidase, stachyose and raffinose would generally seem to be metabolically inert. Fermentation of these sugars in the lower digestive tract, however, may be the source of some unwanted galactose in the diet of galactosemic children, and it is believed to be a component of the flatus factor of various legume foods.

A number of fruits and vegetables have been examined (Lee *et al.*, 1970) using a quantitative paper chromatographic technique, and the concentration and distribution of individual sugars have been tabulated. While many fruits and vegetables were replicated several times, using different cultivars, the results for individual cultivars have been averaged. The difference between species was judged to be of greater significance. Results obtained with various fruit are shown in Table I. The percent total solids are shown so that the contribution of either total or individual sugars to either parameter may be determined; for example, sugars account for 70% of total apple solids, and fructose alone accounts for nearly 40% of the total solids. In addition to the sugars listed in Table I, grapes contained a reducing trisaccharide (maltotriose?) while apples and pears contained a trace of a pentose (xylose?).

The content and distribution of glucose, fructose, and sucrose in vegetables other than legumes are shown in Table II. Some of the vegetables in Table II contained some raffinose and stachyose, such as onion, and is part of the reason for the previous suggestion that the lactose reported in onions and other plants may indeed be the constituent stachyose reducing oligosaccharides galactobiose or melibiose.

The distribution of sugars in fresh and dry legume foods is shown in Table III.

B. Generated Sugars

Sugars and sugar dehydration products are generated during the production of food substances and during the processing and storage of many foods. These are generally described as "reversion" products and sugar anhydrides. Under acid conditions, glucose has been found by Thompson *et al.* (1954) to yield, in varying amounts, the compounds shown in Table IV. All the compounds are reducing or nonreducing disaccharides, with the exception of levoglucosan, which is a sugar anhydride formed by inverting the β-D-glucopyranose ring and the elimination of a molecule of

TABLE I

Free Sugars in Fruit as Percentage Fresh Basis

Fruit	Cultivars	Total solids	Glucose	Fructose	Sucrose	Maltose
Apple, *Pyrus malus*	6	15.96	1.17	6.04	3.78	Trace
Apricot, *Prunus armeniaca*	2	14.44	1.73	1.28	5.84	
Blackberry, *Rubus*	2	15.28	2.48	2.15	0.59	0.66
Blueberry, *Vaccinium corymbosum*	2	15.89	3.76	3.82	0.19	0.08
Currant, *Ribes sativum*	4	17.68	3.33	3.68	0.95	0.64
Gooseberry, *Ribes grossularia*	3	14.81	3.29	3.90	1.21	
Grape, *Vitus labruscana*	4	19.13[a]	6.86	7.84	2.25	1.58
Grape, *Vitis vinifera*	3	17.97[a]	5.35	5.33	1.32	2.19
Peach, *Prunus persica*	5	12.79	0.91	1.18	6.92	0.12
Pear, *Pyrus communis*	3	13.58	0.95	6.77	1.61	0.31
Plum, *Prunus domestica*	3	17.97	3.49	1.53	4.94	0.15
Raspberry (red), *Rubus idaeus*	1	20.67	2.40	1.58	3.68	
Raspberry (black), *Rubus occidentalis*	1	28.22	4.56	4.84	1.90	
Cherry (sour), *Prunus cerasus*	4	15.05	4.30	3.28	0.40	
Cherry (sweet), *Prunus avium*	5	22.39	6.49	7.38	0.22	
Strawberry, *Fragaria chiloensis*	2	9.45	2.09	2.40	1.03	0.07

[a] Soluble solids.

water between the anomeric and the primary methylene OH groups. Levoglucosan is, therefore, 1,6-anhydro-β-D-glucopyranose.

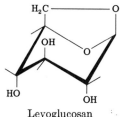

Levoglucosan

The trivial name is derived from the fact that the compound is levorotatory. Ough and Rohwer (1956) found that it is present to the extent of 2.3% in acid hydrolyzates of cornstarch. The compound readily polymerizes, however, and is probably the precursor to the reversion disaccharides isomaltose and gentiobiose. Other reversion products are kojiobiose (2-*O*-α-D-glucopyranosyl-D-glucose) and nigerose (3-*O*-α-D-glucopyranosyl-D-glucose), which are reported (Aso *et al.*, 1961) to account for 3.3% of the total sugars in a beer.

When D-fructose reacts with aqueous acid, a series of dimeric anhydrides are formed which possess a central dioxane ring that seems to be the element of acid stability these compounds possess. The structure di-D-fructose anhydride I has been determined by Lemieux and Nagarajan (1964) to be 1′,2-anhydro-1-(α-D-fructofuranosyl)-β-D-fructofuranoside.

Upon prolonged storage in acid solution, dehydration, fragmentation, and condensation compounds are generated that bear little relationship to the starting compounds. Among these are the saccharinic acids and the furfuraldehydes.

C. Sugars Intentionally Added to Foods

The sugars intentionally added to foods for sweetening, texturizing, preservation, and other purposes are sucrose (cane or beet sugar), dextrose (α-D-glucose), sugar syrups prepared by the hydrolysis of starch, invert sugar prepared by the hydrolysis of sucrose, and "isomerized" glucose syrups. One example of the latter, prepared (Kooi and Smith, 1972) by the action of an enzyme from *S. olivochromogenes* on glucose syrup, yields a product very similar to that produced by the hydrolysis of sucrose, i.e., it contains glucose and fructose in nearly equal amounts. Thus, the sugars intentionally added to foods for various purposes are essentially those which occur naturally in the plant kingdom and in foods.

TABLE II

Free Sugars in Vegetables as Percentage Fresh Basis

Vegetable[a]	Cultivars	Total solids	Glucose	Fructose	Sucrose
Asparagus, *Asparagus officinalis*	2	9.15	0.92	1.30	0.28
Beet,[a] *Beta vulgarus*	2	11.19	0.18	0.16	6.11
Broccoli, *Brassica oleraceae (botrytis)*	3	11.84	0.73	0.67	0.42
Brussels sprout,[a] *Brassica oleracea (gemmifera)*	2	11.45	0.66	0.75	0.41
Cabbage, *Brassica oleracea (capitata)*	3	6.67	1.58	1.20	0.15
Cabbage, *Brassica oleracea (capitata)*, re	1	9.06	2.06	1.74	0.50
Carrot, *Daucus carota*	4	12.00	0.85	0.85	4.24
Cauliflower,[a] *Brassica oleracea (botrytis)*	3	8.05	0.83	0.74	0.67
Celery, *Apium graveolens*	3	8.29	0.49	0.43	0.31
Cucumber, *Cucumis sativus*	4	3.46	0.86	0.86	0.06
Eggplant, *Solanum melongena (esculentum)*	2	8.49	1.51	1.53	0.25
Endive, *Cichorum endivia*	1	5.60	0.07	0.16	0.07
Escarole, *Cichorum endivia*	1	6.15	0.16	0.32	0.10
Kale, *Brassica oleracea (acephala)*	2	9.74	0.27	0.21	
Kohlrabi, *Brassica oleracea (gongylodes)*	1	7.55	1.34	1.24	0.58
Leek,[a] *Allium porrum*	2	11.95	0.98	1.47	1.06
Lettuce, *Lactuca sativa*	4	4.97	0.25	0.46	0.10
Melon, honeydew, *Cucumis melo*	1	12.74	2.56	2.62	5.86
Melon, muskmelon *Cucumis melo (reticulatus)*	2	10.84	1.72	2.03	3.56
Melon, watermelon *Citrullus vulgarus*	2	9.57	1.81	3.54	2.35
Okra, *Hibiscus esculentus*	2	10.70	1.03	1.06	0.75
Onion,[b] *Allium cepa*	4	11.56	2.07	1.09	0.89
Onion, green,[a] *Allium cepa*	1	9.59	0.56	0.76	0.86
Parsley, *Petroselinum hortense*	2	11.28	0.10		0.20
Parsnip,[a] *Pastinaca sativa*	1	20.99	0.18	0.24	2.98

Pepper, *Capsicum frutescens*	3	6.21	0.90	0.87	0.11
Potato, new, *Solanum tuberosum*	5	20.08	0.15	0.09	0.14
Potato, stored at 35°F[a]	5		1.04	1.15	1.69
Pumpkin,[a] *Cucurbita pepo*	2	7.13	1.69	1.43	1.30
Radish, white, *Raphanus sativus*	1	4.40	0.84	0.30	0.22
Radish, red, *Raphanus sativus*	1	5.46	1.34	0.74	0.09
Rhubarb, *Rheum rhaponticum*	2	6.20	0.42	0.39	0.07
Rutabaga, *Brassica napobrassica*	1	6.69	0.38	0.34	0.06
Spinach, *Spinacia oleracea*	3	8.04	0.09	0.04	0.09
Squash, summer, *Cucurbita pepo*	3	5.55	0.77	0.82	1.61
Squash, winter, *Cucurbita pepo*	4	13.08	0.96	1.16	3.03
Sweet corn, *Zea mays*	6	22.69	0.34	0.31	0.06
Swiss chard, *Beta vulgaris (cicla)*	1	9.20	0.17	0.09	
Sweet potato, *Ipomoea batatas Poir*	1	22.53	0.33	0.30	3.37
Tomato, *Lycopersicon esculentum*	5	5.23	1.12	1.34	0.01
Turnip, *Brassica rapa*	1	7.40	1.50	1.18	0.42

[a] Contains traces (0.02–0.20%) raffinose, stachyose, or both.
[b] Contains 0.24%–>1.0% raffinose and stachyose.

TABLE III

Free Sugars in Legumes

Legume	Cultivars	Total solids	Glucose	Fructose	Sucrose	Raffinose	Stachyose
Fava bean,[a] *Vicia faba*	1	16.61		0.18	3.36	0.66	
Lima bean,[a] *Phaseolus lunatus*	5	26.74	0.04	0.08	2.59	0.20	0.59
Pole Lima bean,[a] *Phaseolus lunatus*	1	24.58	0.18		2.26	0.32	0.60
Pole snap bean,[a] *Phaseolus lunatus*	1	10.21	0.48	1.30	0.28	0.26	0.19
Snap bean,[a] *Phaseolus vulgaris*	6	7.79	1.08	1.20	0.25	0.11	0.06
Pea (Alaska)[a] *Pisum sativum*	1	25.54		0.08	3.00	0.06	
(wrinkled)[a] *Pisum sativum*	3	22.77	0.32	0.23	5.27	0.58	0.49
Cow Pea,[a] *Vigna sinensis*	1	39.30	0.08	0.06	1.86	0.10	1.66
Dry bean,[b] *Phaseolus vulgaris*	1				2.40	0.80	3.40
Mung bean,[b] *Phaseolus aureus*	1				1.19	0.40	1.75
Pea bean,[b] *Phaseolus vulgaris*	1				2.55	0.65	3.06
Pea seed,[a] *Pisum sativum*	9		0.24		4.11	1.75	7.96
Soybean,[a] *Glycine max*	1				4.53	0.73	2.73

[a] Sugars as percent fresh basis.
[b] Sugars as percent total bean weight.

TABLE IV

Acid "Reversion" Products of Glucose

Sugar	Definitive name
β-β-Trehalose	β-D-Glucopyranosyl-β-D-glucopyranoside
β-Sophorose	2-*O*-β-D-Glucopyranosyl-β-D-glucopyranose
β-Maltose	4-*O*-α-D-Glucopyranosyl-β-D-glucopyranose
α-Cellobiose	4-*O*-β-D-Glucopyranosyl-α-D-glucopyranose
β-Cellobiose	4-*O*-β-D-Glucopyranosyl-β-D-glucopyranose
β-Isomaltose	6-*O*-α-D-Glucopyranosyl-β-D-glucopyranose
α-Gentiobiose	6-*O*-β-D-Glucopyranosyl-α-D-glucopyranose
β-Gentiobiose	6-*O*-β-D-Glucopyranosyl-β-D-glucopyranose
Levoglucosan	1,6-Anhydro-β-D-glucopyranose

D. Research Needed

Further research is needed on reactions of sugars in foods and the occurrence and distribution of their reaction products. This is particularly true for foods to which sugars have been intentionally added. The chemistry of the sugars at high concentration in solution and at various pH and temperature is only known approximately, and these data are from model studies. Results of such studies can only be applied with caution since each food presents a special case.

References

Acree, T. E., Shallenberger, R. S., Lee, C. Y., and Einset, J. W. (1969). Thermodynamics and kinetics of D-galactose, tautomerism during mutarotation. *Carbohyd. Res.* **10,** 355–360.

Aso, K., Watanabe, T., Sasaki, K., and Motomura, Y. (1961). Studies on beer. I. Sugar composition of beers. *Tohoku J. Agr. Res.* **12,** 261–268.

Englis, D. T., and Hanahan, D. J. (1945). Changes in autoclaved glucose. *J. Amer. Chem. Soc.* **67,** 51–54.

French, D. (1954). The raffinose family of oligosaccharides. *Advan. Carbohyd. Chem.* **9,** 149–184.

Hardinge, M. G., Swarner, J. B., and Crooks, H. (1965). Carbohydrates in foods. *J. Amer. Diet. Ass.* **46,** 197–204.

Kooi, E. R., and Smith, R. J. (1972). Dextrose-levulose syrup from dextrose. *J. Food Technol.* **26,** 57–59.

Lee, C. Y., Shallenberger, R. S., and Vittum, M. T. (1970). Free sugars in fruits and vegetables. *N.Y. Food Life Sci. Bull.* No. 1.

Lemieux, R. U., and Nagarajan, R. (1964). The configuration and conformation of "Di-D-fructose Anhydride I." *Can. J. Chem.* **42,** 1270–1278.

MacLaurin, D. J., and Green, J. W. (1968). Carbohydrates in alkaline systems. I. Kinetics of the transformation and degradation of D-glucose, D-fructose, and D-mannose in 1 *M* sodium hydroxide at 22°C. *Can. J. Chem.* **47**, 3947–3955.

Ough, L. D., and Rohwer, R. G. (1956). Presence of levoglucosan in cornstarch hydrolyzates. *J. Agr. Food Chem.* **4**, 267–269.

Pigman, W., ed. (1957). "The Carbohydrates," 1st ed. Academic Press, New York.

Pigman, W., and Isbell, H. S. (1968). Mutarotation of sugars in solution. Part I. History, basis, kinetics, and composition of sugar solutions. *Advan. Carbohyd. Chem.* **23**, 11–57.

Rules of Carbohydrate Nomenclature. (1963). *J. Org. Chem.* **28**, 281–291.

Shallenberger, R. S. (1973). Sugar structure and taste. *Advan. Chem. Ser.* **117**, 256–263.

Shallenberger, R. S., and Acree, T. E. (1971). Chemical structure of compounds and their sweet and bitter taste. *In* "Handbook of Sensory Physiology" (L. M. Beidler, ed.), Vol. IV, Part 2, pp. 222–277. Springer-Verlag, Berlin and New York.

Thompson, A., Anno, K., Wolfrom, M. L., and Inatome, M. (1954). Acid reversion products from D-glucose. *J. Amer. Chem. Soc.* **76**, 1309–1311.

CHAPTER 6

World Sugar Production and Usage in Europe

ARVID WRETLIND

In many countries sugar—sucrose—is an essential part of the dietary content. Its nutritional and medical significance has been the subject of exhaustive discussions during recent years. The reason for this has been, among other things, the demonstration of the connection between sugar and dental caries and the decrease in essential nutrients in the diet, which occurs when sugar forms an increasing part of the energy supply.

In order to conduct a meaningful discussion on the role of sugar in the diet from a nutritional point of view, it is necessary to base the discussion on a clear idea of production and consumption. Consequently, such a survey is given in what follows. In this connection both total world production of sucrose and consumption in Europe will be dealt with.

A. Sugar Production

The world production of raw sugar has increased sharply since the beginning of the present century, when it was about 14 billion kg, and rose to 73 billion kg in 1970 (Fig. 1). This corresponds to approximately a fivefold increase. During the same period world population has doubled, increasing from 1.7 to 3.5 billion. Thus, the production of white sugar

(100 kg of raw sugar correspond to 90 kg of white sugar) calculated per person and day has increased from 21 to 51 gm, from 1905 to 1970 (Fig. 2). As is evident from Fig. 2, during the First and the Second World War sugar production decreased to the 1905 level.

According to Sukhatme (1961) the average energy supply among the population of the world was estimated at 2400 kcal in around 1960. On the whole, this is probably still the value. This means that the available amount of sugar corresponds to nearly 9 cal %.

Most of the sugar is produced in North and Central America. Then follow, in terms of production volume, Asia, eastern Europe, western Europe, South America, Africa, and Oceania. Figure 3 shows the production volumes. The production of the different regions does not correspond to their actual consumption (Fig. 3). Thus, production exceeds consumption in North and Central America, South America, Africa, and Oceania. The surplus is exported to Asia and to western and eastern Europe.

At the beginning of the twentieth century about the same amount of cane sugar and of beet sugar was produced. Thereafter, the production of cane sugar has risen and was 58% of the total sugar production in the year 1970. During the years 1919–20 and 1944–45 this proportion was higher—78 and 70%, respectively.

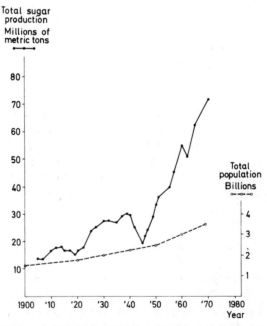

Fig. 1 Production of raw sugar and world population during the twentieth century. The values for production have been taken from Licht and Ahlfeld (1971).

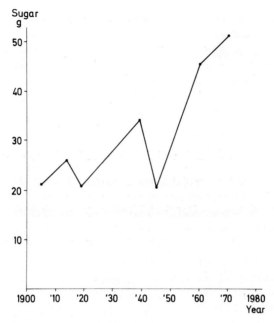

Fig. 2 The world production of sugar calculated per person and day of the total world population. The calculation is based on the values given in Fig. 1.

B. Sugar Consumption in Europe

Sugar consumption in the European countries varies considerably, as shown in Fig. 4 (Production Yearbook, 1972). The highest consumption per capita is in Iceland, followed by Ireland, Holland, Denmark, and England. Consumption in Iceland is about 150 gm per person and day. For the other countries mentioned consumption exceeds 135 gm per person and day. In several countries less than 100 gm per person and day are consumed. The lowest consumption figures are reported for Roumania, Albania, Greece, Portugal, and Bulgaria where the daily intake per person is between 48 and 62 gm.

Changes in sugar consumption in some European countries are presented in Fig. 5. The diagram in Fig. 5 shows that in most of the countries reported there has been a continuous increase in consumption during the last 25 years. In some of the countries with high consumption, however, there seems to have been either stagnation or decrease during the period in question. Thus, Sweden, where until 1950, sugar consumption had continuously increased to a daily amount of 128 gm per person, has since then decreased to 114 gm in 1970. In Denmark sugar consumption has

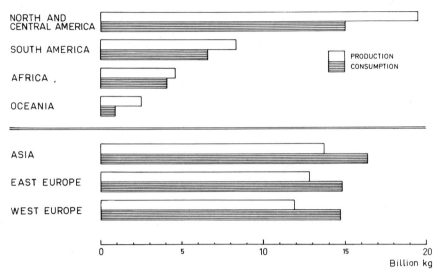

Fig. 3 The production and consumption of sugar in different regions of the world. The values are according to Licht and Ahlfeld (1971).

not increased since the beginning of the 1960's. It seems that, if anything, the data indicate a certain decrease. In Norway and Austria consumption has decreased somewhat in recent years.

Figure 5 also shows the percentage of the total energy consumption, represented by sugar consumption, in the different countries. In Norway, Ireland, Sweden, Denmark, England, and Holland, sugar consumption is between 15 and 18% of the total energy consumption. The lowest percentage of energy from sugar is found in the countries of southern Europe, Greece, Portugal, Yugoslavia and Spain where it varies between 8 and 10 cal %.

Part of the sugar is used by the food industry. Another part is bought and used by households. The portion of the sugar used by the food industry has increased continuously during recent years. This is due, among other things, to the increased use of composite foods, ready-prepared food, etc. In Germany, for example, the proportion of sugar used in the food industry was 35% in the years 1952–53 and rose to 54% in 1968–69 (Cremer, 1972). In Holland 60% is used by the food industry. The corresponding values for England, Finland, and Switzerland are 47, 34, and 34%, respectively. In Italy and Austria only 24 and 27%, respectively, of the sugar consumption are used by the food industries.

The values given above are taken from the official statistics. It is obvious that the values given contain quite a number of errors. It must

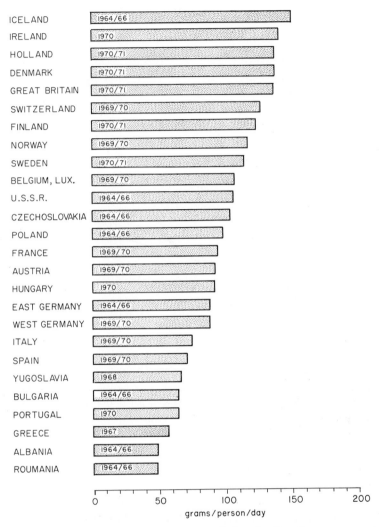

Fig. 4 Consumption of sugar per person and day in the European countries. The amounts shown refer to sugars, syrups, honey, and other sugar products according to Production Yearbook (1972).

be taken into account that some of the given amounts of sugar are not consumed in the countries in question. Thus, a part of the sugar may be used as ingredients of composite foods which are subsequently exported to other countries without full data being recorded. Some of the sugar is fermented, among other things, when added to bread. On the other hand, it seems obvious that the stated values for sugar consumption show

SUGAR
g/person/day

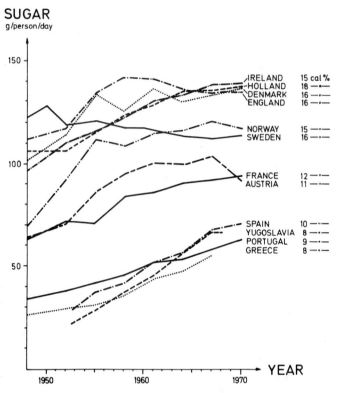

Fig. 5 Changes in the consumption of sugars in some European countries during the last 25 years. The values have been obtained from Production Yearbook (1972).

the magnitude of the amount of sugar consumed and the level of sugar in our diet.

C. Nutritional Aspects of Sugar Consumption

When assessing the role of sugar in our diet from the nutritional and medical point of view, it is necessary to make quantitative estimates of the amount of sugar and of other food included in the diet. From the viewpoint of nutrition, sugar only functions as a readily available source of energy for the body. At the same time this implies that the consumption of other foods, with more or less high contents of essential nutrients, has to be reduced by an amount which corresponds with the energy content of the sugar consumed. Thus, in principle, the higher the sugar level in the diet, the lower its level of essential nutrients. This relationship should be taken into account when it is found that the sugar level in all countries has shown, or still shows, a tendency to increase. The dietary changes in most countries are characterized not only by an increased sugar content but also by a simultaneous increase in the fat content. The simultaneous increase in the content of sugar and of fat was previously pointed out by McGandy *et al.* (1967) and by Périssé *et al.* (1969). Figure 6 shows the interrelations between the sugar and the fat levels in the European countries. From that interrelation it is evident that the higher the sugar consumption, the higher the amount of fat in the diet, and the lower the amount of carbohydrates such as the polysaccharides (starch).

The proportion of sugar in the total amount of carbohydrates in the

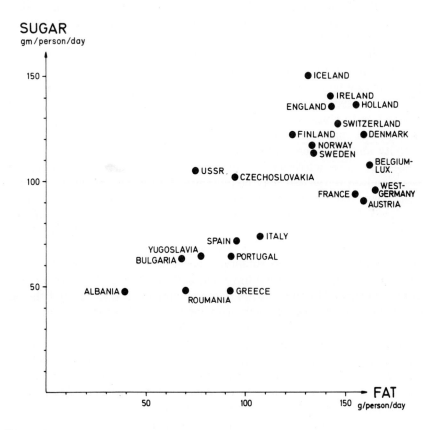

Fig. 6 Interrelation between sugar and fat consumption in some European countries. The values have been obtained from Production Yearbook (1972) and are for the period 1964–66 for Albania, Bulgaria, Czechoslovakia, Denmark, Greece, Iceland, Roumania, Yugoslavia, and the USSR. All other values are for the period 1969–70.

diet has undergone substantial change during this century. At the beginning of the century the average Swedish diet had a sugar content corresponding to less than one-tenth of the total carbohydrates. In 1970, the proportion of sugar had risen to about a third. Similar conditions exist in the rest of the northern European countries. The nutritional and medical consequences of this are obvious but have still not been fully elucidated. Up to now there has been nothing to indicate that this development is in any way advantageous from the nutritional or medical point of view.

Changes in the diet, with special reference to the increasing content of sugar and fat are not in accordance with the changes which can be regarded as desirable from a nutritional viewpoint. This situation has to do with the characteristic features in the development of our modern society, which include a decrease in the physical activity of the population. Thus, the composition of the diet and the dietary habits have not been adjusted to the life of modern man, who has been called *Homo sedentarius* by Passmore (1966): indeed, they are better adapted to physically active *Homo sapiens,* who has been designated *Homo sportivus* (Passmore, 1964). One of the great and important tasks for nutritional research is to elucidate and solve the problems that are concerned with the optimal supply of nutrients for sedentary persons with low physical activity and low energy requirement, or, what are known as low calorie consumers. The nutritional complications of an increasing content of sugar and fat in the diet may be illustrated by the changes in the Swedish diet. Some 200–300 years ago Sweden, as well as many other European countries, had much in common with the present developing countries, with cereals dominating the diet. The greatest changes in the diets during recent centuries have been the same in most other industrialized European countries. Cereal consumption has decreased to a low level. The changes in food consumption have had both positive and negative effects from a nutritional point of view. Greater variety in the diet and the enrichment of a number of foods with vitamins and other nutrients have made it possible to reduce the risk of a lack of essential nutrients. The changes in food habits have resulted in an increase in the sugar and fat content of the Swedish diet up to about 16 and 42%, respectively, of the energy value of the foods consumed. Thus, up to about 58% of the total energy content of the food consumed is derived from sugar and fat. At the beginning of this century these two dietary components accounted for only one-third or less of the energy consumed.

There seems to be an interrelation between the average income and the consumption of both fat and sugar, which has been described by Périssé *et al.* (1969). The greatly increased consumption of sugar and fat, as at present combined with other foods, has involved certain health hazards. Thus, the occurrence of *dental caries* is very common in the Scandinavian countries. This may partly be explained by the food habits and the way in which sugar is used. The high incidence of *obesity,* especially in persons over 40, in the Scandinavian countries indicates overeating, which may easily occur with a diet of high calorie density composed by sugar and fat-rich foods. It is obvious that the Swedish diet, as well as that of many other North-European national diets, is not suitable for low calorie consumers requiring less than 2000 kcal/day. The low calorie consumers run the risk of not obtaining a sufficient amount of essential nutrients. Consequently, in the Scandinavian countries there is a high incidence of *anemia* among women because of a low intake of food iron.

A logical countermeasure against this unsatisfactory situation would be a change in the food habits of the population, such that the national diet per 1500–2000 kcal would have a nutrient content adequate for both a sedentary adult male and a sedentary female of fertile age. Furthermore, the diet should cover the requirements of adolescents and children from 2 years of age. This should help to ensure an adequate intake of nutrients for almost all individuals. As early as 1965 these views were presented by Blix *et al.* in a memorandum prepared for a Swedish governmental committee that was planning the national farming policy. It was further pointed out that, for the realization of the desired dietary changes, it would be necessary to establish a new food policy in its broadest sense. Simple calculations showed that without a reduction in the sugar and fat content it would be practically impossible for a national diet, consisting of conventional foods, to ensure the requirements of the low calorie groups.

The memorandum contained details or recommendations regarding the composition of a revised national diet. All the proposed changes involved a reduced consumption of products rich in sugar or fat, and a greater consumption of cereals, potatoes, low-fat dairy products, vegetables, and fruit. Medical reasons for the changes recommended were put forward. The reduction of sugar and fat, which will mean a diet with low calorie density, might help to counteract a too high supply of energy. The reduction in sugar consumption, particularly in the form of confectionery, should, moreover, be a valuable measure to reduce the incidence of dental caries. The increased content of iron in such a new national diet might contribute to reduce the frequency of iron deficiency anemia in women. In view of the correlation between consumption of saturated fats and the frequency of atherosclerotic heart disease, it might also help in preventing this disease, especially if part of the saturated fat in the diet were at the same time replaced by polyunsaturated fat.

The memorandum did not leave any appreciable traces in the report of the governmental committee previously mentioned. Part of the memorandum was, however, published in 1965 in two Swedish nutrition periodicals, and was thus made available for further discussion (Blix *et al.*, 1965a,b). However, after some time an expert group belonging to the Nutrition Section of the Swedish Medical Society (Svenska Läkaresällskapet) discussed the matter extensively. This resulted in a statement produced in collaboration with nutritionists from the other Scandinavian countries. The statement was published in 1968 under the title "Medical View Points on the National Diet in Scandinavian Countries" (*Medicinska synpunkter på folkkosten i de nordiska länderna*) and appeared as an official recommendation by the Medical Boards in Finland, Norway, and Sweden (Editorial, 1968; Keys, 1968). Adopting the main ideas expressed in the memorandum, it stated in general terms, the principal means of realizing the program outlined. The statement ended with the recommendation

> *that* the dietary energy supply should, in many cases, be reduced to prevent overweight;
> *that* the total fat consumption, at present about 40%, should be decreased to between 25 and 35% of the total calories;
> *that* the use of saturated fat should be reduced and the consumption of polyunsaturated fat should be simultaneously increased;
> *that* the consumption of sugar and products containing sugar should be decreased;
> *that* the consumption of vegetables, fruit, potatoes, skim milk, fish, lean meat, and cereal products should be increased.

With regard to the uniform food habits of families and other groups, it does not seem rational, when planning the diet, to use the detailed Recommended Dietary Allowances of nutrients (Food and Nutrition Board, 1968), which are applied at present. It would be both preferable and more realistic to state the quality of the diet by the minimal amounts of different nutrients that should be contained, per energy unit, or per 1000 kcal, or 10 MJ, in order for a diet to be considered as satisfying the nutritional demands for a national diet (Wretlind, 1967, 1970). In this connection, the recommended contents of nutrients should be so adjusted as to satisfy the nutritional requirements of those groups of persons who have the lowest calorie needs. Such a recommendation is given in Table I. The values have been calculated from the Recommended Dietary Allowances (Food and Nutrition Board, 1968). In a diet meeting that recommendation, the sugar and fat content have to be lower than in our present diet.

TABLE I

Recommended Content of Nutrients per 1000
kcal for Healthy Persons with a low Energy
Requirement[a]

Nutrients	Amount per 1000 kcal
Protein	>32 gm
Fat	27–38 gm
Calcium	> 0.6 gm
Iron	>10 mg
Retinol	> 0.5 mg
Thiamin	> 0.6 mg
Riboflavin	> 0.9 mg
Ascorbic acid	>32 mg

[a] Swedish National Institute of Public Health
(1969).

From the above and other facts, the amount of sugar in the diet should, from a general nutritional point of view, be restricted or moderated. The problem is, what content can be regarded as moderate? There is a place in the diet for sugar as a preservative in jam, marmelade, juice, etc., and as a condiment. Any other use of sugar should, and could, be restricted. It would be reasonable to state or to propose that the diet should not contain more than 10 cal % of sugar.

D. Summary

The total world production of sugar has continuously increased during the beginning of this century, from 14 to 73 billion kg. This means that the amount of white sugar available for consumption has increased from 21 to 51 gm/person/day for the whole world population, corresponding to nearly 9 cal %. In Europe there has been, and in some countries still is, a continuous increase in sugar consumption. In some of these countries the sugar content of the diet has reached a level between 15 and 18 cal %. The increase in sugar consumption, followed by an increased fat intake will, generally speaking, result in a decreased content of essential nutrients and in a reduced consumption of other foods which contain not only energy but also valuable nutrients. The conclusion is that the amount of sugar in a modern diet should be moderate. A maximum level of 10 cal % is proposed. This means that in most of the northern European countries sugar consumption should be reduced.

References

Blix, G. (1968). Cereals in the Swedish diet (in Swedish). *Nord. Cerealkemistfoeren. Nord. Cerealistforbunds Kongr., 1966* p. 23.

Blix, G., Wretlind, A., Bergström, S., and Westin, S. I. (1965a). Swedish national diet. *Var Föda* **17,** 7.

Blix, G., Wretlind, A., Bergström, S., and Westin, S. I. (1965b). Nutritional viewpoints on the Swedish national diet. *Naeringsforskning* **9,** 31.

Cremer, H. (1972). Sugar in nutrition and in Central European diets. *Lect. Symp. Expansion of Sugar through Research, 1972* p. 4.

Editorial. (1968). Medical viewpoints on the national diet in the Scandinavian countries. *Laekartidningen* **65,** 2012.

Food and Nutrition Board. (1968). "Recommended Dietary Allowances," 7th rev. ed., Publ. No. 1694. Nat. Acad. Sci.—Nat. Res. Counc., Washington, D.C.

Keys, A. (1968). Official collective recommendation on diet in the Scandinavian countries. *Nutr. Rev.* **26,** 259.

Licht, F. O., and Ahlfeld, H. (1971). "F. O. Licht's Weltsuckerstatistik 1970/71." Rotzeburg, Germany.

McGandy, R. B., Hegsted, D. M., and Stare, F. J. (1967). Dietary fats, carbohydrates and atherosclerotic vascular disease. *N. Engl. J. Med.* **277,** 186.

Passmore, R. (1964). "Assessment of the Second Report on Caloric Requirements." FAO, Rome.

Passmore, R. (1966). Caloric expenditure in man. *Nutr. Dieta* **8,** 161.

Périssé, J., Sizaret, F., and François, P. (1969). The effect of income on the structure of the diet. *Nutr. Newslett.* **7,** 1.

Production Yearbook. (1972). Vol. 25. Food and Agriculture Organization of the United Nations, Rome.

Sukhatme, P. V. (1961). The world's hunger and future needs in food supplies. *J. Roy. Statist. Soc., Ser. A* **124,** 463.

Swedish National Institute of Public Health. (1969). Desirable content of nutrients per 1000 kcal in the diet. *Var Foeda* **21,** 167.

Wretlind, A. (1967). Nutrition problems in healthy adults with low activity and low calorie consumption. *Symp. Swed. Nutr. Found.* **5,** 114.

Wretlind, A. (1970). Food iron supply. *In* "Iron Deficiency, Pathogenesis, Clinical Aspects, Therapy" (A. Vanotti, L. Hallberg, and H. G. Harwerth, eds.), p. 39. Academic Press, New York.

CHAPTER 7

Level of Use of Sugars in the United States

LOUISE PAGE AND BERTA FRIEND

Sugar, of one kind or another, is found in the diet of most, if not all, individuals. The sugar may be granulated sugar used as an ingredient in prepared food or added to food at the table, it may be in the form of other sweeteners such as syrup, or it may be one or more of the naturally occurring sugars present in a wide variety of foods. One way to find out how much of these sugars Americans are eating is to examine the assortment of foods that make up our national food supply and to calculate the amounts of sugars provided. To do this, estimates of food consumption are used that are based on food disappearing into consumption channels (U.S. Department of Agriculture, Economic Research Service, 1968). This then is food that is consumed—or "used up"—in an economic sense.

Use of these "disappearance" data has advantages and disadvantages. An obvious disadvantage is that they do not tell us what people actually eat. On the other hand, they are useful for giving an overall picture of average per capita consumption at any one time. They are also useful for following trends in consumption over a period of years. Moreover, at this level of consumption, it is possible to account for most of the refined sugar and other sweeteners before they are combined with other ingredients in prepared foods.

This last point is important since at present there is no reliable way to estimate amounts of the different sugars in diets of individuals. This is because existing information on various sugars in food is limited. Only recently has methodology progressed to the point where obtaining quantitative data on individual sugars as part of the carbohydrate fraction in food mixtures is thought to be feasible.

A. Sugars and Other Carbohydrate in the United States Diet

First, what has happened to sugars in relation to carbohydrate in the national food supply (United States diet) during the past 60 years will be pointed out. "Sugars," in the plural, is used to refer to all sugar. This includes naturally occurring sugar found in many foods such as milk and fruit and the sugar contained in syrups and honey as well as refined cane and beet sugar used as such and in food processing.

1. Changes in Level of Total Carbohydrate

The trend in total carbohydrate consumption is indicated in Fig. 1. The trend line is based on a 5-year moving average and represents consumption as a percent of the level reported for the base period of 1909–13. These are the first years for which such statistics are available. For comparison, trend lines for protein, fat, and food energy are also included.

The carbohydrate provided by the national food supply has declined about one-fourth since early in the century. Nearly 500 gm of carbohy-

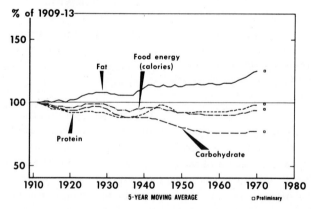

Fig. 1 Per capita civilian consumption of food energy (calories), protein, fat, and carbohydrate, 1909–13 to 1972.

drate per person per day was available then as compared with approximately 380 gm today. Decreased consumption of flour and cereal products is largely responsible for the decline. In 1909–13, the per capita per year estimate for these foods was about 300 lb, now it is 142 lb. The slight upturn in the carbohydrate level within the past 5 years is the result of an increase in use of sugars and other sweeteners combined with an apparent leveling off in the use of grain products.

Carbohydrate consumption has decreased at a faster rate than total food use measured as food energy or calories. The calorie level is now about 3300 kcal, only about 5% less than at the beginning of the century. At that time the national food supply provided about 3500 kcal/person/day. Furthermore, over the past 60 years, the calorie level has never dropped more than 10 percentage points below the levels at the beginning of this time series (5-year moving average). An upward trend in the calorie level can be seen in the past few years following a relatively stable period dating back to the early 1950's. This upturn reflects primarily upward changes in the levels of the other two energy-yielding nutrients—protein and fat rather than changes in carbohydrate. Protein is now near the record level reported in the early 1900's—around 100 gm/person/day. Fat also is at its peak having increased about one-fourth over the past six decades to a level of approximately 155 gm/person/day.

2. Changes in Levels of Sugars and Starch

In Fig. 2 carbohydrate is broken down into total sugars and starch, and trend lines are shown based on percentage changes in consumption

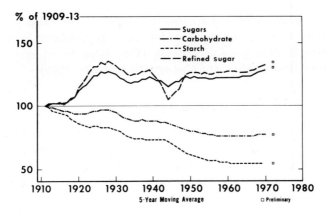

Fig. 2 Per capita civilian consumption of total sugars, refined sugar, starch, and carbohydrate, 1909–13 to 1972.

since the early 1900's. For the purpose of estimating total sugars and starch, it was assumed that the carbohydrate in foods such as fruit and sweeteners was present mainly as sugar and that carbohydrate in foods such as grain products and vegetables was present mainly as starch. A trend line for refined sugar, the largest single component of total sugars, is also shown. Refined sugar refers to cane and beet sugar. A line for total carbohydrate is included as a point of reference.

The decrease in total starch over the years is twice as great as the decline in total carbohydrate, on a percentage basis. The nearly 50% drop in starch is primarily the result of the sharp decline in the use of grain products mentioned earlier. In contrast, use of total sugars is up one-fourth over the early level, reaching a peak in the late 1920's. Use declined somewhat during the depression years of the 1930's and the war years of the 1940's. With these two exceptions, however, consumption showed little variation from the early 1930's to the mid-1960's when the amount in the United States diet started to rise. It is now a few percentage points higher than in the late 1920's. The current level is about 200 gm/person/day. Changes in consumption of refined sugar are largely responsible for the changes noted for total sugars.

Refined sugar use reached a peak in the late 1920's during the prohibition years when part of the sugar may have been used in making illegal alcoholic beverages. The consumption of refined sugar dropped in the early 1930's but remained fairly stable during the next 40 years or so except for a period of relative scarcity during the war years of the 1940's. Only recently has the use of refined sugar again reached the high levels reported in the 1920's. The per capita level is presently about 102 lb/year or about 130 gm/day. The higher current use of refined sugar may be partly the result of the withdrawal of cyclamates from the market in 1970 with their partial replacement by sugar. It may also partly result from the development of new uses for sugar, particularly in convenience foods, among other possible reasons (U.S. Department of Agriculture, Agricultural Stabilization and Conservation Service, 1972).

3. Changes in Proportion of Sugars to Starch

The share of total carbohydrate in the United States diet provided by sugars has increased continuously since the turn of the century (Fig. 3). This is true even though consumption of sugars on a per capita basis only recently has matched levels of the 1920's. The reason for this, of course, is the sharp decline in the amount of starch provided as consumption of grain products dropped. Early in the century, sugars accounted for about one-third of the total carbohydrate. By the late 1950's, sugars

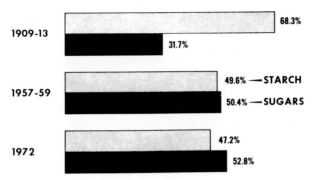

Fig. 3 Proportion of carbohydrate in the national food supply provided by starch and sugars, selected periods.

and starch contributed about equally to the total amount. Now sugars make up a little more than half of the carbohydrate.

4. Changes in Sources of Sugars

Food sources of sugars in the United States diet have been categorized into broad groups as shown in Fig. 4. Sugars, as mentioned earlier, include sugar that occurs naturally in a food such as lactose in milk, sugar that is provided by sweeteners such as syrups, and refined sugar. The current level of sugars, about 200 gm/person/day, supplied by the national diet is equivalent to 7 oz. Of this amount, over two-thirds comes from the sugar and syrups group, one-eighth from dairy products, one-eighth from fruits, and the remainder from other foods. In making these estimates,

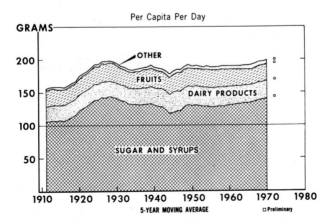

Fig. 4 Sources of sugars in the national food supply, 1909–13 to 1972.

refined sugar and other sweeteners added to processed fruits and sweet-ened condensed milk are included with the fruits and dairy products groups, respectively. This is done for statistical convenience. However, the amounts involved represent only a small proportion of the total sugars—less than 5%.

These broad food groups accounted for roughly the same proportion of the total sugars at the beginning of the century as they do today, even though the total quantity of sugars at that time was about one-fourth less. There has been some variation in the share provided by these groups over the years, but it has not been large.

5. Changes in Proportion of Individual Sugars

An attempt to estimate the amounts of individual sugars in the United States diet has been only partially successful because of inadequate infor-mation about sugars in food. Nevertheless, a rough estimate is given in Table I. About 85–90% of the total sugars in the food supply can be accounted for as individual sugars by calculation; for example, in the 1972 United States diet, sucrose accounted for 61.8% of the total, lactose for 12.5%, glucose for 6.4%, maltose for 2.7%, fructose for 1.7%, and other sugars for 3.2% to give a total of 88% of the 201 gm of total sugars supplied per person per day. By comparison, in 1909–13 sucrose accounted for 64.8% of the total, lactose for 13.6%, fructose for 4%, glucose for 3.5%, maltose for 1.1%, and other sugars for 2.1% to give a total of 89% of the 156 gm of total sugars per capita per day provided at that time. Sucrose, lactose, and fructose are the sugars most likely to have been underestimated in our calculations.

TABLE I

Estimate of Percent of Total Sugars Provided by U.S. Diet Accounted for as Individual Sugars, Selected Periods

| Year | Total sugars (gm) | Individual sugars | | | | | |
		Fructose (%)	Glucose (%)	Lactose (%)	Maltose (%)	Sucrose (%)	Other (%)
1972	201	1.7	6.4	12.5	2.7	61.8	3.2
1947–49	192	2.6	5.6	14.5	1.7	62.0	2.4
1925–29	197	2.9	6.1	11.6	1.3	66.2	2.0
1909–13	156	4.0	3.5	13.6	1.1	64.8	2.1

B. Use of Caloric Sweeteners

The term "caloric sweeteners" excludes substances such as saccharin that provide sweetness but do not yield food energy. Five categories of use generally account for the bulk of sweeteners used by the industry in commercially prepared foods. These are (1) in cereals and bakery products; (2) in processed foods such as canned, bottled, and frozen food (mainly fruits and vegetables) and in jams, jellies, and preserves; (3) in confectionery products; (4) in ice cream and other dairy products; and (5) in beverages (mainly soft drinks). Nonindustry use includes purchase for home consumption, for institutional use (including the military), and for use by commercial eating establishments.

The major sweeteners are refined sugar, corn syrup, and corn sugar or dextrose. These products accounted for over 98% of the sweeteners consumed in 1971 as shown in Fig. 5. The sugar indicated in this figure is refined sugar. Other sweeteners include honey, maple syrup, sorghum syrup, sugarcane syrup, refiners' syrup, and edible molasses.

1. Changes in Use of Sweeteners

Sugar's share of the sweeteners has been decreasing in recent years (Fig. 5), but it is still used in by far the largest quantity. As mentioned earlier, its present per capita per year consumption is about 102 lb. By comparison, the per capita quantity for corn syrup is about 16 lb, and for dextrose about 5 lb.

From 1960 to 1971, sugar's share dropped 4 percentage points—from

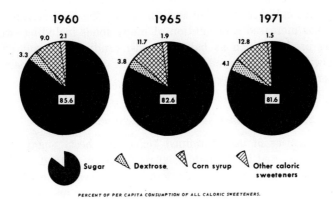

Fig. 5 Share of total caloric sweeteners consumed provided by refined sugar, dextrose, corn syrup, and other sweeteners, selected years.

86 to 82%. However, its use in commercially prepared food, and particularly in soft drinks, increased so that there was a net increase of about 5 lb/person. Corn syrup's share rose from 9 to 13% of the total sweeteners. This reflects primarily an increase in its use in commercially prepared foods, particularly in cereal and bakery products, in processed foods, and in dairy products. The small increase in the share accounted for by dextrose resulted primarily from its increased use in cereals and baked goods and in soft drinks (U.S. Department of Agriculture, Economic Research Service, Food Consumption Section, unpublished information).

The proportion accounted for by the remaining sweeteners is small. Their share dropped slightly from 2.1 to 1.5% in the past 10 years or so. Of these sweeteners, honey is currently used in the largest quantity. Its present use is a little over 1 lb/person/year. However, its per capita consumption has never reached 2 lb. The other sweeteners are now consumed in amounts of 0.2 lb or less each.

From 1960 to 1971, the total consumption of caloric sweeteners, which consists mainly of refined sugar, increased about 10% from roughly 114 to 125 lb/person/year. Several reasons have been suggested for this increase including those mentioned earlier as possible reasons for increased use of refined sugar (U.S. Department of Agriculture, Agricultural Stabilization and Conservation Service, 1972):

The development of new uses for caloric sweeteners, particularly in convenience foods.

A higher proportion of teens and subteens in the population. These are age groups that are likely to consume above average quantities of sweetener-containing foods such as soft drinks.

A higher income, including increased support of USDA's food stamp program, that enhances the demand for food. The food stamp program no doubt has increased consumption of many foods including sweetener-containing products such as bakery goods, candies, and soft drinks.

The withdrawal of cyclamates from the market and their partial replacement with caloric sweeteners.

2. Changes in Use of Refined Sugar

Three categories of use of refined (cane and beet) sugar are shown in Fig. 6: sugar used in the industrial preparation of food products and beverages, sugar used by consumers, and sugar used by institutions and others. Two noteworthy changes have occurred in the use of refined sugar since the beginning of the century: (1) an increase of one-third in the per capita quantity consumed and (2) a shift to the increased use of

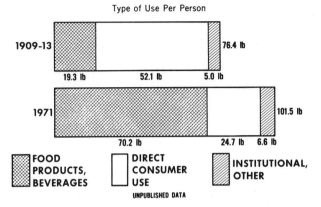

Fig. 6 Per capita per year use of refined sugar in industrially prepared food products and beverages, direct consumer use, and use by institutional and other users, selected periods.

sugar in food products and beverages and a shift away from the household purchase (direct consumer use) of refined sugar as such. Household use is assumed to be synonymous with the purchase of granulated sugar packaged in quantities smaller than 50 lb.

Use in processed food products and beverages has increased more than threefold from nearly 20 to 70 lb, while household purchase has dropped one-half from a little more than 50 to about 25 lb. Currently, food products and beverages account for more than two-thirds of the refined sugar consumed—70 lb out of a little over 100 lb. Moreover, beverages now comprise the largest single industry use of refined sugar, accounting for over one-fifth of the total refined sugar in the United States diet, or nearly 23 lb (Table II). Furthermore, the amount used in beverages has increased nearly sevenfold since early in the century when $3\frac{1}{2}$ lb/person/year was used in these products. Use of refined sugar in beverages is now second only to household use.

The next largest use of refined sugar by the food industry is in cereal and bakery products (Table II). These foods now account for about one-sixth of the per capita total or around 18 lb. Moreover, the per capita quantity going into these products has quadrupled in the past 60 years. Confectionery products and processed foods (which includes fruit products) each account for about 10% of the current per capita quantity of a little more than 100 lb or 10–11 lb each. Dairy products account for about 5% of the total or about 6 lb, and other foods for 2 or 3 lb.

By comparison, six decades ago the manufacture of confectionery products represented the largest industrial use of refined sugar (Table II).

TABLE II

Refined Sugar, Estimated per Capita Consumption by Type of Use, Selected Periods, 1909–13 to 1971[a]

Type of use	1909–13 (lb)	1925–29 (lb)	1935–39 (lb)	1947–49 (lb)	1957–59 (lb)	1965 (lb)	1971 (preliminary) (lb)
In processed foods:							
Cereal and bakery products	4.5	7.7	9.7	12.9	15.4	15.6	17.6
Confectionery products	6.5	8.0	8.2	9.8	9.6	10.4	11.0
Processed fruits and vegetables[b]	3.0	4.6	4.4	9.0	9.8	9.5	10.4
Dairy products	1.5	2.3	2.4	4.6	4.9	5.3	5.8
Other food products[c]	0.3	0.7	1.2	1.5	1.7	2.5	2.6
Total food products	15.8	23.4	25.9	37.8	41.4	43.3	47.4
Beverages (largely in soft drinks)	3.5	5.0	5.2	10.6	12.6	16.9	22.8
Total processed food and beverages	19.3	28.4	31.1	48.4	54.0	60.2	70.2
Other food uses:							
Eating and drinking places[d]	4.5	5.7	6.3	7.7	7.3	6.2	5.5
Household use[e]	52.1	65.0	58.8	37.4	33.1	28.2	24.7
Institutional and other use[f]	0.5	0.9	0.9	1.3	1.0	1.4	1.1
Total	57.1	71.6	66.0	46.4	41.4	35.8	31.3
Total food use	76.4	100.0	97.1	94.8	95.4	96.0	101.5
Nonfood use[g]	0.3	0.4	0.4	0.4	0.7	0.6	0.9
Total consumption	76.7	100.4	97.5	95.2	96.1	96.6	102.4

[a] Prepared by Food Consumption Section, Economic Research Service, U.S. Department of Agriculture.
[b] Canned, bottled, and frozen foods (processed fruit and vegetable products); jams, jellies, and preserves.
[c] Includes miscellaneous food uses such as meat curing, and syrup blending.
[d] Includes hotels, motels, restaurants, cafeterias, and other eating and drinking establishments.
[e] Household use assumed synonymous with deliveries in consumer-sized packages (less than 50 lb).
[f] Largely for military use.
[g] Includes use in pharmaceuticals, tobacco, and other nonfood use.

But this use accounted for less than one-tenth of the per capita total of 76 lb of refined sugar or for $6\frac{1}{2}$ lb. The amount going into cereals was 4.5 lb, into beverages 3.5 lb, and into processed foods around 3.0 lb. The amount used in dairy products was 1.5 lb. Other food products accounted for 0.3 lb.

Little change is noted in the use of refined sugar as such by institutions including the military and other users such as hotels, motels, restaurants, cafeterias, and other eating and drinking establishments (Table II). However, it is reasonable to assume that these users are also consuming more refined sugar now than earlier since they serve more foods that are prepared off the premises.

C. Use of Sweetener-Containing Foods

Finally, data from USDA's nationwide food consumption surveys are presented to show patterns in the use of selected sweetener-containing foods (U.S. Department of Agriculture, Agricultural Research Service, 1956, 1968, 1972). Figure 6 indicates a decreased use by the consumer in recent years of refined sugar as such and an increase in the use of refined sugar added to commercially prepared foods and beverages. This shift in use is supported by information obtained from the 1955 and 1965 surveys of family food consumption as shown in Fig. 7. In these surveys, families reported quantities of food they used in a week. The data shown are for the spring of the year. Use of sugars and sweets (syrups, molasses, honey, jellies, jams, candies, and toppings) was 10% lower in 1965 than in 1955. This decrease largely resulted from a decline in use of refined sugar as such. On the other hand, consumption of certain sweetener-containing foods increased. For example, the use of frozen milk desserts (ice

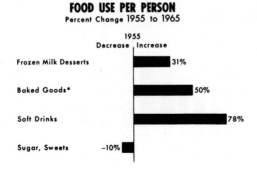

FOOD USE PER PERSON
Percent Change 1955 to 1965

Fig. 7 Change in household use of selected types of foods, 1955–65.

cream, ice milk, and sherbet) increased by about one-third, baked goods (cake, pie, cookies, doughnuts, and the like) increased by one-half, and soft drinks increased by more than three-fourths.

1. Use of Soft Drinks by Individuals

The average intake of soft drinks by men, women, and children of various ages in the 1965 survey is shown in Fig. 8. As mentioned earlier, about one-fifth of the total refined sugar is currently used in beverages, and mainly in soft drinks. In this survey, persons reported on the kinds and amounts of food they ate in a day. As would be expected, children drank more soft drinks when they became older and the quantity reached a peak when they became older teenagers. Young adults reported drinking substantial quantities of soft drinks, but somewhat less than the older teenager. Intake by older adults dropped sharply. Males generally drank more soft drinks than females, but for several of the groups the differences were small. The percent of persons that drank soft drinks was similar for both males and females and was highest for the older teenagers, with 5 or 6 persons out of 10 reporting use of soft drinks in the day.

No consistent differences in intake were noted by income or degree of urbanization, but differences were found by region of the country (Fig. 9). Young children and males over 9 in the South reported using substantially larger quantities of soft drinks, on the average, than their counterparts in the North. Females 9 years and older in the South also used more soft drinks than those in the North, but differences between the two regions were not as large as noted for the males.

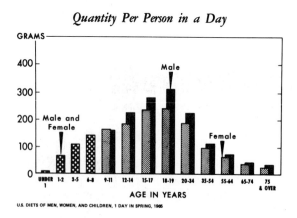

Fig. 8 Average consumption of soft drinks by individuals of different ages.

Quantity Per Male Person in a Day

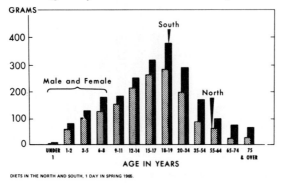

DIETS IN THE NORTH AND SOUTH, 1 DAY IN SPRING 1965.

Fig. 9 Average consumption of soft drinks by children and by males 9 years of age and older in the North and South.

2. Use of Baked Goods by Individuals

The use of baked goods, including both commercially and home prepared, shows a somewhat different pattern among the various groups of individuals than was found for soft drinks (Figs. 8 and 10). Peak use was reported for males 18–19 years of age, but the greatest use by females occurred at a much earlier age—9–11 years. Percentages of persons who reported consuming baked goods showed little variation among all groups above 1 year of age—from 5 to 7 persons out of 10. Such a high proportion of consumers is not surprising in view of the assortment of foods included in this broad classification. Cakes, pies, cookies, sweet crackers, sweet buns, sweet rolls, and so forth were included. However, breads, plain rolls and buns, and most crackers were not included.

Quantity Per Person in a Day

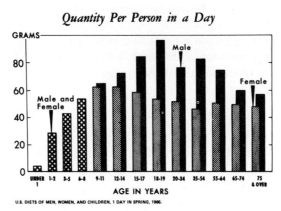

U.S. DIETS OF MEN, WOMEN, AND CHILDREN, 1 DAY IN SPRING, 1965.

Fig. 10 Average consumption of baked goods by individuals of different ages.

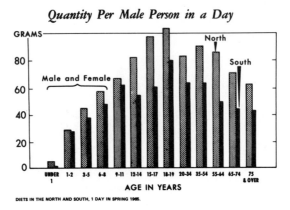

Quantity Per Male Person in a Day

Fig. 11 Average consumption of baked goods by children and by males 9 years of age and older in the North and South.

Regional differences occurred in the use of baked goods as they did for soft drinks (Figs. 9 and 11). Children and males over 9 years of age in the North consumed larger amounts on the average than did those in the South. However, differences among females tended to be considerably smaller. A possible explanation for the regional differences might be found in the fact that bread use—which includes hot breads such as biscuits, muffins, and cornbread—was higher in the South.

3. Use of Sugars and Sweets by Individuals

Some sugar, syrup, honey, molasses, jelly, jam, or candy was included in the day's diet by about 6 or 7 out of every 10 persons in each group over 1 year of age. The use of sugar reported here does not include indirect use such as an ingredient in prepared foods other than in those foods mentioned. The amount of sugars and sweets eaten increased up to age 14 for both boys and girls, then decreased slowly with age. Quantities eaten were larger for males over 9 years than for females of the same ages. Boys 12–14 consumed the largest amounts (Fig. 12).

D. Conclusion

During the past 60 years a marked change has occurred in the total sugar content of the United States diet and in the way refined sugar is entering the diets of individuals:

Total sugar content of the United States diet is up one-fourth with most of the increase occurring between 1909–13 and 1930.

Quantity Per Person in a Day

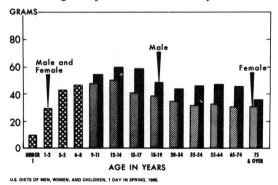

Fig. 12 Average consumption of sugar and sweets by individuals of different ages.

Greater use of refined sugar is largely responsible for this increase.

An increasingly larger proportion of the total refined sugar is being used in prepared foods and beverages before they are brought into the home. The largest increase in use comes from use in soft drinks.

With increasingly more refined sugar and other sweeteners added to foods and beverages before they enter the home, today's consumer appears to have considerably less control over the sugar content of his diet than the consumer of 60 years ago.

References

U.S. Department of Agriculture, Agricultural Research Service. (1956). "Food Consumption of Households in the United States," U.S. Dept. of Agriculture Household Food Consumption Survey 1955, Rep. No. 1. USDA, Washington, D.C.

U.S. Department of Agriculture, Agricultural Research Service. (1968). "Food Consumption of Households in the United States, Spring 1965," U.S. Dept. of Agriculture Household Food Consumption Survey 1965–1966, Rep. No. 1. USDA, Washington, D.C.

U.S. Department of Agriculture, Agricultural Research Service. (1972). "Food and Nutrient Intake of Individuals in the United States, Spring 1965," U.S. Dept. of Agriculture Household Food Consumption Survey 1965–1966, Rep. No. 11. USDA, Washington, D.C.

U.S. Department of Agriculture, Agricultural Stabilization and Conservation Service. (1972). Sugar Report, No. 241. USDA, Washington, D.C.

U.S. Department of Agriculture, Economic Research Service. (1968). "Food Consumption, Prices, and Expenditures," Agr. Econ. Rep. No. 138, Suppl. for 1971. USDA, Washington, D.C.

Sugars in Food:
Recent Technological
Developments

CHAPTER 8

The Chemistry and Technology of Sugars

SIDNEY M. CANTOR

People's food habits with regard both to what they eat and how they eat are related in multiple and complex ways to their cultural heritage. Food habits—or better stated, food behavior—appears to be subject to some change in certain of its components; the economic component is perhaps the most flexible. However, it is remarkable how old food behavior patterns resurface regularly or, to put it another way, how every man's soul food retains an important part of his soulful identity.

The basic nature of sweetness perception puts the "soul" character of sugars in a special class. The early experience of sweetness on the day of birth or, as it has been suggested, even before birth (DeSnoo, 1937), along with its pleasant associations of warmth and security and the regularly repeated reinforcement of the sensation leave a lasting mark not only on the dietary but also on such revealing culture mirrors as language: sweet is good, sugar is a pretty girl, and sugar is money (Cantor, 1953).

Technologies also derive from the cultural heritage and not only do they carry reflections of the art or craft which is their origin, but also they develop an identifiable character as a result of being responsive to the social, behavioral, and economic forces which shape and control them. In other words, technology and craft interact and are modified by complex forces. In a comparable way, usually somewhat later, the technology and its related science interact and are each responsive to modifying forces.

The disciplines of sugar technology and sugar chemistry clearly reflect man's pursuit of sweetness. The pursuit of this pleasant sensation has resulted in an increase in the consumption and in altered patterns of use of certain sugars. In seeking to understand these changes and the relationships among the complex of forces which produced them, perhaps we can find effective countermeasures, if indeed, these altered patterns represent—as feared by many—an increasingly threatening form of malnutrition. Therefore, some historical aspects of sugar industry development which have helped to set patterns of consumption will be briefly examined along with the impact on these patterns of what are generally recognized as critically important recent developments in sugar chemistry and biochemistry and their related technologies.

A. Honey, Cane, and Beet Sugar

1. Honey

Honey was man's first important concentrated sweetener, and the earliest direct evidence of human association is illustrated by a Neolithic cave painting (Lehner and Lehner, 1962). Beekeeping, a highly developed art in many ancient cultures, is probably at least 5000 years old. Honey was so important a sweetener that the bee was domesticated, and manuals of instruction on beekeeping were prepared. Sources and quality differences of honeys were noted, and relative values were established for trading in a number of societies throughout Asia and in Central and South America.

The words for honey and sweetness are similar in ancient Biblical languages and the Bible contains frequent references to "milk and honey" as well as to manna (a sugar-containing tree exudate). Most of these references are associated with "the good life."

Along with the good life association, there also developed superstitions and rituals connected with honey—ways to control the good life—because there was seldom enough for all. Thus, there developed an abundant lore about honey, its natural and healthful qualities. The ease of assimilation of honey by the body, as a result of its pure sugar identity, was—and for that matter still is—emphasized, thus illustrating the persistence of deep, culture-rooted beliefs. This is pointed out not to imply a value judgment but as a setting for a later return to a consideration of honey in this context.

2. Cane Sugar

The sugarcane, which along with the sugar beet is the principal source of sucrose, the major nutritive sweetener in the human dietary, originated in India. It was introduced to Persia in the third century A.D., to Arabia in the fifth century, and to Egypt in the ninth century. The Moslem conquerors carried sugarcane throughout the Mediterranean region, and in 1493 Columbus brought the plant from the Canary Islands to San Domingo, thus establishing it in the West Indies. In the seventeenth century, the British contributed to further progression by bringing cane to Barbados. The continuing movement has resulted in today's cultivation of sugarcane in most of the tropical and semitropical regions around the world (Lehner and Lehner, 1962).

In most of these countries, except for that portion of sugar directed to domestic consumption as a somewhat more refined product, raw sugar—produced by clarifying and evaporating cane juice and crystallizing the crude product—is transported to high consumption areas for refining. The speed with which the sugar cane was carried around the world—barely 500 years—is testimony to the power of the demand for sweetness. What started out as an expensive rarity enjoyed by the wealthy was increased in supply by extending tropical production and utilizing low cost or slave labor for working the cane fields. But even as the cost was reduced, the demand for sweetness increased. On a world basis, total sugar production increased from 21 gm/day per capita to 51 gm/day per capita from 1905 to 1970 (Wretlind, Chapter 6, this volume).

The refining process is carried on essentially to demand in the major consuming areas of the world because sugar at best is a perishable commodity and storing it as raw sugar allows whatever deterioration that occurs to proceed one step away from the last refining operation. Nonetheless, a continuing improvement in raw sugar quality, reducing refining requirements and costs, has been a notable development in cane sugar refining practice.

There is also the problem of storage of the vast quantities of sugar which are processed. In the United States, for example, this is currently over 11 million tons annually. But, in addition, there is the effect of supply on price to be considered. If too much sugar is placed on the market at any one time, the price is likely to fall to a point detrimental to the refiner. As a result, the amount of sugar placed on the market in the United States is controlled under the Sugar Act on a quarterly basis by the Secretary of Agriculture after consultation with both producers and manufacturing consumers. This is part of a much more elaborate control system which protects domestic producers of sugar, that is, both beet and cane processors, by establishing an import duty on raw sugar, quotas for domestic production, and quotas for "favored nation" producers and exporters of raw sugar.

Chemically, cane and beet sugar are sucrose, but for the consumer the white crystalline product is sugar, a word which in popular language applies only to this single substance. To sugar chemists, however, sugar is a generic term describing a large and varied class of organic compounds. Sucrose is structurally unique among sugars. Aside from being the purest organic compound produced in such quantity, it is unique chemically. It is a disaccharide in which one molecule of α-D-glucose (an aldohexose) in the pyranose or six-membered ring form is condensed with one molecule of β-D-fructose (a ketohexose) in the furanose or five-membered ring form through their respective acetal and ketal functions. Thus sucrose is nonreducing but because of the unique carbonyl to carbonyl linkage also demonstrates great acid lability and hydrolyzes readily in acid systems forming a mixture of D-glucose and D-fructose called invert sugar (see Fig. 1).

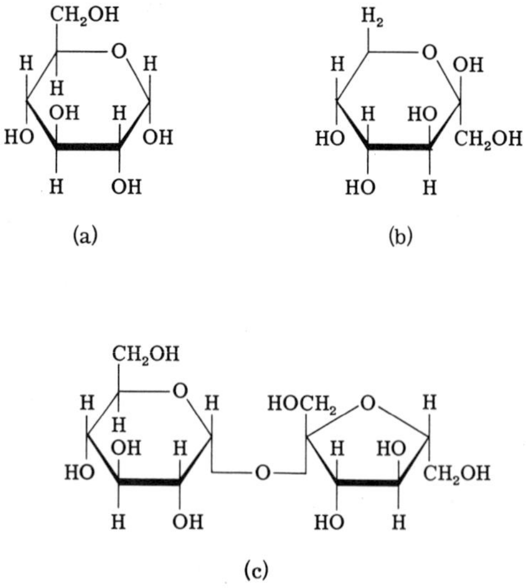

Fig. 1 (a) α-D-Glucopyranose, (b) β-D-fructopyranose, and (c) sucrose (α-D-glucopyranosyl-β-D-fructofuranoside).

3. Beet Sugar

The development of the sugar beet as a temperate zone source of sucrose was literally a research and development achievement (McGinnis, 1951). Although Andreas Margraf, a German chemist, proved that the beet contained sucrose in 1747, it was not until 1799 that one of his students, by publishing promising cost data, attracted the interest of the French government and also the King of Prussia, Frederick Wilhelm III. Motivation for the interest in a new sugar source was the English blockade of European ports as a result of the Napoleonic Wars, which cut off raw cane sugar imports from the West Indies, thus denying the growing demand for sugar to increasingly restless citizens. Early results, unfortunately, were poor and the French devoted their attention to extracting sugar from grapes. Judging by commentary of the times, however, Napoleon was under heavy pressure to provide an acceptable source of sweetness.

By 1811, however, the commercial processing of sugar beets improved to such an extent that Napoleon decreed that 70,000 acres of land be planted to beets and one million francs be appropriated for establishing processing plants. But, before much progress was made, the Battle of Waterloo (1815) intervened, the coastal blockade was lifted and the infant beet sugar industry of Europe collapsed. It did not succeed in rising again in importance until the mid-nineteenth century. Then a combination of process improvement and tariff protection against the slave (but soon to be freed), low-cost labor of the tropical cane areas set up circumstances which allowed the beet industry of Europe to compete (McGinnis, 1951).

In the United States, sugar beet industry development was somewhat slower. Although technologically sound by 1895, outbreak of the Spanish American War in 1898 and the subsequent change in status of Puerto Rico, Cuba, and the Philippines had far reaching effects on the sugar economy. A succession of tariff measures which caused sugar prices to fluctuate wildly, the demands on United States production during World War I, which further complicated the picture, and the dumping of world sugar on the United States market almost destroyed the beet sugar industry and led ultimately to the Sugar Act of 1934. This act built a protective wall around United States domestic production of beet and cane sugar, established quotas, and stabilized prices in a manner constructive to both producers and users.

It is quite clear that the rapid technological development and measures for price control which marked the sugar industry very quickly brought quality sugar within the economic reach of everyone in the industrially developed world; for example, from 1909–13 to 1972 per capita daily sucrose consumption in the United States increased from approximately 101 to 124 gm while total sugars in the diet went from 156 to 201 gm in the same period (Page and Friend, Chapter 7, this volume).

B. Starch-Derived Sweeteners

The year 1811 was critical for another branch of sugar technology because in this year S. Kirchoff, also a German chemist, who was working in St. Petersburg, discovered the acid-catalyzed hydrolysis of starch to D-glucose. Though this discovery was too late to affect Napoleon's decision for a new sugar source, it did mark the start of a starch-derived sweetener industry based on potato starch in Europe and corn (maize) starch in the United States.

The United States corn refining industry flourished largely because the cost of its raw material was buffered by a huge corn crop raised specifically for animal feed. Starch processing developed, in effect, as a by-product industry.

Corn syrup, or glucose syrup as it is known commercially, developed quickly, first as a sugar syrup substitute and honey extender; this was a product surrounded by considerable dispute as to use and identity. But as starch-derived syrups were improved and as food technologists brought some structured development to food formulation, corn syrup flourished. Thus, because of a combination of price factors and functional utility, it has become an important and widely accepted commercial product.

The relationship between corn sweeteners and sucrose has been marked by considerable conflict often joined at food standards hearings, but the relative size of the two industries and the worldwide public image of sugar have fashioned their respective development patterns.

Technological change in the cane sugar refining industry has been associated largely with operating cost problems such as heat utilization, crystallization improvement, decolorization, materials handling, and the production of ever-purer sugar to meet industry demands. This is also true of beet sugar processing although this industry has been more involved in by-product development. An outstanding by-product was monosodium glutamate derived from beet molasses which had a relatively short commercial life because it was succeeded by the less expensive fermentation product based on dextrose. Thus, the beet sugar industry is also essentially a one product industry.

It is noteworthy that sugar demands have been described in increasingly technical terms as major sugar usage has moved from the kitchen to convenience food factories and as the multiple functional roles of sugar have been appreciated. As a result, the sucrose industry in both branches has sought to maintain and extend its markets via sales service channels. The exigencies of competition, however, and the need to improve its products to meet greater technological acceptance, gain wider markets and thus grow, have led the starch industry into an ever-increasing investment in research and new product development.

Out of this came crystalline α-D-glucose hydrate (commercially, dextrose hydrate) in 1921 and later both alpha and the more rapidly soluble beta anhydrous crystal forms of D-glucose. Commercial crystalline dextrose in all its various forms was developed largely in response to the request of pediatricians for an easily assimilable baby food component, but the amount used directly for this purpose now is hardly significant.

Being less sweet than sucrose is a handicap to direct consumer use of dextrose and the more than one billion pounds produced in 1971 reached consumers in indirect ways. On the other hand, glucose syrup, a partial hydrolysis product of starch, is different. It blends well with sucrose-derived syrups, can be handled as a liquid, and offers cost savings. It therefore has become a product of much greater production volume than crystalline dextrose.

Recognition of the limitations of acid-catalyzed starch hydrolysis in terms of increasing demands for functional flexibility led to commercial application of enzyme-catalyzed starch hydrolysis. High maltose syrups were first produced about 1940, after β-amylase at reasonable cost became available. Later, with the commercial development of glucoamylases, high dextrose syrups or so-called liquid dextrose became feasible and was produced in increasing volume.

Acid-catalyzed liquefaction of starch followed by glucoamylase treatment of the resultant low dextrose solution, referred to in the industry as the acid-enzyme process, continued to be the process of choice for dextrose production largely because of cost. More recently, development of lower cost α-amylase (a starch granule breakdown enzyme) not only provided the basis for a totally enzymatic starch hydrolysis process (dual enzyme process) but also gave new maltodextrin products. These substances, which are newcomers to the corn refining industry's increasingly broad spectrum of controlled starch hydrolyzates, are completely soluble, nonhygroscopic, minimum sweetness products. The diversity of hydrolysis products of starch available commercially is shown in Table I.

The proliferating use of these products with and without sucrose and sucrose syrups in formulated foods point to these practices as a source of changing patterns in sugar consumption. In this connection, it is interesting to note that in the United States starch-derived sweeteners repre-

TABLE I

Types and Analyses of Corn Syrups[a]

Component (%)	Type I[b] acid	Type II[c] acid	Type II[c] dual enzyme	Type II[c] acid	Type III[d] acid/enzyme	Type III[d] dual enzyme	Type IV[e] dual enzyme
Moisture	20	19.7	19.7	19	16	17.5	29
Dextrose equivalent	37	42	42	52	62	69	96
Dextrose	15	19	6	28	39	50	93
Maltose	12	14	45	17	28	27	—
Trisaccharides	11	12	15	13	14	8	—
Tetrasaccharides	10	10	2	10	4	5	—
Pentasaccharides	8	8	1	8	5	3	—
Hexasaccharides	6	6	1	6	2	2	—
Higher Saccharides	38	31	30	18	8	5	—

[a] Information supplied by J.M. Newton, Clinton Corn Processing Company, 1972.
[b] From 20 to 38 D.E. Dextrose equivalent or total reducing sugars expressed as dextrose. Below 20 D.E. products are classified as maltodextrins.
[c] From 38 to 58 D.E.
[d] From 58 to 73 D.E.
[e] From 73 D.E. above.

sented approximately 16% of nutritive sweetener distribution in 1972 while sucrose provided 61.8%. In the period 1925–29 by contrast, sucrose provided over 66% of the total (Page and Friend, Chapter 7, this volume).

As noted, product diversification in the sucrose industry has been limited. Liquid sugars were developed as a means of cutting refining costs and offering economies of handling to manufacturing consumers. Ion exchange resin refining was developed as a competitor to crystallization; thus, most developments have been process-oriented.

C. Importance of Sweetness

Invert syrups, items of currently increasing interest as we shall see, contain various amounts of hydrolyzed sucrose as well as sucrose and are old and well-recognized products of the sugar industry. They are used mostly in commercial operations as crystallization inhibitors and food plasticizers. Sucrose is inverted commercially by acid, by cation exchange resins, and also by yeast invertase to provide required functional properties.

Invert syrups providing varied levels of sucrose inversion are sweeter than sucrose at comparable concentrations. This greater sweetness reflects the D-fructose (commercially, levulose) component of the syrup and is an important and critical feature.

The history of sugar chemistry is in large measure the history of stereochemistry and, therefore, a recognition of both the interconvertibility of sugars and the minimal structural differences which result in changes in sugar properties, notably sweetness. Examination of the complex mixture resulting from the action of alkali on D-glucose, for example, demonstrates that any hexose may be expected from any other hexose by such treatment; this includes L- as well as D-hexoses.

The greater sweetness of levulose (D-fructose) over D-glucose in addition to the relative ease of chemical interconvertibility made levulose a target for research efforts in the corn refining industry. Sweetness was its missing asset. The commercial realization of this objective represents its most recent accomplishment. However, because of lack of specificity, the success was not achieved by chemical means but rather by enzymatic isomerization. Isolation of isomerases capable of isomerizing D-glucose to D-fructose by both Japanese and American biochemists and intensive development work in the United States led to commercial introduction of a levulose-containing, starch-derived syrup about 3 years ago.

The motivations for Japanese development of isomerases was ap-

parently not only an extension of the country's acknowledged investment
in enzyme technology but also related, according to authoritative sources,
to a recognition of two factors, namely, the increased expendable income
or affluence of the Japanese population and the related increasing demand
for sweet foods. Both factors are concomitants of increasing industrialization. Japan's hard currency expenditures for raw cane sugar were a cause
of concern; thus, the government, appreciating the country's respectable
domestic starch industry based on the sweet potato, funded a research
program to produce sweeteners based on starch. This appears to be another instance where the pursuit and satisfaction of sweetness affected
national policy.

Some comparative analytical data on levulose-containing syrups are
shown in Table II. The absence of sucrose from the products derived
from starch hydrolyzates is, of course, a major and unique difference.
Now syrups of sweetness comparable to that of invert syrups and independent of sucrose are commercially available.

The development of an independently arrived at common product from

TABLE II

Levulose-Containing Syrups

	Syrup identity			
Component (%)	Dextrose-levulose syrup (isomerized dextrose)[a]	Commercial invert syrups		Isomerose-100[b]
		Medium[a]	Total[a]	
Total solids	71	76	72	71
Levulose	45	28	44	42
Dextrose	53	29	47	50
Sucrose	0	39	6	0
Polysaccharides	<1	3	3	8
D-Psicose	<0.3	<0.2	<0.5	<0.3
D-Mannose	Trace	Trace	Trace	—
Maltose (with trace of maltulose)	—	—	—	2.5
Isomaltose	—	—	—	1.5
Higher polysaccharides (mostly DP 3 and 4)	—	—	—	3–4

[a] From Kooi and Smith (1972).
[b] "Isomerose-100" is a registered trade-mark of Clinton Corn Processing Company.

the starch and sugar industries at the invert syrup level is a notable achievement. While starch- and sucrose-derived syrups are regularly blended to meet cost and functional requirements, there are indications that the cost of an isocaloric unit of sweetness from starch is lower than a comparable unit from sucrose. Thus, in the United States, some adjustment of market shares seems reasonable to expect and the impact on patterns of consumption may be significant.

While the isomerization development is too recent for the readjustment of sweetener sources to be clearly apparent, the acceleration of starch syrup consumption in recent years, in part due to continuing commercialization of the food supply and its accompanying pressure for sweetness cost dilution, provides a basis for projection. In 1972 cane, beet, and corn shares of sweetener market were respectively 57.7, 24.7, and 16.6%. Isomerization of corn-syrup dextrose as an independent source of additional sweeteners can be expected to enhance growth rate. Moreover, other trends must be respected, notably the decline in sucrose from domestic cane sources, the problems experienced by many favored raw sugar suppliers in meeting their United States quotas and the encouragement of increased sugar beet production as a national policy. Accordingly, the comparable share of market figures by the year 2000 could easily reach 38.8, 28.7, and 32.5% (Cantor and Shaffer, 1973).

D. Nonnutritive Sweeteners

No discussion of sugars in the context of technological change and consumption patterns is complete without reference to nonnutritive sweeteners. The importance of the technological achievement of separation of sweetness as a flavor from sweetness associated with calories can hardly be overestimated. At the time of the ban on cyclamates under Food and Drug Administration regulations, the impact on cyclamate-sweetened products in sugar equivalent terms was verging on 5% of United States nutritive sweetener consumption. This figure is testimony to the existence of a problem, whether it is real or imagined.

It is quite likely that the search for acceptable nonnutritive sweeteners has only begun. The recent announcement of the first protein sweetener (Morris and Cagan, 1972) has opened the door to a totally new area of exploration.

The difficulties of achieving significant changes in food behavior patterns associated with sweetness are formidable. Therefore, our ability to formulate foods to satisfy sweetness requirements and, at the same time, manage the problems which are associated with unlimited caloric intake by some of the population, represent a most rational source of dietary control and optimally acceptable solutions to a complex situation. It is important to note that many sugar-containing formulated foods are standardized; that is, allowable limits of ingredients including sugars are set at government-sponsored hearings in which interested commodity producers are invited to testify. Thus canned fruits, catsup, and many other products are fixed with respect both to amount and kind of sugars. This procedure offers a recognizable means of sugar control.

E. Role of Sugar Chemistry

We have reviewed the highlights of technological change in the sweetener industries without particular reference to the impact of research in sugar chemistry and biochemistry. The sugar chemistry of the late nineteenth and early twentieth centuries was essentially organic chemistry-oriented and was concerned chiefly with establishing structures and structural interrelationships, identifying sugars and related products isolated from plant tissues and synthesis. It succeeded in building a disciplinary structure that was essential but was also isolationist in the sense that while it added in a compartmentalized way to the general literature of organic chemistry, new organic chemistry had relatively little impact on sugar chemistry. The situation has improved considerably, and the easing of this isolation promises to continue.

With basic structures well in hand, sugar chemistry is now very much occupied with the dynamics of structure. In this respect, the discipline can be thought of as following the lead of physical organic chemistry. Much of the chemistry of sugars in solution, a subject closely related to nutrition, is still unclear, although work on ring conformation and structural interactions is making substantial contributions.

Looking at the past two or three decades, perhaps the greatest contributions to sugar chemistry which touch on nutrition have been made from outside the field. First, from analytical chemists, the advent of various kinds of column chromatography and then thin layer and gas–liquid chromatography has had great impact. The ability to isolate and identify relatively small amounts of constituents has opened to view avenues of interactions in complex systems and provided new insights to studies of mechanism in many areas of application.

A second and related development is represented by the contributions of biochemists. The great strides in recent years in enzymology have elaborated on the synthetic as well as hydrolytic roles of enzymes, including those operating on carbohydrate substrates. It is noteworthy that a great deal of the inspiration for these developments originated on a "need to know" basis from the starch industry.

A major effect of all of this new information on sugars has been to complicate what were previously thought to be simple systems. In a relatively current treatise on carbohydrate chemistry (Pigman, 1957), the following statement occurs: "Prehistoric man was acquainted with honey, a fairly pure mixture of the three sugars, sucrose, D-fructose, and D-glucose." As promised, this statement brings us full circle to honey. Some recently published data (Siddiqui, 1970) on honey constituents are assembled in Table III. Honey is apparently not so simple and "pure" in terms of limited constituents. The identification of so many additional sugars provides some understanding of the sugar metabolism of the bee and the complex ways in which sugars are synthesized and stored by the insect. It is significant in terms also of the early and continuing popularity of honey to note that present world production is estimated at 500,000 tons (Siddiqui, 1970). The United States production of honey for 1972 was 107,300 tons (U.S. Department of Agriculture, 1973), or not quite 1% of United States refined sugar production.

F. Research Needs

Equipped with this sophisticated chemistry and technology, it would appear that collectively we are reasonably prepared to contribute to the

TABLE III

Average Composition of Canadian Honey (95 samples)[a]

Component	Percentage	Range
Moisture	17.9	15.0–21.8
D-Fructose	37.1	31.1–41.4
D-Glucose	33.7	28.5–40.7
Oligosaccharides[b]	7.4	2.2–15.2
Undetermined	3.9	0.0–10.8

[a] From Siddiqui (1970).

[b] At least 24 oligosaccharides were identified including maltose, kojibiose, isomaltose, nigerose, α,β-trehalose, gentiobiose, laminarabiose, melezitose, maltotriose, sucrose, turanose, 1-kestose, panose, maltulose, and isomaltotriose. A specific sample which contained 3.65% oligosaccharides showed in that fraction: maltose, 29.4%; kojibiose, 8.2%; turanose, 4.7%; isomaltose, 4.4%; sucrose, 3.9%; kestoses, 3.1%; nigerose, 1.7%; α,β-trehalose, 1.1%; and others.

delineation of what might be called the fine structure of sugar metabolism in humans. One conclusion is certain: Advances in sugar chemistry and technology, modified by many other changes ranging from commercial processes to advanced industrialization, urbanization, and other social and political factors, have modified man's nutritive sweetener consumption pattern in significant ways. Moreover, evidence for continuing changes because of new technology and competitive pressures is reasonably clear.

Just as we know more about honey and bees, we need to know more about the realities of sugar consumption by humans. The statistics that we consider as indicative of the human dietary are based largely on disappearance of commercial products from available stores. For the most part, they are not based on an accurate knowledge of what sugars are consumed in a particular food; for example, all the sucrose that goes into carbonated beverages suffers some inversion. Therefore, it is not consumed entirely as sucrose but as mixtures of sucrose, dextrose, and levulose. The same is true of all acidic, sucrose-containing food products. The contribution of starch-derived sweeteners to processed foods is equally important to clarify. Recent revelations on problems of lactose metabolism associated with particular ethnic groups make it mandatory to know not only the lactose content of formulated food products but also the lowest cost means of reducing the lactose content. Current work on *in situ* methods of lactose hydrolysis in milk and on desugaring whey present promising developments in lactose technology. In short, if, as appears to be the case, the different sugars in the dietary follow significantly different metabolic pathways, then it is important to know much more than we do at present about the relative amounts of these sugars in the diet. From the standpoint of sugar chemistry and technology, it seems clear that much of the information is at hand to formulate products which minimize metabolic disfunctions.

We also need to know more quantitatively about the interactions of the numerous factors involved in establishment of dietary patterns of sugar consumption because, in knowing how they developed, we may be able to achieve acceptable methods of change. Thus, behavioral aspects of sugar consumption are critical. The situation appears increasingly to demand interdisciplinary approaches, including those which embrace cultural patterns as well as technological and medical factors. New approaches to systems analysis are available which properly applied can help to increase the effectiveness of corrective measures.

Interdisciplinary communication is difficult. Recently, while visiting a medically oriented laboratory where metabolic work on sugars was in progress, it was surprising to hear one of the workers say that he was puzzled because sucrose and a 1:1 mixture of D-glucose and D-fructose did not behave similarly. Apparently, the unique character of the sucrose structure was being overlooked in the work under way. Thus, while communication is difficult it is through the means of interdisciplinary exchange that the multidimensional forms of the problem are resolved in such a way as to reveal common denominators and set objectives. The communication which develops from such interchanges also serves to emphasize in a coherent and constructive way the vast and varied resources which are required and which are at our disposal for rationalizing and correcting in an acceptable, therefore effective, way the problems associated with sugars in nutrition.

References

Cantor, S. M. (1953). Where it does most good. *Sugar Mol.* **7**, No. 3, 3–6.

Cantor, S. M., and Shaffer, G. E., Jr. (1973). Some aspects of a three commodity sweetener system. Sweetener Symposium, A. C. S. National Meeting, April 1973, Dallas, Tex.

DeSnoo, K. (1937). Das Trinkinde Kind Im Uterus. *Monatsschr. Geburtsh. Gynaekol.* **105**, 88.

Kooi, E. R., and Smith, R. J. (1972). Dextrose-levulose syrup from dextrose. *Food Technol.* **26**, 57–59.

Lehner, E., and Lehner, J. (1962). Folklore and odysseys of food and medicinal plants. "The Story of Sugar," pp. 90–95. Tudor Publ. Co., New York.

McGinnis, R. A. (1951). "Beet Sugar Technology." Van Nostrand-Reinhold, Princeton, New Jersey.

Morris, J. A., and Cagan, R. H. (1972). Purification of monellin, the sweet principle of *Dioscoreophyllum cumminsii. Biochim. Biophys. Acta* **261**, 114–122.

Pigman, W., ed. (1957). "The Carbohydrates," 1st ed., p. 5. Academic Press, New York.

Siddiqui, I. R. (1970). The sugars of honey. *Advan. Carbohyd. Chem. Biochem.* **25**, 285–309.

U.S. Department of Agriculture. (1973). Honey Prod., 1972 Annual Summary. Crop Reporting Board, Statistical Reporting Service.

CHAPTER 9

Current Developments in Industrial Uses for Sugars

E. L. MITCHELL

Industrial uses for sugars are almost identical to uses in the home. We use nutritive sweeteners principally for their sweetening power—just as in the home. There are several other properties which make nutritive sweeteners important both to industry and home use. One is its preservative effect which is vital to the keeping quality of products such as jams, jellies, syrups, and candies. Another useful property of nutritive sweeteners is that they serve as a food for yeasts or other fermenting agents as in bread, pickles, and beer. Nutritive sweeteners provide body as in syrups, candies, and baked goods. Mixtures of sweeteners can provide functional advantages including freezing point, osmotic pressure, and crystallization control, as well as sweetness control.

With respect to current developments there have not been a great number of developments of note, but there are three worth mentioning:

1. The tremendous shift in usage from home use to industrial usage.
2. The shift from 100 lb bags to bulk liquid and bulk dry usage.
3. The most recent availability of high fructose corn syrup.

A. Consumption and Production

In the fourteenth century, one pound of sugar cost the equivalent of a workman's weekly wage. Today a pound of sugar costs about 15

TABLE I

Food Expenditures as a Percentage of Disposable Income[a]

Year	Percent
1932	23
1935	23
1940	22
1945	22
1950	23
1955	22
1960	20
1965	18
1970	16
1971	16

[a] U. S. Department of Agriculture, 1971.

cents and sometimes less on sale. We hear so much today about high food prices; however, in reality food prices are as cheap as they have ever been with respect to disposable income (Table I). From 1932 to 1971 food expenditures have gone from 23 to 16% of disposable income, or a reduction of 30%. Certain portions of the food industry actually need higher prices in order to obtain profits which can be used to plow back into the business to update facilities. This is certainly true in the highly competitive canning business.

The largest producers of sugar in the world today are shown in Table II. Table III lists the principal users of sugar in 1970. Table IV shows the per capita consumption of sugar in those countries with the highest

TABLE II

1970 Production[a]

Country	Amount (tons)	Sugar
U.S.S.R.	10,100,000	Beet
Cuba	6,000,000	Cane
United States (includes Hawaii and Puerto Rico)	5,800,000	Beet and cane
Brazil	5,500,000	Cane
India	4,950,000	Cane

[a] From "Sugar Information, Inc.," Farr Whitlock Dixon, Inc., 1971.

TABLE III

Principal Users of Sugar in 1970[a]

Country	Tons sugar usage	Population	Usage (lb per capita)
U.S.S.R.	12,533,000	246,000,000	90
United States	11,354,000	207,100,000	120
India	4,135,000	560,000,000	11
Brazil	4,134,000	93,000,000	79
China	4,134,000	770,000,000	9

[a] From "Sugar Information, Inc.," Farr Whitlock Dixon, Inc., 1971.

TABLE IV

Per Capita Consumption—1970[a]

Country	Amount (lb)
Ireland	126
Holland	122
Australia	115
United Kingdom	111
New Zealand	110

[a] From "Sugar Information, Inc.," Farr Whitlock Dixon, Inc., 1971.

TABLE V

Sugar Consumption in the United States[a]

Year	Sugar per capita (lb)	Corn sweetener per capita (lb)	Total sweetener per capita (lb)
1830	15		
1900	65		
1925	102		
1948	94		
1956	99	12	111
1960	96	13	109
1965	97	17	114
1970	104	19	123

[a] U. S. Department of Agriculture.

rates USDA excepted. The sugar consumption in the United States is shown in Table V.

B. Changes in Usage

The use of sugar in the United States is divided between industrial use—sugar used by manufacturers of confection, beverages, baked goods, canned goods, etc.—and sugar used in home, restaurant, institution, etc. There is a rapid shift from 1956 until 1968 in the use of sugar from the home to the industrial user (Table VI) and a significant change in the distribution of sweeteners from home use to industrial uses. In spite of population increase, the use of sugar as such continues to decline in the home. All categories of industrial users increased their share of the total sweetener deliveries with the exception of canners and confectioners. The most significant increase is in beverages.

Much of the sugar that was formerly consumed in the household for homemade drinks, cakes, pies, cereals, and the like, is now consumed as convenience foods in the form of cake and cookie mixes, frozen or fresh-baked pies, pre-sweetened cereals, carbonated beverages, etc. The success of many of these convenience foods insures a continuing introduction of new products, which will surely result in further reductions in the use of sugar in the home. Canning of fruits and vegetables in the home has

TABLE VI

Industrial and Nonindustrial Users of Sugar[a,b]

User	1956	1960	1960	1968
Industrial				
Beverage industries	19,042	23,512	31,955	41,278
Bakers	22,510	26,569	31,419	35,974
Confectioners	21,068	22,925	27,152	29,905
Canners	16,987	18,496	20,817	22,705
Ice cream and dairy industries	7,346	8,732	11,981	13,408
Other food uses	6,807	7,272	11,744	12,574
Total industrial use	90,877	110,471	138,185	159,472
Nonindustrial use	81,646	78,108	74,245	72,358
Total use	172,523	188,579	212,430	231,830
Industrial (%)	52	59	65	68

[a] Use in 1000 hundredweights. (Figures include sugar and corn sweeteners.)
[b] U. S. Department of Agriculture.

declined in favor of greater use of commercially canned and frozen foods. This shift has been accelerated by the declining rural population and the corresponding increase in urban dwellers.

C. Use of Corn Sweeteners

During the past 25 years we have seen a striking increase in the use of corn sweeteners—either corn syrup or dextrose—in industrial use. This trend actually started during World War II when supplies of sugar became tight and were rationed. Many industries substituted corn sweeteners for some of the sucrose to make their supplies of sugar stretch further. Experience during this period made it apparent that a portion of the sugar used in some products could be replaced with a sucrose–corn syrup blend or sucrose–dextrose blend without seriously sacrificing product quality or acceptibility with the added advantage of a savings in cost.

By 1956, 12% of all industrial deliveries were in the form of corn sweeteners. This increased to 18% in 1960 and 20% in 1967. It is readily apparent that deliveries of corn sweeteners increased markedly during this period. Chief gainers were bakeries, ice cream and dairy product businesses, and canners. Much of the increased use by the baking industry can be attributed to the development of a new type of corn syrup of extra high dextrose equivalent of 70%. Because of increased fermentability, this syrup was well received by the baking industry.

Now the corn industry has come up with a new development which may greatly increase the use of corn sweeteners by industrial users. The production of a corn syrup equal to sucrose in sweetness has long been a research goal of the corn industry. The fermentation Research Institute of the Japanese government has developed a process utilizing an isomerizing enzyme to convert dextrose into fructose or levulose. Clinton Corn Processing Company acquired the exclusive patent and sublicensing rights for the process in the United States and are now marketing a product—Isomerose 100 brand high fructose corn syrup. This product consists of 50% dextrose, 42% levulose, and 8% disaccharides and other carbohydrates. It is much sweeter than any other corn syrup on the market at this time and for some items may have equivalent sweetness to sucrose and for others will approach the sweetness of sucrose.

The product is now available from Clinton Corn Processing Company at Clinton, Iowa, and A. E. Staley Manufacturing Company, Morrissville, Pennsylvania. Other manufacturers of corn syrups are considering additional facilities. Availability of high fructose corn syrup must be keeping some sugar executives awake at night because this syrup will surely increase the corn sweetener share of the industrial market.

One other factor of importance in industrial use of sweeteners has been the rapid shift away from 100 lb bags. In 1956, 75% of industrial sugar deliveries were granulated in 100 lb bags. By 1968 the delivery in bags had decreased to 32.5% with 67.5% of the industrial deliveries in bulk— either in liquid or granulated form.

D. Usage in the Canning Industry

For many years in the canning industry there was an association between the quality of fruit and the amount of sugar used. The better quality or fancy fruit always received the highest amount of sugar—fancy or extra heavy syrup. It was found, however, that this amount of sugar tended to mask the fruit flavor and produced a very sweet product high in calories. One-half cup serving equaled 127 cal for cling peaches. Today the majority of fruits are packed in choice or heavy syrup where the calorie content is 99 cal per one-half cup of cling peaches. There is some trend to light syrup where the calorie content of one-half cup of cling peaches is 73 cal. Fruits are now being packed in juices, or sweetened juices. Pineapple packed in pineapple juice may today be the most popular packing medium for this product. Grapefruit in grapefruit juice is gaining in popularity. White grape juice, apple juice, and pear juice are also used as packing mediums for apricots, peaches, pears, fruit cocktail, etc. The public desire for canned products is definitely trending toward less sweet products with fewer calories. Calcium cyclamate was very popular in fruit products and probably accounted for 6–8% of the fruit pack before its removal. Saccharin packs, juice packs, water packs, and low sugar packs have been made to replace cyclamate packs. The volume of these products in total is probably less than 50% of the volume of cyclamate-sweetened fruits previously produced. Thus, with respect to canned fruits, we have not found a substitute or substitutes as well liked as calcium cyclamate.

Whereas in sugar granular size may determine end use, in corn syrups dextrose content, or dextrose equivalent (D.E.), determines end use. Syrups are available from 26 D.E. to 96 D.E. The lower D.E. is used for coffee whitener, whereas the higher D.E. is used principally for bakery products and fermentation processes. Many other uses fall in between— ice cream, bakery products, canned foods, jams, jellies and confections.

A brief mention of industrial prices today for sweeteners: Sugar is $11.90 F.O.B. basing point, or near this point. Corn syrups are in the midst of a very strong competitive price war and prices are extremely depressed. Corn syrups on a dry basis F.O.B. factory are priced at about $4.70 per 100 lb. With this price advantage it can readily be seen why the corn sweetener proportion is constantly rising. The high fructose syrups are not priced this cheaply—$9.56 on a dry solids basis, F.O.B. factory—but this still represents a considerable savings over sucrose.

Now that the wet corn milling industry has enjoyed success with high fructose syrup, it is safe to predict that research laboratories are busy with efforts to make further changes in the starch or dextrose molecule. One would also expect that this threat to the sugar industry will encourage renewed effort in their research resulting in new and improved industrial sweeteners.

CHAPTER 10

New Carbohydrate Sweeteners

CARL AMINOFF

Sucrose has been the totally dominating carbohydrate sweetener for a very long time. With a world production over 70 million tons this product constitutes a considerable part of the daily calorie intake, especially in the developed countries.

Only the starch-based glucose products have been able to conquer a significant part of the market for sweeteners. The division line between sucrose and the starch sweeteners has, however, been clear and very stable. Fluctuations in price differences between sucrose and glucose syrup or dextrose have brought about remarkably little substitution of one for the other. The reason for this lack of interchangeability has mainly been the lower sweetness values of the starch sweeteners.

Lately, this stable picture has begun to show unmistakable signs of change. Behind the currents of change are several factors of varied nature. A few of them can easily be identified. There is a growing tendency to look upon sucrose as a not-so-beneficial part of the diet, especially at the levels of per capita intake reached in most developed countries. These doubts encourage a search for other alternatives even if the differences in nutritive value might be marginal. Another driving force is the development of new technologies which enhance the possibilities of manufacturing alternative products along alternative routes or processes. A third factor is the distinct possibility that the increase in world sugar production is lagging behind the increase in world sugar consumption. Estimates of decreasing amounts of sugar in stock and a nervously fluctuating world market price are strong symptoms, which point in the direction of considerably higher sugar price levels in the seventies than in the sixties.

A. Technological Advances

1. Enzyme Technology

The development of new enzymatic processes is currently one of the fastest moving frontiers of food technology. New enzymes are isolated, their stability and purity is improved, and novel applications are devised. A case in point is the development of isomerase, an enzyme capable of transforming glucose into fructose. The first patent was granted in the United States in 1960, but no immediate industrial application followed. Japanese institutions did some fundamental development work and a small industrial production of fructose-containing starch syrup came into existence in Japan. Much more significant, however, is the late development in the United States of three big plants which have already begun production.

Because of abundant and cheap cornstarch and the Sugar Act, which fixes the domestic sugar prices at a fairly high level, the temptation to produce invert sugar from cornstarch is particularly great in the United States. Thus, the United States sugar industry is in danger of losing a big slice of the considerable market for liquid sugar. This is the first serious shift in the long established balance between sucrose and starch-based industries. For the first time some almost identical products can be made by both industries.

The starch industry has benefited from enzymatic processes before. Acid hydrolysis of starch is becoming more and more a thing of the past, as enzymes have taken over with improvements in quality and yield of the products as a result. The hydrolysis can also be steered enzymatically in the direction of high maltose content, and from maltose one Japanese company is already producing maltitol, a new carbohydrate sweetener.

In enzyme technology still another epoch-making development may be just around the corner. Until now enzyme preparations have been mixed as suspensions or solutions with the material to be treated. Subsequently the enzymes or the cell material has to be removed, a process which can be both difficult and costly. Now the possibility of fixing the pure enzymes on inert carrier materials has become the subject of intensive study within a great number of research and development organizations. A breakthrough in this technology would have great significance for the economy of enzymatic processes. One can visualize a series of reactors with different fixed enzymes, which convert starch in one continuous process to invert sugar solution.

Finally, if a recombining of glucose and fructose into sucrose could be achieved by enzymatic processing, consequences would be shattering, not only for the sugar industry but also for the cane and beet producing areas of the world.

2. Separation Processes

The individual mono- and disaccharides have very similar chemical and physical properties. To separate them from each other out of mixtures has thus been difficult. In those comparatively few industrial applications where such separation has been the target, the rather crude device of chemically altering one saccharide component while leaving the other intact has been the only way. Selective oxidation of glucose to gluconic acid leaving fructose intact or selective precipitation of calcium-fructosate leaving glucose intact are examples of this. Such processes are inherently costly and yield-sensitive.

However, various analytical chromatographic methods have been well established for a long time and have been capable of precise separation of such mixtures. One of the chromatographic methods, namely, ion-exchange chromatography has lately been scaled up into a viable industrial process. The inventions and know-how involved are still the property of very few companies, but some patents have already been made public. Such a situation will never persist for long. This new unit process will spread in the industry and although its first application has been only the separation of glucose–fructose mixtures, it is in principle applicable to most soluble carbohydrate and polyol mixtures. The ion exclusion variety of this process is a tool to separate ionic impurities from the sugars or polyols.

B. New Carbohydrate Sweeteners

The carbohydrates referred to below are not new in the scientific sense. They have all been known chemical compounds for a long time. However, as industrial products for food purposes they are new (see Fig. 1). Besides sucrose and glucose, sorbitol is produced in substantial quantities, about 250,000 tons/year, including captive use, of which about one-third ends up in foods. This has been going on for a long time and will thus be outside the scope of this chapter. Neither is the fructose-containing

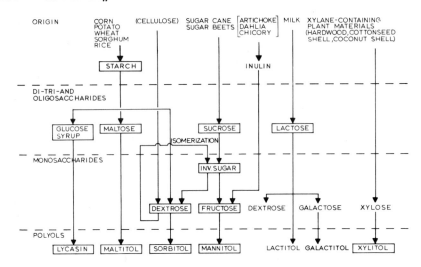

Fig. 1 Industrial carbohydrates for food. The materials in frames are in industrial production for food purposes. The raw materials in brackets have been in industrial use earlier but are not economical at present.

starch syrup a radically new product. It is an invert sugar produced from a new raw material.

1. Maltitol

A Japanese company produces on a small scale, presumably a few hundred tons, maltitol, the polyol obtained by hydrogenation of maltose. This product has about 90% of the sweetness of sucrose and thus 50% more than sorbitol, which it resembles. It is even more hygroscopic than sorbitol and is so far produced only as a liquid.

For maltitol the Japanese company makes a noncaloric claim based on animal tests. Reports on tests in humans are so far lacking. In view of the fact that Lycasin, the hydrogenation product of glucose syrup, which contains some maltitol, has been found in Sweden to be almost completely metabolized by humans, this claim seems rather doubtful. On the other hand, if it were true, maltitol would necessarily cause osmotic diarrhea. Thus, the whole concept of a noncaloric soluble carbohydrate or carbohydrate-derived polyol seems to be self-defeating. If it is not, it causes diarrhea. The same company also has a patent for the production of lactitol, the hydrogenation product of lactose and the same noncaloric claim is made for that product. Lactitol, however, has a low sweetness value and would thus be less attractive as a sweetener than maltitol.

2. Fructose

Fructose or levulose, the sweetest of all known sugars, already has a long production history. The first production methods started from inulin-containing plants like Jerusalem artichoke, dahlia, or chicory. These were superseded by methods involving the precipition of calcium-fructosate from invert sugar solutions. Another method which is still practiced derives from microbiological oxidation of glucose in invert to gluconic acid and the subsequent recovery of fructose from the liquid remaining after precipitation of calcium gluconate.

The most economical method today, however, is separation of glucose and fructose by ion-exchange chromatography from invert sugar solutions. Variations of this new method have drastically lowered the price of fructose during the last 3 years and opened up food use markets for fructose in certain European countries. It can be estimated that the world production of pure fructose in the mid-sixties was about 2000 tons/year. The corresponding figure at present is about 8000 tons/year.

Fructose has a spectrum of properties that makes it a very interesting carbohydrate sweetener. A short listing of these would be:

Sweeter than sucrose
Higher solubility than sucrose
Ability to improve fruit flavors
Metabolized without initial need for insulin and does not induce
    insulin release
Less cariogenic than sucrose

These properties, together with the new possibilities for production available, justify a prediction that fructose will capture a growing part of the total carbohydrate sweetener market. In Finland, where comparatively cheap fructose has been marketed a few years, the consumption of fructose in 1972 has risen to 0.5% of the consumption of sucrose.

3. Xylitol

Another promising carbohydrate sweetener is xylitol, the polyol obtained by hydrogenation of xylose. Xylose can be produced from xylane-containing plant materials. The present world production which can be estimated at about 1000 tons/year derives partly from prehydrolysis of hardwood in connection with the production of cord-cellulose in Japan and Russia and partly from plants using cottonseed hulls and coconut shells as raw material.

Like fructose, xylitol has first been used for parenteral nutrition and subsequently as a sweetener in diets for diabetics. The latter use has been mainly in Germany and Russia. Xylitol is as sweet as sucrose, and because of its higher heat of solution it has a pleasant cool taste. What makes xylitol attractive aside from the above mentioned uses is the promise it holds for production of noncariogenic confectionery products. A new plant using birch residues from the wood-working industry, which will considerably increase the world production of xylitol and make a lower priced product available, is under construction in Finland.

C. Summary

The traditionally stable, even stagnant picture of carbohydrate sweetener production is changing today. The new techniques enabling starch-based industries to produce fructose-containing products is upsetting the long established balance between starch and sucrose-based industries.

At the same time new promising carbohydrate sweeteners like pure fructose and xylitol are emerging. It surely is more than a coincidence that these two substances are given so much attention in this book. It would, however, be short-sighted to believe that only these new carbohydrates will be important in the future. The production potential in terms of available technology for many other carbohydrates is today clearly greater than 10 or even 5 years ago.

It would be pointless to list all the possible saccharides and their corresponding polyols that could be produced in quantity at reasonable prices if the potential of use were there. It is a challenge to the scientists, whether medical researchers, nutritionists, or food technologists, to uncover what useful properties still lie undiscovered in this large category of compounds so abundantly present in nature.

References to the Relevant Patent Literature

1. Enzymatic Isomerization of Glucose to Fructose

USA Pat 2 950 228	R. Marshall: Enzymatic process. 23.08.1960
3 623 953	W. Cotter, N. Lloyd et al.: Method of isomerizing glucose syrups. 30.11.1971
Brit. Pat 1 103 394	Anon: A method of manufacturing syrup containing fructose by use of an enzymatic process. 14.02.1968

2. Insoluble Enzymes

USA Pat 3 519 538	R. Messing and H. Weetall: Chemical coupled enzymes. 07.07.1970
Fr. Pat 2 020 527	Anon. (Corning Glass Works): Enzymes stabilisés et leur procédé de préparation. (= Insoluble enzymes and a method for their preparation) 17.07.1970
German Pat 1 944 418	R. Messing and H. Weetall: Stabilized insoluble enzyme preparations (Stabilizierte Enzymepräparate und Verfahren zu seiner Herstellung). 23.04.1970
Fr. Pat 2 001 336	R. Messing (Corning Glass Works): Stabilization of enzymes by adsorption on an inorganic support 26.09.1969
Brit. Pat 1 183 260	M. Lilly, G. Kay et al.: Insolubilized enzymes. 04.03.1970 (Re issue 01.05.1972)
German Pat 1 959 169	S. Amotz: Enzymic reactions with insoluble enzymes. 04.06.1970

3. Fructose Ion-Exchange Chromatography Separations

USA Pat 3 044 904	G. Serbia: Separation of dextrose and levulose. 17.12.1968
3 044 905	L. Lefèvre: Separation of fructose from glucose using cation exchange resin salts. 17.07.1962
3 483 031	K. Lauer: Method of recovering pure glucose and fructose from sucrose or sucrose-containing invert sugars. 09.12.1969
USA Pat 3 416 961	C. B. Mountfort, B. Cortis-Jones and R. Wickham: Process for the separation of fructose and glucose. 17.12.1968
3 692 582	A. Melaja: Procedure for the separation of fructose from the glucose of invert sugar. 19.09.1972
3 184 334	R. Sargent: Separation of dextran from fructose using ion-exchange resins. 18.05.1965
Brit. Pat 1 178 892	Anon(Anheuser-Busch): Improved process of purifying isomerized liquors. 21.01.1970
German Pat 2 036 525	K. Lauer, H-G. Budka and G. Stoeck: Verfahren zur chromatographischen Auftrennung von Mehrstoffgemischen. 03.02.1972 (= A method for the chromatographic separation of mixtures of several components)

4. Lycasin

USA Pat 3 329 507	E. Conrad: Process for preparing non-fermentable sugar substitute and product therefore. 04.07.1967

5. *Maltitol and Lactitol*

German Pat 2 034 700   M. Mitsuhashi et al.: Süssmittel für Nahrungsmittel und Getränke. 18.03.1971
(= Sweetening agent for foodstuffs and drinks)

Brit. Pat 1 252 371   Anon (Hayashibara Co): Improvements in and relating to the production of maltitol. 03.11.1971

1 250 952   Anon (Hayashibara Co): Improvements in and relating to the preparation of foodstuffs. 27.10.1971

1 253 300   K. Hayashibara: Improvements in and relating to the preparation of foodstuffs. 10.11.1971

6. *Xylitol*

USA Pat 3 558 725   S. Kohno, I. Yamatsu and S. Ueyama: Preparation of xylitol. 26.01.1971

Japan Pat 70/37 817   Y. Takahira: Preparation of crystalline xylitol by saccharification of cellulosic materials. 30.11.1970

German Pat 2 005 851   H. Buckl, R. Fahn and C. Hofstadt: Verfahren zur Gewinnung von Xylit aus xylanhaltigen Naturprodukten. 18.11.1971
(= A method for the preparation of xylitol from natural products which contain xylane)

2 047 898   G. Jaffe, W. Szkrybald and P. Weinert: Verfahren zur Herstellung von Kohlenhydraten. 22.04.1971 (A method for the preparation of carbohydrates)

Digestion and Absorption
of Sugars

CHAPTER 11

Hydrolysis and Absorption of Carbohydrates, and Adaptive Responses of the Jejunum

ROBERT H. HERMAN

The gastrointestinal tract serves to hydrolyze dietary carbohydrates (polysaccharides) into their constituent components (monosaccharides) and then provides an absorptive surface whereby various monosaccharides can enter the body from the gastrointestinal lumen. The movement of sugars across the small intestinal epithelial cells occurs through an active transport mechanism in the case of glucose and galactose. Other sugars are handled differently. The exact mechanisms of sugar transport by the small intestine are as yet unknown. However, at least, some properties of these mechanisms are known.

Although the bulk of intraluminal sugars are absorbed into the portal venous blood a portion of these sugars are metabolized by the mucosal epithelial cells of the small intestine thereby maintaining their viability enabling the absorptive systems to function. The small bowel can also utilize blood glucose. The small intestine can also metabolize dietary amino acids and is capable of transforming some of these amino acids into glucose since all of the gluconeogenetic enzymes are present in the small intestinal mucosal cells. However, this gluconeogenetic mechanism is probably not physiologically significant in terms of the maintenance of blood glucose levels.

The hormonal controls that regulate the utilization of glucose and other sugars by the small intestine are as yet unclear. Although adenyl cyclase is present in the small intestine of man and animals, it is not yet clear what hormones significantly affect the small intestinal metabolic pathways via adenyl cyclase.

Despite the lack of knowledge concerning the exact role of hormones in regulating metabolic pathways in the small intestine it has been shown that dietary substances can affect the level of activities of various jejunal enzymes. Certain enzymes of the jejunal epithelial cells undergo adaptive responses to dietary substances. The exact mechanism of this adaptive response to dietary substances is unknown but would appear to involve the protein synthetic pathway. Various dietary sugars affect the activities of glycolytic, gluconeogenetic, and folic acid-metabolizing enzymes. The adaptive response of enzymes would appear to ensure that these various enzymes are present in sufficient quantity to handle the dietary input. When dietary input is lacking it would be a waste of energy and amino acids to manufacture enzymes that would be excessive for the amount of sugar to be handled. In this sense dietary substances serve as environmental stimuli to which certain enzymes of the jejunum can respond. As the dietary environment changes, the jejunal enzymes respond accordingly. There is evidence now accumulating which suggests that when jejunal enzymes fail to respond to dietary sugars significant clinical gastrointestinal illness occurs.

A. Hydrolysis of Carbohydrates by Digestive Enzymes

The major dietary carbohydrates of man consist of starch, sucrose, and lactose. Starch is a polysaccharide composed of a complex linkage of glucose residues. Sucrose and lactose are disaccharides which are hydrolyzed by their respective disaccharidases. The disaccharidases will not be discussed here in any great detail since this subject is treated by Rosensweig (Chapter 12, this volume). Other dietary sugars are glucose and fructose which may occur as the free sugars. As the free sugars they are not affected by digestive enzymes but are absorbed intact through the small intestine. Small amounts of various other sugars such as ribose, mannose, glucosamine, fucose, galactosamine, trehalose, stachyose, and raffinose occur in dietary nucleic acids, glycoproteins, mucopolysaccharides, etc., but seem to be of little nutritional and metabolic consequence as far as we know today. Although the metabolic pathways entered into by these minor dietary sugars are interesting, no attempt will be made to discuss these pathways. In general, these minor dietary sugars are not significantly utilized by the body. Rather, various of these sugars are derived from glucose through a complex sequence of transformations.

1. Starch Hydrolysis

(a) Starch Structure

Starch (amylum) is derived from plant seeds, especially cereal grains, tubers, fruits, roots, and stem piths. The starch occurs in granules from 2 to 150 μ in diameter (Geddes, 1969) which vary in appearance depending on their origin. Starch consists of amylose and amylopectin (Samec and Blinc, 1938; Meyer, 1942; Greenwood, 1960) which are associated with each other in a crystal lattice (Doi, 1965; but see Geddes, 1969). Amylose is a linear polymer of glucose units linked by α-1,4-glucosidic bonds (Meyer *et al.*, 1940; Geddes, 1969). Amylopectin is a branched polymer of glucose having the same α-1,4-glucosidic bonds with the branch points consisting of α-1,6-glucosidic bonds occurring every 25–27 units (Meyer and Bernfeld, 1940; Geddes, 1969). Various figures are given for the proportion of amylose and amylopectin in native starch: 22–27% amylose and 73–78% amylopectin. The molecular weight of amylose is 10^5 to 10^6 and that for amylopectin is 10^7 to 10^8 (Geddes, 1969). Amylopectin would appear to have a randomly branched structure ("ramified tree structure") (Meyer and Bernfeld, 1940; Larner *et al.*, 1952; Geddes, 1969). "Waxy" cereal starches (e.g., corn) contain only amylopectin (Geddes, 1969). Interestingly, tuber starches (e.g., potato starch) may have phosphate attached in the ester form at the C-6 position with 90% of this linked to amylopectin (Schoch, 1942a,b). Cereal starches may have phosphorus present as phosphatides which are easily extractable (Posternak, 1935).

The major sources of starch which serve as dietary staples are different in various parts of the world. Most familiar are wheat, oats, rye, barley, potatoes, corn, and rice. Less well known starch-containing foods that are important around the world are millet, sweet potato, cassava, taro, arrowroot, yam, breadfruit, plantain, and sago. Table I lists these various sources. Peas and lentils also have significant amounts of starch.

(b) The Action of α-Amylase on Starch

Starch is hydrolyzed by the action of α-amylase (α-1,4-glucan 4-glucanohydrolase), an enzyme that is present in saliva and pancreatic juice.

TABLE I

Major Sources of Dietary Starch

Starch source	Plant part	Generic name	Major geographical areas of use	Comment
Wheat	Seed	*Triticum sativum, T. vulgare,* and other species	Worldwide: N. America, Europe, Middle East, India, China, N. Africa, Asia, Australia, Argentina	Grass family
Rice	Seed	*Oryza sativa*	Worldwide: Europe, Asia, N. America, Middle East, Egypt, Italy, Ethiopia, Brazil, India, Africa, Spain	Grass family
Corn (maize)	Seed	*Zea mays*	N. America, Central and S. America, Europe, India, S. Africa	Grass family, Indian corn
Oats	Seed	*Avena sativa* and other species	N. America, Europe, S. Africa, Argentina	
Rye	Seed	*Secale cereale*	N. America, N. Asia, Europe, Argentina, Korea, Japan, Manchuria; temperate and cool regions	Least important of the grain crops
Barley	Seed	*Hordeum vulgare* and *H. distichon* and other species	N. America, Europe, Middle East, N. Africa, India, Tibet; temperate regions and subtropics	Grass family
Potato	Tuber	*Solanum tuberosum*	Worldwide: N. America, Central and S. America, Europe, China, N. Africa, Japan, E. Indies; temperate regions	
Millet	Seed	*Panicum miliaceum* and other genera	Ethiopia, Africa, Far East, Middle East, Europe, Asia	Grass family
Sweet Potato	Tuberous root	*Ipomoea batatas*	U.S., Central and S. America, U.S.S.R., Pacific Islands, Japan; tropical and subtropical areas	

TABLE I (*Continued*)

Starch source	Plant part	Generic name	Major geographical areas of use	Comment
Cassava (Manioc)	Root	*Manihot esculenta* and *M. aipi*	Africa, Nigeria, S. America, Brazil, W. tropical Africa, Malay archipelago	*M. esculenta* (bitter cassava) contains cyanide; source of tapioca
Taro (Eddo)	Tuber	*Colocasia anti-quorum* var. *esculenta*	Polynesia, Pacific Islands esp. Hawaii	Source for poi, Arum family
Arrowroot	Tuber	*Maranta arundi-nacea*	W. Indies, Guiana, Brazil, E. Indies, Australia, S. Africa	
Yam	Tuber	*Dioscorea* species	W. Indies, E. Indies, India; tropical and subtropical areas	
Breadfruit	Fruit	*Artocarpus incisa* and *A. altilis*	Polynesia, Pacific Islands, S. Pacific, Malayan archipelago, W. Indies, Mexico, Central America, Brazil	From bread-fruit tree; seedless form more valuable
Plantain	Fruit	*Musa para-disiaca*	Africa, tropics	Fruit similar to banana but cooked and eaten like potato
Sago	Pith	*Metroxylan loeva, M. rumphii, M. sagu* and other genera	E. Indies	From sago palm

There are species differences in salivary gland content of amylase. The dog, for example, lacks salivary amylase (Davenport, 1961). Through the action of α-amylase starch is hydrolyzed to maltose, maltotriose, isomaltose, some branched tri-, tetra-, penta- and hexasaccharides and traces of glucose (Bernfeld, 1951; Meyer and Bernfeld, 1941; Meyer and Gonon, 1951; Whelan and Roberts, 1953; Whelan, 1953). Prolonged action of α-amylase produces glucose in quantities never exceeding 16–19%. With hydrolysis to completion with α-amylase starch yields glucose, mal-

tose, and isomaltose (Meyer and Gonon, 1951). Highly purified salivary
α-amylase has been prepared (Whelan and Roberts, 1953). This enzyme
formed only maltose and maltotriose. The α-1,6 linkages of amylopectin
are not hydrolyzed but can be bypassed by α-amylase. Human pancreatic
α-amylase has also been purified (Fischer *et al.*, 1950). Human pancreatic
and salivary α-amylases have a pH optimum of 6.9 and are not activated
by calcium. Salivary amylase is activated by chloride ion. Hog pancreatic
α-amylase has a pH optimum of 6.9, a molecular weight of 45,000, is
not stimulated by calcium but is activated by chloride ion (Danielsson,
1947), and has been found to exist in two isozyme forms (Cozzone *et
al.*, 1970). Salivary α-amylase action is destroyed by the low pH found
in the stomach. Soluble starch is split in the intestine not only by pancreatic α-amylase in the chyme but also by α-amylase sorbed onto the
surface of the epithelial villi (Ugolev, 1965), perhaps facilitated by the
presence of a polysaccharide-containing material or glycocalyx covering
the membrane of the microvilli (Crane, 1969; Bennett and Leblond,
1970). Crystalline human pancreatic and salivary α-amylases and pig
pancreatic α-amylase are the same with respect to the content of nitrogen,
phosphorus, sulfur, pH optimum, chloride activation, electrophoretic
mobility, absorption spectrum, and the ratio of saccharifying and dex-
trinizing power. Human amylases, both pancreatic and salivary, differ
from pig pancreatic α-amylase with regard to specific activity, pH stability, and solubility at pH 8.0 (Bernfeld *et al.*, 1950).

2. Disaccharide Hydrolysis

The maltose and isomaltose resulting from α-amylase action on starch
are hydrolyzed by maltase (α-glucosidase) (Dahlqvist, 1962a,b) and isomaltase (oligo-1,6-glucosidase) (Larner and McNickle, 1954, 1955;
Dahlqvist *et al.*, 1963a), both of which are present in the brush border
of the small intestinal mucosal cells (Ugolev, 1965; Borgström and
Dahlqvist, 1958; Borgström *et al.*, 1957; Dahlqvist and Borgström, 1961;
Miller and Crane, 1961; Crane, 1962, 1966; Jos *et al.*, 1967a,b). Electron
microscopy has demonstrated the presence of knobs on the microvillus
portion of the brush border of the mucosal cellular surface which had
sucrase and maltase activities (Johnson, 1967, 1969; Benson *et al.*, 1971;
Eichholz, 1969). Disaccharidase activity is highest in the apical half or
two-thirds of the villus but there is no activity in the crypts (Dahlqvist
and Nordstrom, 1966). Human and rat intestinal β-galactosidases have
been isolated and separated (Asp and Dahlqvist, 1968; Kraml *et al.*,
1969; Asp *et al.*, 1969, 1970; Alpers, 1969; Fluharty *et al.*, 1971; Asp,
1971). Although isomaltase is able to act on the dextrin intermediates

that result from the action of α-amylase on starch, this probably does not occur physiologically since isomaltase is a brush border enzyme and probably only hydrolyzes isomaltose as it comes in contact with the brush border. Figure 1 summarizes the digestion of starch.

B. Absorption of Sugars by the Small Intestine

1. Glucose and Galactose

Glucose and galactose are actively transported across the small intestinal surface and appear in the portal venous blood. Glucose must cross against a concentration gradient, and thus energy is required. The exact molecular mechanism is unknown. However, certain pertinent facts have emerged from the *in vitro* study of animal intestine preparations and *in vivo* human and animal studies.

In the duodenum the rate of glucose absorption is 6–20 gm/hr while in the jejunum 25–40 gm/hr can be absorbed from a 45-cm length. Because of the high rate of absorption of glucose in the duodenum, the amount of glucose entering into the jejunum is approximately 5% of that entering the duodenum (Davenport, 1961). The rates of absorption of various sugars in the rat vary in the sequence galactose > glucose > fructose > mannose > xylose > arabinose. This sequence is the same for man, rabbit, and the dog. From the *in vitro* work on everted hamster gut of Wilson and his co-workers (Wilson, 1962; Wilson and Landau, 1960; Wilson and Vincent, 1955; Landau and Wilson, 1959; Wilson and Wiseman, 1954; Crane, 1960; Alvarado, 1966; Barnett *et al.*, 1968; Wilson and Crane, 1958) the requirements for active transport of sugars have been determined. Table II lists the various sugars that are actively transported and passively diffuse across the intestinal wall. From these data the structural requirements for the active transport of sugars have

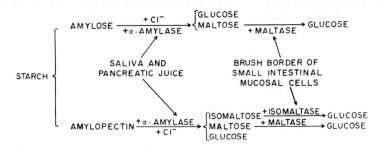

Fig. 1 The hydrolysis of starch to glucose.

TABLE II

Sugars That Are Actively Transported and Passively Diffuse
across the Intestinal Wall

Actively transported sugars	Passively diffusing sugars
D-Glucose	6-*O*-Methylglucose
α-Methyl-D-glucoside	Gold thioglucose
1-Deoxy-D-glucose	2-Deoxy-D-glucose
D-Glucoheptulose	3-*O*-Ethylglucose
2-C-hydroxymethyl-D-glucose	3-*O*-Propylglucose
3-Deoxy-D-glucose	3-*O*-Butylglucose
3-*O*-Methyl-D-glucose	3-*O*-Methyl-D-fructose
D-Galactose	D-Gulose
4-*O*-Methyl-D-galactose	2-Deoxy-D-galactose
D-Allose	D-Mannoheptulose
6-Deoxy-D-glucose	D-Glucosamine
6-Deoxy-D-galactose	2-*O*-Methyl-D-glucose
6-Deoxy-6-fluoro-D-glucose	*N*-Acetyl-D-glucosamine
7-Deoxy-D-glucoheptulose	L-Galactose
1,5-Anhydro-D-glucitol	6-Deoxy-L-mannose
	6-Deoxy-L-galactose
Passively diffusing sugars	1-Deoxy-D-mannose
	1,4-Sorbitan
D-Fructose	Mannitol
D-Mannose	Sorbitol
L-Sorbose	Glycerol
D-Lyxose	1,5-Anhydro-D-mannitol
D-Ribose	2,4-Di-*O*-methyl-galactose
L-Xylose	3-*O*-Hydroxyethyl-D-glucose
D-Xylose	1,4-Anhydro-D-glucitol
D-Talose	6-Deoxy-6-iodo-D-galactose
L-Arabinose	L-Glucose
D-Arabinose	

been formulated. Figure 2 lists these requirements. There must be at least 6 carbons in a D-pyranose ring structure and there must be a hydroxyl group at carbon-2. Pentoses are not actively transported. Substituted groups cannot be too large. The proper steric orientation is necessary. Inversion of the hydroxyl group at carbon-3 to form allose or at carbon-4 to form galactose does not alter the active transport of the sugar. Inversion of the hydroxyl groups at both carbons-3 and -4 to form gulose results in a nonactively transported sugar. Figure 3 compares the structures of some of these sugars. The rate of absorption of a sugar by passive diffusion is about one-tenth that of the rate of absorption of a sugar by active transport.

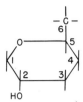

Fig. 2 Structural requirements for the active transport of sugars: (1) At least six carbons must be present, (2) there must be a hydroxyl group at carbon-2, (3) D-pyranose ring structure, (4) substituent groups cannot be too large, and (5) proper steric orientation is necessary.

Active transport can be inhibited by lack of oxygen or the presence of various metabolic inhibitors such as cyanide, 2,4-dinitrophenol, or iodoacetate.

Glucose and galactose compete for the same transport system. Each inhibits the rate of transport of the other when both are present. Mannose and glucosamine will inhibit galactose transport, but fructose has no effect. In the dog 7–17% of absorbed glucose is metabolized to lactate, while the bulk of the glucose appears as glucose in the portal venous blood. The formation of an intermediate phosphorylated form of glucose does not appear to be involved in the transport of glucose (Landau and Wilson, 1959). There are data to suggest that the carrier protein for glucose has considerable specificity for the hydroxyl group at carbon-6 (Ber-

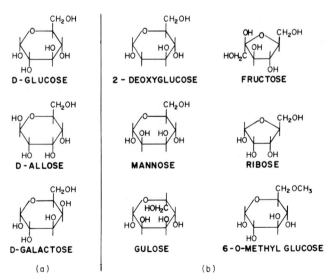

Fig. 3 Comparison of structures of some sugars that are (a) actively transported with (b) sugars that passively diffuse.

lin, 1970; Barnett *et al.*, 1968) as well as for the hydroxyl groups at carbons-1, -2, and -5.

Sodium ions are required for the active transport of sugars by *in vitro* preparations of small intestine (Bihler and Crane, 1962; Csaky and Thale, 1960; Csaky and Zollicoffer, 1960; Riklis and Quastel, 1958). The brush border membrane is the site of action of the sodium ions (Crane, 1962; Bihler *et al.*, 1962). It is assumed that the cell membrane contains a glucose binding carrier substance which binds glucose at one site and Na^+ at a separate site. The Na^+ is assumed to be necessary in order for the glucose-binding site to have the proper conformation to bind glucose (Crane *et al.*, 1965). Since K^+ as well as Li^+, Rb^+, Cs^+, and NH_4^+ inhibit both glucose and Na^+ influx (Bosackova and Crane, 1965), it is suggested that these ions bind at the Na^+-binding site and alter the degree of glucose binding through another conformational change in the carrier (Crane *et al.*, 1965). An efflux of K^+ through the same mechanism would facilitate the release of the glucose by the carrier on the intracellular side of the membrane. In this regard it has been shown (Csaky and Ho, 1966) that high concentrations of K^+ enhance the absorption of glucose but not galactose or 3-O-methylglucose. K^+ alone stimulated glucose transport at higher glucose concentrations. It was concluded that K^+ enhanced the intracellular metabolic disappearance of glucose thus creating a larger extracellular–intracellular concentration gradient which increased the rate of glucose entry into the cell. The active transport of Na^+ out from and K^+ into the cell could be handled by a separate system via a brush border, ouabain-sensitive Na^+-K^+-dependent membrane ATPase (Taylor, 1962; Rosenberg and Rosenberg, 1968; Parkinson *et al.*, 1972; Quigley and Gotterer, 1969a) thereby providing Na^+ necessary for glucose transport (Fig. 4) (see Crane, 1962; Crane *et al.*, 1965; Schultz and Zalusky, 1964). On the other hand, it is suggested (Crane, 1969) that the inward diffusion of Na^+ via the glucose carrier provides the energy for glucose absorption via the same carrier. This has been called gradient-coupled transport. The internal concentration of Na^+ can be kept low by the active transport of Na^+ out of the cell via another Na^+-K^+-dependent ATPase system (Parkinson *et al.*, 1972; Quigley and Gotterer, 1969b) on the serosal side of the cell. In the physiological state there would have to be a source for Na^+ on the mucosal, brush border side of the cell. Glucose in aqueous solutions is absorbed, *in vivo*, relatively rapidly. Thus, one would have to postulate that the cell itself provides Na^+ if none is present in the lumen from dietary sources if gradient-coupled transport is of physiological importance. With the influx of Na^+ there is an associated absorption of Cl^- and water.

From another viewpoint, sodium transport and water absorption are

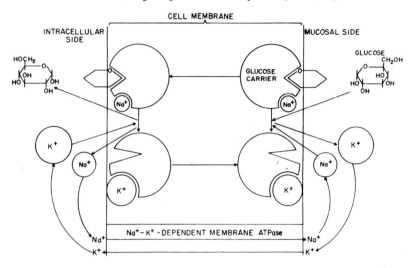

Fig. 4 Hypothetical scheme for the active transport of glucose in the small intestine.

stimulated by the presence of glucose or galactose. It has been suggested that the resulting increased sodium level within the cell stimulates the Na^+-K^+-dependent membrane ATPase at the serosal surface with a resulting net increase in the rate of sodium transport (Fordtran *et al.*, 1968; Goldner *et al.*, 1969; Schultz and Zalusky, 1964; Schultz and Curran, 1970a,b). The sugar, 3-*O*-methylglucose, which is not metabolized, facilitates sodium transport while fructose, which is metabolized, has no effect on sodium transport. It may well be that physiologically glucose may enhance sodium transport and that with large amounts of glucose sodium is not necessary for, or may have no apparent effect on, glucose transport (Crane *et al.*, 1965; Crane, 1965). In that case we have the model shown in Fig. 5. It may well be as has been suggested (Crane *et al.*, 1965) that membrane ATPases at both the brush border and serosal sides are involved in the total mechanism.

Phlorizin inhibits glucose transport by acting on the mucosal surface (Schultz and Zalusky, 1964) competing with substrates (Alvarado, 1967) although it does not enter into the cell (Stirling, 1968). Ouabain inhibits Na^+ transport at the serosal surface by inhibiting the Na^+-K^+-dependent membrane ATPase (Fig. 6) (Taylor, 1962; Schultz and Zalusky, 1964; Skou, 1965; Csaky *et al.*, 1961). In this connection it is interesting to note that oral glucose-electrolyte solutions have been used therapeutically in cholera to cause increased Na^+ absorption. Such therapy has been successful in maintaining the cholera patient in good electrolyte balance (Hirschhorn *et al.*, 1968; Pierce *et al.*, 1969). On the other hand, studies

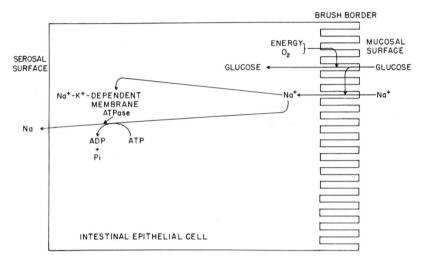

Fig. 5 Stimulation of sodium transport by glucose at mucosal surface, and stimulation of sodium transport via Na⁺-K⁺-dependent membrane ATPase at serosal surface.

of man, rat, and dog ileum *in vivo* showed that glucose absorption proceeded normally even when intraluminal Na^+ was at very low levels (Saltzman *et al.*, 1972). Nevertheless, it was considered theoretically possible that the microenvironment of the brush border had a high Na^+ concentration even though the luminal Na^+ concentration was quite low.

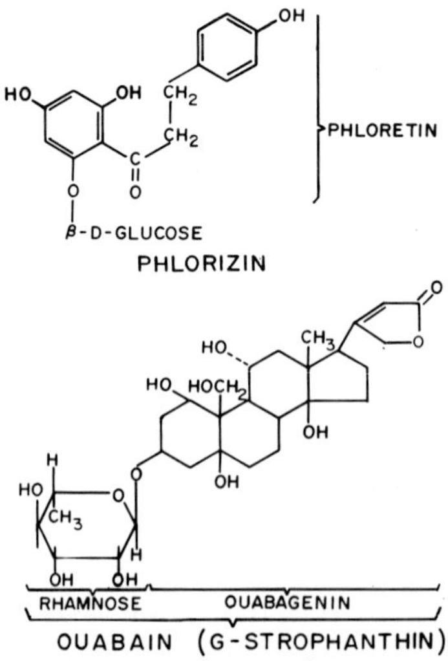

Fig. 6 Structures of phlorizin and ouabain.

Another interpretation is that glucose is transported into the cell and thence into a space between the cells, the lateral intercellular space. The glucose produces osmotic pressure which draws water across the tight junction, the point of attachment of adjoining epithelial cells on the mucosal side of the intestine (Fordtran *et al.,* 1968). As the water enters NaCl follows as a result of solvent drag. This is shown in Fig. 7A. The presence of mannitol prevents net water absorption but does not prevent Na^+ transport. Thus, it is further postulated that water moves across the tight junction and back out through the cell through Na^+ impermeable channels so that there is no net water absorption (Fig. 7B). The magnitude of the effect of solvent drag, however, does not seem to account for the degree of stimulation of sodium transport by glucose (Schultz and Curran, 1970a; Sladen and Dawson, 1969).

Various drugs affect Na^+ transport by the small intestine. Ethacrynic acid inhibits Na^+, water, and glucose absorption in the hamster small intestine (Binder *et al.,* 1966). It probably acts by inhibiting Na^+ absorption but conceivably may inhibit a Na^+-K^+-dependent membrane ATPase as it does in red blood cells (Hoffman and Kregenow, 1966). Chlorothiazide inhibits Na^+ and water absorption in the hamster small intestine, while probenecid inhibits glucose absorption without affecting Na^+ or water movement (Binder *et al.,* 1966). Phlorizin (Fig. 6) inhibits glucose absorption by the intestine (Crane, 1960). Phlorizin inhibits the absorption of those sugars that are actively transported but not those that are passively absorbed. Phlorizin appears to affect the active transport

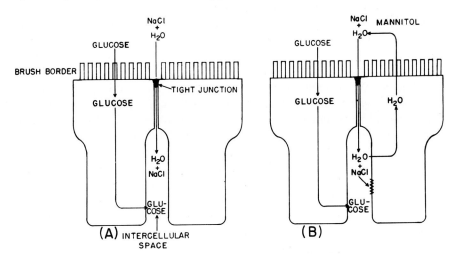

Fig. 7 The stimulation of sodium transport by glucose via solvent drag: (A) Action of glucose causing solvent drag and (B) movement of water through brush border.

mechanism of glucose. Phloretin, the aglycone of phlorizin, is much less active than phlorizin (Fig. 6) (Jervis *et al.*, 1956). It has been shown that at 37°C D-glucose binds to the brush border of hamster jejunum in preference to mannose or L-glucose after the brush borders had been disrupted with tris (hydroxymethyl)-aminomethane (Faust *et al.*, 1967). This binding was inhibited by 0.1 mM phlorizin, 1 mM HgCl$_2$, and binding was decreased by a temperature of 23°C.

Sodium is able to influence the rate of transport of many solutes in many different types of cells. Thus, the interaction of Na$^+$ and glucose in the intestine would appear to be a special case of a more general phenomenon. The effect of Na$^+$ on the transport of other substances in various cells has been extensively reviewed (Crane, 1965; Schultz and Curran, 1970b).

2. Fructose

Fructose is absorbed across and into intestinal epithelial cells of man and rat but not by an active transport mechanism (Crane, 1960; Wilson, 1962). In the rat small intestine about 10% of fructose is converted to glucose and about 60% to lactate. The remaining 30% diffuses passively into the portal venous blood. In the guinea pig about 55–80% of fructose is converted to glucose, 10% to lactate, and the remaining 35–10% passes into the portal vein unchanged. In normal man fructose tolerance tests regularly elevate blood glucose and fructose levels with the blood glucose resulting from conversion of fructose to glucose by the liver. In the rare individuals with essential fructosuria as a result of a deficiency of fructokinase a fructose tolerance test elevates blood fructose and results in fructosuria. The transformation of fructose into glucose and lactic acid in the small intestine occurs via the well-known glycolytic pathway shown in Fig. 8 (White and Landau, 1965; Herman and Zakim, 1968). Thus, fructose may be converted to glucose and lactic acid via the hexokinase and fructokinase portions of the glycolytic pathway. Fructose is converted to glucose by rat (Kiyasu and Chaikoff, 1957), guinea pig (Kiyasu and Chaikoff, 1957; Fridhandler and Quastel, 1955), and hamster intestines (Wilson and Vincent, 1955). Guinea pig intestine appears to convert fructose to glucose to a greater degree than rat intestine (Kiyasu and Chaikoff, 1957) presumably because of a very low level of glucose-6-phosphatase in rat jejunum (Ginsburg and Hers, 1960; Hers and de Duve, 1950). Fructose is metabolized to lactate in rats and guinea pigs (Kiyasu and Chaikoff, 1957; Sherratt, 1968). The metabolism of fructose by the intestine is reviewed elsewhere (Herman *et al.*, 1972).

In man, up to 80–90% of fructose is absorbed from the jejunum as

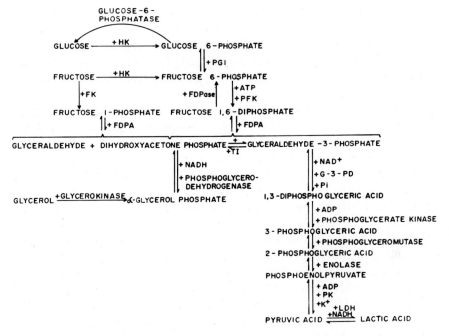

Fig. 8 Schematic outline of the glycolytic pathway.

fructose. There is only minor conversion of glucose to lactate during absorption (Cook, 1969). However, the conversion of some fructose to glucose by the human intestine has been demonstrated (Ockerman and Lundborg, 1965; White and Landau, 1965). In the latter study, the tissue was obtained at the time of surgery from patients with a variety of intestinal lesions from areas adjacent to the resection site that were grossly free of pathological processes. In the former study, sugars were introduced by tube and mesenteric blood was sampled at the time of surgery. It was shown in this study that galactose is not converted to glucose by the human small intestine. More recently, data have been obtained to suggest that the epithelial cells of the small intestine of the rat (Guy and Deren, 1971) and rabbit ileum (Guy and Deren, 1971; Schultz and Strecker, 1970) have a special membrane mechanism that allows for the rapid entry of fructose. This mechanism was not dependent on Na^+ nor sensitive to phlorizin.

3. Disaccharides

Disaccharides may be absorbed intact across the small intestine if the concentration is high, as may occur in primary disaccharidase deficiency, or if there is damage to the mucosal cells with a secondary disaccharidase deficiency or perhaps injury to the cell membrane resulting in a loss of selective permeability. Maltose has a relatively slow rate of absorption and usually is hydrolyzed more rapidly than it can be absorbed (Wilson, 1962). It has been shown that 10 gm of lactose is the largest amount that can be fed to men and nonmenstruating women without lactosuria occurring. Menstruating women can tolerate 20 gm or more without developing lactosuria (Watkins, 1928). Normal individuals given 20 gm each of sucrose or lactose excrete no more than 20 mg of lactose nor more than 10 mg of sucrose in 3 hr (Fischer *et al.*, 1965). In various gastrointestinal diseases (Gryboski *et al.*, 1963) and in sprue (Santini *et al.*, 1957) sucrosuria and lactosuria occur. Some patients with various gastrointestinal diseases such as regional enteritis, idiopathic steatorrhea, amyloidosis, ulcerative colitis, duodenal ulcer, cirrhosis, and miscellaneous other diseases had sucrosuria and/or lactosuria after test meals containing sucrose or lactose. Sucrosuria and/or lactosuria has been found in patients with various gastrointestinal and several types of neoplastic diseases (Fischer *et al.*, 1965). In three patients with complete removal of the tumors the disacchariduria disappeared. These data indicate that in various disease conditions disacchariduria after oral administration of disaccharides may occur without there necessarily being disease of the gastrointestinal tract or florid disaccharidase deficiency. Clearly, the significance and mechanism of disacchariduria remains to be determined. Lactosuria may occur in individuals with celiac disease and secondary lactase deficiency after the administration of oral lactose (Weser and Sleisenger, 1965). Disacchariduria can be used to detect disaccharidase deficiencies such as lactase (Dahlqvist *et al.*, 1963b; Holzel *et al.*, 1962; Kern *et al.*, 1963; Weijers and van de Kamer, 1962) and sucrase deficiencies (Greene *et al.*, 1972). Sucrose absorption *in vitro* by the intestine is dependent on the concentration on the mucosal side (Fridhandler and Quastel, 1955).

C. Adaptive Responses of Jejunal Glycolytic and Other Enzymes to Dietary Sugars

The administration of various sugars to normal human volunteers on an isocaloric constant diet results in changes in jejunal glycolytic enzymes. In humans the jejunal tissue was obtained perorally using an intestinal biopsy capsule (Crosby and Kugler, 1957). Thus, in man (Rosensweig *et al.*, 1968b) and in rats (Stifel *et al.*, 1968b) fructokinase, fructose 1,6-diphosphate aldolase and fructose 1-phosphate aldolase activities were highest in order on the following diets: fructose > sucrose > glucose > carbohydrate-free > fasting. Hexokinase activities were highest in order on the following diets: glucose > sucrose > fructose > carbohydrate-free > fasting (Table III). Similar results were obtained in rats (Table IV) (Stifel *et al.*, 1968b). Such adaptive changes occur within 24 hr (Rosensweig *et al.*, 1969b), occurring as early as 2 hr with a maximum at 6–12 hr with effects persisting at 24 hr (Rosensweig *et al.*, 1969a). Folic acid–metabolizing enzymes are also affected by dietary sugars (Table IV) (Stifel *et al.*, 1970). Adaptive changes have been shown in rat jejunum with regard to galactose-metabolizing enzymes with the enzyme activities being highest on a galactose diet (Stifel *et al.*, 1968a) (Table V). Similar results have been found in human volunteers (Rosensweig *et al.*, 1968a).

The adaptive effects of dietary sugars on jejunal glycolytic enzymes, fructosediphosphatase, and folate-metabolizing enzymes would appear to involve protein synthesis since inhibitors of protein synthesis inhibit the effects of dietary sugars in rat jejunum (Stifel *et al.*, 1971) (Table VI).

Other authors have demonstrated the effect of diet on jejunal enzymes and other physiological parameters. Fed rats compared to fasted rats have a heavier intestine with the mucosa being a greater percentage of

TABLE III

Effect of Dietary Sugars on Human Jejunal Glycolytic Enzymes[a]

	Diets				
Enzymes	Fructose	Sucrose	Glucose	Carbohydrate free	Fasting
Fructokinase[b]	3.24	2.50	1.59	0.84	0.28
Fructose 1-phosphate aldolase[b]	3.52	3.45	2.12	1.29	0.70
Hexokinase[b]	0.32	0.37	0.55	0.20	0.17
Fructose 1,6-diphosphate aldolase[b]	3.60	2.96	2.68	1.23	0.65
Phosphofructokinase[c]	3.52	—	2.34	1.50	1.16
Pyruvate kinase[c]	10.28	—	5.12	3.16	2.95
Fructosediphosphatase[c]	0.50	—	0.46	0.62	0.88

[a] Mean values given as μmoles of substrate metabolized/min/gm wet weight of mucosa.
[b] Rosensweig *et al.*, 1968b.
[c] Rosensweig *et al.*, 1970a. Data of subject 3.

TABLE IV

Effect of Dietary Sugars on Rat Jejunal Glycolytic Enzymes

Enzymes	Diets				
	Fructose	Sucrose	Glucose	Casein	Fasting
Fructokinase[a]	41.54	27.32	15.39	7.28	4.50
Hexokinase[a]	1.69	2.93	4.18	1.33	0.91
Aldose reductase[a]	5.20	5.69	7.00	4.15	3.44
Fructose 1-phosphate aldolase[a]	69.09	45.15	16.41	9.20	6.02
Fructose 1,6-diphosphate aldolase[a]	45.84	40.79	45.31	22.47	10.01
Phosphofructokinase[b]	27.62	—	15.83	9.84	3.97
Pyruvate kinase[b]	86.97	—	39.38	18.03	8.25
Glycerol-3-phosphate dehydrogenase[b]	120.60	—	96.35	81.00	65.23
Fructosediphosphatase[b]	16.41	—	10.15	21.33	27.18
Glutamate formimino-transferase[c]	0.068	—	0.048	0.018	0.007
Serine hydroxymethyl-transferase[c]	0.381	—	0.324	0.273	0.094
Methylene tetrahydrofolate dehydrogenase[c]	0.820	—	0.724	0.552	0.349
Formyltetrahydrofolate synthetase[c]	0.030	—	0.029	0.067	0.057

[a] Stifel *et al.*, 1968b. Mean values given as nmoles of substrate metabolized/min/mg of protein in footnotes *a* and *b*.

[b] Stifel *et al.*, 1969.

[c] Stifel *et al.*, 1970. Mean values given as increases in A_{355nm}.

TABLE V

Effect of Dietary Sugars on Rat Jejunal Galactose-Metabolizing Enzymes[a]

Enzymes	Diets					
	Galactose	Fructose	Sucrose	Glucose	Casein	Fasting
Galactokinase	20.77	11.61	9.98	6.90	4.69	2.68
Uridyl-transferase	75.94	39.81	34.60	26.80	17.66	8.99
Epimerase	41.10	31.08	29.95	24.19	18.22	11.09

[a] Mean values given as nmoles of substrate metabolized/min/mg of protein. Data from Stifel *et al.*, 1968a.

TABLE VI

Effects of Inhibitors of Protein Synthesis on the Adaptive Effects of Diet on
Certain Rat Jejunal Glycolytic and Folate-Metabolizing Enzymes[a]

Enzymes	Inhibitor of protein synthesis	Diets		
		Fasting	Casein	Fructose
Pyruvate kinase	None	12.47	26.33	69.13
	Actinomycin	—	10.82	26.42
	Ethionine	—	16.61	46.98
Fructose 1,6-diphosphate aldolase	None	11.37	19.43	39.43
	Actinomycin	—	12.10	16.41
	Ethionine	—	13.01	24.87
Serine hydroxymethyl-transferase	None	0.015	0.034	0.055
	Actinomycin	—	0.019	0.029
	Ethionine	—	0.022	0.041
Methylene tetrahydro-folate dehydrogenase	None	0.038	0.060	0.150
	Actinomycin	—	0.030	0.062
	Ethionine	—	0.041	0.111

[a] Mean values given as nmoles of substrate metabolized/min/mg of protein. Actinomycin was given intraperitoneally for 3 days at a dose of 25 μg/100 gm body weight, and ethionine was similarly administered at a dose of 100 mg/100 gm body weight. Data from Stifel *et al.*, 1971.

the total weight (McManus and Isselbacher, 1970). There was evidence to suggest that the mucosal cells of the fed rats were larger than those of the fasted rats and that there was increased deoxyribonucleic acid synthesis and proliferative activity. Rats fed *ad libitum* had higher activities of jejunal hexokinase and pyruvate kinase than fasted rats, while fasted rats had higher jejunal fructosediphosphatase activities than fed rats (Anderson and Zakim, 1970). It has been found that in obese and lean men fasting for 5 days there is a decrease in the absorption of glucose, water, sodium, chloride, and folic acid as well as a decline in the activities of jejunal alkaline phosphatase, sucrase, and maltase (Billich *et al.*, 1972).

These data are consistent with the model of Jacob and Monod (Pardee *et al.*, 1959) as shown in Fig. 9. It is postulated that a regulatory gene causes a repressor protein to be synthesized which in turn inhibits the action of the operator gene. Glucose or other sugars (e.g., fructose and galactose) bind the repressor protein so that it cannot continue to bind to the operator gene. With the inhibition of the operator gene lifted the operator gene enables the structural gene to synthesize messenger ribonucleic acid (mRNA) which then engages in protein synthesis of the

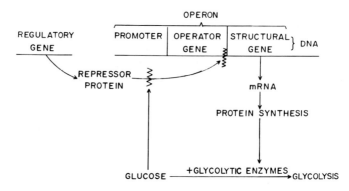

Fig. 9 Postulated mechanism of the adaptive effect of dietary glucose on jejunal glycolytic enzymes.

glycolytic enzymes. In this way, substrate sugars cause an increase in the enzymes involved in their metabolism. It is thought that the inhibition of a gluconeogenetic enzyme such as fructosediphosphatase occurs through a similar mechanism. In this case the formation of a glucose-repressor protein complex occurs and the complex then inhibits the operator gene adjacent to the structural gene for fructosediphosphatase.

In this way, substrate sugars cause an increase in the enzymes involved in their metabolism. If this mechanism were of physiological importance, failure of the adaptation of jejunal glycolytic enzymes should lead to gastrointestinal disease. Over the past several years more than 30 patients have been studied with unexplicable diarrhea and/or a dumping syndrome (Rosensweig *et al.*, 1970b,c, 1972). These patients had little or no adaptation of jejunal glycolytic enzymes to sugars and indeed were symptomatic with dietary sugar. Removal of the offending sugars from their diets in many cases, but not in all, resulted in marked improvement. The exact biochemical defect involved in the adaptive failure of these patients is as yet unknown, and the mechanism of symptom production is as yet in the speculative stage of study.

D. Research Needs

The areas of digestion, absorption, and metabolism of sugars by the intestine are extensive and complex. The influence of carbohydrates on gastrointestinal motility, gastric emptying, and gastrointestinal hormones is, as yet, a largely unexplored field.

A more unresolved area of investigation is that of glucose transport. The exact mechanism of glucose and sodium transport is unknown. How the transport of each is coupled to that of the other still remains speculative. Isolation and characterization of the glucose and sodium carriers is a major task that is essential to solution of the problems in this area. The relationship of cation-sensitive membrane ATPases to glucose and sodium transport remains to be clarified. Isolation and characterization of intestinal membrane ATPases is necessary but this is a formidable task.

A detailed examination of the metabolic pathways of the small intestine has yet to be carried out. The exact pathways that exist and the localization along the length of the gut are certainly not completely known. The exact mechanism whereby dietary sugars affect jejunal enzymes remains to be elucidated. All of the known steps in protein synthesis should be examined. There are limitations in studying the biochemistry of the human intestine since only small amounts of mucosa can be obtained by peroral biopsy. Yet, such studies coupled with those of animals should determine the steps of protein synthesis that occur in human intestine. It is hoped that intestinal mucosa from patients with various types of intestinal diseases can then be tested for all of the steps of protein synthesis and some clue as to the nature of the disease can be obtained. This is, of course, especially a necessary course to be taken in those patients whom we have studied who lack the normal adaptation of jejunal enzymes to dietary sugars. Equally important is the determination of the mechanism whereby dietary sugars cause gastrointestinal symptoms in these patients. Certainly the gastrointestinal tract contains appreciable quantities of histamine, serotonin, prostaglandins, and kinin-like proteins. Some or all of these substances could be involved in causing diarrhea and/or the symptoms of dumping.

A large amount of relatively easily obtainable data needs to be tabulated. The effect of various sugars on the various intestinal enzymes of different species must be studied. The effect of nutritional deficiencies on jejunal enzyme activities needs to be determined.

A great deal of work thus remains to be done. At least we now are aware of the metabolic functions of the small intestinal mucosal cells instead of only considering the intestine as a conduit and absorptive surface for dietary nutrients.

References

Alpers, D. H. (1969). Separation and isolation of rat and human intestinal β-galactosidases. *J. Biol. Chem.* **244,** 1238.

Alvarado, F. (1966). D-Xylose active transport in the hamster small intestine. *Biochim. Biophys. Acta* **112,** 292.

Alvarado, F. (1967). Hypothesis for the interaction of phorizin and phloretin with membrane carriers for sugars. *Biochim. Biophys. Acta* **135**, 483.

Anderson, J. W., and Zakim, D. (1970). The influence of alloxan-diabetes and fasting on glycolytic and gluconeogenic enzyme activities of rat intestinal mucosa and liver. *Biochim. Biophys. Acta* **201**, 236.

Asp, N.-G. (1971). Human small-intestinal β-galactosidases. Separation and characterization of three forms of an acid β-galactosidase. *Biochem. J.* **121**, 299.

Asp, N.-G., and Dahlqvist, A. (1968). Rat small-intestinal β-galactosidases. Separation by ion-exchange chromatography and gel filtration. *Biochem. J.* **106**, 841.

Asp, N.-G., Dahlqvist, A., and Koldovsky, O. (1969). Human small-intestinal β-galactosidases. Separation and characterization of one lactase and one hetero β-galactosidase. *Biochem. J.* **114**, 351.

Asp, N.-G., Dahlqvist, A., and Koldovský, O. (1970). Small intestinal β-galactosidase activity. *Gastroenterology* **58**, 591.

Barnett, J. E. G., Jarvis, W. T. S., and Munday, K. A. (1968). Structural requirements for active intestinal sugar transport. *Biochem. J.* **109**, 61.

Bennett, G., and Leblond, C. P. (1970). Formation of cell coat material for the whole surface of columnar cells in the rat small intestine, as visualized by radio-autography with L-fucose-³H. *J. Cell Biol.* **46**, 409.

Benson, R. L., Sacktor, B., and Greenawalt, J. W. (1971). Studies on the ultrastructural localization of intestinal disaccharidases. *J. Cell Biol.* **48**, 711.

Berlin, R. D. (1970). Specificities of transport systems and enzymes. *Science* **168**, 1539.

Bernfeld, P. (1951). Enzymes of starch degradation and synthesis. *Advan. Enzymol.* **12**, 379.

Bernfeld, P., Duckert, F., and Fischer, E. H. (1950). The amylolytic enzymes. XV. Properties of the α-amylase of human pancreas. Comparison with the other crystalline α-amylases. *Helv. Chim. Acta* **33**, 1064.

Bihler, I., and Crane, R. K. (1962). Studies on the mechanism of intestinal absorption of sugars. V. The influence of several cations and anions on the active transport of sugars, *in vitro,* by various preparations of hamster small intestine. *Biochim. Biophys. Acta* **59**, 78.

Bihler, I., Hawkins, K. A., and Crane, R. K. (1962). Studies on the mechanism of intestinal absorption of sugars. VI. The specificity and other properties of Na⁺-dependent entrance of sugars into intestinal tissue under anaerobic conditions *in vitro. Biochim. Biophys. Acta* **59**, 94.

Billich, C., Bray, G. A., Gallagher, T. F., Jr., Hoffbrand, A. V., and Levitan, R. (1972). Absorptive capacity of the jejunum of obese and lean subjects. *Arch Intern. Med.* **130**, 377.

Binder, H. L., Katz, L. A., Spencer, R. P., and Spiro, H. M. (1966). The effect of inhibitors of renal transport on the small intestine. *J. Clin. Invest.* **45**, 1954.

Borgström, B., and Dahlqvist, A. (1958). Cellular localization, solubilization and separation of intestinal glycosidases. *Acta Chem. Scand.* **12**, 1997.

Borgström, B., Dahlqvist, A., Lundh, G., and Sjövall, J. (1957). Studies of intestinal digestion and absorption in the human. *J. Clin. Invest.* **36**, 1521.

Bosačková, J., and Crane, R. K. (1965). Studies on the mechanism of intestinal absorption of sugars. VIII. Cation inhibition of active sugar transport and ²²Na influx into hamster small intestine, *in vitro. Biochim. Biophys. Acta* **102**, 423.

Cook, G. C. (1969). Absorption products of D(−)-fructose in man. *Clin. Sci.* **37**, 675.

Cozzone, P., Pasero, L., Beaupoil, B., and Marchis-Mouren, G. (1970). Characterization of porcine pancreatic isoamylases. Chemical and physical studies. *Biochim. Biophys. Acta* **207**, 490.

Crane, R. K. (1960). Intestinal absorption of sugars. *Physiol. Rev.* **40**, 789.

Crane, R. K. (1962). Hypothesis for mechanism of intestinal active transport of sugars. *Fed. Proc., Fed. Amer. Soc. Exp. Biol.* **21**, 891.

Crane, R. K. (1965). Na⁺-dependent transport in the intestine and other animal tissues. *Fed. Proc., Fed. Amer. Soc. Exp. Biol.* **24**, 1000.

Crane, R. K. (1966). Enzymes and malabsorption: A concept of brush border membrane disease. *Gastroenterology* **50**, 254.

Crane, R. K. (1969). A perspective of digestive-absorptive function. *Amer. J. Clin. Nutr.* **22**, 242.

Crane, R. K., Forstner, G., and Eichholz, A. (1965). Studies on the mechanism of the intestinal absorption of sugars. X. An effect of Na⁺ concentration on the apparent Michaelis constants for intestinal sugar transport, *in vitro. Biochim. Biophys. Acta* **109**, 467.

Crosby, W. H., and Kugler, H. W. (1957). Intraluminal biopsy of the small intestine. The intestinal biopsy capsule. *Amer. J. Dig. Dis.* **2**, 236.

Csáky, T. Z., and Ho, P. M. (1966). The effect of potassium on the intestinal transport of glucose, *J. Gen. Physiol.* **50**, 113.

Csáky, T. Z., and Thale, M. (1960). Effect of ionic environment on intestinal sugar transport. *J. Physiol. (London)* **151**, 59.

Csáky, T. Z., and Zollicoffer, L. (1960). Ionic effect on intestinal transport of glucose in the rat. *Amer. J. Physiol.* **198**, 1056.

Csáky, T. Z., Hartzog, H. G., III, and Fernald, G. W. (1961). Effect of digitalis on active intestinal sugar transport. *Amer. J. Physiol.* **200**, 459.

Dahlqvist, A. (1962a). The intestinal disaccharidases and disaccharide intolerance. *Gastroenterology* **43**, 694.

Dahlqvist, A. (1962b). Specificity of the human intestinal disaccharidases and implications for hereditary disaccharide intolerance. *J. Clin. Invest.* **41**, 463.

Dahlqvist, A., and Borgström, B. (1961). Digestion and absorption of disaccharides in man. *Biochem. J.* **81**, 411.

Dahlqvist, A., and Nordström, C. (1966). The distribution of disaccharidase activities in the villi and crypts of the small-intestinal mucosa. *Biochim. Biophys. Acta* **113**, 624.

Dahlqvist, A., Auricchio, S., Semenza, G., and Prader, A. (1963a). Human intestinal disaccharidases and hereditary disaccharide intolerance. The hydrolysis of sucrose, isomaltose, palatinose (isomaltulose), and a 1,6-α-oligosaccharide (isomalto-oligosacchride) preparation. *J. Clin. Invest.* **42**, 556.

Dahlqvist, A., Hammond, J. B., Crane, R. K., Dunphy, J. V., and Littman, A. (1963b). Intestinal lactase deficiency and lactose intolerance in adults. *Gastroenterology* **45**, 488.

Danielsson, C.-E. (1947). Molecular weight of α-amylase. *Nature (London)* **160**, 899.

Davenport, H. W. (1961). "Physiology of the Digestive Tract," p. 175. Yearbook Publ., Chicago, Illinois.

Doi, K. (1965). Formation of amylopectin granules in gelatin gel as a model of starch precipitation in plant plastids. *Biochim. Biophys. Acta* **94**, 557.

Eichholz, A. (1969). Fractions of the brush border. *Fed. Proc., Fed. Amer. Soc. Exp. Biol.* **28**, 30.

Faust, R. G., Wu, S.-M. L., and Faggard, M. L. (1967). ᴅ-Glucose: Preferential binding to brush borders disrupted with tris(hydroxymethyl) aminomethane. *Science* **155**, 1261.

Fischer, E. H., Duckert, F., and Bernfeld, P. (1950). Amylolytic enzymes. XIV. Isolation and crystallization of α-amylase of human pancreas. *Helv. Chim. Acta* **33**, 1060.

Fischer, R. A., Rosoff, B. M., Altshuler, J. H., Thayer, W. R., and Spiro, H. M. (1965). Disacchariduria in malignant disease. *Cancer* **18**, 1278.

Fluharty, A. L., Lassila, E. L., Porter, M. T., and Kihara, H. (1971). The electrophoretic separation of human β-galactosidases on cellulose acetate. *Biochem. Med.* **5**, 158.

Fordtran, J. S., Rector, F. C., Jr., and Carter, N. W. (1968). The mechanisms of sodium absorption in the human small intestine. *J. Clin. Invest.* **47**, 884.

Fridhandler, L., and Quastel, J. H. (1955). Absorption of sugars from isolated surviving intestine. *Arch. Biochem. Biophys.* **56**, 412.

Geddes, R. (1969). Starch biosynthesis. *Quart. Rev., Chem. Soc.* **23**, 57.

Ginsburg, V., and Hers, H. G. (1960). On the conversion of fructose to glucose by guinea pig intestine. *Biochim. Biophys. Acta* **38**, 427.

Goldner, A. M., Schultz, S. G., and Curran, P. F. (1969). Sodium and sugar fluxes across the mucosal border of rabbit ileum. *J. Gen. Physiol.* **53**, 362.

Greene, H. L., Stifel, F. B., and Herman, R. H. (1972). Dietary stimulation of sucrase in a patient with sucrase isomaltase deficiency. *Biochem. Med.* **6**, 409.

Greenwood, C. T. (1960). Physiochemical studies on starches. XXII. The molecular properties of the component of starches. *Starke* **12**, 169.

Gryboski, J. D., Thayer, W. R., Gabrielson, I. W., and Spiro, H. M. (1963). Disacchariduria in gastrointestinal disease. *Gastroenterology* **45**, 633.

Guy, M. J., and Deren, J. J. (1971). Selective permeability of the small intestine for fructose, *Amer. J. Physiol.* **221**, 1051.

Herman, R. H., and Zakim, D. (1968). Fructose metabolism. I. The fructose metabolic pathway. *Amer. J. Clin. Nutr.* **21**, 245.

Herman, R. H., Stifel, F. B., Greene, H. L., and Herman Y. F. (1972). Intestinal metabolism of fructose. *Acta Med. Scand., Suppl.* **542**, 19.

Hers, H. G., and deDuve, C. (1950). Le système hexosephosphatasique. II. Repartition de l'activité glucose-6-phosphatasique dans les tissues. *Bull. Soc. Chim. Biol.* **32**, 20.

Hirschhorn, N., Kinzie, J. L., Sachar, D. B., Northrup, R. S., Taylor, J. O., Ahmad, Z., and Phillips, R. A. (1968). Decrease in net stool output in cholera during intestinal perfusion with glucose-containing solutions. *N. Engl. J. Med.* **279**, 176.

Hoffman, J. F., and Kregenow, F. M. (1966). Characterization of new energy-dependent cation transport processes in red blood cells. *Ann. N.Y. Acad. Sci.* **137**, 566.

Holzel, A., Mereu, T., and Thomson, M. L. (1962). Severe lactose intolerance in infancy. *Lancet* **2**, 1346.

Jervis, E. L., Johnson, F. R., Sheff, M. F., and Smyth, D. H. (1956). Effect of phlorizin on intestinal absorption and intestinal phosphatase. *J. Physiol. (London)* **134**, 673.

Johnson, C. F. (1967). Disaccharidase: Localization in hamster intestine brush borders. *Science* **155**, 1670.

Johnson, C. F. (1969). Hamster intestinal brush-border surface particles and their function. *Fed. Proc., Fed. Amer. Soc. Exp. Biol.* **28**, 26.

Jos, J., Frézal, J., Rey, J., and Lamy, M. (1967a). Histochemical localization of intestinal disaccharidases: Application to peroral biopsy specimens. *Nature* *(London)* **213,** 516.

Jos, J., Frézal, J., Rey, M. L., and Weymann, R. (1967b). La localisation histochimique des disaccharidases intestinales par un nouveau procède. *Ann. Histochim.* **12,** 53.

Kern, F., Jr., Struthers, J. E., Jr., and Attwood, W. L. (1963). Lactose-intolerance as a cause of steatorrhea in an adult. *Gastroenterology* **45,** 477.

Kiyasu, J. Y., and Chaikoff, I. L. (1957). On the manner of transport of absorbed fructose. *J. Biol. Chem.* **224,** 935.

Kraml, M., Koldovský, O., Heringová, A., Jirsova, V., Kácl, K., Ledvina, M., and Pelichova, H. (1969). Characteristics of β-galactosidase in the mucosa of the small intestine of infant rats. Physiochemical properties. *Biochem. J.* **114,** 621.

Landau, B. R., and Wilson, T. H. (1959). The role of phosphorylation in glucose absorption from the intestine of the golden hamster. *J. Biol. Chem.* **234,** 749.

Larner, J., and McNickle, C. M. (1954). Action of intestinal extracts on "branched" oligosaccharides. *J. Amer. Chem. Soc.* **76,** 4747.

Larner, J., and McNickle, C. M. (1955). Gastrointestinal digestion of starch. I. The action of oligo-1,6-glucosidase on branched saccharides. *J. Biol. Chem.* **215,** 723.

Larner, J., Illingworth, B., Cori, G. T., and Cori, C. F. (1952). Structure of glycogens and amylopectins. II. Analysis by stepwise enzymatic degradation. *J. Biol. Chem.* **199,** 641.

McManus, J. P. A., and Isselbacher, K. J. (1970). Effect of fasting versus feeding on the rat small intestine. *Gastroenterology* **59,** 214.

Meyer, K. H. (1942). Newer investigation on starch. *Tech.-Ind. Schweiz. Chem.-Ztg.* **25,** 37.

Meyer, K. H., and Bernfeld, P. (1940). Starch. V. Amylopectin. *Helv. Chim. Acta* **23,** 875.

Meyer, K. H., and Bernfeld, P. (1941). Starch. XIX. Degradation of amylose by α-amylase. *Helv. Chim. Acta* **24,** 359E.

Meyer, K. H., and Gonon, W. F. (1951). Researches on Starch. LI. Degradation of amylopectin by α-amylases. *Helv. Chim. Acta* **34,** 308.

Meyer, K. H., Wertheim, M., and Bernfeld, P. (1940). Starch. IV. Methylation and determination of terminal groups of maize amylose and amylopectin. *Helv. Chim. Acta* **23,** 865.

Miller, D., and Crane, R. K. (1961). The digestive function of the epithelium of the small intestine. II. Localization of disaccharide hydrolysis in isolated brush border portion of intestinal epithelial cells. *Biochim. Biophys. Acta* **52,** 293.

Ockerman, P. A., and Lundberg, H. (1965). Conversion of fructose to glucose by human jejunum. Absence of galactose-to-glucose conversion. *Biochim. Biophys. Acta* **105,** 34.

Pardee, A. B., Jacobs, F., and Monod, J. (1959). The genetic control and cytoplasmic expression of "inducibility" in the synthesis of beta-galactosidase by *E. coli*. *J. Mol. Biol.* **1,** 165.

Parkinson, D. K., Ebel, H., DiBona, D. R., and Sharp, G. W. G. (1972). Localization of the action of cholera toxin on adenyl cyclase in mucosal epithelial cells of rabbit intestine. *J. Clin. Invest.* **51,** 2292.

Pierce, N. F., Sack, R. B., Mitra, R. C., Banwell, J. G., Brigham, K. L., Fedson,

D. S., and Mondal, A. (1969). Replacement of water and electrolyte losses in cholera by an oral glucose-electrolyte solution. *Ann. Intern. Med.* **70**, 1173.

Posternak, T. (1935). Phosphorus in starch. *Helv. Chim. Acta* **18**, 1351.

Quigley, J. P., and Gotterer, G. S. (1969a). Distribution of (Na⁺-K⁺)-stimulated ATPase activity in rat intestinal mucosa. *Biochim. Biophys. Acta* **173**, 456.

Quigley, J. P., and Gotterer, G. S. (1969b). Properties of a high specific activity (Na⁺-K⁺)-stimulated ATPase from rat intestinal mucosa. *Biochim. Biophys. Acta* **173**, 469.

Riklis, E., and Quastel, J. H. (1958). Effects of cations on sugar absorption by isolated surviving guinea pig intestine. *Can. J. Biochem. Physiol.* **36**, 347.

Rosenberg, I. H., and Rosenberg, L. E. (1968). Localization and characterization of adenosine triphosphatase in guinea pig intestinal epithelium. *Comp. Biochem. Physiol.* **24**, 975.

Rosensweig, N. S., Stifel, F. B., and Herman, R. H. (1968a). Dietary regulation of the galactose-metabolizing enzymes in human jejunum. *J. Lab. Clin. Med.* **72**, 1009.

Rosensweig, N. S., Stifel, F. B., Herman, R. H., and Zakim, D. (1968b). The dietary regulation of the glycolytic enzymes. II. Adaptive changes in human jejunum. *Biochim. Biophys. Acta* **170**, 228.

Rosensweig, N. S., Stifel, F. B., Herman, R. H., and Zakim, D. (1969a). Time response of diet-induced changes in human jejunal glycolytic enzymes. *Fed. Proc., Fed. Amer. Soc. Exp. Biol.* **28**, 323.

Rosensweig, N. S., Stifel, F. B., Zakim, D., and Herman, R. H. (1969b). Time response of human jejunal glycolytic enzymes to a high sucrose diet. *Gastroenterology* **57**, 143.

Rosensweig, N. S., Herman, R. H., and Stifel, F. B. (1970a). Dietary regulation of glycolytic enzymes. VI. Effect of dietary sugars and oral folic acid on human jejunal pyruvate kinase, phosphofructokinase and fructosediphosphatase activities. *Biochim. Biophys. Acta* **208**, 373.

Rosensweig, N. S., Herman, R. H., and Stifel, F. B. (1970b). Familial glucose intolerance: A failure of jejunal glycolytic enzyme adaptation to dietary glucose. *Gastroenterology* **58**, 990.

Rosensweig, N. S., Herman, R. H., Stifel, F. B., and Hagler, L. (1970c). The intestinal maladaptation syndrome: A new approach to "functional" gastrointestinal disease, *J. Clin. Invest.* **49**, 81a.

Rosensweig, N. S., Herman, R. H., Stifel, F. B., Hagler, L., Greene, H. L., and Herman, Y. F. (1972). Gastrointestinal disease associated with a failure of jejunal glycolytic enzymes. *Annu. Meet. Gastroenterol. Res. Group, 1972*, Dallas, Texas, May 24–27.

Saltzman, D. A., Rector, D. A., Jr., and Fordtran, J. S. (1972). The role of intraluminal sodium in glucose absorption *in vivo*. *J. Clin. Invest.* **51**, 876.

Samec, M., and Blinc, M. (1938). Newer results of studies on starch. *Kolloid-Beih.* **47**, 371.

Santini, R., Jr., Perez-Santiago, E., Martinez-De Jesus, J., and Butterworth, C., Jr. (1957). Evidence of increased intestinal absorption of molecular sucrose in sprue. *Amer. J. Dig. Dis.* **2**, 663.

Schoch, T. J. (1942a). Noncarbohydrate substances in the cereal starches. *J. Amer. Chem. Soc.* **64**, 2954.

Schoch, T. J. (1942b). Fractionation of starch by selective precipitation with butanol. *J. Amer. Chem. Soc.* **64**, 2957.

Schultz, S. G., and Curran, P. F. (1970a). Stimulation of intestinal sodium absorption by sugars, *Amer. J. Clin. Nutr.* **23**, 437.

Schultz, S. G., and Curran, P. F. (1970b). Coupled transport of sodium and organic solutes. *Physiol. Rev.* **50**, 637.

Schultz, S. G., and Strecker, C. K. (1970). Fructose influx across brush border of rabbit ileum. *Biochim. Biophys. Acta* **211**, 586.

Schultz, S. G., and Zalusky, R. (1964). Ion transport in isolated rabbit ileum. II. The interaction between active sodium and active sugar transport. *J. Gen. Physiol.* **47**, 1043.

Sherratt, H. S. A. (1968). The metabolism of the small intestine. Oxygen uptake and L-lactate production along the length of the small intestine of the rat and guinea pig. *Comp. Biochem. Physiol.* **24**, 745.

Skou, J. C. (1965). Enzymatic basis for active transport of Na^+ and K^+ across cell membrane. *Physiol. Rev.* **45**, 596.

Sladen, G. E., and Dawson, A. M. (1969). Interrelationships between the absorptions of glucose, sodium and water by the normal human jejunum. *Clin. Sci.* **36**, 119.

Stifel, F. B., Herman, R. H., and Rosensweig, N. S. (1968a). Dietary regulation of galactose-metabolizing enzymes: Adaptive changes in rat jejunum. *Science* **162**, 692.

Stifel, F. B., Rosensweig, N. S., Zakim, D., and Herman, R. H. (1968b). Dietary regulation of glycolytic enzymes. I. Adaptive changes in rat jejunum. *Biochim. Biophys. Acta* **170**, 221.

Stifel, F. B., Herman, R. H., and Rosensweig, N. S. (1969). Dietary regulation of glycolytic enzymes. III. Adaptive changes in rat jejunal pyruvate kinase, phosphofructokinase, fructosediphosphatase and glycerol-3-phosphate dehydrogenase. *Biochim. Biophys. Acta* **184**, 29.

Stifel, F. B., Herman, R. H., and Rosensweig, N. S. (1970). Dietary regulation of glycolytic enzymes. VII. Effect of diet and oral folate upon folate metabolizing enzymes in rat jejunum. *Biochim. Biophys. Acta* **208**, 381.

Stifel, F. B., Herman, R. H., and Rosensweig, N. S. (1971). Dietary regulation of glycolytic enzymes. XI. Effect of inhibitors of protein synthesis on the adaptation of certain jejunal glycolytic and folate-metabolizing enzymes to diet and sex steroids. *Biochim. Biophys. Acta* **237**, 484.

Stirling, C. E. (1968). High-resolution radioautography of phlorizin-^{3}H in rings of hamster intestine. *J. Cell Biol.* **35**, 605.

Taylor, C. B. (1962). Cation-stimulation of an ATPase system from the intestinal mucosa of the guinea pig. *Biochim. Biophys Acta* **60**, 437.

Ugolev, A. M. (1965). Membrane (contact) digestion. *Physiol. Rev.* **45**, 555.

Watkins, O. (1928). Lactose metabolism in women. *J. Biol. Chem.* **80**, 33.

Weijers, H. A., and van de Kamer, J. H. (1962). Diarrhea caused by deficiency of sugar splitting enzymes. II. *Acta Paediat. (Stockholm)* **51**, 371.

Weser, E., and Sleisenger, M. H. (1965). Lactosuria and lactase deficiency in adult celiac disease. *Gastroenterology* **48**, 571.

Whelan, W. J. (1953). The enzymic breakdown of starch. *Biochem. Soc. Symp.* **11**, 17.

Whelan, W. J., and Roberts, P. J. P. (1953). The mechanism of carbohydrase action. Part II. α-amylolysis of linear substrates. *J. Chem. Soc., London* p. 1298.

White, L. W., and Landau, B. R. (1965). Sugar transport and fructose metabolism in human intestine *in vitro*. *J. Clin. Invest.* **44**, 1200.

Wilson, T. H. (1962). "Intestinal Absorption," p. 69. Saunders, Philadelphia, Pennsylvania.

Wilson, T. H., and Crane, R. K. (1958). The specificity of sugar transport by hamster intestine. *Biochim. Biophys. Acta* **29**, 30.

Wilson, T. H., and Landau, B. R. (1960). Specificity of sugar transport by the intestine of the hamster. *Amer. J. Physiol.* **198**, 99.

Wilson, T. H., and Vincent, T. N. (1955). Absorption of sugars *in vitro* by the intestine of the golden hamster. *J. Biol. Chem.* **216**, 851.

Wilson, T. H., and Wiseman, G. (1954). The use of sacs of everted small intestine for the study of the transference of substances from the mucosal to the serosal surface. *J. Physiol (London)* **123**, 116.

CHAPTER 12

Adaptive Effects of Dietary Sugars on Intestinal Disaccharidase Activity in Man

NORTON S. ROSENSWEIG

Dietary carbohydrate constitutes almost one-half of the calories of the average diet in the United States. Much of this carbohydrate is in the disaccharide form. The common disaccharides (lactose or milk sugar, sucrose or table sugar, and maltose, the breakdown product of starch digestion) are hydrolyzed in the small intestine to their component monosaccharides (glucose and galactose from lactose, glucose and fructose from sucrose, and two molecules of glucose from maltose). This hydrolysis is carried out by the disaccharidases lactase, sucrase, and maltase, respectively, which are located in the brush border of the mucosal epithelial cells. Most of this digestion occurs in the upper part of the small intestine. If one or more of these enzymes is low or absent, the corresponding disaccharide will not be hydrolyzed when ingested. The intact disaccharide is too large to be absorbed, and it therefore remains in the intestinal lumen. In the lumen it acts osmotically to draw water into the intestine and cramps, bloating, and diarrhea ensue.

During the last decade, it has become clear that the amount and type of carbohydrate that is ingested has an effect upon the enzyme activity of the small intestine of animals and man. Blair *et al.* (1963) and Deren *et al.* (1967) demonstrated that diets with a high carbohydrate content increased intestinal sucrase and maltase activities in rats when these diets were compared with isocaloric carbohydrate-free diets. This suggested that dietary content could alter the enzymatic capacity of the small intestinal epithelium. However, the effect of these sugars on intestinal enzyme activity in man was not known.

The recent technical development of instruments capable of performing peroral intestinal biopsies safely in man has revolutionized the study of gastrointestinal enzymology in humans. This development coupled to the report by Dahlqvist (1964) of an appropriate assay for measuring disaccharidase activity in human biopsy specimens has made it possible to study the effect of dietary sugars and other substances on intestinal enzyme activity in man.

A. Disaccharidase Adaptation

1. Sucrose vs. Glucose Feeding

The impetus of animal studies plus the development of appropriate techniques led to an investigation of the effect of dietary substances on human intestinal enzymes. The role of specific dietary sugars in the regulation of jejunal disaccharidase activity in man has recently been delineated (Rosensweig and Herman, 1968a). In a series of studies, normal nonfasted volunteers, with no history of disaccharide intolerance, were fed isocaloric liquid diets on a metabolic ward. Only the carbohydrate content was changed. Protein was in the form of sodium or calcium caseinate and fat as corn oil. Initially, diets containing glucose as the sole source of carbohydrate were compared with isocaloric diets in which sucrose was the sole source of carbohydrate calories. In each subject, each diet was fed for at least one week and jejunal biopsies were performed at regular intervals. At least three biopsies were obtained on each diet.

In each of the seven subjects, sucrase and maltase activities were significantly higher on the sucrose diet than on the glucose diet regardless of which diet was fed first (Fig. 1). By contrast, lactase activity was not changed on the two diets (Fig. 2). This demonstrated that dietary sucrose, as compared with glucose, has a specific effect on jejunal sucrase and maltase activities. In one individual, lactose, maltose, and galactose were compared with glucose. No increase in activity was demonstrated compared with glucose.

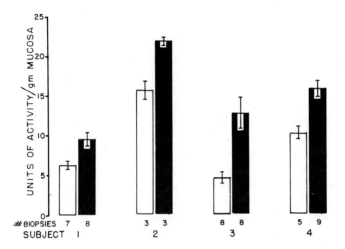

Fig. 1 Mean sucrase activity in four normal subjects on isocaloric glucose (□) and sucrose (■) diets. The standard error of the mean is in brackets. [Reprinted from Rosensweig and Herman (1968, p. 2255), with permission of the Rockefeller University Press, New York.]

When expressed as activity per gram wet weight of mucosa or per gram of protein, the effect of sucrose on enzyme activity is clearly demonstrable. In addition, the fact that lactase activity was unchanged on the two diets afforded an opportunity to express the data in another

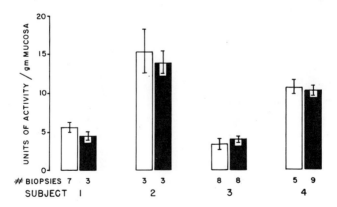

Fig. 2 Mean lactase activity in four normal subjects on isocaloric glucose (□) and sucrose (■) diets. The standard error of the mean is in brackets. [Reprinted from Rosensweig and Herman (1968, p. 2256) with permission of the Rockefeller University Press, New York.]

manner; namely, the ratios sucrase to lactase (S/L) and maltase to lactase (M/L).

The S/L and M/L ratios are employed to minimize possible variations in enzyme activity resulting from differences in the depth of the biopsy. It is known that some variability of disaccharidase levels is often seen with peroral biopsy specimens in humans. This may be attributed to location of the biopsy, variations in the depth of the biopsy, and differing proportions of epithelial cells in the biopsy specimens. Since disaccharidase activity is found only in the brush border of mature villus epithelial cells, any variations in disaccharidase activity that result from the depth of the biopsy should have a similar effect on all disaccharidases. A specific effect of a dietary sugar or a drug on one disaccharidase, however, will lead to disproportionate changes in the activity of the disaccharidases. Since lactase activity did not change on the glucose and sucrose diets, lactase activity was taken as a common denominator from which to compare changes in sucrase and maltase activities. Therefore, for each biopsy, S/L and M/L ratios as well as units per gram of tissue

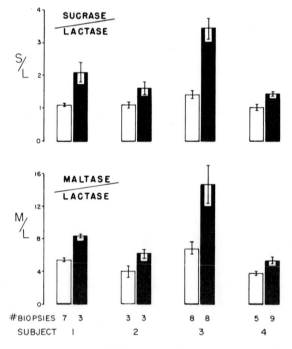

Fig. 3 Mean sucrase to lactase (S/L) and maltase to lactase (M/L) ratios in four normal subjects on isocaloric glucose (□) and sucrose (■) diets. The standard error of the mean is in brackets. [Reprinted from Rosensweig and Herman (1968, p. 2257) with permission of the Rockefeller University Press, New York.]

were computed. When expressed as ratios as well as units of activity, sucrose feeding produced an increase in both S/L and M/L ratios compared with glucose feeding (Fig. 3). This showed that sucrase and maltase activities were increased relative to lactase activity.

2. Effect of Fructose

From these original studies it was not known if the intact sucrose molecule was necessary to obtain these enzyme changes. Accordingly, the effect of dietary fructose, compared to glucose, was studied. The fructose feeding produced results similar to sucrose (Fig. 4). This suggested that fructose is the active principle in the sucrose molecule.

The studies that had been performed at that time permitted certain conclusions. These findings demonstrated that specific dietary sugars could control certain enzyme activity in the human small bowel in a specific manner. Furthermore, this enzyme activity could be controlled by substances other than substrate. Fructose is an end product of sucrose hydrolysis by sucrase and yet it has the capacity to increase sucrase activity. Therefore, it is most likely the active principle in the sucrose molecule.

This finding that the regulation of certain enzyme activities in man is not substrate-dependent is consistent with data in bacteria which dem-

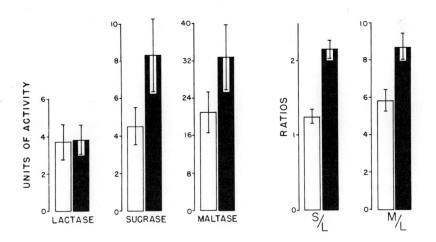

Fig. 4 Mean disaccharidase activities and ratios in one normal subject on iso-caloric glucose (□) and fructose (■) diets. The standard error of the mean is in brackets. [Reprinted from Rosensweig and Herman, (1968, p. 2259), with permission of the Rockefeller University Press, New York.]

onstrates that enzyme substrate may have no inducer properties (Monod, 1966). Likewise, the most potent inducers may not be a substrate of the enzyme.

3. Maltase Isoenzymes

The demonstration that sucrose or fructose feeding increases maltase as well as sucrase activity can be attributed to earlier reports that there are several maltase isoenzymes in man, two of which have sucrase activity (Dahlqvist, 1962; Auricchio *et al.*, 1965). Rosensweig and Herman (1968a) postulated that only the maltase isoenzymes with sucrase activity would be increased by sucrose feeding. In recent studies, this hypothesis has been confirmed (Schmitz *et al.*, 1972). These workers found that sucrase and maltase activities were 50% of normal in children with hereditary fructose intolerance who had been treated with a fructose-free diet. The maltases with sucrase activity (heat labile) were significantly lower than the controls, whereas the maltase isoenzymes without sucrase activity (thermoresistant maltases) were not significantly lower than the controls.

4. Dose Response

From the original observations of Rosensweig and Herman (1968a) it appeared that sucrose and fructose might be the only sugars that could regulate sucrase and maltase activities. However, it became readily apparent that sucrase and maltase activity did not disappear on a fructose- or sucrose-free diet. Other factors had to be involved in maintaining enzyme activity. In an investigation of the dose response of these enzymes to dietary sugars, Rosensweig and Herman (1970) showed that increasing the amount of dietary glucose from 0 to 80% of the total calories, with a constant amount of calories ingested, increased sucrase and maltase activities and the S/L and M/L ratios. Therefore, glucose is also able to regulate sucrase and maltase activities. When compared with increasing amounts of sucrose, however, the increase in activity was greater with sucrose (Fig. 5).

If not only fructose but also all carbohydrate is removed from the diet, disaccharidase activity falls, but not to zero. The factors maintaining activity are not understood. It may be endogenous or exogenous, constitutive or inductive. Limited data suggest that refeeding a carbohydrate-free diet to fasting obese subjects does not increase sucrase activity whereas there is an increase with feeding carbohydrate (Rosensweig and Herman, 1968b). It is possible that the blood sugar acts to maintain this activity.

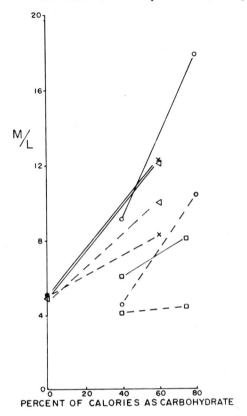

Fig. 5 Maltase to lactase (M/L) ratios in four normal subjects on isocaloric diets in which the carbohydrate portion of the diet ranged from 0 to 80% of the total calories. In each subject, high and low glucose (- - -) diets were compared with similar high and low sucrose (——) diets. Subject: 1, 0; 2, Δ; 3, □; and 4, ×. [Reprinted from Rosensweig and Herman (1970, p. 1375), with permission of the American Society for Clinical Nutrition, Inc., Bethesda.]

Indirectly supporting this possibility is the finding by Olsen and Rogers (1971) that disaccharidase activity is increased in diabetes. Additional studies are needed to delineate the factors other than dietary sugars that are needed to regulate disaccharidase activity.

It seems that dietary protein is not one of these controlling factors. Rats fed protein-free diets did not exhibit lowered disaccharidase activity (Solimano *et al.*, 1967; Prosper *et al.*, 1968). However, dietary carbohydrate was not constant in these studies. There is no definitive study of this point in man. However, one normal volunteer subject was fed a high protein diet for 9 days followed by a protein-free diet for 9 days. The carbohydrate content was kept constant and the diets were isocaloric.

No change was observed in mean disaccharidase activities or ratios (Rosensweig and Herman, 1970).

5. Time Response

In the initial studies, the sugars were fed for 1–2 weeks and the adaptive response was measured during this period. Additional studies were undertaken which measured the exact time required for this adaptive response (Rosensweig and Herman, 1969a). Normal subjects were fed glucose diets followed by isocaloric sucrose diets, and serial biopsies were obtained on changing the diets. There was no change in activity at 1 day (Fig. 6). At 2 days, a change in activity had begun and by 5 or 6 days it was complete. When a carbohydrate-free diet was next fed, it took 2–5 days for the enzyme activity to fall. A similar response was

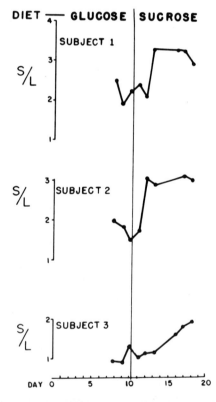

Fig. 6   Serial sucrase to lactase (S/L) ratios in three normal subjects on consecutive isocaloric glucose and sucrose diets. [Reprinted from Rosensweig and Herman (1969a, p. 502), with permission of the Williams and Wilkins Co., Baltimore.]

obtained when the diet was changed from sucrose to glucose or glucose to fructose. After 5 days there was no further change, even when the diet was continued for 9 weeks.

This 2–5 day response is quite similar to the estimated turnover time for the small intestinal epithelial cell in man (Bertalanffy and Nagy, 1961; Lipkin *et al.*, 1963; Shorter *et al.*, 1964; MacDonald *et al.*, 1964). On the basis of these findings, it has been suggested that the sucrose–fructose effect on disaccharidase activity occurs at the crypt cell level where disaccharidase activity is absent. The effect of the sugar becomes measurable as the crypt cell (with new information) migrates up the villus, matures, and produces an increased amount of disaccharidase activity.

To date, this hypothesis has not been tested. The studies reported above record the time response of enzyme activity to dietary sugars. However, they do not measure the mechanism of this change in activity. Additional studies are needed to prove that enzyme induction accounts for these changes and, further, that the inductive mechanism is mediated via the crypt cell and intestinal cell turnover.

B. Lactase Deficiency and Lactose Feeding

The role of dietary sugars in the regulation of intestinal enzyme activity takes on potential clinical significance when dealing with the problem of lactase deficiency and milk intolerance. This condition of lactase deficiency, which is more correctly termed "hypolactasia," is present in large numbers of adults all over the world. Its possible relationship to the ingestion of milk has occupied investigators for several years.

In independent reports, Cook and Kajubi (1966) and Bayless and Rosensweig (1966) noted marked differences in the incidence of hypolactasia among different African tribes and between American blacks and whites. Both laboratories suggested this condition was genetically controlled. Since there is a high correlation in some parts of the world between milk consumption and the incidence of hypolactasia (Cook and Kajubi, 1966; Bayless and Rosensweig, 1967; Kretchmer *et al.*, 1971), it was proposed by others that lactase deficiency was secondary to lactose deprivation (Bolin and Davis, 1970). However, there is no direct evidence in man to support this hypothesis. In rats, lactose feeding has yielded conflicting results. Some observers (Fischer, 1957) have found no effect of lactose feeding, while others (Cain *et al.*, 1969; Bolin *et al.*, 1971) have shown that lactose feeding increases lactase.

In man, several approaches have failed to show a correlation between lactase deficiency and the ingestion of lactose. One approach has been to feed lactose or milk to lactase deficient patients. Three separate and independent studies in which one or more quarts of milk or the lactose equivalent were fed daily to lactase-deficient patients for a time period ranging from one month to one year all showed no increase in lactase activity after lactose feeding (Cuatrecasas *et al.*, 1965; Keusch *et al.*, 1969; Gilat *et al.*, 1972). Another approach was to feed lactose-free, but otherwise complete, diets to normal volunteers and measure lactase activity. In two such studies (Knudsen *et al.*, 1968; Rosensweig and Herman, 1969b) lactose deprivation for as long as 2 months produced no decrease in lactase activity. In addition, Kogut *et al.* (1967) showed that galactosemic patients on a lactose-free diet for many years showed no increase in the incidence of lactose intolerance.

Lastly, Rosensweig and Herman (1968a, and unpublished observations) fed up to 450 gm/day of lactose to two normal subjects without any increase in lactase activity. These studies were short-term (2 weeks), and the lactose diets were compared with similar glucose diets. Therefore, lactose feeding or deprivation in man has not been shown to alter human small intestinal lactase activity.

Most of these feeding experiments have been short-term studies. It is possible that an effect of lactose feeding or deprivation takes many years to develop. However, the mechanism by which such an effect would occur is vague. Also, this possibility is opposite to the experience with the intestinal enzymes which do adapt to dietary sugars. This adaptation usually occurs in 1–5 days (Rosensweig and Herman, 1969a; Rosensweig *et al.*, 1969a).

It is pertinent to consider another aspect of this subject. The concept that lactase deficiency results from lactose deprivation implies that only lactose can regulate lactase activity, i.e., a one substrate–one inducer hypothesis. Experience with bacteria and with mammalian systems, however, suggests that regulation of enzyme activity can occur with nonsubstrate inducer substances as well as substrate; for example, fructose, the end product of sucrose hydrolysis, regulates sucrase activity. Other intestinal enzymes are regulated by hormones, drugs, folic acid, and other substances as well as sugars (Stifel *et al.*, 1968, 1969; Rosensweig *et al.*, 1968, 1969b,c; Lufkin *et al.*, 1972; Zakim *et al.*, 1969). Similarly, if lactase activity is regulated by substances other than lactose, then lactose deprivation alone would not necessarily lead to hypolactasia.

To date, the possible effect of other substances on lactase activity has not been well studied. However, there is a suggestion in rats that glucose feeding may increase lactase activity (Bolin *et al.*, 1969). One normal human volunteer subject showed a significant increase in lactase activity with a high glucose diet but others did not (N. S. Rosensweig and R. H. Herman, unpublished observations). If substances other than lactose can be shown to regulate lactase activity, this would render unlikely the hypothesis that hypolactasia is a result of lactose deprivation only.

C. Significance

With the demonstration of dietary regulation of intestinal enzyme activity, it is clear that a convenient model for the study of enzyme regulation in man has been developed. If the regulatory factors in the normal are understood, then potentially this information will be useful in the treatment of a variety of enzyme disorders. Theoretically, oral fructose should be useful in the treatment of sucrase deficiency.

Recent preliminary evidence suggests that fructose may be useful in the treatment of sucrase deficiency. Greene *et al.* (1972) fed large amounts of fructose to a sucrase-deficient girl with chronic diarrhea. There was an increase in sucrase activity, but it did not return to normal. However, clinically, the child became asymptomatic with continued fructose therapy.

It is not known if this type of sucrase deficiency is typical of all patients with sucrase-isomaltase deficiency. Other patients must be treated with fructose before any definitive conclusions can be drawn about this form of therapy. Quite possibly, only a few selected mild cases will benefit from this form of therapy. Regardless, through the studies in normal volunteers and those in selected patients, the concept of dietotherapy of certain intestinal conditions has been established.

D. Research Needed

The studies reported to date provide good evidence for the dietary regulation of human intestinal enzyme activity. However, these must be looked upon as beginning studies. Relatively little is known about this regulation by diet in various disorders. The effect of hormones and drugs in disease states likewise needs further investigation. Currently investigators often look for the underproduction or overproduction of an enzyme to explain some diseases. The studies reported herein lead to an additional concept. The enzymes may be present but they may not be under proper regulation. Preliminary studies (Rosensweig *et al.*, 1972) suggest that such an entity of failure to adapt to dietary sugars does indeed exist. Many more studies are needed before the full significance of this regulatory failure can be determined. With the development of these newer approaches to disease states, it may be possible to understand the etiology and pathogenesis of many intestinal and metabolic disorders that are currently ill-defined.

Acknowledgments

This work was supported in part by Research Grant AM 14122 from the National Institutes of Health, U.S. Public Health Service.

The secretarial assistance of Miss Josephine Mamahit is gratefully acknowledged.

References

Auricchio, S., Semenza, G., and Rubino, A. (1965). Multiplicity of human intestinal disaccharidases. II. Characterization of the individual maltases. *Biochim. Biophys. Acta* **96**, 498–507.

Bayless, T. M., and Rosensweig, N. S. (1966). A racial difference in the incidence of lactase deficiency in healthy adult males. *J. Amer. Med. Ass.* **197**, 968–972.

Bayless, T. M., and Rosensweig, N. S. (1967). Topics in clinical medicine: Incidence and implications of lactase deficiency and milk intolerance in white and negro populations. *Johns Hopkins Med. J.* **121**, 54–64.

Bertalanffy, F. D., and Nagy, K. P. (1961). Mitotic activity and renewal rate of the epithelial cell of the human duodenum. *Acta Anat.* **45**, 362–370.

Blair, D. G. R., Yakimets, W., and Tuba, J. (1963). Rat intestinal sucrase. II. The effects of rat age and sex and of diet on sucrase activity. *Can. J. Biochem. Physiol.* **41**, 917–929.

Bolin, T. D., and Davis, A. E. (1970). Primary lactase deficiency: Genetic or acquired? *Amer. J. Dig. Dis.* **15**, 679–692.

Bolin, T. D., Pirola, R. C., and Davis, A. E. (1969). Adaptation of intestinal lactase in the rat. *Gastroenterology* **57**, 406–409.

Bolin, T. D., McKern, A., and Davis, A. E. (1971). The effect of diet on lactase activity in the rat. *Gastroenterology* **60**, 432–437.

Cain, G. D., Moore, P., Jr., Patterson, M., and McElveen, M. A. (1969). The stimulation of lactase by feeding lactose. *Scand. J. Gastroenterol.* **4**, 545–550.

Cook, G. C., and Kajubi, S. K. (1966). Tribal incidence of lactase deficiency in Uganda. *Lancet* **1**, 725–730.

Cuatrecasas, P., Lockwood, D. H., and Caldwell, J. R. (1965). Lactase deficiency in the adult: A common occurrence. *Lancet* **1**, 14–18.

Dahlqvist, A. (1962). Specificity of the human intestinal disaccharidases and implications for hereditary disaccharide intolerance. *J. Clin. Invest.* **41**, 463–470.

Dahlqvist, A. (1964). A method for assay of intestinal disaccharidases. *Anal. Biochem.* **7**, 18–25.

Deren, J. J., Broitman, S. A., and Zamcheck, N. (1967). Effect of diet upon intestinal disaccharidases and disaccharide absorption. *J. Clin. Invest.* **46**, 186–195.

Fischer, J. E. (1957). Effects of feeding a diet containing lactose upon β-D-galactosidase activity and organ development in the rat digestive tract. *Amer. J. Physiol.* **188**, 49–53.

Gilat, T., Russo, S., Gelman-Malachi, E., and Aldor, T. A. M. (1972). Lactase in man: A nonadaptable enzyme. *Gastroenterology* **62**, 1125–1127.

Greene, H. L., Stifel, F. B., and Herman, R. H. (1972). Dietary stimulation of sucrase in a patient with sucrase-isomaltase deficiency. *Biochem. Med.* **6,** 409–418.

Keusch, G. T., Troncale, F. J., Thavaramara, B., Prinyanont, P., Anderson, P. R., and Bhamarapravathi, N. (1969). Lactase deficiency in Thailand: Effect of prolonged lactose feeding. *Amer. J. Clin. Nutr.* **22,** 638–641.

Knudsen, K. B., Welsh, J. D., Kronenberg, R. S., *et al.* (1968). Effect of a nonlactose diet on human intestinal disaccharidase activity. *Amer. J. Dig. Dis.* **13,** 593–597.

Kogut, M. D., Donnell, G. N., and Shaw, K. N. F. (1967). Studies of lactose absorption in patients with galactosemia. *J. Pediat.* **71,** 75–81.

Kretchmer, N., Ransone-Kuti, O., Hurwitz, R., Dungy, C., and Alakija, W. (1971). Intestinal absorption of lactose in Nigerian ethnic groups. *Lancet* **2,** 392–395.

Lipkin, M., Sherlock, P., and Bell, B. (1963). Cell proliferation kinetics in the gastrointestinal tract of man. II. Cell renewal in stomach, ileum, colon, and rectum. *Gastroenterology* **45,** 721–729.

Lufkin, E. G., Stifel, F. B., Herman, R. H., Rosensweig, N. S., and Hagler, L. (1972). Effect of testosterone on jejunal pyruvate kinase activities in normal and hypogonadal males. *J. Clin. Endocrinol. Metab.* **34,** 586–591.

MacDonald, W. C., Trier, J. S., and Everett, N. B. (1964). Cell proliferation and migration in the stomach, duodenum and rectum in man: Radioautographic studies. *Gastroenterology* **46,** 405–417.

Monod, J. (1966). From enzymatic adaptation to allosteric transitions. *Science* **154,** 475–483.

Olsen, W. A., and Rogers, L. (1971). Jejunal sucrase activity in diabetic rats. *J. Lab. Clin. Med.* **77,** 838–842.

Prosper, J., Murray, R. L., and Kern, F., Jr. (1968). Protein starvation and the small intestine. II. Disaccharidase activities. *Gastroenterology* **55,** 223–228.

Rosensweig, N. S., and Herman, R. H. (1968a). Control of jejunal sucrase and maltase activity by dietary sucrose or fructose in man: A model for the study of enzyme regulation in man. *J. Clin. Invest.* **47,** 2253–2262.

Rosensweig, N. S., and Herman, R. H. (1968b). On the effect of fasting and refeeding. *Gastroenterology* **55,** 746–747.

Rosensweig, N. S., and Herman, R. H. (1969a). Time response of jejunal sucrase and maltase activity to a high sucrose diet in normal man. *Gastroenterology* **56,** 500–505.

Rosensweig, N. S., and Herman, R. H. (1969b). Diet and disaccharidases. *Amer. J. Clin. Nutr.* **22,** 99–102.

Rosensweig, N. S., and Herman, R. H. (1970). The dose response of jejunal sucrase and maltase activities to isocaloric high and low carbohydrate diets in man. *Amer. J. Clin. Nutr.* **23,** 1372–1377.

Rosensweig, N. S., Stifel, F. B., Herman, R. H., and Zakim, D. (1968). The dietary regulation of the glycolytic enzymes. II. Adaptive changes in human jejunum. *Biochim. Biophys. Acta* **170,** 228–234.

Rosensweig, N. S., Stifel, F. B., Zakim, D., and Herman, R. H. (1969a). Time response of jejunal glycolytic enzymes to a high sucrose diet. *Gastroenterology* **57,** 143–146.

Rosensweig, N. S., Herman, R. H., Stifel, F. B., and Herman, Y. F. (1969b). The regulation of human jejunal glycolytic enzymes by oral folic acid. *J. Clin. Invest.* **48,** 2038–2045.

Rosensweig, N. S., Stifel, F. B. and Herman, R. H. (1969c). Effect of phenobarbital on human jejunal glycolytic, gluconeogenetic, and pentose phosphate path enzymes. *Clin. Res.* **17,** 596.

Rosensweig, N. S., Herman, R. H., Stifel, F. B., Hagler, L., Greene, H. L., Jr., and Herman, Y. F. (1972). Gastrointestinal disease associated with a failure of adaptation of jejunal glycolytic enzymes. *Gastroenterology* **62**, 802.

Schmitz, J., Odievre, M., and Rey, J. (1972). Specificity of the effects of a fructose-free diet on the activity of intestinal α-glucosidases in man. A study in hereditary fructose intolerance. *Gastroenterology* **62**, 389–392.

Shorter, R. G., Moertel, C. G., Titus, J. L., and Reitemeier, R. J. (1964). Cell kinetics in the jejunum and rectum of man. *Amer. J. Dig. Dis.* **9**, 760–763.

Solimano, G., Burgess, E. A., and Levin, B. (1967). Protein-calorie malnutrition: Effect of the deficient diets on enzyme levels of jejunal mucosa of rats. *Brit. J. Nutr.* **21**, 55–68.

Stifel, F. B., Rosensweig, N. S., Zakim, D., and Herman, R. H. (1968). Dietary regulation of the glycolytic enzymes. I. Adaptive changes in rat jejunum. *Biochim. Biophys. Acta* **170**, 221–227.

Stifel, F. B., Herman, R. H., and Rosensweig, N. S. (1969). Dietary regulation of glycolytic enzymes. III. Adaptive changes in rat jejunal pyruvate kinase, phosphofructokinase, fructose-diphosphatase and glycerol-3-phosphate dehydrogenase. *Biochim. Biophys. Acta* **184**, 29–34.

Zakim, D., Herman, R. H., Rosensweig, N. S., and Stifel, F. B. (1969). Clofibrate-induced changes in the activity of human intestinal enzymes. *Gastroenterology* **56**, 496–499.

CHAPTER 13

Enzyme Deficiency and Malabsorption of Carbohydrates

ARNE DAHLQVIST

Until recently very little concrete information has been available about impaired carbohydrate digestion and absorption as a cause of gastrointestinal disease. From a more speculative point of view this matter caused some attention in the beginning of the twentieth century. Littman and Hammond, in a review article of 1965, wrote: "Pediatricians and internists in Vienna in the early part of this century wrote extensively about 'fermentative diarrhea,' and presented evidence in some cases that certain carbohydrates were poorly tolerated and that their removal from the diet would be followed by cessation of the diarrhea."

But essentially nothing happened in this field, and the "fermentative diarrhea" was more or less forgotten until the second half of the 1950's. At this time the possible occurrence of disaccharide intolerance received new interest both from the clinical and biochemical point of view. The cause of this multiple interest is difficult to explain, but a contributing factor probably was that Crane and his collaborators had developed a completely new model for the events occurring during the active transport of certain monosaccharides (for review, see Crane, 1960, 1967; Wilson, 1962) thereby focusing interest on sugar digestion and absorption in general. Rather independently from each other, several groups described what apparently must be congenital enzyme defects in the digestion of

187

different disaccharides. Holzel *et al.* (1959), Weijers *et al.* (1960, 1961), and Prader *et al.* (1961) described patients with congenital lactase or sucrase deficiency. One patient, earlier reported by Durand (1958) and initially interpreted as a case of congenital lactase deficiency, probably was instead a case of the disease later classified by Holzel *et al.* (1962) and named "severe lactose intolerance." This disease is not caused by lactase deficiency, but rather by increased permeability of the stomach for lactose (and probably other molecules of corresponding size) with subsequent toxic effects on certain tissues (Berg *et al.*, 1969).

At the same time the present author performed a biochemical study of the intestinal disaccharidases and their specificity, using experimental animals (mainly pigs). These studies were collected into a doctoral thesis (Dahlqvist, 1960a). They revealed a picture of the enzymes involved which was quite different from that accepted earlier. In these studies a large number of different separation methods were used, but the most efficient one in the case of the intestinal α-glucosidases (to which most of the intestinal disaccharidases belong) was heat inactivation (Dahlqvist, 1959a, 1960a). This method was therefore selected for a comparative study of the disaccharidases in the *human* small intestine, which provided the necessary enzymological basis for discussions about the defect in the patients (Dahlqvist, 1962a,b, 1963). The human enzymes were somewhat similar to those in the pig, but there were a few minor differences.

Still, the enzyme defects in disaccharidase deficiency syndromes had only been demonstrated indirectly with blood sugar curves and fecal analysis after oral administration of the sugars. Intestinal intubation had, however, been practiced for some time [in fact, it was intubation studies in the human that first told us that the disaccharidases are not secreted but act on the cellular surface of the mucosal epithelium of the villi (Borgström *et al.*, 1957)]. More recently, methods for oral biopsy of the intestinal mucosa had been developed. A method for disaccharidase activity assay was therefore devised, which was sensitive enough to use with a few milligrams of mucosa (wet weight) (Dahlqvist, 1964). [The method was later simplified and improved (Dahlqvist, 1968, 1970).] With this method a study was performed in Chicago (Dahlqvist *et al.*, 1963b) in which it was attempted to find some relationship between mucosal activity of the disaccharidases and the oral tolerance for disaccharides in single adult patients. Much to the surprise of the author it was found quite easy to find adult subjects with lactase deficiency, as revealed both from the enzyme assay and from the oral tolerance test, with very good agreement in results between these two methods. Quite independently the syndrome of lactase deficiency in adults was described during the same year by a Swiss group (Haemmerli *et al.*, 1963) and an American group

(Kern *et al.*, 1963). These initial papers have been followed by a large number of clinical reports and reviews by these and other investigators. Lactase deficiency in adults was later found to be present in about 6% of the adults in North America, but even more frequently in other populations [so far it seems only to have lower frequency in Sweden and Denmark and possibly among the Hamites (Cook, 1972)]. The majority of other human subjects over the world lose their intestinal lactase before they grow up to be adults, quite independently of whether they drink milk or not. The amount of publications on this subject that have appeared during the last decade is enormous, and no attempts at a literature survey will be made. It is, however, fascinating to recall that for many years prior to the 1960's no attention was paid to the possible occurrence of disaccharidase deficiency in patients with gastrointestinal symptoms. Around 1960 disaccharidase deficiency was found to occur in different forms as very rare "inborn errors of metabolism." Only a few years later lactase deficiency was found to be rather common among adults, first interpreted as a disease, but later recognized as the normal condition in most populations. Those few humans who are lactase-persistant, i.e., continue to have high intestinal lactase activity throughout adult life are, in fact, the deviant group, rather than those who become deficient in lactase.

A. Physiology of Carbohydrate Digestion and Absorption

Carbohydrates contribute a considerable fraction of the available energy in our diet. Approximately 50% of the calories are obtained from carbohydrates, and in countries with a less developed economy this figure may be increased to 90% since carbohydrates provide relatively cheap food. The only exception seems to be people living under arctic conditions such as the Greenland Eskimos, where the carbohydrate consumption until recently has been practically zero for obvious reasons.

Most of the carbohydrate is consumed as disaccharides or polysaccharides, which require digestion into monosaccharides before absorption can occur in the intestine. The average consumption of different carbohydrates based on statistical figures for production, export, and import in Sweden during 1959 are shown in Table I. Later studies performed by the double-portion technique have indicated somewhat lower figures for the carbohydrate consumption, expecially for sucrose (Borgström *et al.*, 1970), but there are certain sources of error with this technique that make it undesirable to adjust the figures as yet. The consumption in the United States is rather similar to that in Sweden.

TABLE I

Average Daily Consumption of Different
Carbohydrates in Sweden

Carbohydrate	Amount (gm)
Starch[a]	135
Sucrose	140
Lactose	30
Monosaccharides	5–20

[a] Includes maltose and other hydrolytic products.

For the hydrolysis of these carbohydrates we possess a number of hydrolytic enzymes, namely, amylase, produced by the salivary glands (but rapidly inactivated by the HCl in the stomach), the pancreas, and probably to some extent by the small intestinal mucosa, and a number of disaccharidases, produced by the small intestinal mucosa.

The amylase is an α-amylase (*endo*-amylase), and it is present in such large amounts in the small intestinal content that it will hydrolyze the dietary starch within minutes (Dahlqvist and Borgström, 1961). The pancreatic amylase seems to play the most important role. The small intestinal mucosa also has γ-amylase (glucamylase) activity (Dahlqvist and Thomson, 1963), but this activity may be a side effect of the maltases. Sometimes weak maltase activity has been described in the pancreatic juice of different species. This maltase activity most probably does not represent anything but a side effect of the amylase. All commercially available maltose preparations contain significant amounts of maltotriose, and this trisaccharide is known to be slowly hydrolyzed by α-amylase.

The hydrolysis of disaccharides, those present in the food as such as well as the disaccharides and oligosaccharides formed from starch by the amylase, thus seems to be confined to the small intestine. It has also been long known that the small intestinal mucosa has powerful hydrolytic activity with disaccharides as substrates.

It was a widespread textbook concept (and still is to some extent) that the disaccharidases and other small intestinal enzymes were secreted with the "succus entericus." In a study of the digestion in the human intestine, using the intubation-nonabsorbable marker technique (Borgström *et al.*, 1957) we therefore attempted to measure the disaccharidase activity as an indicator of the admixture of succus entericus. Very little disaccharidase activity was found in the samples collected, and therefore a more detailed study was performed (Dahlqvist and Borgström, 1961). The results were clear-cut: The disaccharidase activity of the small intestinal content was too weak to be of any importance for the digestion, and the conclusion was drawn that the intestinal disaccharidases exert their physiological function while still attached to the mucosal cells. In animal experiments, Miller and Crane (1961b,c) arrived at identical conclusions. These authors also developed a brilliant technique for the isolation of the brush border membranes covering the musocal surface of the epithelial cells. The disaccharidases were located in these membranes (Miller and Crane, 1961a).

A histochemical staining method was also developed, which was based on a coupled reaction sequence leading to the precipitation of an insoluble formazan (Dahlqvist and Brun, 1962). The reaction sequence is shown in Fig. 1. This method initially yielded a more diffuse staining as a result of diffusion-adsorption phenomena. It was, however, later improved by Lojda (1965) in Prague and by Jos *et al.* (1967) in Paris, and then yielded results consistent with those of Miller and Crane (1961a). A detailed review on the localization has been published (Dahlqvist, 1967).

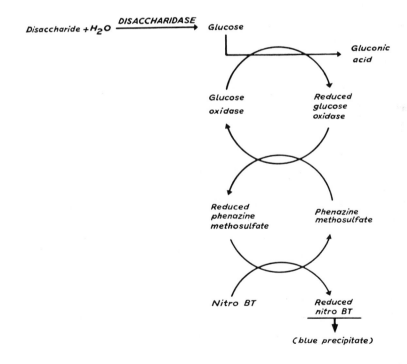

Fig. 1 Reaction sequence in the disaccharidase activity staining method of Dahlqvist and Brun (1962).

Occasionally the staining method has also been used for the clinical diagnosis of disaccharidase deficiency.

The histochemical staining method indicated higher disaccharidase activity in the villi than in the crypts of the small intestinal mucosa. This is also what would be expected if the enzymes act in the absorbing cells instead of being secreted. More quantitative data on the enzyme distribution were obtained by a microdissection technique subsequently devel-

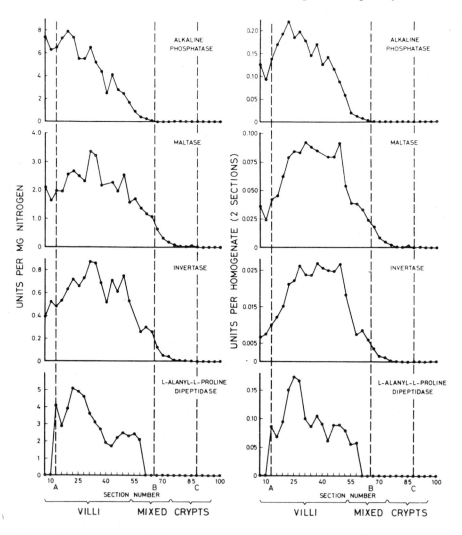

Fig. 2 Localization of alkaline phosphatase, disaccharidases and dipeptidases in the small intestine, as studied by the microdissection technique of Nordström *et al.* (1968).

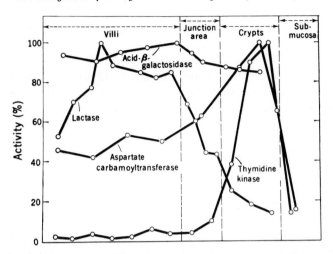

Fig. 3 Localization of one digestive enzyme (lactase), one lysosomal enzyme (acid β-galactosidase), and two enzymes involved in cell regeneration, using the same technique as in Fig. 2. (Fortin-Magana *et al.*, 1970. Copyright 1970 by the American Association for the Advancement of Science.)

oped (Dahlqvist and Nordström 1966; Nordström *et al.*, 1968). All the digestive enzymes seem to be located exclusively in the villi (Fig. 2). It is now generally agreed that the crypts act as cell proliferation centra, which is also reflected in their enzyme activity (Fig. 3).

B. Specificity of the Enzymes

The specificity of the α-amylase, which is the most important enzyme for the hydrolysis of starch (and glycogen) in food, has been well known for a long time. The main hydrolysis products are maltose, small oligosaccharides (α-limit dextrins) containing $\alpha\text{-}1 \to 4$ and $\alpha\text{-}1 \to 6$ glucosidic links and a small fraction of free glucose.

Concerning the intestinal disaccharidases, there was once believed to be one enzyme for each substrate that could be hydrolyzed. Subsequent studies of enzymes from other sources revealed that a single enzyme could often hydrolyze several different substrates, provided that they contained an identical glycon structure, and from this structure they were named "α-glucosidases," "β-glucosidases," "α-galactosidases," · etc. This classification originated from the brilliant studies of yeast enzymes by Weidenhagen in the 1930's.

Neither of these theories, however, covers the specificity of the intestinal disaccharidases. The intestinal brush borders contain at least five different α-glucosidases, with partly overlapping specificity, and one β-galactosidase (Borgström and Dahlqvist, 1958; Dahlqvist, 1958, 1959a,b,c, 1960a–e; Dahlqvist and Borgström, 1959). This first study was performed with mucosa from pigs, but later human mucosa was studied with heat inactivation experiments (Dahlqvist, 1962a) and then with different forms of column chromatography (Dahlqvist and Telenius, 1969).

After the initial characterization of the intestinal disaccharidases, a provisional numbering system was proposed. Other authors have used different orders of numbering, however, and this can very easily cause confusion. It is therefore time to abandon these numbering systems and name the enzymes according to their main substrate. In Table II the different enzymes are listed together with information about the approximate importance of each enzyme for the hydrolysis of different substrates in a crude homogenate of the mucosa.

The lactase is a β-galactosidase, and possibly also a β-glucosidase, while all the other five disaccharidases are α-glucosidases. As seen from the table, there are no less than four different enzymes with maltase activity. Therefore, maltase deficiency as a genetic disease is not likely to occur. The presence of only one of the maltases would be sufficient for the digestion of fairly large amounts of maltose.

One other research group has described the further fractionation of

TABLE II

Disaccharidases in the Human Small Intestinal Mucosa[a]

Enzyme	Substrates	
Isomaltase	Isomaltose	(>95%)
	Maltose	(∼50%)
Sucrase	Sucrose	(100%)
	Maltose	(∼25%)
Two heat-stable maltases	Maltose	(∼25%)
	Isomaltose	(<5%)
Trehalase	Trehalose	(100%)
Lactase	Lactose	(>95%)
	Hetero-β-galactosides	(<50%)
	Cellobiose	(100%?)

[a] All enzymes located in the brush border.

sucrase and lactase into two components each. Both these fractionations have, however, later been revealed as artifacts, caused by interaction with the Sephadex used for the columns. One positive thing appeared from these experiments, however. It became apparent that the isomaltase and sucrase, although enzymatically quite different, exist in some form of complex, bound to each other. There is also some genetic relationship between these two enzymes since the inborn sucrase deficiency always seems to be combined with isomaltase deficiency (Auricchio *et al.*, 1963).

In addition to the brush border enzymes there are in the mucosa other enzymes with disaccharidase or glycosidase activity. These have different localization and probably do not take part in the digestion of the food. Asp *et al.* (1969; Asp, 1971) and Gray and Santiago (1969) have isolated and studied one β-galactosidase with acid pH optimum (probably lysosomal) which hydrolyzes both lactose and hetero-β-galactosides, and one soluble β-galactosidase with neutral pH optimum, which hydrolyzes certain hetero-β-galactosides, but *not* lactose. These enzymes are named "acid β-galactosidase" and "hetero-β-galactosidase," respectively.

An acid α-glucosidase has also been demonstrated in the intestine.

C. Clinical Diagnosis of Impaired Carbohydrate Utilization

The ingestion of substantial amounts of carbohydrate, which for one reason or another cannot be absorbed and utilized in the body, will lead to what is called "fermentative diarrhea," i.e., diarrhea with usually liquid and acid stools. In small infants it is often possible to demonstrate the unabsorbed carbohydrate in the stools, but later in life the bacteria of the large intestine consume the sugar so effectively that only lactic acid is found in the stools.

In addition to chronic or intermittent diarrhea, the patient often reports abdominal pain, borborygmia, flatulence, and possibly other abdominal symptoms which may sometimes cause considerable diagnostic difficulties (Gudmand-Høyer *et al.*, 1969, 1970). The degree of severity of the symptoms varies with the amount of sugar ingested, but constitutional factors play an important role also. Small infants usually get proportionally more intense symptoms than do adults. In a recently performed double-blind study the reaction of lactase-deficient adults given varying doses of lactose was studied (Andersson *et al.*, 1973). In the subjects studied a dose of up to 5 gm of lactose did not result in increased symptoms, but 10 gm or more did. It made no difference whether the sugar was fed in milk or in water. There are a few subjects who react to much smaller amounts of lactose, but these subjects are exceptional (Gud-mand-Høyer *et al.*, 1970; A. C. Frazer, personal communication, 1967).

It has already been mentioned that in small infants fecal analysis, which should of course be performed *before* the patient receives a carbo-hydrate-restricted diet, may give valuable information (Durand *et al.*, 1962; Ford and Haworth, 1963; Kerry and Anderson, 1964). In adult pa-tients, however, this method is of very limited value. It is important to note that it is the liquid part of the stools that should be analyzed for sugar, and possibly for pH and lactic acid. In adult patients, when disac-charide intolerance is suspected, the main diagnostic tool is the oral toler-ance test.

Oral tolerance tests with starch have been used as a measurement of the amylase efficiency (Weijers *et al.*, 1961; Prader *et al.*, 1961). This method, however, is not satisfactory for the demonstration of pancreatic dysfunction (Nugent and Millhon, 1958). For diagnosis of amylase de-ficiency, assay of the amylase activity in intestinal content, collected by intubation, is recommended instead (Borgström *et al.*, 1957). A simple reducing sugar test with a 3,5-dinitrosalicylate reagent is excellent for this purpose (Dahlqvist, 1962c). As far as the author is aware, no patient with specific amylase deficiency has, however, so far been described (al-though specific deficiencies in some other pancreatic enzymes do occur). In cases of general pancreatic dysfunction, the fat balance studies as rou-tinely performed in hospitals probably give more valuable information from a clinical point of view (van de Kamer *et al.*, 1949).

In contrast, oral tolerance tests with disaccharides have been used all over the world and are to be regarded as the regular test for the diagnosis of disaccharide intolerance. A control test with the corresponding mono-saccharides should be performed to exclude general malabsorption or glu-cose–galactose malabsorption. In most cases the result of an oral toler-ance test with a disaccharide is easy to interpret. In some cases, however, difficulties occur. In these cases the test should be complemented with enzyme assay in a mucosal biopsy, if possible, or a new tolerance test should be performed.

The details in the performance of these tests do, however, vary consid-erably in different clinics and laboratories with regard to calculation of dosage, intervals for blood collection, evaluation of the limit between nor-mal and pathological evaluation of blood sugar, etc. These differences become especially important when simplified tolerance tests are used, e.g., for screening of the frequency of lactase deficiency in a population. These problems were discussed during a recent symposium "Intestinal Enzyme Deficiencies and their Nutritional Implications," sponsored by the Swedish Nutrition Foundation (Dahlqvist *et al.*, 1973). Since many scientists with considerable experience of such tests were present, it was decided to give a joint recommendation on the performance of the lactose tolerance test, which might be used as a standard. This recommendation is cited below (see also Dahlqvist, 1972):

Recommendations on the Performance of Lactose Tolerance Tests

During a symposium arranged by the Swedish Nutrition Foundation at Saltsjöbaden (Stockholm), August 14–16, 1972, the following proposal for a protocol on methodology for the performance of lactose tolerance tests was endorsed by the participants.

Note: It is appreciated that laboratories with long experience in the field may wish to continue with a technique which is not exactly similar to the recommendations.

1. The patient should ideally be fasted overnight, or alternatively for at least 6 hr.

2. The amount of lactose used for the tolerance test should be as follows:
   for adults: 50 gm;
   for infants and children: 2 gm/kg body weight (maximum 50 gm).

Note: A calculation of lactose amount per body surface area (e.g., 50 gm/sq m) does not seem to give such an advantage that it makes up for the extra work involved. It may have a value when comparing the situation in different age groups.

3. Lactose should be given as a 10% solution/suspension in water and preferably at room temperature.

4. Capillary (not venous) blood should be obtained and analyzed for glucose. Determination with glucose oxidase is recommended, but it is recognized that the use of automatic equipment often makes a ferricyanide method necessary.

5. Blood samples should be taken as follows:
   Before lactose administration: 2 samples with 5 or 10 min interval (mean value to be used);
   After lactose administration: at 15, 30, 45, and 60 min. A sample at 90 min may sometimes give additional information but is not regarded as obligatory.

6. A blood glucose rise by 25 mg/100 ml or more indicates efficient hydrolysis and absorption of lactose. A rise below 20 mg/100 ml suggests low lactase activity. A value between 20 and 25 mg/100 ml should be regarded as a borderline. If diarrhea occurs (cf. below), the patient probably has low lactase activity.

7. Ideally, a "flat" lactose tolerance test should be followed by a glucose plus galactose test, in order to exclude impaired monosaccharide absorption or increased peripheral glucose uptake in the tissues.

8. A "flat" lactose tolerance test is strong evidence of low lactase activity. However, it does not necessarily mean that the patient is intolerant to lactose. This can only be evaluated by elucidation of signs and symptoms. The subject should be advised to carefully report his or her experience in this respect, not

only during the period of blood sampling but also during the day and night following the test.

   9. Prevalence studies may be performed with a simplified procedure:

      Fasting time, dose and mode of administration of lactose, should be performed as described above. Blood samples should be taken at 0 (single sample), 15, 30, and 45 min. If for practical reasons only three blood samples can be obtained they should be taken at the following times: 0, 20, and 40 min.

   The claim for glucose-galactose test can rarely be upheld in prevalence studies.

   Also in prevalence studies it is of fundamental importance to register any clinical signs of intolerance which may appear after the lactose load, both during the time of blood sampling and during the remainder of the day.

The performance of tolerance tests with other disaccharides does not differ essentially from that with lactose.

If glucose–galactose malabsorption is suspected, it is not equally easy to establish the diagnosis with only oral tolerance tests since a flat curve after feeding these monosaccharides may also be the result of a general malabsorption as in celiac disease. The different methods available have been discussed in an earlier review by Dahlqvist *et al.* (1968). The most adequate method is to feed a solution containing equal amounts of glucose and fructose to a patient under intestinal intubation and to recover samples for analysis at different degrees of absorption. In a normal subject the sugars are rapidly absorbed. In a patient with general malabsorption the sugars are absorbed slowly. In both cases, however, glucose is absorbed more rapidly than fructose. Only in a patient with glucose–galactose malabsorption is fructose absorbed more rapidly than glucose (Lindquist and Meeuwisse, 1963; Fig. 4).

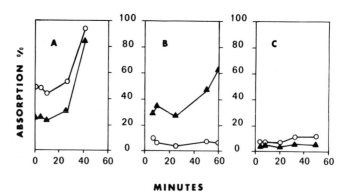

MINUTES

Fig. 4   The absorption of fructose and glucose in an equimolar mixture studied by the intubation technique in (A) a control subject, (B) a patient with glucose-galactose malabsorption, and (C) a patient with celiac disease. Symbols: (○) glucose and (▲) fructose. (Lindquist and Meeuwisse, 1963; Dahlqvist *et al.*, 1968.)

D. Disaccharidase Activity Assay

Assay of disaccharidase activity utilizing a nonreducing substrate such as sucrose is easily performed by using a reducing sugar method for assaying the monosaccharides formed on hydrolysis. Unfortunately, very few disaccharides are nonreducing, and measurement of their hydrolysis with conventional methods is much more complicated (Dahlqvist, 1960a).

A number of synthetic substrates (heteroglycosides) yielding colored or fluorescent hydrolysis products have been synthesized, and for some purposes these substrates are very useful. They are not recommended for the assay of intestinal disaccharidase activity since the lysosomal and soluble enzymes in the mucosa usually interfere more than when disaccharides are used (Asp, 1971).

When glucose oxidase became available, yielding a simple method for sensitive and accurate measurement of glucose in the presence of other sugars, it was hoped to add a simplified disaccharidase assay method. But it was soon found that the glucose oxidase preparations contained powerful contaminant disaccharidases which were practically impossible to remove from the glucose oxidase.

In the same year it was independently found in three different laboratories that these contaminant enzymes could be specifically inhibited by tris buffer (Dahlqvist, 1961; Sols and de la Fuente, 1961; White and Subers, 1961). This enabled the development of a method for intestinal disaccharidase activities which could also be used for such small amounts of mucosa as could be obtained by peroral biopsy (Dahlqvist, 1964).

Some years later glucose oxidase preparations with low disaccharidase activity prepared from another organism appeared. A simplified one-step procedure was then described in which the hydrolysis and color development occurred simultaneously (Messer and Dahlqvist, 1966). There are, however, certain disadvantages with the one-step method, and therefore it was later abandoned in favor of an improved and simplified version of the original two-step method (Dahlqvist, 1968, 1970).

Whichever method is used, it is important to note that the homogenate should *not* be centrifuged before analysis, since a considerable fraction of the activity may be lost. Unfortunately, this point was not expressed clearly enough in one of the author's papers (Dahlqvist, 1968), which sometimes has led to misunderstandings.

The biopsy is usually taken at the ligament of Treitz (the duodenojejunal flexure). The maximal disaccharidase activity in the small intestine does not occur until in the middle-distal part of the jejunum. The difficulties in getting oral biopsies from the deeper parts of the small intestine and the disadvantage of taking the biopsy at a site where the activity is not yet maximal are counterbalanced by the fact that the ligament of Treitz can be used as a landmark, giving rather reproducible localization. Most published results have therefore been obtained with biopsies from this location.

If the biopsy is not analyzed immediately, it should be wrapped in parafilm to prevent excessive drying and stored at −20°C. When duplicate biopsies were stored in this way, no decrease was found in any of the disaccharidase activities even after storage for 2 years (A. Dahlqvist and G. Meeuwisse, unpublished observation). After the biopsy has been homogenized the stability is limited and the analysis should be completed within 24–48 hr. During this time the homogenate is stored in crushed ice (on the bench) or in the refrigerator (overnight). If it is frozen, part of the protein will precipitate.

The disaccharidase activities are expressed as units (μmoles of substrate hydrolyzed per minute) per gram protein. Some authors use units per gram wet weight of the mucosa, but there is a risk that partial drying of the biopsy may disturb the results.

Some disaccharidase activities found in normal subjects are seen in Table III. In adult subjects with persistant lactase the activity found of this disaccharidase is 9–98 units/gm of protein (mean 44 units) (Dunphy *et al.*, 1965). This seems to be approximately the same activity as is found in infants. In adults with lactase deficiency there is a residual lactase activity of 1–5 units/gm of protein, the mean residual activity

TABLE III

Disaccharidase Activities Measured in Intestinal Biopsies from Presumably Normal Subjects of Some Different Populations[a,b]

	Units/gram of protein (mean)				
Activity	American controls $n = 30$	American controls $n = 23$	Swedes $n = 37$	Indian $n = 30$	Thai $n = 74$
Maltase	363	375	258	224	292
Sucrase	88	108	69	81	79
Trehalase	43	—	24	19	—
Lactase	36	38	29	3.4	2.1

[a] From Berg *et al.* (1970), Keusch *et al.* (1969), Swaminathan *et al.* (1970), and Dunphy *et al.* (1965).

[b] In two of the investigations the trehalase activity was not measured. The Indian and Thai adults are lactase-deficient.

seems to be somewhat different in various populations (Asp, 1971). A considerable fraction (up to 90%) of this residual lactase activity is not caused by brush border lactase, but by the lysosomal acid β-galactosidase (Asp, 1971). However, there is also a small residue of brush border lactase when assay methods are used which permit the specific measurement of each of these two enzymes (Asp and Dahlqvist, 1972; Asp *et al.*, 1971, 1973).

Whether there is also a residual activity in the congenital forms of disaccharidase deficiency is not known for sure. Some authors have reported weak residual activity, others not, but the sensitivity limits of the methods and the nature of the activity found have usually not been studied.

In spite of the residual activity occurring in at least the adults with lactase deficiency, it is in most cases quite easy to interpret the results of disaccharidase activity assays in mucosal biopsies. When a patient has a specific deficiency, the activity affected is usually reduced to less than 10% of the average. This method also makes it possible to distinguish between specific disaccharidase deficiencies and general deficiencies in which all the activities are low (the latter condition is usually accompanied by marked morphological changes, both in the dissection microscope and in histological sections). Assay of the disaccharidase activities in biopsy preparations is therefore often of value for the clinical diagnosis. It also opens a more direct way for the demonstration of the enzyme defect than do tolerance tests, and in addition enables enzymological characterization.

E. Clinical Forms of Enzyme (or Carrier) Deficiency

1. Specific Enzyme Defects

a. Pancreatic

Amylase deficiency would be expected to occur but has not yet been found.

b. Small Intestinal

  i. Adult lactase deficiency
 ii. Congenital lactase deficiency
iii. (Congenital) sucrase-isomaltase deficiency
 iv. (Congenital) trehalase deficiency

v. Glucose–galactose malabsorption
vi. Fructose malabsorption—not yet found

2. General (Secondary) Enzyme Defects

a. Pancreatic

Can result from congenital disease (e.g., cystic fibrosis) or acquired disease with impairment of pancreatic glandular tissue or ducts.

b. Small Intestinal

In diseases with villous "atrophy," as celiac disease, the disaccharidase activities are strongly reduced and also the monosaccharide transport is impaired.

Pancreatic amylase seems to be a single enzyme; furthermore, the salivary amylase appears to be identical in structure with pancreatic amylase (Meyer *et al.*, 1948). Therefore, amylase deficiency would be expected to occur as a genetic defect. It has not been found, however. The small intestinal wall has powerful amylase activity. This activity may partly result from pancreatic amylase since the pancreatic enzymes have been reported to be absorbed onto the mucosal surface (Ugolev, 1965). Although it cannot be said whether the small intestine also produces α-amylase, and whether this α-amylase is similar to that from the pancreas, it is known, however, that the small intestine does produce γ-amylase (Dahlqvist and Thomson, 1963), an enzyme which probably is identical with one of the maltases. The possibility thus exists that the small intestine by itself is able to hydrolyze dietary starch, and if such is the case pancreatic amylase deficiency will not give clinical symptoms.

How the adult lactase deficiency was first found in a few subjects (too many to be adult patients with congenital lactase deficiency), and then found to be the rule in most parts of the world, has been outlined above. Initially, the problem was to explain why the lactase disappeared after childhood in some subjects. Possibilities such as a permanent biochemical defect after an acute intestinal disease were discussed. Now the problem has changed so that we must instead explain why a few ethnic groups have developed the ability to retain high intestinal lactase activity throughout adult life.

One thing is sure, the lactase activity is *not* regulated by adaptation to lactose in the diet of the single individual. Rosensweig *et al.* (Rosensweig and Herman, 1969; Rosensweig, 1973) have studied the effect of different sugars on the intestinal disaccharidase activity in man. Certain sugars increased the activity of sucrase and other α-glucosidases some-what, but the lactase activity was not changed either by lactose or by any other sugar.

Rather, a selection *ad modum* Darwin has been responsible. Both the North Europeans and the Hamites, the two groups that have developed lactase persistence, have for several thousands of years been heavily dependent on cattle breeding for their living. Of course, it was then a considerable advantage to be able to consume milk also in adult life. Several long articles have been written about this (Simoons, 1969, 1970; Bolin and Davis, 1970; McCracken, 1970).

In adult lactase deficiency, the lactase activity is not zero, but if a sensitive enough method is used a residual activity of 1–5 units of lactase per gram of protein is found. This is about 10% or less of the "normal" activity. The accurate measurement of such low activities offers certain difficulties since rather concentrated homogenates have to be used. These then interfere with the color formation in the glucose oxidase reaction (Asp *et al.*, 1967) so that too low values are found. To avoid these difficulties, Dahlqvist and Asp (1971) have developed a fluorimetric method based on another enzyme, galactose dehydrogenase (utilizes NAD^+ as coenzyme). This method is more sensitive, and concentrated homogenates do not interfere. (It should be noted that this method is not needed if the problem is only to differentiate between high or low lactase activity since then the glucose oxidase reaction works equally well.) This new method gave more accurate measurements of the residual value in lactase deficiency. Both methods register, however, the activity of the acid β-galactosidase in addition to the brush border lactase. Therefore, a new method was devised which permitted the separate assay of each of these two enzymes in the homogenate (Asp and Dahlqvist, 1972). With one additional substrate the hetero-β-galactosidase, which does not hydrolyze lactose, was also measured.

The results of measurements of each of the three intestinal β-galactosidases are seen in Fig. 5. It can be seen that there is a statistically significant difference between the populations in residual lactase activity. All biopsies were frozen in dry ice and transported to Lund for analysis; thus, this difference seems to be real, and not one resulting from varied techniques or standards in the laboratories. There was, however, detectable residual activity of brush border lactase in all the samples.

Whether there is also a residual enzyme activity in the *congenital* forms of disaccharidase deficiency is not yet known. Some authors have reported residual activity, others have not, but the sensitivity limits and specificity of the assay methods used have not been defined.

The congenital lactase deficiency was described by Holzel *et al.* (1959)

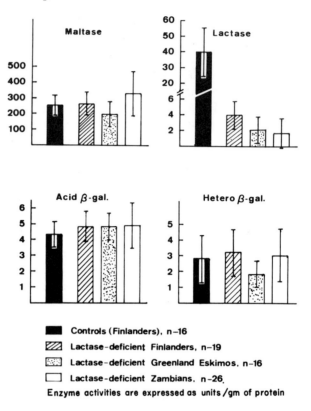

Fig. 5   Activity of maltase (as control), brush border lactase, acid β-galactosidase, and hetero-β-galactosidase in one high-lactase group of Finlanders (control group), one group of lactase-deficient Finlanders, one group of lactase-deficient Greenland Eskimos, and one group of lactase-deficient Zambian Africans. The frequencies of lactase deficiency among the adults, the populations from which the lactase-deficient patients are obtained, are: Finlanders, 15%; Greenland Eskimos, 90%; and Zambians, 100%.

and later described by a large number of other authors. The earlier report by Durand (1958), and the later distinction between lactase deficiency and the "severe lactose intolerance" with lactosuria have been discussed above. In patients with disaccharidase deficiency, none or only small amounts of disaccharides are excreted in the urine, but in severe lactose intolerance there is marked lactosuria.

Sucrase–isomaltase deficiency is in fact a congenital defect in which *two* enzymes are missing. Sucrose intolerance was first described by Weijers *et al.* (1960, 1961) and Prader *et al.* (1961). Patients were also shown to lack isomaltase (isomaltose represent the branching links in starch) (Auricchio *et al.*, 1963). Later, a large number of patients with sucrase deficiency were described, and in all of those tested for isomaltase activity, this enzyme was found to be absent also. Nevertheless, intestinal sucrase and isomaltase are two different enzymes, which has been shown both by heat inactivation (Dahlqvist, 1962a) and by substrate competition experiments (Dahlqvist *et al.*, 1963a).

Also structurally, these two enzymes are strongly interrelated. It has been shown that in solubilized homogenates the sucrase and isomaltase activities are bound to each other to some form of complex (Kolinska and Semenza, 1967), but the nature of the power that keeps them together is not known.

Congenital maltase deficiency will probably not occur, since in addition to the sucrase and the isomaltase (both of which also have maltase activity) there are two more maltases in the mucosa, and these will be powerful enough for the digestion of dietary maltose or starch. A patient with maltase deficiency described by Weijers *et al.* (1961) most probably had a secondary (general) disaccharidase deficiency.

Trehalase deficiency would be expected to occur since there is in the intestine a specific trehalase which is the only intestinal enzyme that can hydrolyze trehalose. The author once predicted that trehalase deficiency would probably never be found clinically since our food contains so little of this sugar. Trehalose is chiefly present in fungi and insects.

In spite of this, Bergoz (1972) recently demonstrated isolated trehalase deficiency in a woman, probably as a genetic defect. This woman was very fond of mushrooms and ate large quantities of them. Afterward she often had diarrhea. Apparently, trehalase deficiency only gives symptoms if the patient has somewhat unusual consumption habits according to our standards. In some other populations people often eat "manna" (the name "trehalose" came from *Trehala manna* where it was found) in large quantities. In some microorganisms which are potential protein sources for humans in the future there are also large amounts of trehalose. It is therefore not impossible that with changing food habits trehalase deficiency may become more important than it is at present.

The congenital forms of disaccharidase deficiency belong to the "inborn errors of metabolism" and as other such diseases they are very rare. There are no figures available for calculation of their frequency, but surely they are *not* more common than phenylketonuria or galactosemia which occur in one of 10,000–40,000 births.

Glucose–galactose malabsorption is not an enzyme defect but a carrier defect. The sodium-dependent glucose–galactose carrier in the small intestinal brush border, postulated by Crane (1960, 1967), is missing. We do not know the structure of this carrier, but it appears quite probable that it at least is partially a protein, like the enzymes. Glucose–galactose malabsorption was described by Lindquist *et al.* (1962, 1963) in Sweden, and at the same time by Laplane *et al.* (1962) in France. Later, Meeuwisse (1970) made a very careful study of this disease. More than 20 patients have now been described. Glucose and galactose are absorbed very slowly in this disease, and mucosal biopsies incubated *in vitro* are not able to accumulate these sugars to a concentration above that of the medium, which biopsies from normal subjects can do (Meeuwisse and Dahlqvist, 1968). The absorption of other substances such as amino acids is normal as is the absorption of fructose, which in these patients is absorbed more rapidly than glucose and galactose (see above).

This indicates that there is a specific transport mechanism for fructose in the intestine, and that concept also as received support from other experiments. It has been shown that fructose, in contrast to previously accepted belief, is transported actively (Gracey *et al.*, 1970). One would therefore except that in the future a "fructose malabsorption" will also be found, but as far as the author knows a patient with this problem has not yet been reported.

The general enzyme defects both in the pancreas and in the small intestine are secondary to some disease of other origin which severely disturbs all the functions of the affected tissue or tissues (Dahlqvist *et al.*, 1964; Arthur *et al.*, 1966). In these cases it is, of course, the primary disease that should be diagnosed and cured. An especially disturbing reaction is that severe protein deficiency seems to cause lactase deficiency (for review, see Dahlqvist and Lindquist, 1971). For the treatment of protein deficiency, skim milk powder is used since it contains proteins with high nutritional value. The large amount of lactose in milk often leads to diarrhea, however; and it has also been demonstrated that in the initial phase of treatment the nitrogen balance becomes negative (Graham and Paige, 1973). Work is going on in different laboratories to develop a suitable method to remove the lactose. Since these patients lack not only protein but also calories, a process which hydrolyzes lactose into monosaccharides is preferable to such processes which remove the sugar, and therefore the first methods have been concentrated on (Dahlqvist *et al.*, 1973).

F. The Greenland Eskimos

In a recent investigation, performed as a Danish–Swedish cooperation (Asp *et al.*, 1973), biopsies from a number of Greenland Eskimos were analyzed for disaccharidase activities. The Eskimos were regarded as an especially interesting group since they have for a very long time had a diet essentially without any carbohydrates. The disaccharidase activities in biopsies from 19 Eskimos are given in Table IV. As expected, nearly all are lactase-deficient. The 2 subjects who had high lactase activity were both closely related to a Dane, who had moved to Greenland and married an Eskimo. The mean maltase, isomaltase, sucrase, and trehalase activities were somewhat lower than in most other published materials, but if one looks more carefully at the table, one will find that the lower mean values for these enzymes are caused by a few subjects, in which they are extremely low, while the other subjects have more normal values. Thus, 3 of the 19 subjects were practically without sucrase or isomaltase activity. As in the earlier described infants with sucrase–isomaltase deficiency, these two enzymes thus follow each other. The

TABLE IV

Disaccharidase Activities in Jejunal Biopsies from Adult Greenland Eskimos[a,b]

Results of oral lactose tolerance test	Case no.	Disaccharidase activity (units/gm of protein)				
		Maltase	Isomaltase	Sucrase	Trehalase	Lactase (total)
Lactose	1	56	16	16	15	1.4
malabsorption	2	226	95	62	10	2.3
	3	145	76	61	13	1.7
	4	185	67	59	21	1.6
	5	334	105	82	8.7	3.0
	6	204	86	58	15	2.9
	7	43	2.2	0.05	14	1.9
	9	319	130	100	42	3.6
	10	92	1.5	0.6[c]	20	4.0
	11	132	41	29	17	1.5
	12	240	82	67	7.7	2.3
	13	154	55	46	2.7	8.2
	14	159	69	54	6.0	2.0
	16	56	1.4	0.02	28	4.3
	17	180	59	64	11	2.4
	18	192	67	56	9.1	2.1
	19	307	82	59	6.0	3.8
No lactose	8	290	105	75	1.5	21
malabsorption	15	236	69	60	14	21
Mean ($n = 19$)		187	64	50	14	(5)

[a] From Asp *et al.* (1973).

[b] Most of the biopsies with very low isomaltase, sucrase, or trehalase activity have been reassayed with a more sensitive method (hexokinase glucose 6-phosphate DH NADP; fluorimetric assay of the reduced coenzyme).

[c] Not analyzed with the more sensitive method.

somewhat higher value for residual isomaltase than for residual sucrase in these 3 subjects fits with our knowledge of the intestinal enzyme specificity (Dahlqvist and Telenius, 1969) as does the rather low maltase activity in these 3 subjects, since the remaining maltase activity is caused by the two heat-stable maltases.

Two of the nineteen Eskimos had extremely low trehalase activity, and three others had moderately low trehalase.

It thus looks as if the "inborn errors" with absence of one or another of the intestinal disaccharidases, which occur as very rare diseases elsewhere, are rather common findings in the Greenland Eskimos. It is easy to understand that these enzyme defects do not give any symptoms as long as no carbohydrates are consumed, but it is more difficult to explain why these mutants should have become so frequent among the Eskimos. This would imply that they have some advantage over those who are not disaccharidase-deficient (are "better fit for life").

G. Future Research

The demand for a method for preparation of lactose-free milk on a large scale has already been mentioned. One might also want some method for substituting the enzyme(s) which is(are) lacking in the other patients. Attempts have been made with concentrated enzyme preparations taken together with the meals. These experiments have not been very successful, however, probably mainly because the mixing conditions are not ideal, and the ionic conditions in the gastrointestinal tract are not quite optimal for the enzymes used. Therefore, large amounts of these enzymes must be fed, which makes the method wasteful and makes it necessary for the patient to take rather bulky enzyme preparations. One would therefore prefer a method that could make the intestine produce the missing enzyme. Rosensweig *et al.* (1971; Rosensweig, 1972) have successfully adapted a sucrase-deficient patient by feeding fructose. They found that the intestinal sucrase activity, which can be increased by feeding sucrose, also can be increased by fructose, but not by glucose. To the patient with sucrase deficiency they fed fructose for some time (fructose did not cause diarrhea, which sucrose would have done), and thereafter the patient was tolerant for at least a moderate amount of sucrose. It is possible that this was an especially favorable patient, however, since the residual sucrase activity present prior to fructose treatment was higher than is usually reported in patients of this kind.

Another method to make the intestine produce the missing enzyme could be to transplant a piece of normal intestine in the patient. We have a large reserve capacity in the small intestine, and such a transplantation would therefore not do any harm to the donor. This idea apparently is something for the surgeons to work with.

One may also attempt not to make a surgical transplantation but a gene transplantation, i.e., to make the patient's mucosal epithelial cells take up a normal gene or deoxyribonucleic acid, which can initiate the synthesis of a normal enzyme.

The author would like to have more convincing data concerning the development of lactase persistence in a few populations and an explanation for the high frequency of isolated disaccharidase deficiencies in the Greenland Eskimos. Finally, it would be interesting to get more detailed knowledge about the complex formation between the sucrase and the isomaltase and about the structure of the specific sugar carrier or carriers in the small intestine.

References

Andersson, H., Dotevall, G., Dahlqvist, A., and Isaksson, B. (1973). Dietary treatment of lactose intolerance in adults. *Swed. Nutr. Found. Symp.: Intestinal Enzyme Deficiencies and their Nutritional Implications, 1972.*

Arthur, A. B., Clayton, B. E., Cotton, D. G., Seakins, J. W. T., and Platt, J. W. (1966). Importance of disaccharide intolerance in the treatment of coeliac disease. *Lancet* **1,** 172–174.

Asp, N.-G. (1971). Small-intestinal β-galactosidases. Characterization of different enzymes and application to human lactase deficiency. Ph.D. Dissertation, University of Lund.

Asp, N.-G., and Dahlqvist, A. (1972). Human small-intestinal β-galactosidases. Specific assay of three different enzymes. *Anal. Biochem.* **47,** 527–538.

Asp, N.-G., Koldovský, O., and Hosková, J. (1967). Use of the glucoseoxidase method for assay of disaccharidase activities in the small intestine—a limitation. *Physiol. Bohemoslov.* **16,** 508–511.

Asp, N.-G., Dahlqvist, A., and Koldovský, O. (1969). Human small-intestinal β-galactosidases. *Biochem. J.* **114,** 351–359.

Asp, N.-G., Berg, N. O., Dahlqvist, A., Jussila, J., and Salmi, H. (1971). The activity of three different small-intestinal β-galactosidases in adults with and without lactase deficiency. *Scand. J. Gastroenterol.* **6,** 755–762.

Asp, N.-G., Cook, G. C., and Dahlqvist, A. (1973). Activities of brush border lactase, acid β-galactosidase, and hetero-β-galactosidase in the intestine of lactose intolerant Zambian Africans. *Gastroenterol.* **64,** 405–410.

Auricchio, S., Dahlqvist, A., Mürset, G., and Prader, A. (1963). Isomaltose intolerance causing decreased ability to utilize dietary starch. *J. Pediat.* **62,** 165–176.

Berg, N. O., Dahlqvist, A., Lindberg, T., and von Studnitz, W. (1969). Severe familial lactose intolerance—a gastrogen disorder? *Acta Paediat. Scand.* **58,** 525–527.

Berg, N. O., Dahlqvist, A., Lindberg, T., and Nordén, Å. (1970). Intestinal dipeptidases and disaccharidases in celiac disease in adults. *Gastroenterology* **59,** 575–582.

Bergoz, R. (1973). Intestinal trehalase deficiency. *Swedish Nutr. Found. Symp.: Intestinal Enzyme Deficiencies and their Nutritional Implications, 1973,* Almqvist–Wiksell, Uppsala.

Bolin, T. D., and Davis, A. E. (1970). Primary lactase deficiency: Genetic or acquired? *Amer. J. Dig. Dis.* **15,** 679–692.

Borgström, B., and Dahlqvist, A. (1958). Cellular localization, solubilization and separation of intestinal glycosidases. *Acta Chem. Scand.* **12,** 1997–2006.

Borgström, B., Dahlqvist, A., Lundh, G., and Sjövall, J. (1957). Studies of intestinal digestion and absorption in the human. *J. Clin. Invest.* **36,** 1521–1536.

Borgström, B., Dahlqvist, A., Dencker, I., Krabisch, L., Krasse, B., Lindstrand, K., Nordén, Å., Åkesson, B., and Övrum, P. (1970). Erfarenheter från en kostundersökning med dubbelportionsmetoden inom Dalby kommun under 1968–69. *Näringsforskning* **2,** 40–54.

Cook, G. C. (1973). Incidence and clinical features of specific hypolactasia in man. *Swed. Nutr. Found. Symp.: Intestinal Enzyme Deficiencies and their Nutritional Implications, 1973,* Almqvist–Wiksell, Uppsala.

Crane, R. K. (1960). Intestinal absorption of sugars. *Physiol. Rev.* **40,** 789–825.

Crane, R. K. (1967). Gradient coupling and the membrane transport of water-soluble compounds—a general mechanism? *Protides Biol. Fluids, Proc. Colloq.* **15,** 227–235.

Dahlqvist, A. (1958). Characterization of intestinal invertase as a glucosido-invertase. I. Action on different substrates. *Acta Chem. Scand.* **12,** 2012–2020.

Dahlqvist, A. (1959a). Studies on the heat inactivation on intestinal invertase, maltase and trehalase. *Acta Chem. Scand.* **13,** 945–953.

Dahlqvist, A. (1959b). The separation of intestinal invertase and three different intestinal maltases on TEAE-cellulose by gradient elution, frontal analysis and mutual displacement chromatography. *Acta Chem. Scand.* **13,** 1817–1827.

Dahlqvist, A. (1959c). Specificity of a purified hog intestinal maltase fraction. Competitive inhibition of maltase activity by other substrates. *Acta Chem. Scand.* **13,** 2156–2158.

Dahlqvist, A. (1960a). Hog intestinal α-glucosidases solubilization, separation and characterization. Ph.D. Dissertation, University of Lund.

Dahlqvist, A. (1960b). Characterization of three different hog intestinal maltases. *Acta Chem. Scand.* **14,** 1–8.

Dahlqvist, A. (1960c). Characterization of hog intestinal trehalase. *Acta Chem. Scand.* **14,** 9–16.

Dahlqvist, A. (1960d). Characterization of hog intestinal invertase as a glucosido-invertase. III. Specificity of purified invertase. *Acta Chem. Scand.* **14,** 63–71.

Dahlqvist, A. (1960e). Hog intestinal isomaltase activity. *Acta Chem. Scand.* **14,** 72–80.

Dahlqvist, A. (1961). Determination of maltase and isomaltase activities with a glucose-oxidase reagent. *Biochem. J.* **80,** 547–551.

Dahlqvist, A. (1962a). Specificity of the human small-intestinal disaccharidases and implications for hereditary disaccharide intolerance. *J. Clin. Invest.* **41,** 463–470.

Dahlqvist, A. (1962b). The intestinal disaccharidases and disaccharide intolerance. (Editorial.) *Gastroenterology* **43,** 694–696.

Dahlqvist, A. (1962c). A method for the determination of amylase in intestinal content. *Scand. J. Clin. Lab. Invest.* **14,** 145–151.

Dahlqvist, A. (1963). Specificity of the human intestinal disaccharidases and implications for hereditary disaccharide intolerance. *In* "Malabsorption Syndromes" (H. Schön, ed.), pp. 47–51. Karger, Basel.

Dahlqvist, A. (1964). Method for assay of intestinal disaccharidases. *Anal. Biochem.* **7**, 18–25.

Dahlqvist, A. (1967). Localization of the small-intestinal disaccharidases. *Amer. J. Clin. Nutr.* **20**, 81–88.

Dahlqvist, A. (1968). Assay of intestinal disaccharidases. *Anal. Biochem.* **22**, 99–107.

Dahlqvist, A. (1970). Assay of intestinal disaccharidases. *Enzymol. Biol. Clin.* **11**, 52–66.

Dahlqvist, A. (1972). Enzymdefekter i tunntarmen—betydelse för nutritionen. *Näeringsforskning* **16**, 153–164 (with English summary).

Dahlqvist, A., and Asp, N.-G. 1971. Accurate assay of low levels of lactase in the intestinal mucosa with a fluorimetric method. *Anal. Biochem.* **44**, 654–657.

Dahlqvist, A., and Borgström, B. (1959). Characterization of intestinal invertase as a glucosido-invertase. II. Studies on transglycosylation by intestinal invertase. *Acta Chem. Scand.* **13**, 1659–1667.

Dahlqvist, A., and Borgström, B. (1961). Digestion and absorption of disaccharides in man. *Biochem. J.* **81**, 411–418.

Dahlqvist, A., and Brun, A. (1962). A method for the histochemical demonstration of disaccharidase activities: Application to invertase and trehalase in some animal tissues. *J. Histochem. Cytochem.* **10**, 294–302.

Dahlqvist, A., and Lindquist, B. (1971). Lactose intolerance and protein malnutrition. *Acta Paediat. Scand.* **60**, 488–494.

Dahlqvist, A., and Nordström, C. (1966). The distribution of disaccharidase activities in the villi and crypts of the small-intestinal mucosa. *Biochim. Biophys. Acta* **113**, 624–626.

Dahlqvist, A., and Telenius, U. (1969). Column chromatography of human small-intestinal maltase, isomaltase and invertase activities. *Biochem. J.* **111**, 139–146.

Dahlqvist, A., and Thomson, D. (1963). Separation and characterization of two rat-intestinal amylases. *Biochem. J.* **89**, 272–277.

Dahlqvist, A., Auricchio, S., Semenza, G., and Prader, A. (1963a). Human intestinal disaccharidases and hereditary disaccharide intolerance. The hydrolysis of sucrose, isomaltose, palatinose (isomaltulose), and a 1,6-α-oligosaccharide (isomalto-oligosaccharide) preparation. *J. Clin. Invest.* **42**, 556–562.

Dahlqvist, A., Hammond, J. B., Crane, R. K., Dunphy, J. V., and Littman, A. (1963b). Intestinal lactase deficiency and lactose intolerance in adults: Preliminary report. *Gastroenterology* **45**, 488–491.

Dahlqvist, A., Hammond, J. B., Crane, R. K., Dunphy, J. V., and Littman, A. (1964). Assay of disaccharidase activities in peroral biopsies of the small-intestinal mucosa. *Acta Gastro-Enterol. Belg.* **27**, 543–555.

Dahlqvist, A., Lindquist, B., and Meeuwisse, G. (1968). Disturbances of the digestion and absorption of carbohydrates. *In* "Carbohydrate Metabolism and its Disorders" (F. Dickens, P. J. Randle, and W. J. Whelan, eds.), Vol. 2, pp. 199–222. Academic Press, New York.

Dahlqvist, A., Borgström, B., and Hambraeus, L. (1973). "Intestinal Enzyme Deficiencies and the Nutritional Implication." Almqvist-Weksell, Uppsala.

Dahlqvist, A., Mattiasson, B., and Mosbach, K. (1973). Enzymic hydrolysis of lactose in milk using polymer-entrapped lactase. *Biotechnol. Bioeingen.* **15**, 395–402.

Dunphy, J. V., Littman, A., Hammond, J. B., Forstner, G., Dahlqvist, A., and

Crane, R. K. (1965). Intestinal lactase deficit in adults. *Gastroenterology* **49**, 12–21.

Durand, P. (1958). Lactosuria idiopatica in una paziente con diarrea cronica ed acidosi. *Minerva Pediat.* **10**, 706–711.

Durand, P., Martino, A. M., and Lamedica, G. M. (1962). Diagnosis of carbohydrate intolerance by stool chromatography. *Lancet* **2**, 374–375.

Ford, J. D., and Haworth, J. C. (1963). The fecal excretion of sugars in children. *J. Pediat.* **63**, 988–990.

Fortin-Magana, R., Huwitz, R., Herbst, J. J., and Kretchmer, N. (1970). Intestinal enzymes: Indicators of proliferation and differentiation in the jejunum. *Science* **167**, 1627–1628.

Gracey, M., Burke, U., and Oshin, A. (1970). Intestinal transport of fructose. *Lancet* **2**, 827–828.

Graham, G. G., and Paige, D. M. (1973). The nutritional implications of low intestinal lactase activity in children. *Swed. Nutr. Found. Symp.: Intestinal Enzyme Deficiencies and their Nutritional Implications, 1973*, Almqvist–Wiksell, Uppsala.

Gray, G. M., and Santiago, N. A. (1969). Intestinal β-galactosidases. I. Separation and characterization of three enzymes in normal human intestine. *J. Clin. Invest.* **48**, 716–728.

Gudmand-Høyer, E., Dahlqvist, A., and Jarnum, S. (1969). Specific small-intestinal lactase deficiency in adults. A report of 18 patients. *Scand. J. Gastroenterol.* **4**, 377–386.

Gudmand-Høyer, E., Dahlqvist, A., and Jarnum, S. (1970). The clinical significance of lactose malabsorption. *Amer. J. Gastroenterol.* **53**, 460–473.

Haemmerli, U. P., Kistler, H. J., Ammann, R., Auricchio, S., and Prader, A. (1963). Lactase-Mangel der Dünndarmmucosa als Ursache gewisser Formen erworbener Milchintoleranz beim Erwachsenen. *Helv. Med. Acta* **30**, 693–705.

Holzel, A., Schwarz, V., and Sutcliffe, K. W. (1959). Defective lactose absorption causing malnutrition in infancy. *Lancet* **1**, 1126–1128.

Holzel, A., Mereu, T., and Thomson, M. L. (1962). Severe lactose intolerance in infancy. *Lancet* **2**, 1346–1348.

Jos, J., Frézal, J., Rey, J., and Lamy, M. (1967). Histochemical localization of intestinal disaccharidases: Application to peroral biopsy specimens. *Nature (London)* **213**, 516–518.

Kern, F., Struthers, J. E., and Attwood, W. L. (1963). Lactose intolerance as a cause of steatorrhea in an adult. *Gastroenterology* **45**, 477–480.

Kerry, K. R., and Anderson, C. M. (1964). A ward test for sugar in faeces. *Lancet* **1**, 981–982.

Keusch, G. T., Troncale, F. J., Thavaramara, B., Prinyanont, P., Anderson, P. R., and Bhamarapravathi, N. (1969). Lactase deficiency in Thailand: Effect of prolonged lactose feeding. *Amer. J. Clin. Nutr.* **22**, 638–640.

Kolinska, J., and Semenza, G. (1967). Studies on intestinal sucrase and on intestinal sugar transport. V. Isolation and properties of sucrase-isomaltase from rabbit small intestine. *Biochim. Biophys. Acta* **146**, 181–195.

Laplane, R., Polonovski, C., Etienne, M., Debray, P., Lods, J.-C., and Pissarro, B. (1962). L'intolérance aux sucres a transfert intestinal actif. Les rapports avec l'intolérance au lactose et le syndrome coeliaque. *Arch. Fr. Pediat.* **19**, 895–944.

Lindquist, B., and Meeuwisse, G. (1963). Intestinal transport of monosaccharides in generalized and selective malabsorption. *Acta Paediat. Scand., Suppl.* **146**, 110–115.

Lindquist, B., Meeuwisse, G., and Melin, K. (1962). Glucose-galactose malabsorption. *Lancet* **2,** 666.

Lindquist, B., Meeuwisse, G., and Melin, K. (1963). Osmotic diarrhoea in genetically transmitted glucose-galactose malabsorption. *Acta Paediat. (Stockholm)* **52,** 217–218.

Littman, A., and Hammond, J. B. (1965). Diarrhoea in adults caused by deficiency in intestinal disaccharidases. *Gastroenterology* **48,** 237–249.

Lojda, Z. (1965). Some remarks concerning the histochemical detection of disaccharidases and glucosidases. *Histochemie* **5,** 339–360.

McCracken, R. D. (1970). Lactase deficiency: An example of dietary evolution. *Curr. Anthropol.* (Unpublished manuscript).

Meeuwisse, G. (1970). Glucose-galactose malabsorption. An inborn error of carrier-mediated transport, Ph.D. Dissertation, University of Lund.

Meeuwisse, G., and Dahlqvist, A. (1968). Glucose-galactose malabsorption. A study with biopsy of the small-intestinal mucosa. *Acta Paediat. Scand.* **57,** 273–280.

Messer, M., and Dahlqvist, A. (1966). A one-step ultramicro method for the assay of intestinal disaccharidases. *Anal. Biochem.* **14,** 376–392.

Meyer, K. H., Fischer, E. H., Bernfeld, P., and Duckert, F. (1948). Purification and crystallization of human pancreatic amylase. *Arch. Biochem.* **18,** 203–205.

Miller, D., and Crane, R. K. (1961a). A procedure for the isolation of the epithelial brush border membrane of hamster small intestine. *Anal. Biochem.* **2,** 284–286.

Miller, D., and Crane, R. K. (1961b). The digestive function of the epithelium of the small intestine. I. An intracellular locus of disaccharide and sugar phosphate ester hydrolysis. *Biochim. Biophys. Acta* **52,** 281–292.

Miller, D., and Crane, R. K. (1961c). The digestive function of the epithelium of the small intestine. II. Localization of disaccharide hydrolysis in the isolated brush border portion of intestinal epithelial cells. *Biochim. Biophys. Acta* **52,** 293–298.

Nordström, C., Dahlqvist, A., and Josefsson, L. (1968). Quantitative determination of enzymes in different parts of the villi and crypts of rat small intestine. Comparison of alkaline phosphatase, disaccharidases and dipeptidases. *J. Histochem. Cytochem.* **15,** 713–721.

Nugent, F. W., and Millhon, W. A. (1958). Clinical evaluation of starch tolerance test. *J. Amer. Med. Ass.* **168,** 2260–2262.

Prader, A., Auricchio, S., and Mürset, G. (1961). Durchfall infolge hereditären Mangels an intestinaler Saccharaseaktivität (Saccharose-Intoleranz). *Schweiz. Med. Wochenschr.* **91,** 465–476.

Rosensweig, N. S. (1973). The influence of dietary carbohydrates on intestinal disaccharidase activity in man. *Swed. Nutr. Found. Symp.: Intestinal Enzyme Deficiencies and their Nutritional Implications, 1973,* Almqvist–Wiksell, Uppsala.

Rosensweig, N. S., and Herman, R. H. (1969). Diet and disaccharidases. *Amer. J. Clin. Nutr.* **22,** 99–102.

Rosensweig, N. S., Herman, R. H., and Stifel, F. B. (1971). Dietary regulation of small-intestinal enzyme activity in man. *Amer. J. Clin. Nutr.* **24,** 65–69.

Simoons, F. J. (1969). Primary adult lactose intolerance and the milking habit: A problem in biological and cultural interrelations. I. Review of the medical research. *Amer. J. Dig. Dis.* **14,** 819–836.

Simoons, F. J. (1970). Primary adult lactose intolerance and the milking habit. II. A culture historical hypothesis. *Amer. J. Dig. Dis.* **15,** 695–710.

Sols, A., and de la Fuente, G. (1961). Hexokinase and other enzymes of sugar

214 *Arne Dahlqvist*

metabolism in the intestine. *In* "Methods in Biochemical Research" (J. Quartel, ed.) pp. 302–309. Yearbook Publ., Chicago, Illinois.

Swaminathan, N., Mathan, V. I., Baker, S. J., and Radakrishnan, A. N. (1970). Disaccharidase levels in jejunal biopsy specimens from American and South Indian control subjects and patients with tropical sprue. *Clin. Chim. Acta* **30**, 707–712.

Ugolev, A. M. (1965). Membrane (contact) digestion. *Physiol. Rev.* **45**, 555–595.

van de Kamer, J. H., ten Bokkel Huinink, H., and Weijers, H. A. (1949). Rapid method for the determination of fat in feces. *J. Biol. Chem.* **177**, 347–355.

Weijers, H. A., van de Kamer, J. H., Mossel, D. A. A., and Dicke, W. K. (1960). Diarrhoea caused by deficiency of sugar-splitting enzymes. *Lancet* **2**, 296–297.

Weijers, H. A., van de Kamer, J. H., Dicke, W. K., and Ijsseling, J. (1961). Diarrhoea caused by deficiency of sugar-splitting enzymes. I. *Acta Paediat.* **50**, 55–71.

White, J. W., Jr., and Subers, M. H. (1961). A glucose oxidase reagent for maltase assay. *Anal. Biochem.* **2**, 380–384.

Wilson, T. H. (1962). "Intestinal Absorption." Saunders, Philadelphia, Pennsylvania.

CHAPTER 14

Effect of Carbohydrates on Intestinal Flora

PAUL GYÖRGY

A. Symbiosis between Animals and Man

Man and animals live in symbiosis with microbes; in particular, the intestinal flora may act as a modifying environmental factor and as such may influence growth, development, and the metabolic processes of the host organism. The question whether its presence is harmful, beneficial, or even indispensable for live processes of the higher organism harboring it has been raised as early as 1885 by Pasteur. The distinction between saccharolytic and bacterial action in the intestine, as first brought into focus by Pasteur, has been applied by clinicians for pathological conditions and by Metchnikoff (1903) even to the problem of longevity. According to Metchnikoff, saccharolytic acid producing intestinal flora, such as composed by lactobacilli, by virtue of its suppressive action on proteolytic, putrefying microbes, should promote longer life. In contrast, putrefaction was considered to be the cause of many ills and the source of "intestinal autointoxication." Bouchard (1894) reasoned that if the intestinal material is toxic and under usual circumstances no toxemia occurs, the kidney must eliminate toxic products. Other authors introduced the concept of neutralization, detoxification of harmful substances originating in the intestine as products of bacterial putrefaction. In retrospect, the sweeping conclusions of the early proponents of intestinal autointoxication with their generalization of very few, poorly documented biochemical findings, were certainly devoid of convincing experimental and clinical scientific proof (György, 1971).

A more scientific cornerstone in the edifice of symbiosis between host organism and intestinal flora was laid by the systematic bacteriological study of the intestinal microflora by Escherich (1886), followed by Tissier (1900), Moro (1900, 1905), and others. It was Escherich and his followers who first established the fact that the composition of the intestinal flora is primarily influenced by the food ingested and, in final analysis, by the milieu in the intestinal lumen, acting as culture medium, with selective capacity for bacterial inhabitants of the intestine. Theoretically, it is impossible to produce a particular intestinal flora by the ingestion of the microorganism even in large amount, yet the same task may be easily achieved by establishing the required favorable cultural milieu. In consequence, the effort to establish acidophilus flora in the intestine, as often advocated, solely by the administration of a culture of *Lactobacillus acidophilus*, even after frequently repeated ingestion, must be doomed to failure. This applies also to inclusion in the diet of yogurt, sour milk treated with *Lactobacillus bulgaricus* and recommended by Metchnikoff for prolongation of life. Yogurt is only part of the diet consumed per day and, in consequence, the intestinal milieu will support the growth not only of *L. bulgaricus* but of other bacteria as well.

B. Differences between Human and Cow's Milk

The most impressive illustration between food and intestinal flora is offered by infants fed either human milk or cow's milk formulas. In contrast to the acid reaction of the feces of normal breast-fed infants, the pH of the feces of those infants given the usual cow's milk formulas fall in the neutral or alkaline range. Unlike the mixed intestinal flora of infants on cow's milk formulas, the healthy breast-fed infants are characterized by the prevalence of a particular subspecies of *Lactobacillus*, i.e., *bifidus*. According to the recent nomenclature it is called *Bifidobacterium bifidum* (Mata and Wyatt, 1971). The old name *"Lactobacillus"* is preferable since it emphasizes the relationship to lactose in milk and the production in large amounts of lactic and acetic acid and in trace amounts of formic and succinic acid, determining the acid reaction under experimental and dietary conditions.

TABLE I

Differences in Human and Cow's Milk

Content	Human milk (%)	Cow's milk (%)
Protein	1.1	3.5
Lactose	6.6	4.0
Fat	3.5	3.5
Ash	0.2	0.7

The difference of the intestinal flora of breast-fed infants compared with bottle fed (cow's milk formulas) infants was originally related by pediatricians mainly to the differences in protein content: about 3.5% in cow's milk and 1% in human milk (Table I). Attempts with the use of dilute cow's milk formulas with added carbohydrate and more elaborate adaptations to human milk have been only moderately successful, especially with regard to the intestinal flora.

A gram-stained fecal smear obtained from healthy breast-fed infants appear to be almost uniform, as if it would represent a pure culture of gram-positive rods characteristic for *L. bifidus*. On the average, the proportion of gram-positive rods is around 98% with only slight variation in normal breast-fed infants.

Lactobacillus bifidus is a gram-positive straight or curved rod and nonmotile. One end may be bulbous or ricket shaped. One or both ends may appear to be split longitudinally to give the effect of two short branches. This appearance led to the term "bifid." The bifid character usually becomes more prominent in cultures (Fig. 1) than in fecal smears. In cultures and sometimes even in fecal smears, *L. bifidus* may turn gram-negative, either in spots, or throughout the whole length of the bacterium.

Lactobacillus bifidus is anaerobic, or at least micro-aerophilic, and easily mutates into aerobic straight rods called *L. parabifidus*. On agar plate bifid colonies appear as opaque, glistening buff in color and often mucoid, whereas the unbranched *L. parabifidus* straight rods form small translucent colonies. The taxonomic position of *L. parabifidus* as a mutant of *L. bifidus* to *L. acidophilus* has not yet been clearly determined. Studies of metabolic products in cultures of *L. bifidus* furnished no evidence of the presence of a specific antibiotic which would act in the intestine against possible contaminants of the *L. bifidus* flora.

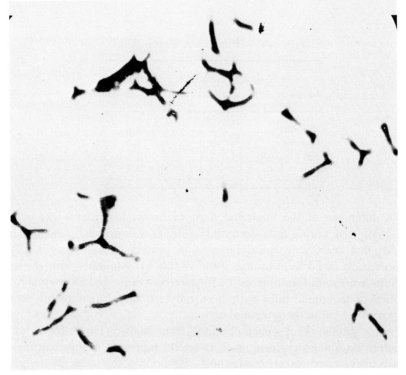

Fig. 1 Lactobacillus bifidus culture.

C. Bifidus Growth Factor

In the course of further bacteriological studies a specific variant of *L. bifidus* was found which on primary isolation on plates containing the usual medium had shown only very scant growth. Upon further propagation on plates or in liquid culture this variant occurred only after addition of human milk to the medium (Fig. 2). This variant was and is still called *L. bifidus* var. *Pennsylvanicus*. In contrast to human milk, cow's milk was practically devoid of this property indicating a ratio of activity (on the average) of 1:40 for cow's milk to human milk. Thus with regard to this microbiological growth factor called "bifidus factor" (György et al., 1952; György, 1953) the range of activity of mature cow's milk is in a definitely lower order of magnitude than that of mature human milk (Table II). In further studies the activity of human milk was compared with that of the milk of other species (Table III) (György et al., 1952).

The average activity was highest for human colostrum, closely followed by rat colostrum, followed by mature human milk and then cow's colos-

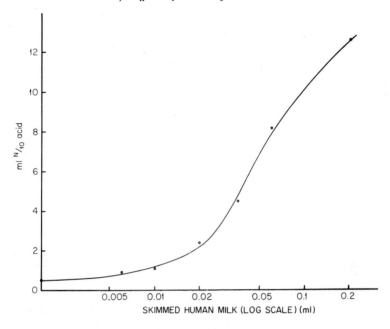

Fig. 2 Microbiological growth effect of human milk.

trum. The milk of such ruminants as cow, sheep, and goat has shown only very slight activity, less than mare's and sow's milk. Observations on the distribution of the bifidus growth factor in man are contained in Table IV (György, 1953).

The high activity of meconium, semen, gastric juice, Pseudomucinous ovarian cyst fluid, amniotic fluid and tears, in addition to that of colostrum, appears to be rather similar to the distribution of blood group poly-

TABLE II

Bifidus Growth Factor in Human and Cow's Milk

	Supplement (ml/10 ml of medium)	Acid production in 40 hr (ml of 0.1 N)
None	—	0.4
Human milk, skimmed	0.02	2.2
	0.06	7.9
	0.2	12.5
Cow's milk, skimmed	0.1	0.6
	0.3	0.8
	1.0	2.3

TABLE III

Activity of Human Milk as Compared with Milk from Other Species

Species	Activity	One unit (ml)	Relative activity
Guinea pig	0	—	—
Cow	(+)	2.5	2.5
Sheep	(+)	2.5	2.5
Goat	(+)	2.5	2.5
Mare	+	0.5	12
Sow	+	0.4	15
Cow colostrum	++	0.15	40
Rat	++	0.13	45
Rat colostrum	++	0.03	200
Human	++	0.06 (0.02–0.15)	100
Human colostrum	++	0.02 (0.01–0.03)	300

saccharides in human secretions. During the purification process of blood group polysaccharides from meconium concentrates were simultaneously obtained with high blood group and microbiological activity. It is tempting to speculate that meconium may play a part in the first propagation of *L. bifidus* in the intestinal content of the newborn infant. The first inoculum of the sterile intestine of the newborn infant may originate from the vagina of the mother, with its rich bifidus flora at term (Harrison *et al.*, 1953).

TABLE IV

Distribution of Bifidus Factor in Man

Factor	Unit (ml)
Colostrum	0.01–0.03 (average 0.02)
Milk	0.02–0.15 (average 0.06)
Gall bladder mucus[a]	0.01
Pseudomucinous ovarian cyst fluid[a]	0.02
Cervix uteri mucus[a]	0.02
Meconium[a]	0.02
Bronchial mucus[a]	0.02
Colon mucus[a]	0.03
Semen	0.07
Saliva	0.03–3.0 (average 0.5)
Gastric juice	0.45
Tears	1

[a] Ten percent dry weight.

Chemically the bifidus factor belongs in the group of N-containing carbohydrates. In human milk the presence of a great variety of oligo- and polysaccharides has been demonstrated (Gauhe *et al.*, 1954). Their total quantity in fresh human milk may be estimated to be around 0.4% which is by no means a negligible amount with respect to total solids. The bifidus factor in human milk is present in low molecular dialyzable and high molecular nondialyzable form (György *et al.*, 1954a). The dialyzable fraction in milk was found to be between 40 and 75%, the nondialyzable 25 and 60%. In contrast, the bifidus active blood group is completely nondialyzable.

The active bifidus factor contains N-acetylhexosamine, chiefly glucosamine (in smaller amount galactosamine), glucose, and galactose (lactose). Lacto-N-fructopentose is the largest single compound of the active bifidus factor (Fig. 3), in addition to smaller active and inactive di- and

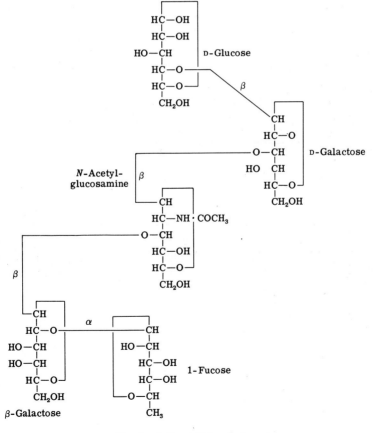

Fig. 3 Active bifidus factor.

trisaccharides, even the alkyl derivatives of N-acetylglucosamine. Unexpectedly, very high activity was encountered with β-alkyl compounds, in particular with the ethyl and n-propyl homologues. The activity of β-methyl-N-acetyl-β-D-glucosaminide was greatly enhanced (György and Rose, 1955) in the pure β-ethyl and β-propyl-N-acetylglucosaminide. The chemical composition of the bifidus factor is very similar to that of the blood group substance, with the exception of glucose (lactose) which is missing in the blood group polysaccharide. In spite of this difference, blood group mucoids act as the bifidus factor and promote the growth of *L. bacillus* var. *Pennsylvanicus* (Springer and György, 1953).

An enzyme present in a cell-free extract from *L. bifidus* var. *Pennsylvanicus* inactivates the various forms of bifidus factor in human milk, also in the blood group, with release of N-acetyl-D-hexosamine and the other constituent monosaccharides (György *et al.*, 1954b).

D. Sialids

In addition to the neutral oligo- and polysaccharides found in the bifidus factor, human milk also contains acidic oligo- and polysaccharides which occur again in dialyzable low molecular form and also in non-dialyzable high molecular form (György, 1958). The acidic constitutent of these N-containing saccharides has been identified as acetylneuraminic (sialic) acid (Hoover *et al.*, 1954; Zilliken *et al.*, 1955). The content of sialic acid in human milk compared with cow's milk is about 40:1, thus in the same proportion as for the neutral bifidus factor. Sialic acid, when present as an additional constituent of the bifidus factor, is most commonly chemically bound at the open end of the carbohydrate chain (Nicolai, 1971). (The acidic oligosaccharides are often found on the surface of erythrocytes and thrombocytes.) Combined with polypeptide chains they are called glycoproteins and become nondialyzable. With the sialic acid in end position, the compounds are inactive as bifidus factor. After treatment with the enzyme neuraminidase the sialic end "breaks off" and the newly restored bifidus factor chain shows the usual activity as growth stimulus for *L. bifidus* var. *Pennsylvanicus*, combined with the production of lactic, acetic, formic, and other acids (Nicolai, 1971).

Sialic acid was long recognized as receptor of the influenza virus group. As a reliable test for the reaction of receptor and virus, the inhibition of hemagglutination is used *in vitro* (Hirst, 1941). This test was carried out in our laboratory with positive results using human milk, and the fraction of acidic glycoproteins as receptors with interaction of active Lee virus or active PR8 virus or vibrio cholerae filtrate. Thus, the presence of normal hemagglutinin inhibitor in human milk has been demonstrated (Silver *et al.,* 1956).

In the light of all these observations on the effect of carbohydrates on intestinal flora, differentiation should be made between three groups of carbohydrates: (a) lactose, (b) bifidus factor, and (c) glycoproteins (sialids). With negligible amounts of factors b and c in cow's milk, the latter could serve as comparison for the effect of lactose alone, whereas in human milk all three carbohydrates are well represented and their effects may differ from that of lactose alone.

E. Effects of Fecal pH

In special studies (Barbero *et al.,* 1952) the pH and relative occurrence of gram-positive bacilli have been determined in the stools of 236 infants fed exclusively either human or cow's milk. The latter was given in the form of the usual formulas of evaporated milk or as commercial brands of "adapted" milk. Such commerical formulas are at present given to the great majority of bottle-fed infants (80%). The valid scientific foundation for its use is the effort to make cow's milk chemically closer (better adapted) to human milk.

The average fecal pH of the breast-fed infants was 5.5. In infants fed evaporated cow's milk modified with maltose-dextrin type of sugar, the fecal pH was 6.7. The flora of breast-fed infant's stool was comprised of over 90% gram-positive bacilli. The stool of the infant on cow's milk was variable in its bacterial flora. The feeding of brands of adapted milk failed to produce stools with an acidity and bacterial flora similar to those found in the stools of breast-fed infants, although with few brands some minor approximation was noted. These changes were related since the end of the nineteenth century to the proximate composition of breast and cow's milk (Table I).

In special investigations it has been shown that the specific bifidus factor of human milk materially enhances the propagation of *L. bifidus* in the intestinal flora of the normal human infant. This effect is observed after feeding of human milk as well as after that of purified fractions of the bifidus factor added to a cow's milk formula, which in its protein and lactose ratio closely resembles human milk. It is of special interest that from this *L. bifidus* flora the particular strain of *L. bifidus* var. *Pennsylvanicus* may easily be recovered, whereas it is absent or very rarely found in the feces of infants fed the usual or adapted cow's milk formulas without the addition of the specific bifidus factor (György, 1957a,b). It is further noteworthy that *L. bifidus* var. *Pennsylvanicus* probably through accelerated initial growth stimulation via the bifidus factor will inhibit the growth of *Escherichia coli* and that of regular strains of *L. bifidus*. The low fecal pH in general suppresses the growth of coliform, shigella, and other proteolytic, putrefying, harmful intestinal bacteria. This observation applies not only to infants but also to children and adults (Hentges and Freter, 1962; Maier *et al.*, 1973).

F. Acrodermatitis Enteropathica

Skepticism against the overestimation of the difference in the intestinal flora between breast-fed and artificially fed infants is also countered by reference to the very interesting, but not well-known, disease entity named "acrodermatitis enteropathica." The disease is characterized by vesicular dermatitis around the body orifices and the distal parts of the extremities with multiple paronychia on hands and feet, diarrheal attacks, and other digestive dysfunction.

The disease usually appears after weaning and without treatment is usually fatal. Brandt, in 1936, stated: "Among all therapeutic experiments that have been instituted, the treatment with mothers' milk is the only one that has any demonstrable effect." After addition of mother's milk to the diet there has been an increase of the baby's weight from a previously almost arrested growth, and, in particular, the general condition has shown a distinct improvement.

The nature of this intriguing condition is unknown. It is probably based on an inborn metabolic error with its original site in the intestinal flora, and in further consequence, on the ability of the organism to detoxify bacterial products absorbed from the lumen of the intestine. The claim that the intestinal antiseptic, diodoquin, has proved to be an effective remedy in the treatment of acrodermatitis enteropathica (Dillaha and Lorenc, 1953) appears to be in accord with the above hypothesis. More recently, a defective interconversion of unsaturated acids, in particular the synthesis of the unsaturated fatty acids, has been considered a basis of the metabolic disorder (Cash and Berger, 1969) which is neutralized by a human milk diet.

G. Specific Role of *L. Bifidus flora*

The specific role of the *L. bifidus* flora, and in particular that of the oligo-polysaccharide bifidus factor, is still shrouded in mystery.

It is easier to characterize the function of the glycoprotein sialids as receptors in the prevention of viral infections, especially concerning the influenza group, but also the virus of polymyelitis and in general those containing neuraminidase.

Human milk contains large quantities of a glycoprotein, lactofcrrin with iron binding capacity. Concentrates with not fully iron-saturated capacity had a powerful bacteriostatic effect on *E. coli* 0111/B4. The bacteriostatic property of the milk is abolished if the iron binding proteins are saturated with iron. Resistance to *E. coli* has been demonstrated in suckled guinea pigs. The exact chemical nature of the active free milk-lactoferrin is not yet known. Its presence in late cow's milk is not yet documented. On the other hand, the practical applications on *E. coli in vitro* and *in vivo* (guinea pigs) make it probable that the iron binding proteins in human milk may play an important part in resistance to infantile enteritis caused by *E. coli* (Bullen *et al.*, 1972).

In conclusion, the positive effect of the carbohydrates of human milk, some of which are practically missing in cow's milk, is a further impressive example for the validity of the teleological statement: "Human milk is for the human infant, cow's milk for the calf." There cannot be any doubt about it (cf. György, 1971).

References

Barbero, G. J., Runge, G., Fisher, D., Crawford, M. N., Torres, F. E., and György, P. (1952). Investigations on the bacterial flora, pH and sugar content in the intestinal tract of infants *J. Pediat.* **40,** 152–163.

Bouchard, C. J. (1894). "Lectures on Auto-intoxication in Disease or Self-poisoning of the Individual," p. 302. Davis, Philadelphia, Pennsylvania.

Brandt, T. (1936). Dermatitis in children with disturbances of general conditions and absorption of food elements. *Acta Dermato-Venereol.* **17,** 513.

Bullen, J. J., Rogers, H. J., and Leigh, L. (1972). Iron-binding proteins in milk and resistance to *Escherichia coli* infection in infants. *Brit. Med J.* **1,** 69–75.

Cash, R., and Berger, C. K. (1969). Acrodermatitis enteropathica: Defective metabolism of unsaturated fatty acids. *J. Pediat.* **74,** 717.

Dillaha, C. J., and Lorenc, A. T. (1953). Enteropathic acrodermatitis (Danbolt). Successful treatment with diodoquin (diodohydroxyquinidine). *AMA Arch. Dermatol. Syphilol.* **67,** 324.

Escherich, T. (1886). "Die Darmbakterien des Säuglings und ihre Beziehungen zur Physiologie der Verdauung," p. 186. Enke, Stuttgart.

Gauhe, A., György, P., Hoover, J. R. E., Kuhn, R., Rose, C. S., Ruelius, H. W., and Zilliken F. (1954). Bifidus factor. IV. Preparations obtained from human milk. *Arch. Biochem. Biophys.* **48,** 214–224.

György, P. (1953). A hitherto unrecognized biochemical difference between human milk and cow's milk. *Pediatrics* **11,** 98–100.

György, P. (1957a). Nutrition and intestinal flora in man. *Ann. Nutr. Aliment.* **11,** A189–A203.

György, P. (1957b). Development of intestinal flora in the breast-fed infant. *Mod. Probl. Pediat.* **2,** 1–12.

György, P. (1958). *N*-Containing saccharides in human milk. *Chem. Biol. Mucopolysaccharides, Ciba Found. Symp., 1957* pp. 140–154.

György, P. (1971). Biochemical aspects of human milk. *Amer. J. Clin. Nutr.* **24,** 970–975.

György, P., and Rose, C. S. (1955). Microbiological studies on the growth factor for L. *bifidus* var. *Pennsylvanicus. Proc. Soc. Exp. Biol. Med.* **90,** 219–223.

György, P., Kuhn, R., Norris, R. F., Rose, C. S., and Zilliken, F. (1952). A hitherto unrecognized biochemical difference between human milk and cow's milk. *Amer. J. Dis. Child.* **84,** 482.

György, P., Hoover, J. R. E., Kuhn, R., and Rose, C. S. (1954a). Bifidus factor. III. The rate of dialysis. *Arch. Biochem. Biophys.* **48,** 209–213.

György, P., Rose, C. S., and Springer, G. F. (1954b). Enzymatic inactivation of bifidus factor and blood group substances. *J. Lab. Clin. Med.* **43,** 543–552.

Harrison, W., Stahl, R. C., Magavran, J., Sanders, M., Norris, R. F., and György, P. (1953). The incidence of *Lactobacillus bifidus* in vaginal secretions of pregnant and non-pregnant women. *Amer. J. Obstet. Gyecol.* **65,** 352–357.

Hentges, D. J., and Freter, R. (1962). *In vivo* and *in vitro* antagonism of intestinal bacteria against *Shigella flexneri*. I. Correlation between various tests. *J. Infec. Dis.* **110,** 30–37.

Hirst, G. K. (1941). The agglutination of red cells of allantoic fluid of chick embryos infected with influenza virus. *Science* **94,** 22–23.

Hoover, J. R. E., Braun, G. A., and György, P. (1954). Neuraminic acid in mucopolysaccharides of human milk. *Arch. Biochem. Biophys.* **47,** 216–217.

Maier, B. R., Onderdonk, A. B., Baskett, R. C., and Hentges, D. J. (1973). *Amer. J. Clin. Nutr.* (in press).

Mata, L. J., and Wyatt, R. G. (1971). Host resistance to infection. *Amer. J. Clin. Nutr.* **24,** 976–986.

Metchnikoff, E. (1903). "Essai sur la nature humaine, essai de philosophie optimiste," p. 399. Masson, Paris.

Moro, E. (1900). Uber die nach gramfärbbaren Bacillen des Säuglings. *Wien. Klin. Wochenschr.* **13,** 114.

Moro, E. (1905). Morphologische und biologische Untersuchungen uber die Darmbakterien des Säuglings. *Jahrb. Kinderheilk.* [N.S.] **61,** 687–734 and 870–899.

Nicolai, von H. (1971). Investigations on the biological function of sialic acid containing oligosaccharides in human milk. Thesis, Bonn.

Pasteur, L. (1885). Observations relative à la note précédente de M. Duclaux. *C.R. Acad. Sci.* **100,** 61.

Silver, R. K., Braun, G., Zilliken, F., Werner, G. H., and György, P. (1956). Factors in human milk interfering with influenza virus activities. *Science* **123,** 932–933.

Springer, G. F., and György, P. (1953). Blood group mucoids in growth promotion of *Lactobacillus bifidus* var. *Pennsylvanicus. Fed. Proc., Fed. Amer. Soc. Exp. Biol.* **12,** 272.

Tissier, H. (1900). Recherches sur la flôre intestinale normale et pathologique du nourrisson. Thesis, Paris.

Zilliken, F., Braun, G. A., and György, P. (1955). Gynaminic acid. A naturally occurring form of neuraminic acid in human milk. *Arch. Biochem. Biophys.* **54,** 564–566.

Metabolism of Sugars

CHAPTER 15

The Metabolism of Polyols

OSCAR TOUSTER

Strictly speaking, polyols are not sugars, but are merely polyalcohols. However, the legitimacy for including polyols in the sugar field stems from biochemical relationships. Polyols are formed from, and are converted to, sugars. The background for the development of xylitol as a nutrient will be given and its metabolism related to that of other metabolites. In general, we can say that the polyol field is of relatively recent vintage; for example, a 1945 review of polyols (Carr and Krantz, 1945) stated that nothing was known about the metabolism of pentitols. Of course, some polyols had been found naturally a long time ago, and some enzymes have long been known to utilize polyols as substrates. However, in the last decade or two, with the development of chromatographic techniques and enzymology, and the continued interest in elucidating the carbohydrate metabolism of mammals, a great deal has been learned about the physiological significance of several polyols.

In the account below, the reader is referred to the review by Touster and Shaw (1962) for additional information and references.

A. Survey of the Occurrence and Metabolism of Polyols by Animals

It should be mentioned in passing that the metabolism of polyols is a vast subject with interesting and important enzymatic and genetic as-

TABLE I

Polyols in Animal Metabolism

Occurrence in tissues

Sorbitol—fetal blood, seminal vesicles and plasma, erythrocytes, brain, nerve, kidney, aorta, lens of alloxan–diabetic rats or rats given cataractogenic dose of D-xylose

Xylitol—lens of rats given cataractogenic dose of D-xylose

Dulcitol—various tissues after cataractogenic dose of D-galactose

Occurrence in urine

Erythritol
D- and L-Arabitol
Sorbitol
D-Mannitol

Utilization by mammals *in vivo*

High—xylitol, ribitol, and sorbitol

Variable—D-mannitol (oral dose moderately utilized; parenteral—poorly)

Poor—D- and L-arabitol, dulcitol

---

pects, but for obvious reasons this topic will not be dealt with. In addition, it should be mentioned that polyols occur widely in nature, many of them having been isolated from a variety of plants and lower animals. This fact is not a trivial or irrelevant matter, since we ingest such materials.

Comments will be focused on the behavior of polyols in animals to serve as background for subsequent papers in this section. Table I summarizes the occurrence of polyols in animal tissues. Sorbitol occurs in fetal blood, and many studies indicate that it is the precursor of fructose, which is especially high in the fetal blood of ungulates. There is general agreement that sorbitol produced from glucose is the precursor of fructose in male accessory organs. Sorbitol occurs in many other mammalian tissues, and its formation has been demonstrated in these tissues (for review, see Winegrad *et al.*, 1972). All these studies clearly indicate that sorbitol is a normal metabolic intermediate. In addition, it has been known for some time that sorbitol accumulates in the lens of diabetic rats or of rats being fed cataractogenic doses of xylose (van Heyningen, 1969). Finally, it should be noted that in Chapter 28, Gabbay presents interesting work on sorbitol metabolism in nervous tissue.

In regard to xylitol, which shall be dealt with in greater detail later in connection with its being a normal metabolite, we find that xylose feeding not surprisingly leads to deposition of the pentitol in the lens (van Heyningen, 1969). Dulcitol (or galactitol) has been found in various tissues by van Heyningen (1959) after cataractogenic doses of galactose.

Extensive analyses of urine have been carried out. In 1960, the isolation of erythritol from normal human urine was reported (Touster *et al.*, 1960). The origin of this tetritol is unknown, but it is of interest that aldose reductase, a widely distributed enzyme which can catalyze the reduction of erythrose to erythritol, is especially active with these two substrates (Moonsammy and Stewart, 1967). Earlier, arabitol was found in human urine (Touster and Harwell, 1958). Pitkänen and Pitkänen (1964) reported that mannitol and sorbitol occur in normal human urine and confirmed our finding of L-arabitol and erythritol. The two hexitols were also detected in human urine by Ingram *et al.* (1971).

Administered xylitol, ribitol, and sorbitol are very efficiently metabolized by animals, whereas the utilization of the arabitols and of dulcitol is very poor (reviewed in Touster and Shaw, 1962). D-Mannitol is unusual in that the route of administration is so decisive, oral doses being utilized much better than parenteral doses (Nasrallah and Iber, 1969).

The utilization of polyols by living organisms has been found to be initiated by one of the following reactions (Fig. 1): (a) oxidation to the ketose, (b) oxidation to an aldose, or (c) phosphorylation to the polyol phosphate. The direct phosphorylation route has not been encountered in mammals, but, as indicated in Fig. 1 for xylitol and sorbitol, the aldose and ketose conversions are common. Reactions indicated by "a" are catalyzed by enzymes commonly called polyol dehydrogenases, which utilize NAD or NADP as coenzymes. The xylitol-D-xylulose and sorbitol-D-fructose reactions are probably catalyzed by the same enzyme, or at least by closely related enzymes. On the other hand, the glucose–sorbitol reaction is usually catalyzed by the NADP-dependent aldose reductase, a type of enzyme responsible for the deposition of polyols in the lens and perhaps for polyol formation from aldoses in such diverse

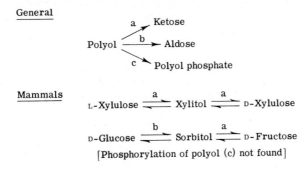

Fig. 1 Enzymatic reactions of polyols.

tissues as the pancreas, placenta, male accessory organs, aorta, and nervous tissue (for review, see Winegrad *et al.,* 1972).

B. Xylitol Metabolism and the
Glucuronate-Xylulose Cycle

Knowledge of the role of xylitol as a normal metabolite stemmed from studies of essential pentosuria, in which it was attempted to discover the route of formation and utilization of L-xylulose, the characteristic urinary sugar in this genetic abnormality. L-Xylulose incubated with tissue preparations was found not to undergo direct conversion to other common sugars or sugar phosphates. Instead, it was reduced to xylitol by a remarkably specific NADP-linked polyol dehydrogenase which was originally found in liver mitochondria (Hollmann and Touster, 1957) but which we now know is largely a soluble cytoplasmic enzyme (Arsenis and Touster, 1969). As shown in Fig. 2, the xylitol is then dehydrogenated by an NAD-linked enzyme, commonly known as sorbitol dehydrogenase (or more systematically, L-iditol dehydrogenase), at the other end of the molecule to yield D-xylulose, which is then phosphorylated to D-xylulose 5-phosphate for metabolism via the pentose phosphate pathway (McCormick and Touster, 1957). The other possible reduction product of L-xylulose, L-arabitol, is in fact produced in small amount by pentosuric subjects, presumably as a "detoxication product" of accumulated L-xylulose (Touster and Harwell, 1958).

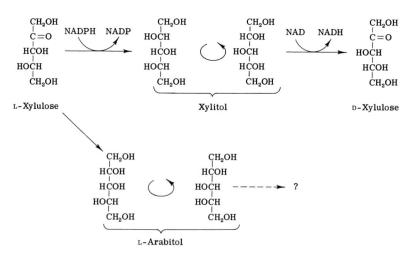

Fig. 2   Xylitol.

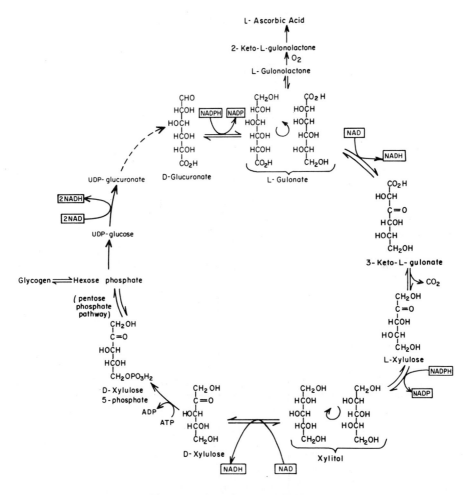

Fig. 3 Carbohydrate metabolism.

As shown in Table II, ^{14}C-xylitol is readily metabolized by animals (McCormick and Touster, 1957). Glycogen, as well as labeled CO_2, is produced, the glycogen having an isotope distribution pattern in the glucose carbon chain consistent with the metabolism of the xylitol through the pentose phosphate pathway.

This whole area of carbohydrate metabolism is summarized by the cycle shown in Fig. 3, which was formulated in 1957 (McCormick and Touster, 1957; Burns and Kanfer, 1957). Several features may be noted:

1. It begins and ends with the ubiquitous glycolytic pathway.
2. It is responsible for ascorbic acid synthesis as well as L-xylulose

TABLE II

Distribution of ^{14}C in Liver Glycogen after Administration of [1-^{14}C]-D-Ribose and [1-^{14}C]Xylitol[a]

Substance injected	Experimental animal	% administered ^{14}C in glycogen	Relative % ^{14}C of glucose carbon atom				
			1	2	3	4 + 5	6
[1-^{14}C]D-Ribose (4.30 μCi, 10.4 mg)	Rat, ♀ (356 gm)	7.1	69.5	2.4	28.5	2.9	1.4
[1-^{14}C]Xylitol (6.70 μCi, 10 mg)	Guinea pig, ♀ (765 gm)	23.8	67.1	0.4	25.6	2.0	2.1
[1-^{14}C]Xylitol (6.70 μCi, 10 mg)	Rat, ♂ (474 gm)	11.8	67.1	0.6	25.1	2.2	3.3

[a] From McCormick and Touster (1957).

synthesis, L-gulonate constituting a branch point. All primates, the guinea pig, flying mammals, and insects (Gupta *et al.*, 1972) have a genetic enzymatic defect in the microsomal oxidase responsible for ascorbic acid formation from L-gulonolactone. The defect in L-xylulose utilization is rare, however, occurring as a recessive defect only in Jews and Arabs (Touster, 1960; Hiatt, 1972).

3. The defect in L-xylulose reduction has until recently only been inferred from the massive accumulation of the pentose in pentosuric individuals, and from studies which ruled out a kidney defect (Freedberg *et al.*, 1959; Bozian and Touster, 1959), since it has been impossible to obtain a sample of liver from a pentosuric person with which to demonstrate the defect. (All available evidence indicates that these people are suffering from an abnormality, not a disease; they are therefore understandably reluctant to donate their tissues for pure research.) It was fortunate that Asakura (Asakura *et al.*, 1967) found a few years ago that erythrocytes contain the two xylitol dehydrogenases already discovered in animal liver. This finding facilitated the work of Wang and van Eys (1970), which indicates that pentosuric individuals have a normal NAD-linked xylitol dehydrogenase but a mutant NADP-linked xylitol dehydrogenase with markedly reduced capacity to catalyze the reduction of L-xylulose to xylitol; that is, the pentosuric erythrocytic NADP-linked xylitol dehydrogenase has a higher K_m (by a factor of 10–20), or lower affinity, for NADP than the normal enzyme. Therefore, it is evident that the enzyme is synthesized by the pentosuric but in a modified, less active form.

4. The cycle has been demonstrated in all animals tested, yet we may not as yet know its main function. The early steps are obviously concerned with the synthesis of nucleotide sugars involved in the biosynthesis of mucopolysaccharides, glycoproteins, glycolipids, and glucuronides formed as detoxication products, and most animals produce ascorbic acid by this route. The crucial importance of the latter steps in the cycle has not been established. In a recent collaborative study with Hankes, Politzer, and Anderson (Hankes *et al.*, 1969), we have obtained evidence that the cyclic polyol inositol is exclusively metabolized in man by conversion to D-glucuronate and further metabolism through the xylulose pathway. Pentosuric subjects, however, accumulated the isotopic label of the administered inositol in urinary L-xylulose.

Pentosuric individuals are apparently at no disadvantage from the insufficiency of L-xylulose reduction [e.g., they have a normal life span (Lasker, 1955)]. Perhaps compensatory mechanisms have come into play; for example, examination of the cycle shows that it could serve a transhydrogenation role in that several steps utilize NADPH while others produce NADH. Yet, we know that other metabolic processes can also effect this overall change in the two coenzymes. Winegrad has reported that this C_6 oxidation pathway is elevated in diabetic rats (Winegrad and Shaw, 1964) and that plasma L-xylulose levels are elevated in diabetes mellitus (Winegrad and Burden, 1966). The significance of these interesting observations remains to be established.

5. From the fact that pentosuric individuals normally excrete several grams of L-xylulose each day, and from experiments on the extent of augmentation of this excretion on feeding the precursor D-glucuronolactone, we can make a rough estimate that the carbohydrate flux through the cycle is between 5 and 15 gm/day for man (Hollmann and Touster, 1964). In other words, several percent of the carbohydrate oxidized daily by man may be handled by this process.

6. The cycle is enhanced by various drugs and hydrocarbons (for review, see Hollmann and Touster, 1964). Many of the substances that induce microsomal oxidative and detoxication enzymes increase the conversion of hexose to ascorbic acid in rats and the production of L-xylulose in pentosurics. While certain enzymes are elevated by these drugs (Hollmann and Touster, 1962), the actual mechanism of the stimulation of the cycle is not really evident as yet.

C. Xylitol in Nutrition

Professor Konrad Lang of Mainz is mainly responsible for developing xylitol for nutritive purposes (Lang, 1969). With the knowledge that xylitol is a normal metabolite, has a sweet taste, and is not particularly expensive, he extended research on the pentitol in a number of areas. Moreover, he emphasized a particular advantage of xylitol, namely, the fact that, lacking a carbonyl group, it does not interact with amino acids and can therefore be sterilized in the presence of protein hydrolysates for infusion therapy. The work of a number of investigators, particularly in Germany and Japan, led to the following views (for review, see Horecker *et al.*, 1969) :

(a) Xylitol is essentially a nontoxic substance.

(b) Several species, including man, handle rather large doses of xylitol efficiently. Humans apparently can utilize several hundred grams of xylitol per day. Little xylitol is found in the urine after large doses are administered.

(c) The utilization of xylitol is insulin independent, a finding suggesting its use in the treatment of diabetics. (In this regard, although xylitol stimulates insulin release in dogs, little change in plasma insulin is induced by xylitol in humans.)

(d) Xylitol is antiketogenic, an effect of obvious relevance to its proposed use in diabetics.

Parenteral xylitol, as well as sorbitol, is first acted upon by the cytoplasmic sorbitol dehydrogenase (L-iditol dehydrogenase) of the liver. With xylitol, this is the rate-limiting reaction in its utilization (Bässler, 1969). Little xylulose is excreted after xylitol administration. [Sorbitol is, of course, oxidized to D-fructose by this dehydrogenase and then further metabolized as discussed by Hue (Chapter 22) and Hers (Chapter 21).]

Xylitol has gained wide acceptance in parenteral nutrition in Japan and has been widely tested in Germany. In America, there has been skepticism about replacing the inexpensive and safe glucose with any other sugar or sugar derivative. Comparative studies of glucose, fructose, sorbitol, and xylitol have been made abroad in recent years (Lang and Fekl, 1971), to which reference will be made in subsequent papers of this volume. The uricemia and bilirubinemia following the administration of xylitol is also observed with fructose and sorbitol and is considered by some investigators to be a minor matter (see, e.g., Förster *et al.*, 1970). Moreover, reference should also be made to a disturbing report from Australia (Thomas *et al.*, 1972) which has, unfortunately, detrimentally influenced attitudes toward xylitol, which may in fact be an innocent bystander. This point will also be discussed subsequently.

D. Concluding Remarks

It is often difficult to make the transition from experimental studies to wide clinical and nutritional application. Xylitol has presented us with this problem. The papers in this volume will not only focus on a variety of specific aspects of the metabolism and use of xylitol but, it is hoped, they may also provide the basis for making more perceptive evaluations of the overall value of xylitol as a safe nutrient for the human.

References

Arsenis, C., and Touster, O. (1969). Nicotinamide adenine dinucleotide phosphate-linked xylitol dehydrogenase in guinea pig liver cytosol. *J. Biol. Chem.* **244,** 3895–3899.

Asakura, T., Adachi, K., Minakami, S., and Yoshikawa, H. (1967). Non-glycolytic sugar metabolism in human erythrocytes. I. Xylitol metabolism. *J. Biochem. (Tokyo)* **62,** 184–193.

Bässler, K. H. (1969). Adaptive processes concerned with absorption and metabolism of xylitol. *In* "Metabolism, Physiology, and Clinical Uses of Pentoses and Pentitols" (B. L. Horecker, K. Lang, and Y. Takagi, eds.), pp. 190–196. Springer-Verlag, Berlin and New York.

Bozian, R. C., and Touster, O. (1959). Essential pentosuria: Renal or enzymic disorder. *Nature (London)* **184**, 463–464.

Burns, J. J., and Kanfer, J. (1957). Formation of L-xylulose from L-gulonolactone in rat kidney. *J. Amer. Chem. Soc.* **79**, 3604–3605.

Carr, C. J., and Krantz, J. C., Jr. (1945). Metabolism of the sugar alcohols and their derivatives. *Advan. Carbohyd. Chem.* **1**, 175–192.

Förster, H., Meyer, E., and Ziege, M. (1970). Erhohung von serumharnsäure und serumbilirubin nach hochdosierten infusionen von sorbit, xylit, und fructose. *Klin. Wochenschr.* **48**, 878–879.

Freedberg, I. M., Feingold, D. S., and Hiatt, H. H. (1959). Serum and urine L-xylulose in pentosuric and normal subjects and in individuals with pentosuric trait. *Biochem. Biophys. Res. Commun.* **1**, 328–332.

Gupta, S. D., Chaudhuri, C. R., and Chatterjee, I. B. (1972). Incapability of L-ascorbic acid synthesis by insects. *Arch. Biochem. Biophys.* **152**, 889–890.

Hankes, L. V., Politzer, W. M., Touster, O., and Anderson, L. (1969). Myo-inositol catabolism in human pentosurics. The predominant role of the glucuronate-xylulose-pentose pathway. *Ann. N.Y. Acad. Sci.* **165**, 564–576.

Hiatt, H. H. (1972). Pentosuria. *In* "The Metabolic Basis of Inherited Disease" (J. B. Stanbury, J. B. Wyngaarden, and D. S. Fredrickson, eds.), 3rd ed., Chapter 5, pp. 119–130. McGraw-Hill, New York.

Hollmann, S., and Touster, O. (1957). The L-xylulose-xylitol enzyme and other polyol dehydrogenases of guinea pig liver mitochondria. *J. Biol. Chem.* **225**, 87–102.

Hollmann, S., and Touster, O. (1962). Alterations in tissue levels of uridine diphosphate glucose dehydrogenase, uridine diphosphate glucuronic acid pyrophosphatase and glucuronyl transferase induced by substances influencing the production of ascorbic acid. *Biochim. Biophys. Acta* **62**, 338–352.

Hollmann, S., and Touster, O. (1964). "Non-glycolytic Pathways of Metabolism of Glucose," p. 107. Academic Press, New York.

Horecker, B. L., Lang, K., and Takagi, Y., eds. (1969). "Metabolism, Physiology, and Clinical Use of Pentoses and Pentitols." Springer-Verlag, Berlin and New York.

Ingram, P., Applegarth, D. A., Sturrock, S., and Whyte, J. N. C. (1971). Sorbitol and mannitol in human urine. *Clin. Chim. Acta* **35**, 523–524.

Lang, K. (1969). Utilization of xylitol in animals and man. *In* "Metabolism, Physiology, and Clinical Uses of Pentoses and Pentitols" (B. L. Horecker, K. Lang, and Y. Takagi, eds.), pp. 151–156. Springer-Verlag, Berlin and New York.

Lang, K., and Fekl, W. (1971). Xylit in infusionstherapie. *Z. Ernährungswiss., Suppl.* **11.**

Lasker, M. (1955). Mortality of persons with xyloketosuria. *Hum. Biol.* **127,** 294–300.

McCormick, D. B., and Touster, O. (1957). The conversion *in vivo* of xylitol to glycogen via the pentose phosphate pathway. *J. Biol. Chem.* **229**, 451–461.

Moonsammy, G. I., and Stewart, M. A. (1967). Purification and properties of brain aldose reductase and L-hexonate dehydrogenase. *J. Neurochem.* **14**, 1187–1193.

Nasrallah, S. M., and Iber, F. L. (1969). Mannitol absorption and metabolism in man. *Amer. J. Med. Sci.* **258**, 80–88.

Pitkänen, E., and Pitkänen, A. (1964). Polyhydric alcohols in human urine. II. *Ann. Med. Exp. Fenn.* **42,** 113–116.

Thomas, D. W., Edwards, J. B., Gilligan, J. E., Lawrence, J. R., and Edwards, R. G. (1972). Complications following intravenous administration of solutions containing xylitol. *Med. J. Aust.* **1,** 1238–1246.

Touster, O. (1960). Essential pentosuria and the glucuronate-xylulose pathway. *Fed. Proc., Fed. Amer. Soc. Exp. Biol.* **19,** 977–983.

Touster, O., and Harwell, S. (1958). The isolation of L-arabitol from pentosuric urine. *J. Biol. Chem.* **230,** 1031–1041.

Touster, O., and Shaw, D. R. D. (1962). Biochemistry of the acyclic polyols. *Physiol. Rev.* **42,** 181–225.

Touster, O., Hecht, S. O., and Todd, W. M. (1960). The isolation of crystalline erythritol from normal human urine. *J. Biol. Chem.* **235,** 951–953.

van Heyningen, R. (1959). Formation of polyols by the lens of the rat with 'sugar' cataract. *Nature (London)* **184,** 194–195.

van Heyningen, R. (1969). Metabolism of xylose by the lens of the eye. *In* "Metabolism, Physiology, and Clinical Uses of Pentoses and Pentitols" (B. L. Horecker, K. Lang, and Y. Takagi, eds.), pp. 109–121. Springer-Verlag, Berlin and New York.

Wang, Y. M., and van Eys, J. (1970). The enzymatic defect in essential pentosuria. *N. Engl. J. Med.* **282,** 892–896.

Winegrad, A. I., and Burden, C. L. (1966). L-Xylulose metabolism in diabetes mellitus. *N. Engl. J. Med.* **274,** 298–305.

Winegrad, A. I., and Shaw, W. N. (1964). Glucuronic acid pathway activity in adipose tissue. *Amer. J. Physiol.* **206,** 165–168.

Winegrad, A. I., Clements, R. S., and Morrison, A. D. (1972). Insulin-independent pathways of carbohydrate. *In* "Handbook of Physiology" (Amer. Physiol. Soc., J. Field, ed.), Sect. 7, Vol. I, pp. 457–471. Williams & Wilkins, Baltimore, Maryland.

CHAPTER 16

The Metabolism of Xylitol

E. R. FROESCH AND A. JAKOB

A. The Hepatic Metabolism of Xylitol

A scheme of xylitol metabolism of the liver is presented in Fig. 1. The liver cell membrane has the interesting property of being permeable to substances to which other cell membranes are relatively impermeable. This is true for the two polyols, sorbitol and xylitol. Xylitol seems to

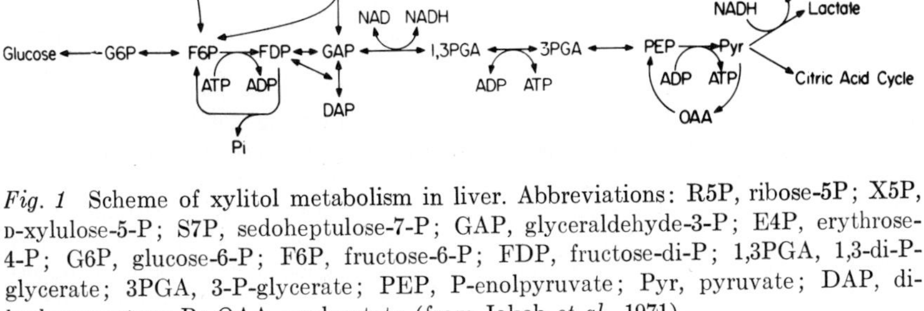

Fig. 1 Scheme of xylitol metabolism in liver. Abbreviations: R5P, ribose-5P; X5P, D-xylulose-5-P; S7P, sedoheptulose-7-P; GAP, glyceraldehyde-3-P; E4P, erythrose-4-P; G6P, glucose-6-P; F6P, fructose-6-P; FDP, fructose-di-P; 1,3PGA, 1,3-di-P-glycerate; 3PGA, 3-P-glycerate; PEP, P-enolpyruvate; Pyr, pyruvate; DAP, di-hydroxyacetone-P; OAA, oxalacetate (from Jakob *et al.*, 1971).

enter the hepatic cell without difficulty although nothing is known about the characteristics of xylitol uptake by the liver cell. Xylitol is first oxidized to D-xylulose by the NAD-xylitol dehydrogenase causing the NADH-NAD ratio to increase (Smith, 1962; Williamson *et al.*, 1971). The next step is the phosphorylation of D-xylulose to D-xylulose 5-phosphate by D-xylulose-kinase (Ashwell, 1962). D-Xylulose 5-phosphate is an intermediate of the pentose phosphate shunt and it is metabolized to fructose 6-phosphate and glyceraldehyde phosphate by this pathway. The stoichiometry is such that three xylitol molecules yield two molecules of fructose 6-phosphate and one molecule of glyceraldehyde phosphate (Touster and Shaw, 1962). Fructose 6-phosphate can readily be converted to glucose and glycogen; glyceraldehyde phosphate either to glucose, glycogen, or lactate.

As shall be seen from the following results, it is likely that most of the xylitol oxidized to D-xylulose and phosphorylated to D-xylulose 5-phosphate is rapidly converted to glucose and that only small quantities are converted to lactate (Jakob *et al.*, 1971). This is illustrated quite clearly by Fig. 2 which shows the results of a liver perfusion with xylitol and lactate. As can be seen more than half of the xylitol taken up by the liver is released into the perfusion medium as glucose. Less than 50% of xylitol is oxidized and lactate formation is very small. When

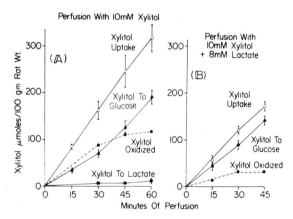

Fig. 2 Effect of lactate on xylitol metabolism by the perfused rat liver. (A) Conversion of xylitol to glucose was calculated by multiplying measured glucose production by 1.2. Likewise, xylitol conversion to lactate was obtained by multiplying the measured lactate production by 0.6. The amount of xylitol oxidized over each time interval was calculated by subtracting xylitol conversion to glucose and lactate from the measured xylitol uptake. (B) Xylitol conversion to glucose was calculated from the difference between the total glucose production and [^{14}C]glucose formed from [^{14}C]lactate. The amount of xylitol oxidized was obtained from the difference between xylitol uptake and xylitol conversion to glucose (from Jakob *et al.*, 1971).

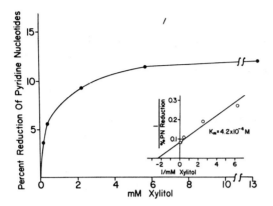

Fig. 3 Titration of pyridine nucleotide fluorescence increase with xylitol concentration in the perfused rat liver. The inset expresses the reciprocal of the fluorescence change against the reciprocal of the xylitol concentration in the perfusion fluid (from Jakob *et al.,* 1971).

the liver is perfused with xylitol and lactate simultaneously, the proportion of xylitol which is converted to glucose increases, whereas less xylitol is oxidized as a result of the preferential oxidation of lactate. Figure 3 shows the effect of xylitol on the redox state of the liver. The reduction of the pyridine nucleotides in the liver cell increases with the concentration of xylitol in the perfusion medium and reaches a plateau at about 6 mM xylitol. When these data are plotted according to Lineweaver-Burk one gets a relatively straight line and an apparent K_m of 4.2×10^{-4} M which corresponds to the K_m of the cytosolic polyol dehydrogenase. The increased reduction of the cytosol during perfusion with xylitol is reflected by an increase of the lactate–pyruvate ratio from 13 to 67.

B. Extrahepatic Metabolism of Xylitol

It has been claimed by several authors that xylitol is also metabolized by various tissues other than liver and kidney (reviewed by Lang, 1971; Horecker *et al.,* 1969). As we will see later, it is very difficult to ascertain from results of *in vivo* experiments whether one is really following the metabolism of xylitol itself or the metabolism of glucose after the liver has converted xylitol to glucose. Crofford *et al.* (1965) have shown that adipose tissue metabolizes small amounts of sorbitol. We were interested to see whether xylitol was also metabolized by adipose tissue *in vitro*. Figure 4 shows the results of an *in vitro* experiment in which four concentrations of xylitol (namely, 10, 30, 90, and 270 mg%) were presented to adipose tissue. The specific activity of [^{14}C]xylitol was the same in all flasks. As can be seen from this figure, adipose tissue does, indeed, take up xylitol and incorporate it into the various metabolites in a concentration-dependent manner.

When the xylitol concentration was further increased to 800 and 1600 mg% there was a further increase of xylitol metabolism. Whether this increase of xylitol metabolism with an increasing concentration of xylitol is dependent on transport or on the first enzymatic steps responsible for xylitol metabolism is entirely unclear. It was argued that if sorbitol dehydrogenase were the first enzyme oxidizing D-xylitol to D-xylulose there might possibly be a competition between xylitol and sorbitol if adipose tissue is simultaneously incubated with both polyols. The results shown in Fig. 5 demonstrate quite clearly that added cold sorbitol does not compete with xylitol; 30 mg% of xylitol was present in all incubation flasks, whereas sorbitol was added in three concentrations, i.e., 10, 50, and 250 mg%. As is apparent from this figure, xylitol incorporation into total lipids and xylitol oxidation to ^{14}CO$_2$ were not affected by sorbitol.

In contrast to Crofford *et al.* (1965), we have not detected significant activities of sorbitol dehydrogenase in homogenates of adipose tissue. It is, therefore, conceivable that other enzymes are responsible for the metabolism of xylitol and sorbitol in this particular tissue. In any case, it must be concluded from Figs. 4 and 5 that xylitol metabolism in adipose tissue is very small and probably of no physiological significance.

For comparison, the results in Fig. 6 are shown, where fructose metabolism of adipose tissue *in vitro* was studied at a fructose concentration of 200 mg%. Incorporation of [^{14}C]fructose into ^{14}CO$_2$ and total lipids was 5 μmoles compared to 0.4 μmoles of [^{14}C]xylitol at a concentration of 270 mg%. Figure 6 also shows that fructose uptake of adipose tissue is slightly insulin-dependent in the absence of glucose, although much less than glucose metabolism, and that xylitol has no influence on the metabolism of fructose by this tissue.

C. Comparison of the Metabolism of Xylitol, Sorbitol, Fructose and Glucose in Normal Animals

Glucose substitute sugars are often administered to patients because they are believed to be utilized independently of insulin. According to the scheme of hepatic xylitol metabolism (Fig. 1) and the results presented in Fig. 2, xylitol is rapidly converted to glucose by rat liver *in vitro*. If this also occurs *in vivo*, xylitol utilization by extrahepatic tissues must depend primarily on the availability of insulin. It is also known

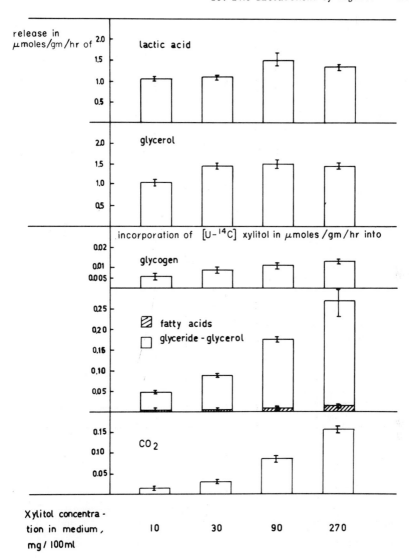

Fig. 4 The effect of xylitol concentration on its metabolism by adipose tissue *in vitro*. Pooled epididymal adipose tissue was incubated in the presence of four xylitol concentrations, the specific activity of [U-14C]xylitol being constant. After a 2-hr incubation the incorporation of 14C into total lipids and glycogen as well as glycerol and lactic acid release into the medium were determined. For methods see Froesch and Ginsberg (1962). The bars and brackets give the mean of the results of two flasks with the range.

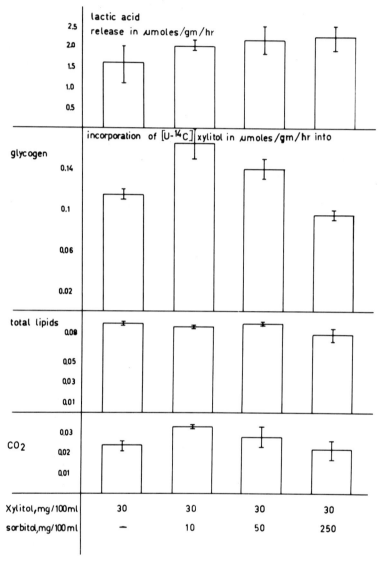

Fig. 5 The effect of various concentrations of sorbitol on the metabolism of [¹⁴C]xylitol by adipose tissue *in vitro*. Cold sorbitol was added to the medium containing 30 mg% of [U-¹⁴C]xylitol. The bars give the mean of the results of two flasks and the brackets the range (unpublished data).

from *in vitro* experiments with fructose that hepatic conversion to glucose is extremely rapid. Therefore, the following experiments were set up.

Normal fed rats were anesthetized with Diazepam subcutaneously and then injected intravenously with 20 mg of cold fructose together with 1 μCi of [^{14}C]fructose. One group of rats received simultaneously with fructose 18 mU of insulin, whereas the other group was given anti-insulin serum neutralizing 20 mU of insulin. The results of these experiments are shown in Fig. 7.

Plasma glucose was, of course, significantly higher in the anti-insulin

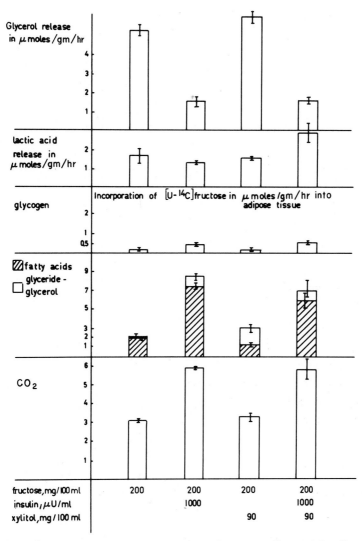

Fig. 6 Fructose metabolism of adipose tissue *in vitro;* effects of insulin and of xylitol. Pooled epididymal adipose tissue was incubated *in vitro* during 2 hr (unpublished data).

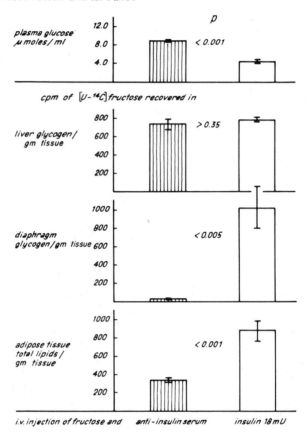

Fig. 7 *In vivo* metabolism of fructose. Metabolism of [¹⁴C]fructose with a load of 20 mg of fructose administered intravenously together with 18 mU of insulin or anti-insulin serum. Seven rats were used in each group and the means and the SEM are given (from Froesch *et al.*, 1971).

serum treated rats than in insulin treated rats. The incorporation of [¹⁴C]fructose into liver glycogen was unaffected by anti-insulin serum and insulin. The incorporation of carbon-14 into total lipids of adipose tissue was stimulated by insulin about $2\frac{1}{2}$-fold. There was almost no incorporation of carbon-14 in the anti-insulin serum treated rats into diaphragm glycogen, a metabolic index markedly enhanced by insulin. These results were interpreted to mean that (1) the acute formation of liver glycogen is not dependent on the acute release of insulin, (2) adipose tissue can take up carbon-14 from fructose and convert it to total lipids, and (3) the stimulation of the incorporation of carbon-14 into total lipids is not the result of an insulin stimulation of fructose uptake but rather of glucose uptake originating in the liver from [^{14}C]fructose. This latter interpretation is also based on the data obtained with [^{14}C]sorbitol. After the injection of [^{14}C]sorbitol almost no carbon-14 was incorporated into diaphragm glycogen and adipose tissue total lipids in the absence of insulin. Insulin markedly stimulated both processes (Fig. 8).

Since it is well known that even *in vitro* fructose incorporation into adipose tissue and diaphragm is not stimulated by insulin in the presence of glucose (Froesch, 1965) this must mean that fructose was rapidly converted to [^{14}C]glucose and that the stimulatory effect of insulin on carbon-14 incorporation was on glucose transport and incorporation into adipose tissue and diaphragm rather than on fructose transport.

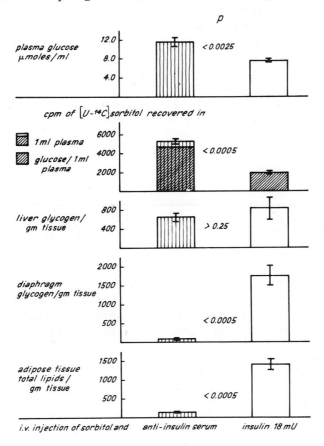

Fig. 8   In vivo metabolism of sorbitol. Metabolism of [^{14}C]sorbitol with a load of 20 mg of sorbitol administered intravenously together with 18 mU of insulin or anti-insulin serum. Ten rats were used in each group and the means and the SEM are given (from Froesch *et al.,* 1971).

Similar experiments were carried out with [^{14}C]sorbitol and [^{14}C]xylitol, and essentially the same results were obtained as with [^{14}C]fructose (Fig. 8). The advantage of using [^{14}C]sorbitol and [^{14}C]xylitol was also that we were able to separate [^{14}C]glucose in the blood from these polyols by the glucosazone formation. Whereas fructose yields the same glucosazone as glucose, sorbitol and xylitol do not. In this way the formation of [^{14}C]glucose can be accurately assessed. Again, the blood sugar of the rats injected with anti-insulin serum was much higher than the blood sugar of the rats given insulin. Almost all carbon-14 present in the serum was recovered in the form of the glucosazone which means that 30 min after sorbitol injection almost 100% of the remaining carbon-14 activity is in the form of [^{14}C]glucose.

Since insulin stimulates the uptake of glucose by various tissues the amount of [^{14}C]glucose present in the blood of the rats treated with insulin was much lower than the [^{14}C]glucose content of the blood of anti-insulin serum treated rats. Liver glycogen formation was not significantly influenced by insulin, whereas there was almost no incorporation of carbon-14 into diaphragm and adipose tissue in the absence of insulin. Both metabolic indices were markedly stimulated by insulin. The interpretation of these results is again that insulin did not stimulate sorbitol incorporation into these tissues but rather the incorporation of [^{14}C]glucose which had been produced by the liver from [^{14}C]sorbitol.

Figure 9 shows similar results, this time obtained after the injection of [^{14}C]xylitol. The results resemble those obtained with sorbitol in every respect. There may be one exception regarding the incorporation of [^{14}C]xylitol into liver glycogen which was significantly higher in this particular experiment than with sorbitol or fructose. Also, there was a fairly significant stimulatory effect of insulin on the incorporation of carbon-14 into liver glycogen. In other experiments not shown here this observation was verified, but the insulin effect was quantitatively less pronounced. In direct comparative experiments the incorporation of [^{14}C]xylitol into liver glycogen was about twice that of fructose and sorbitol incorporation.

In another series of experiments the metabolism of sorbitol and xylitol was compared in normal rats as a function of time. The same metabolic indices as in the foregoing experiments were followed (Fig. 10). The rats were killed after 5, 10, 15, and 30 min, respectively, to get a better idea of the half-life of sorbitol and xylitol and of their turnover. The initial counts per minute per milliliter of plasma and their extrapolation to time zero indicates that the distribution volumes of xylitol and sorbitol are very similar in the rat. The disappearance of [^{14}C]xylitol and [^{14}C]sorbitol was calculated from the difference between the total counts per minute

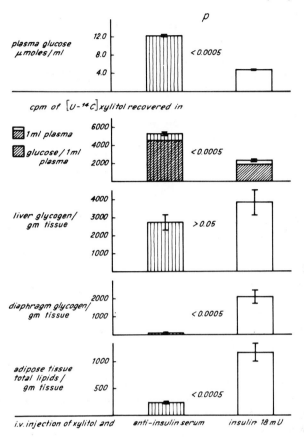

Fig. 9 *In vivo* metabolism of xylitol. Metabolism of [¹⁴C]xylitol with a load of 20 mg of xylitol administered intravenously together with 18 mU of insulin or anti-insulin serum. Ten rats were used in each group and the means and the SEM are given (from Froesch *et al.*, 1971).

in plasma and the counts per minute recovered in the glucosazones. As may be seen from the results presented in Fig. 11, both xylitol and sorbitol disappear extremely rapidly and the calculated half-life is 165 sec for both polyols. Five minutes after the injection of these substrates more than 50% of the carbon-14 was present in the blood in the form of glucose, and after 10 min more than 70%. After 15 min the total amount of carbon-14 in the plasma was almost identical with the carbon-14 recovered in the plasma glucose. This shows clearly that even in the presence of insulin, sorbitol and xylitol are extremely rapidly converted to glucose by the liver. After 5 min almost no carbon-14 was recovered from the total lipids of adipose tissue and diaphragm glycogen, presumably

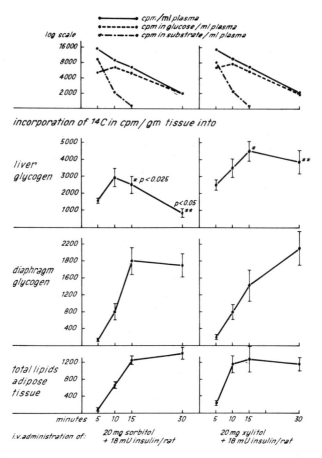

Fig. 10 Comparison of the metabolism of an intravenously administered load (20 mg) of [^{14}C]sorbitol with that of [^{14}C]xylitol as a function of time. Each point represents the mean of the results of 5 rats; p values indicate the significance of the difference between sorbitol and xylitol incorporation into liver glycogen at 15 and 30 min, respectively (from Froesch *et al.*, 1971).

because the conversion of [^{14}C]xylitol and [^{14}C]sorbitol to glucose had not yet occurred to a sufficient extent. There is a steep rise of these two metabolic indices once [^{14}C]glucose accumulates in the blood. We find again a clear-cut difference in the behavior of the liver glycogen insofar as radioactivity in the liver glycogen reaches a maximum already 10 min after sorbitol injection followed by a decrease, whereas xylitol incorporation into liver glycogen increases up to 15 min and remains significantly higher than sorbitol incorporation. This finding substantiates the results of the other experiments shown so far. We cannot offer an explanation for this phenomenon.

In another series of experiments diabetic rats were used instead of normal fed rats (Fig. 12). These were injected with 20 mg and 1 μCi of either fructose, sorbitol, or xylitol alone or together with 18 mU of insulin. As expected, no counts are incorporated into diaphragm glycogen of diabetic rats from all three substrates in the absence of insulin. In contrast, adipose tissue takes up fructose and incorporates significantly greater quantities of fructose into adipose tissue than of xylitol and sorbitol.

The response of the diabetic liver to insulin is different from that of the normal liver. In the liver of normal animals the effect of insulin is counteracted by hyperglycemic and insulin antagonistic hormones. In contrast, a significant stimulation of glycogen synthesis by insulin oc-

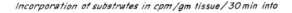

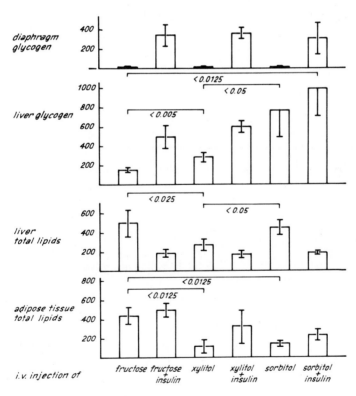

Fig. 11 Metabolism of ^{14}C-labeled fructose, sorbitol, and xylitol administered with a load of 20 mg of substrate with and without 18 mU of insulin by streptozotocin-diabetic rats. Five rats were used in each group and the means and the SEM are given (from Froesch *et al.,* 1971).

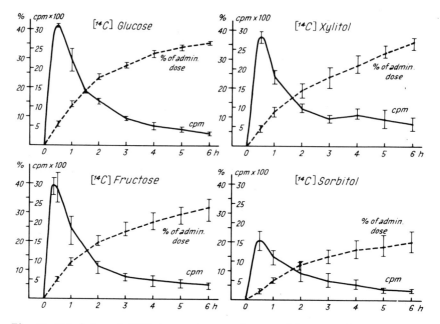

Fig. 12 Exhalation of $^{14}CO_2$ in counts per minute and in accumulated percent of administered dose after U-^{14}C-labeled glucose, xylitol, fructose, and sorbitol administration. Twenty milligrams and 2 μCi of each substrate together with 18 mU of insulin were intravenously administered to two 24-hr fasted rats. $n = 3$ or 4; brackets indicate the range (from Keller and Froesch, 1971).

curred in diabetic animals. In the case of fructose this effect of insulin on liver glycogen synthesis was statistically significant (Fig. 12). It is of interest that the diabetic liver seems to prefer sorbitol for liver glycogen synthesis. In this particular experiment liver glycogen synthesis was clearly greater with [^{14}C]sorbitol than with either [^{14}C]xylitol or [^{14}C]fructose.

Further experiments were designed to find out whether any of the four substrates—glucose, xylitol, fructose, and sorbitol—are preferentially utilized by normal and diabetic rats. Rats were put in a glass cage, flushed with air which left the cage through an FHT 50 A analytic respirometer from Frieseke and Hoepfner GmbH, Erlangen, for continuous registration of $^{14}CO_2$ exhalation. The rats were injected with 20 mg of each of these sugars together with 2 μCi of the uniformly labeled sugar. Exhalation of $^{14}CO_2$ was monitored for 6 hr, at which time the rats were killed. The total amount of $^{14}CO_2$ expired after 6 hr was approximately 35% of the given dose, in the case of glucose, xylitol, and fructose (Fig. 12) and approximately 20% in the case of sorbitol, of which a greater amount was lost in the urine. Diabetic rats exhaled during the same time only between 11 and 18% of the same dose of either one of these sugars (Fig. 13). Exhalation of CO_2 was the same whether [^{14}C]glucose or [^{14}C]xylitol was administered. A somewhat larger percentage of [^{14}C]fructose was exhaled as $^{14}CO_2$.

In Fig. 14 the utilization of [^{14}C]xylitol is compared with that of [^{14}C]glucose in normal and diabetic rats after intravenous injection of 20 mg of each sugar with 2 μCi of uniformly labeled sugar. During 6 hr, 38% of glucose and 39% of xylitol, respectively, were oxidized to $^{14}CO_2$ by normal rats. In comparison, 11% of glucose and 15% of [^{14}C]xylitol were expired by diabetic rats. Normal rats excreted 4% of the total dose of carbon-14 administered as [^{14}C]glucose in the urine of which only a very small portion was in the form of [^{14}C]glucose. In contrast, 15% of the total doses of [^{14}C]xylitol was lost in the urine of normal rats. The amounts of carbon-14 recovered in skeletal muscle, liver, and in the extracellular space were small and similar for both sugars throughout all experiments.

Diabetic rats lost 55% of the total dose of [^{14}C]glucose in the urine, compared to a total loss of 46% of the total dose of [^{14}C]xylitol; 27%

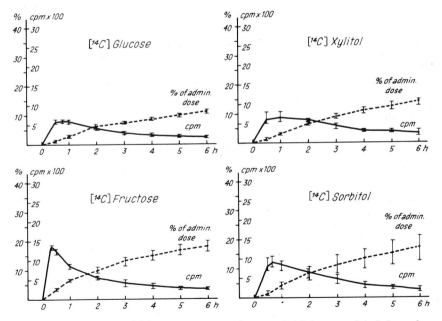

Fig. 13 Exhalation of $^{14}CO_2$ as counts per minute and total accumulated dose after U-^{14}C-labeled glucose, sorbitol, xylitol, and fructose administration. Twenty milligrams and 2 μCi of each substrate were intravenously administered to two streptozotocin-diabetic rats. $n = 3$ or 4; brackets indicate the range (from Keller and Froesch, 1971).

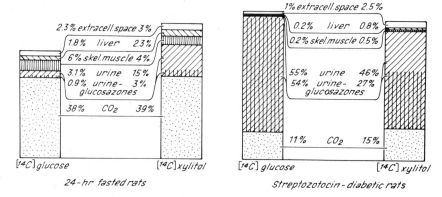

Fig. 14 Recovery of carbon-14 in $^{14}CO_2$, urine, glycogen of skeletal muscle, total lipids, and glycogen of liver and in the extracellular space 6 hr after intravenous administration of 20 mg and 2 μCi of glucose and xylitol in 24-hr fasted and in streptozotocin-diabetic rats (from Keller and Froesch, 1971).

was lost in the form of [^{14}C]glucose as demonstrated by the formation of glucosazones from urine. The peripheral distribution of [^{14}C]glucose and [^{14}C]xylitol was again similar.

The results of this last experiment can be interpreted only in the following way. The oxidation of both sugars to $^{14}CO_2$ was very similar. This therefore proves in an unequivocal manner that xylitol was first converted to glucose before oxidation, with the possible exception of some direct hepatic oxidation of [^{14}C]xylitol. Almost 30% of the total dose of xylitol was excreted in the urine as [^{14}C]glucose, which again conclusively shows that most of the xylitol is rapidly converted to glucose that cannot be taken care of adequately by the diabetic, insulin-deficient rat.

The metabolism of xylitol, fructose, and sorbitol differs from that of glucose with respect to the initial steps by which they enter the glycolytic pathway of the liver. These steps are independent of insulin which led to the misunderstanding that the entire utilization of fructose, sorbitol, and xylitol is also insulin-independent (Lang, 1971).

As shown in these experiments, all three glucose substitutes are extremely rapidly converted to glucose by the liver. The utilization of glucose is clearly insulin-dependent in those tissues which were investigated, i.e., the liver, muscle, and adipose tissue. The rapid conversion of these polyols and of fructose to glucose does not contradict the finding that xylitol, sorbitol, and fructose do not produce significant hyperglycemia in normal animals and in man. In rats, the glucose turnover is very rapid and glucose homeostasis well maintained. Other gluconeogenetic substrates such as amino acids and glycerol are also rapidly converted to glucose but do not markedly elevate blood sugar. Hyperglycemia is prevented by two mechanisms. First, the uncontrolled phosphorylation of D-xylulose and of fructose leads to a rapid increase of the intracellular concentration of all glycolytic intermediates so that glycogenolysis and glucogeogenesis from other substrates are inhibited. These sugars are now the main precursor of hepatic glucose and compete with lactate and amino acids for gluconeogenesis. Second, a small secondary increase of insulin secretion may occur which may be barely detectable in the peripheral blood, but which, delivered into the portal vein, acts on the liver to maintain the blood sugar more or less constant. Whereas fructose may be utilized to a small extent prior to its conversion to glucose by adipose tissue, this is not the case for sorbitol and xylitol.

The use of sorbitol and xylitol for parenteral nutrition offers no advantage but leaves the physician in the mistaken belief that he does not harm the patient because these polyols are easily "utilized" without extra insulin. Furthermore, there is the considerable danger of lactic acidosis if large amounts of fructose, sorbitol, or xylitol are administered.

Acknowledgment

The original work from the Metabolic Unit cited in this chapter was supported by a grant from the Schweizerische Nationalfonds (3.856.69).

Dr. Jakob was a recipient of a fellowship from the Schweizerische Nationalfonds. His work was carried out at the Johnson Research Foundation and was supported by a grant AM-15120 from the U.S. Public Health Service and by grants-in-aid from the American Diabetes Association, the American Heart Association, and the American Medical Association Education and Research Foundation.

References

Ashwell, G. (1962). D-Xylulokinase from calf liver. *In* "Methods in Enzymology," (S. P. Colowick and T. O. Kaplan, eds.), Vol. 5, pp. 208–211. Academic Press, New York.

Crofford, O. B., Jeanrenaud, B., and Renold, A. E. (1965). Effect of insulin on the transport and metabolism of sorbitol by incubated rat epididymal adipose tissue. *Biochim. Biophys. Acta* **111**, 429–539.

Froesch, E. R. (1965). Fructose metabolism of adipose tissue of normal and diabetic rats. *In* "Handbook of Physiology" (Amer. Physiol. Soc., J. Field, ed.), Sect. 5, pp. 281–293. Williams & Wilkins, Baltimore, Maryland.

Froesch, E. R. and Ginsberg, J. L. (1962). Fructose metabolism of adipose tissue. Comparison of fructose and glucose metabolism in epidydimal adipose tissue of normal rats. *J. Biol. Chem.* **237**, 3317.

Froesch, E. R., Zapf, J., Keller, U., and Oelz, O. (1971). Comparative study of the metabolism of U-^{14}C-fructose, U-^{14}C-sorbitol and U-^{14}C-xylitol in the normal and in the streptozotocin-diabetic rat. *Eur. J. Clin. Invest.* **2**, 8–14.

Horecker, B. L., Lang, K., and Takagi, Y., eds. (1969). "Metabolism, Physiology, and Clinical Uses of Pentoses and Pentitols." Springer-Verlag.

Jakob, A., Williamson, J. R., and Asakura, T. (1971). Xylitol metabolism in perfused rat liver. Interactions with gluconeogenesis and ketogenesis. *J. Biol. Chem.* **246,** 7623–7631.

Keller, U., and Froesch, E. R. (1971). Metabolism and oxidation of U-^{14}C-glucose, xylitol, fructose and sorbitol in the fasted and in the streptozotocin-diabetic rat. *Diabetologia* **7,** 349–356.

Lang, K. (1971). Xylit, Stoffwechsel und klinische Verwendung. *Klin. Wochenschr.* **49,** 233–245.

Smith, M. G. (1962). Polyol dehydrogenases. 4. Crystallization of the L-iditol dehydrogenase of sheep liver. *Biochem. J.* **83,** 135–144.

Touster, O., and Shaw, D. R. (1962). Biochemistry of the acyclic polyols. *Physiol. Rev.* **42,** 181–225.

Williamson, J. R., Jakob, A., and Refino, C. (1971). Control of the removal of reducing equivalents from the cytosol in perfused rat liver. *J. Biol. Chem.* **246,** 7632–7641.

CHAPTER 17

Comparative Metabolism of Xylitol, Sorbitol, and Fructose

H. FÖRSTER

It is scarcely possible to eliminate sugars widely used as sweetening agents from human nutrition. Since consumption of sucrose is contraindicated in the diabetic state, fructose, sorbitol, or xylitol can be used instead of sucrose under this condition. Sugars are especially important for intravenous nutrition to meet caloric demands, because polysaccharides cannot be used parenterally. Glucose is certainly the most important carbohydrate in parenteral nutrition, and the indications for its use are wide. Glucose, however, is metabolized preferentially by the peripheral tissues in the presence of insulin, whereas fructose, sorbitol, and xylitol are taken up to a greater part by the liver and are insulin-independent. The preferential metabolism of these three substances in liver is the cause for some effects common to them. Under certain conditions with impaired glucose tolerance, such as diabetes mellitus, the postoperative state, or heavy burns, insulin is additionally required if glucose is used. Especially in these cases fructose, sorbitol, and xylitol would be better suited for intravenous therapy than glucose.

A. Metabolic Pathways

Sorbitol is metabolized via fructose in the liver (Fig. 1). It is oxidized by sorbitol dehydrogenase involving the reduction of NAD (Hollmann,

259

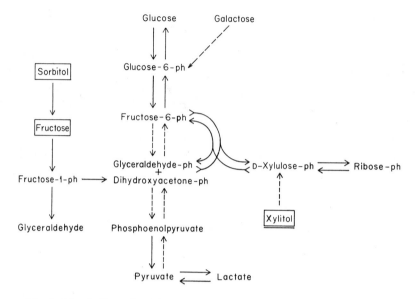

Fig. 1 Metabolism of sorbitol, fructose, and xylitol in the liver.

1961). Fructose is phosphorylated by hepatic fructokinase and ATP. Fructose 1-phosphate and ADP are formed. This product is split to dihydroxyacetone phosphate and glyceraldehyde by hepatic 1-phosphofructaldolase.

While dihydroxyacetone phosphate is part of the glycolysis, glyceraldehyde is, to our knowledge, mainly either phosphorylated directly by thiokinase or it is oxidized to glyceric acid and subsequently phosphorylated by glyceric acid kinase (Heinz *et al.*, 1968). Between 80 and 90% of administered fructose is metabolized in this manner in the liver of man. In the absence of glucose, fructose is phosphorylated in peripheral tissues by means of hexokinase to fructose 6-phosphate. Because of the high K_m for fructose and the low K_m for glucose, glucose is preferentially utilized by this enzyme (Leuthardt and Stuhlfauth, 1960). There is a possibility that fructose is phosphorylated by hexokinase in a state where glucose is excluded from the peripheral tissues, e.g., in the absence of insulin.

Xylitol is oxidized to D-xylulose by the same enzyme which initiates sorbitol metabolism. The intramitochondrial xylitol dehydrogenase is probably not involved in the metabolism of exogenous xylitol (Kupke and Lamprecht, 1967). D-Xylulose is phosphorylated to D-xylulose-5-phosphate which is further metabolized by means of the transaldolase and transketolase reactions of the pentose phosphate cycle to glyceraldehyde phosphate and fructose 6-phosphate. Therefore, xylitol is transformed to components of the glycolysis pathway as a whole, whereas the fate of one part of the 1-phosphofructaldolase reaction, the glyceraldehyde, is not quite clear.

The complete metabolism of all three substances is catalyzed by enzymes which are localized entirely in the cytoplasmic compartment (Bässler, 1971; Förster, 1972; Hollmann, 1961; Kupke and Lamprecht, 1967; Leuthardt and Stuhlfauth, 1960). The products of the metabolism of all three substances are channeled mainly into glycolysis, which is also located in the cytoplasm. Since more than 80% of fructose, sorbitol, and xylitol is metabolized in the liver, the main product should be glucose which is subsequently released to the circulation. This expectation was established by the experiments of Keller and Froesch (1972). In experiments with human subjects and experimental animals, these authors demonstrated that all three glucose precursors are rapidly and extensively converted to glucose in an insulin-independent manner (Keller and Froesch, 1972). These results are in accord with the results of our own laboratory (Förster, 1972; Förster *et al.*, 1973; Mehnert *et al.*, 1964).

B. Experiments in Isolated Perfused Rat Liver

As shown in Fig. 2 fructose as well as xylitol and sorbitol are taken up rapidly by the isolated perfused rat liver. However, glucose is not metabolized by this isolated organ preparation if livers are taken from fasting animals. Surprisingly, until now we have not found any possibility

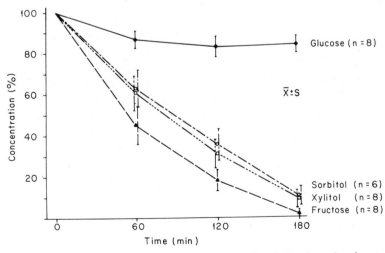

Fig. 2 Fructose, xylitol, and sorbitol taken up by isolated perfused rat liver.

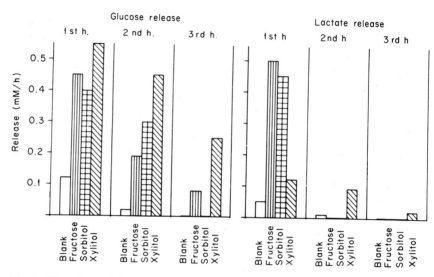

Fig. 3 The effect of fructose, sorbitol, and xylitol on release of glucose and lactate by the isolated perfused rat liver.

to induce the isolated liver to take up glucose from the perfusion solution; for example, addition of insulin to the perfusion solution was without effect (Förster *et al.*, 1974).

In the isolated liver preparation xylitol is metabolized mainly to glucose, whereas addition of fructose and sorbitol leads to an increase in lactate concentrations also (Fig. 3). During the 3-hr experimental period, 60–70% of the xylitol added was transformed to glucose, whereas the values for fructose and sorbitol were 45–50%. Using fructose and sorbitol, 15–25% of the substances taken up by the liver was released in the form of lactate, whereas this was the case for only 5–10% of the added xylitol. As sorbitol and xylitol are metabolized in the cytoplasmic compartment by action of the polyoldehydrogenase, cytoplasmic NAD is reduced concomitantly. The consequence is a rise in the lactate-pyruvate ratio, which can be measured in the isolated liver perfusion also (Fig. 4). A very high lactate-pyruvate ratio is achieved by additions of xylitol and sorbitol, as well as by ethanol. Fructose has only little effect, whereas glucose is without an effect. At the end of the third hour, when sorbitol and xylitol are almost completely taken up from the perfusion solution, a rapid fall in the ratio is seen. Obviously there is a very effective mechanism for transporting reducing equivalents from the cytosol to the mitochondria. On the other hand, the utilization of ethanol is not complete and the ratio remains elevated.

Glucose 6-phosphate is formed in the liver during gluconeogenesis from fructose, sorbitol, and xylitol. After hydrolysis this glucose 6-phosphate can contribute to regulation of blood glucose concentration. It can also be transformed to UDP-glucose via glucose 1-phosphate and stored as glycogen. Glycogen synthesis is minimal in the isolated perfused rat liver. Only after the addition of xylitol is some glycogen formed in this isolated organ preparation (Förster *et al.*, 1972e). In the living animal there is a correlation between blood glucose concentration and glycogen storage (Förster *et al.*, 1972c). Elevation of blood glucose concentration over 130–150 mg/100 ml (i.e., 8–9 mM) leads to glycogen storage (Fig. 5), which increases further with the increasing blood glucose concentration up to 600 mg/100 ml (i.e., 25–30 mM). As glucose is given intravenously to unrestrained animals, blood glucose concentration equals portal glucose concentration, whereas during intestinal glucose absorption, portal blood glucose concentration in rats is about 100 mg/100 ml higher than venous blood glucose concentration (Förster *et al.*, 1972d).

In the whole animal glycogen storage is also achieved by intravenous administration of fructose, sorbitol, and xylitol, as well as of glucose (Fig. 5). A part of this stored glycogen stems from the glucose newly synthesized from the glucose precursors. In the experiment shown in Fig. 5, 0.4 gm of glucose or glucose precursors per kilogram of body weight per

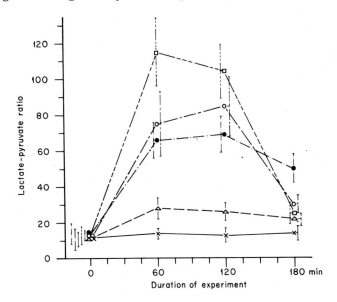

Fig. 4 The effect of fructose, sorbitol, xylitol, or ethanol on lactate-pyruvate ratio using the isolated perfused rat liver ($\overline{X} \pm S$). (×) Basal ($n = 10$), (□) basal, 2.4 mmole xylitol ($n = 10$), (○) basal, 2.0 mmole sorbitol ($n = 6$), (△) basal, 2.0 mmole fructose ($n = 6$), (●) basal, 2.0 mmole ethanol ($n = 6$).

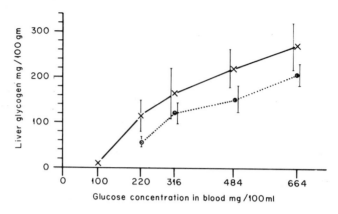

Fig. 5 Relation between blood glucose concentration and liver glycogen deposition during 2 hr intravenous glucose infusion in starving rats (n = 8–15). $\bar{X} \pm$ S. ($\times$) Chemically determined and ($\bigcirc$) ^{14}C-labeled.

hour were infused intravenously for 2 hr. About 50% of the carbohydrates administered were stored in the form of glycogen; the effects of xylitol, fructose, and sorbitol were obviously more distinct than the effect of glucose. In general, maximum glycogen deposition when using these glucose precursors is up to 50% higher than when using glucose. Additionally, glycogen deposition is achieved by a lower blood glucose concentration, as seen from the left side of Fig. 5. The correct values for the transformation of carbohydrates to glycogen would be lower. Using ^{14}C labeled glucose only 35–60% of the ^{14}C glycogen stems from the infused carbohydrates (Förster *et al.,* 1972e). The remaining glycogen obviously stems from gluconeogenesis from other precursors, e.g., protein.

When the infusion rate was doubled, the glucose concentration was raised, especially in those animals which received glucose infusion (Fig. 6). However, glycogen deposition was raised mainly in the animals with infusion of fructose, sorbitol, and especially with xylitol. The glycogen deposition occurred despite nearly normal blood glucose concentrations. The differences in glycogen deposition between animals having received glucose infusions and animals having received infusions of glucose precursors are statistically significant, as well as are the differences in the elevation of blood glucose concentration.

Moreover, the glycogen deposition in streptozotocin-diabetic animals was significant only, under the experimental conditions used, when fructose, sorbitol, or xylitol were administered (Fig. 7). In the course of glucose infusion in diabetic animals no glycogen was stored in the liver. These results confirm earlier observations of Bässler and Heesen (1963) with alloxan-diabetic animals. On the other hand, glycogen deposition

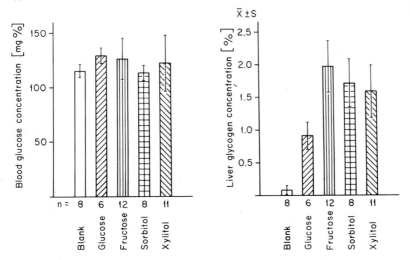

Fig. 6 Blood glucose concentration (second hour) and liver glycogen concentration following 2 hr intravenous infusion of glucose, fructose, sorbitol, and xylitol (0.15 gm/hr) in rats.

in diabetic animals is significantly lower as compared to nondiabetic animals. In all cases the rise in blood glucose concentration was highest after glucose administration, whereas it was much lower following infusions of glucose precursors (Fig. 8).

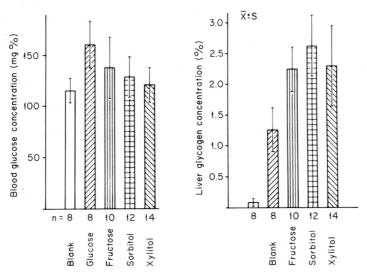

Fig. 7 Blood glucose concentration (second hour) and liver glycogen concentration following 2 hr intravenous infusion of glucose, fructose, sorbitol, and xylitol (0.3 gm/hr) in rats.

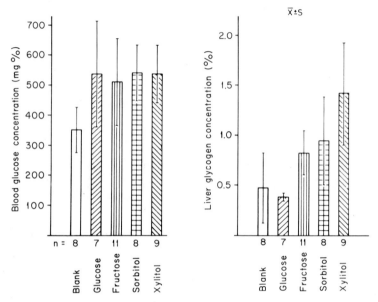

Fig. 8 Blood glucose concentration (second hour) and liver glycogen concentration following 2 hr intravenous infusion of glucose, fructose, sorbitol, and xylitol (0.3 gm/hr) in streptozotocin-diabetic rats.

C. Experiments in Humans

Studies on isolated perfused rat liver and with experimental animals were complemented and extended by experiments on human volunteers (Förster, 1972, 1973; Förster *et al.*, 1970). We have infused 1.5 gm/kg of body weight of the respective substance in 20% solution within 20 min. Except for glucose infusions blood glucose concentration was never significantly increased. Obviously the slow transformation to glucose as well as the partial glycogen deposition and some other factors, perhaps reduction of gluconeogenesis, contributed to this effect. (See also Mehnert *et al.*, 1970.)

On the other hand, we were able to reproduce the other effects shown by rat liver perfusion studies such as changes in pyruvate or lactate concentrations. The changes in blood concentration were diminished partly because the estimations were performed using blood from the cubital vein and not with blood from the hepatic veins. Therefore, dilution effects have to be considered.

As shown in Fig. 9 lactate levels were most elevated following fructose infusions. However, sorbitol was almost as effective, and glucose also caused a significant rise in blood lactate concentration. Confirming the liver perfusion studies, lactate was only slightly elevated following xylitol infusions. On the other hand, the lactate-pyruvate ratio was influenced most by xylitol infusions.

Unlike the results with the isolated liver, glucose was also effective in the human subjects. This might result from peripheral effects of glucose and glucose-mediated insulin secretion. On the other hand, in the living organism glucose uptake and metabolism obviously occur also by the liver in contrast to the isolated perfused organ. The changes in lactate concentrations and in lactate-pyruvate ratio were altogether minimal as compared with the results of liver perfusion studies where manifold changes were observed.

Additional experiments were performed to study the acute effect of the reduction of liver cytoplasmic NAD on a very sensitive parameter. The galactose metabolism in liver is limited by the conversion of UDP-galactose to UDP-glucose by hexose epimerase (Förster *et al.*, 1973). The co-

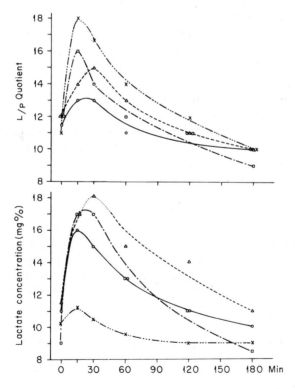

Fig. 9 Lactate-pyruvate quotient and lactate concentration in fasting voluntary subjects following intravenous infusion of (○) glucose, (△) fructose, (□) sorbitol, or (×) xylitol (1.5 gm/kg of body weight).

enzyme of this reaction is probably NAD which must be kept in the oxidized state to be effective. Neither sorbitol nor xylitol effectively in- hibited galactose metabolism in healthy volunteers. These results favor- ably complement the efficient gluconeogenesis in perfused rat liver and the glycogen deposition in the experimental animal. The results taken together also demonstrate that the reduction in cytoplasmic NAD achieved by sorbitol or xylitol does not influence metabolism in the same manner as the same effect of ethanol. Following ethanol administration in human subjects, lactate-pyruvate level is also elevated. However, glu- coneogenesis is diminished (Lieber, 1967) as well as galactose metabolism retarded during ethanol metabolism (Forsander, 1966). Both effects are in contrast to the effect of the polyalcohols. Moreover, in isolated per- fused rat liver, addition of ethanol to the perfusion solution inhibits con- version of fructose, xylitol, and sorbitol to glucose. Under these experi- mental conditions, xylitol metabolism is especially inhibited, whereas the uptake of fructose is not influenced significantly. One may therefore con- clude that there are compartmental differences in the metabolism of poly- alcohols as compared with ethanol.

Some years ago it was reported that intravenous administration of fructose causes an increase in serum uric acid (Förster *et al.*, 1967), an effect which was confirmed shortly thereafter by other authors (Perheen- tupa and Raivio, 1967). It has also been shown that intravenous infusion of sorbitol and xylitol leads to an increase in serum uric acid concentra- tion, whereas glucose and galactose are without effect (Fig. 10) (Förster *et al.*, 1970). Oral administration of fructose, sorbitol, xylitol, and also of sucrose increased serum uric acid concentration as well (Förster and

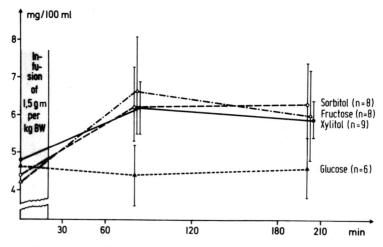

Fig. 10 Serum uric acid in healthy volunteers (according to Förster *et al.*, 1970).

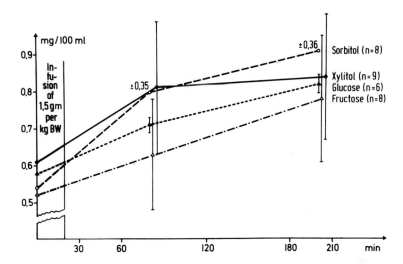

Fig. 11 Serum bilirubin in healthy volunteers (according to Förster *et al.*, 1970).

Ziege, 1971; Förster *et al.*, 1972a). Since sucrose intake of 200 gm/day by human volunteers caused an increase in serum uric acid concentration, this effect should be very common to North American nutrition. Schumer and Edwards confirmed the observations in the case of intravenous administration of xylitol (Edwards and Edwards, 1971; Schumer, 1971). However, these authors interpreted this well-established increase in uric acid concentration as an adverse reaction, not noticing that the same effect is evoked by fructose, sorbitol, or sucrose as well.

The other so-called adverse reaction drawn attention to by Schumer was an increase in serum bilirubin concentration following intravenous xylitol administration (Schumer, 1971; Thomas *et al.*, 1972). Using healthy volunteers (Fig. 11), we were able to demonstrate a small increase in serum bilirubin concentration following rapid intravenous infusion of 1.5 gm/kg of various carbohydrates. The increase was identical regardless of whether glucose, fructose, xylitol, or sorbitol was used (Förster *et al.*, 1970). However, 24 hr later the serum bilirubin concentration had returned to the initial level again. An increase in serum bilirubin concentration following intravenous glucose infusion was already reported more than 20 years ago (Popper and Schaffner, 1961).

Despite the conviction that this increase in serum bilirubin concentration could not be an adverse reaction, animal experiments were immediately undertaken (Förster *et al.*, 1970). Using rats again, we performed 72-hr intravenous infusions of carbohydrates at the rate of 5.6 gm/day, i.e., 0.7 gm/kg/hr (Fig. 12). At the end of the infusion period blood was drawn and aspartate aminotransferase, alanine aminotransferase, and

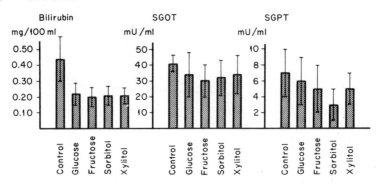

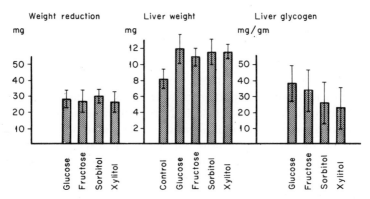

Fig. 12 Effect of 72 hr lasting intravenous infusions of some carbohydrates in rats on some parameters; D = 5.6 gm/day.

	Aspartate aminotransferase (mU/ml)	Alanine aminotransferase mU/ml
Sorbitol	7.2–6.0	4.8–5.0
Xylitol	10.0–9.2	5.0–5.2
Fructose	10.8–9.2	5.4–6.0
Glucose	10.2–9.9	5.5–5.0

bilirubin were determined. In all instances the values were lower than in control animals. However, liver glycogen concentration was raised in the animals receiving the continuous infusions as compared to values obtained from animals starving the same amount of time. The increase in liver glycogen concentration was almost independent of the substance used for infusion. This was true despite the stress situation of the animals which were restrained for the whole time. The findings of these animal studies are inconsistent with any degree of hepatotoxicity of the carbohydrates used. In all groups of animals the mortality during infusions was of the same magnitude.

One can summarize these experimental results by saying that xylitol, sorbitol, and fructose are used as glucose precursors by the liver after oral or intravenous administration. Furthermore they should have advantages also, as compared to glucose, in being the most often used carbohydrate in intravenous nutrition. Summarizing earlier experimental results, it can be stated that blood glucose concentration is increased only slightly following administration of sorbitol, fructose, or xylitol (Bässler, 1971; Förster, 1972; Lang, 1971; Leuthardt and Stuhlfauth, 1960; Mehnert *et al.*, 1970). On the other hand, when using glucose for infusion under several situations, insulin would be required to diminish extreme hyperglycemia. In spite of these widely known results it is a common view that fructose, sorbitol, and xylitol are influencing metabolic events only by way of transformation to glucose and consequently by insulin secretion. The following experiments were undertaken to test this hypothesis.

Using the 72-hr intravenous infusion of carbohydrates in rats, the influence on urinary urea excretion was tested. In the control group which received 0.23% saline, urea excretion increased on the first day of the experiment (Fig. 13). Urea excretion was only slightly reduced on the

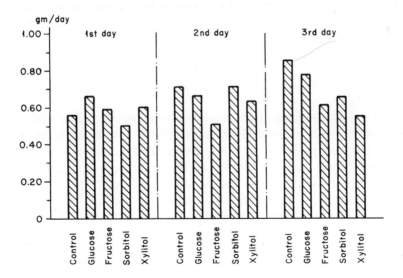

Fig. 13 Urea excretion following intravenous infusion of various carbohydrates ($d = 5.6$ gm/day) in rats for 72 hrs.

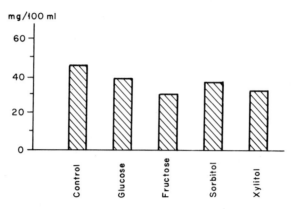

Fig. 14 Serum urea concentration following infusion of glucose and sorbitol.

second and on the third day of the experiment, when glucose was added to the infusion solutions. Sorbitol was as effective as glucose but fructose, and especially xylitol, were much better suited for the reduction of urea excretion. This effect is also corroborated by the serum urea concentration. In comparison to the control animals, serum urea concentration is diminished to the same degree following infusion of glucose and sorbitol (Fig. 14). However, fructose and xylitol have a twofold effect on this parameter. These results show that the effect of the sugars and sugar alcohols on renal urea excretion is an extra-renal one, being connected with protein metabolism. In these groups of experiments blood glucose concentration was raised most during glucose infusion, whereas infusions of xylitol, sorbitol, and fructose had only little effect on blood glucose concentration. Serum insulin concentration (IRI) was only elevated following glucose infusions. It was hardly influenced by the infusions of the glucose precursors.

These results conclude that at least in experimental animals the nitrogen sparing effect of sorbitol and especially of xylitol and fructose is a direct one, not mediated by the primary metabolization to glucose of these substances. Another interesting result of these experiments is the fact that urea excretion rises day by day in the control group. Fructose and xylitol infusion have hardly any effect on urea excretion on the first day, but these substances prevent the increase in excretion on the following days.

Despite the high infusion rate of the substances of 5.6 gm/day, i.e., about 0.7 gm/kg/hr, the urinary excretion of the substances was low. For the polyols, sorbitol and xylitol, it amounted up to 20% of the doses given intravenously. In the case of fructose only half as much was excreted in the urine. Infusion of glucose or glucose precursors was not accompanied by glycosuria.

D. Experimental Model for Extreme Hepatic Gluconeogenesis

To gain more information for the benefit of the various carbohydrates, we looked for a model to perform studies under conditions of extreme hepatic gluconeogenesis, e.g., the phlorizin diabetes studied extensively by Graham Lusk 60 years ago (Lusk, 1912). The glycoside phlorizin blocks renal tubular reabsorption of glucose and in this way leads to renal glycosuria. We administered phlorizin by means of continuous intravenous infusion with or without addition of carbohydrates. Previously we tested the minimum phlorizin dose required to cause maximum glycosuria. The renal glucose loss in phlorizin diabetic rats measured more than 1 gm/24 hr. Therefore, more than 2 gm of protein were metabolized to glucose. Evaluated on body weight the glucose excretion amounted to 3–5 gm/kg/day, and the additional nitrogen loss caused by phlorizin was 2–3 gm/kg of body weight per day.

Glucose excretion was elevated when glucose or glucose precursors were infused additionally (Fig. 15) as shown earlier by Lusk, who created

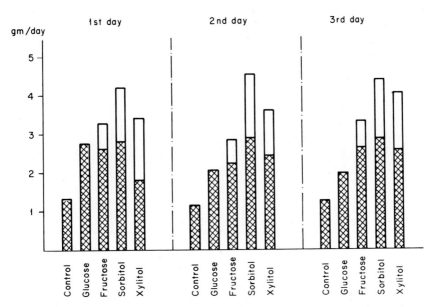

Fig. 15 Sugar excretion following intravenous infusions of various carbohydrates (d = 5.6 gm/day) and phlorizin (d = 0.03 gm/day) in rats for 72 hrs.

the term "extraglucose" for the additional glucose excretion (Lusk, 1912). Glucose loss was highest following sorbitol infusion and least during xylitol infusion. However, even under these extreme conditions the greater part of the infused substances was obvioulsy metabolized by the organism. On the basis of the control experiments, the loss in the form of the substance or in the form of glucose in most cases was less than 50%. The utilization was best using glucose, but it was not much worse using fructose. After the infusion of 5.6 gm of sorbitol per day, approximately 1.5 gm of sorbitol and additionally 1.5 gm of glucose were excreted by the kidneys. The renal excretion of urea was significantly reduced to nearly the same amount by any of these substances in the first 2 days of the experiment (Fig. 16). On the last day no differences in the nitrogen excretion were to be seen in the phlorizin-diabetic animals.

As compared with the infusion series without addition of phlorizin, the "extra nitrogen" excretion caused by phlorizin was prevented by glucose or by the glucose precursors on the first day. In the further course of the experiment, in contrast to the infusions without phlorizin, the effect of the sugar infusions was reduced; nitrogen excretion on the last day was the same in all cases and also was not much different from nitrogen excretion in the control group receiving phlorizin alone. On the third day

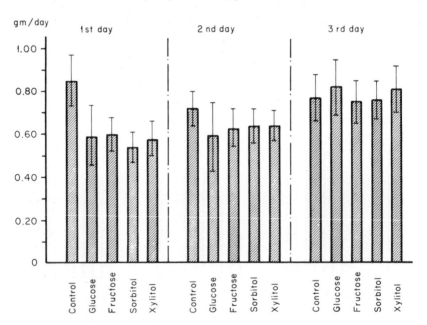

Fig. 16 Renal excretion of urea following intravenous infusions of various carbohydrates (D = 5.6 g/day) in rats for 72 hrs.

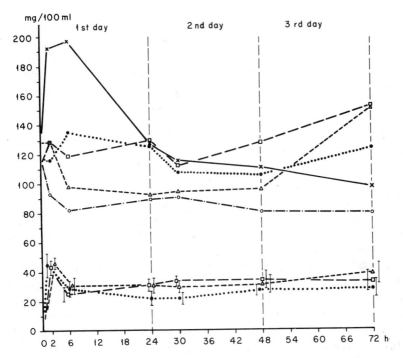

Fig. 17 Blood glucose concentration and substrate concentration during intravenous infusion of various carbohydrates (D = 5.6 gm/day). (□) Sorbitol, (△) fructose, (●) xylitol, (×) glucose, and (○) control.

of the experiment, the urea excretion of the controls receiving only an infusion of 0.23% NaCl was also identical with the maximum urea excretion in the phlorizin series.

When evaluating blood glucose concentrations in the phlorizin-diabetic animals, a rise was seen already in the initial period of the experiment before the infusions were started (Fig. 17). These elevated basal values were obviously caused by the restraining of the rats in the metabolic cages. At the end of the experimental period the blood glucose concentration rose in experiments with sorbitol as well as with fructose infusions, whereas the blood glucose concentration with xylitol infusion was nearly identical with glucose infusion. Obviously, as a result of the stress situation of the animals, blood glucose concentration was also significantly elevated in the experiments with infusion of 0.23% saline. The blood concentrations of sorbitol, xylitol, and fructose were nearly identical over the whole infusion perod of 72 hr. In contrast to this, renal fructose excretion was much less than sorbitol excretion, xylitol excretion being intermediate.

Renal loss of sorbitol, xylitol, and fructose was nearly doubled in the experiments with phlorizin. Since the blood concentrations of these substances were lowered in parallel to increased excretion, this is obviously a renal effect of phlorizin. It can be assumed that renal excretion of fructose, sorbitol, and xylitol is enhanced by phlorizin by way of a mechanism which is similar to that which causes renal glucose loss.

The low blood concentrations of sorbitol, xylitol, and fructose showed an effective elimination of these substances from the bloodstream of the experimental animal. Renal loss was less than 20% despite the high infusion rate of approximately 0.7 gm/kg of body weight per hour. In human subjects, xylitol and sorbitol were significantly less utilized than fructose when continuous infusion studies were performed (Bässler, 1971; Berg, *et al.*, 1973). On the other hand, the elimination rates of fructose, sorbitol, and xylitol in human subjects are fairly good and comparable to that of glucose.

The different mechanism of action of glucose versus fructose, sorbitol, and xylitol is also demonstrated by the influence of the rapid infusion of these substances on the concentration of free fatty acids in human volunteers (Fig. 18). During rapid glucose infusion the concentration of free fatty acids is depressed within the infusion period by more than 50%, and there is a continuous increase to the preinfusion level within the next hours. However, following the infusion of the other three substances the nadir in free fatty acid concentration is reached only 30 min

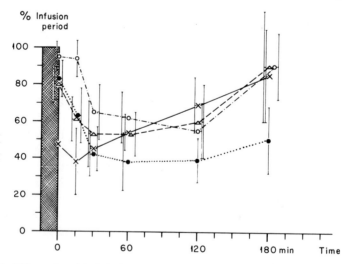

Fig. 18 Effect of quick infusion of carbohydrates on the level of free fatty acids in plasma (d = 1.5 g/kg of body weight, t = 15 min). $\bar{X} \pm$ S. ($\triangle$) Fructose, n = 8; ($\times$) glucose, n = 10; ($\bullet$) xylitol, n = 7; and ($\bigcirc$) sorbitol, n = 5.

after termination of the infusion; the maximum depression lasts for a longer time.

There might be different mechanisms responsible for the depression of serum free fatty acids. It is assumed that the rapid glucose effect is mediated by the measurable glucose-induced insulin secretion. This effect is therefore especially a peripheral one concerning inhibition of fatty acid mobilization. Because fructose and sorbitol are metabolized to more than 80% by the liver, their relatively slow effect might be predominantly hepatic, where an increased esterification of circulating free fatty acids might be achieved by the two substances (Bässler *et al.*, 1966; Gericke *et al.*, 1968; Kupke and Lamprecht, 1967). Finally, the xylitol effect is a combination of the stimulated hepatic esterification and the diminished peripheral release of fatty acids. In this high dose, xylitol stimulates insulin secretion in man. On the other hand, xylitol is metabolized predominantly by the liver. This combined mechanism leads to a deep and long-lasting depression of the concentration of free fatty acids by administration of xylitol. The increased incorporation of labeled fructose into hepatic glycidglycerol was already demonstrated (Gericke *et al.*, 1968; Kupke and Lamprecht, 1967). Another well-known effect of the polyalcohols and of fructose is decreased ketogenesis. This effect could also be mediated by the increased esterification of the fatty acids and by a decreased metabolism of the fatty acids in the liver (Bässler 1963; Bässler *et al.*, 1966; Haydon, 1969).

The extensive investigations performed with human volunteers, with the experimental animal, and with the isolated perfused rat liver have shown good utilization of fructose, sorbitol, and xylitol by the liver. Fortunately, one of the main products of their metabolism is glucose, which is produced in large quantities and is continuously required, especially by the brain and the blood cells. Therefore, the hepatic glucose production is not an artifact following the application of sorbitol, xylitol, and fructose. It is a vital function for the whole body glucose homeostasis. During fasting periods most of the glucose required is contributed by the liver. Under these conditions when glycogen stores are depleted, glucose is formed from proteins. Under these conditions a nitrogen sparing effect is shown by glucose as well as by fructose, sorbitol, and xylitol. There is no indication that this effect of fructose, sorbitol, or xylitol is mediated via gluconeogenesis and insulin secretion. On the contrary, fructose and xylitol were more effective than glucose on inhibition of protein catabolism despite the fact that blood glucose concentration was scarcely elevated.

In the postoperative state, following heavy burns, in diabetes, in renal disease, and perhaps under some other conditions, the glucose utilization by the liver is disturbed. This hepatic glucose intolerance is also shown by the isolated perfused rat liver, which cannot be made to take up glucose from the perfusion solution. On the other hand, also under these conditions, the liver metabolism requires energy and some extrahepatic tissues require glucose also. This need for energy in the liver which amounts to 30% of the basal metabolic rate and for glucose in extrahepatic tissues is met best by the application of glucose precursors which are to a great part transformed to glucose by the liver and which have the additional advantage of reducing nitrogen catabolism. The application of xylitol, sorbitol, or fructose causes contemporary glycogen storage in the liver of diabetic animals, whereas glucose is without effect. Following the application of these substances hyperglycemia is generally not seen.

In contrast to some speculations published by other authors which are based on very incomplete material, we were unable to find so-called adverse reactions by any of these substances. Apart from a rise in lactate concentration and an increase in lactate-pyruvate ratio which are common metabolic consequences of carbohydrate metabolism, increases in uric acid concentration and in serum bilirubin concentration were said to be adverse reactions. However, uric acid production is stimulated by fructose, sorbitol, and xylitol as well. Since an oral intake of sucrose has the same effect, most North Americans would suffer from this adverse reaction.

Rapid intravenous infusions of fructose, sorbitol, and xylitol in high dosage of 1.5 gm/kg of body weight were followed by an increase in serum bilirubin concentration. However, the same effect was also caused by glucose. In extensive studies using experimental animals, there were no signs of adverse reactions caused by infusions of sorbitol, xylitol, and fructose. However, the rat, like man, does not respond with hypoglycemic reactions to administration of xylitol. Therefore, experiments with other animals like the dog where hypoglycemic reactions are caused by xylitol are not to be compared to the situation with the human subject or even with the rat.

References

Bässler, K. H. (1971). Die Rolle der Kohlenhydrate in der parenteralen Ernährung. *Z. Ernaehrungswiss. Suppl.* **10,** 57.

Bässler, K. H., and Dreiss, G. (1963). Antiketogene Wirkung von Xylit bei alloxandiabetischen Ratten. *Klin. Wochenschr.* **41,** 593.

Bässler, K. H., Fingerhut, M., and Czok, G. (1966). Hemmung der Fettsäureoxydation als ein Faktor bei der antiketogenen Wirkung von Zuckern and Zucheralkoholen. *Klin. Wochenschr.* **44,** 899.

Bässler, K. H., and Heesen, D. (1963). Die Bildung von Leber und Muskelglykogen aus Sorbit, Xylit und Glucose bei gesunden und alloxandiabetischen Ratten. *Klin. Wochenschr.* **41**, 595.

Berg, G., Matzkies, F., Bickel, H., and Zeilhofer, R. (1973). Säure-Basen-Haushalt bei Dauerinfusion von Xylit, Fructose, Glucose und Kohlenhydratmischungen. *Z. Ernaehrungswiss. Suppl.* **15**, 47.

Edwards, R. G., and Edwards, J. B. (1971). *Technicon Symp.*, Frankfurt, 7.-9.6.1971.

Forsander, O. A. (1966). The galactose tolerance test as a measurement of the redox potential of the liver. *Scand. J. Clin. Lab. Invest* **18**, Suppl. 92, 143.

Förster, H. (1972). Grundlagen für die Verwendung der drei Zucker austauschstoffe Fructose, Sorbit und Xylit. *Med. Ernaehrung* **13**, 7.

Förster, H. (1974). The effect of intravenous carbohydrates on various parameters in blood. *Z. Ernaehrungswiss.* **1**, 24.

Förster, H., Hoffman, H. and Hoos, I. (1973). Stoffwechselwirkungen verschiedener Kohlenhydrate und deren Bedeutung für die Infusionstherapie. *Z. Ernaehrungsw. Suppl.* **15**, 28.

Förster, H., and Ziege, M. (1971). Anstieg der Serumharnsäurekonzentration nach oraler Zufuhr von Fructose, Sorbit und Xylit. *Z. Ernaehrungswiss.* **10**, 524.

Förster, H., Mehnert, H., and Alhough, I. (1967). Anstieg der Serumharnsäure nach Fructose. *Klin. Wochenschr.* **45**, 436.

Förster, H., Meyer, E., and Ziege, M. (1970). Erhöhung von Serumharnsäure und Serum-bilirubin nach hochdosierten Infusionen von Sorbit, Fructose und Xylit. *Klin. Wochenschr.* **48**, 878.

Förster, H., Boecker, S., and Ziege, M. (1972a). Anstieg der Konzentration der Serumharnsäure nach akuter und nach chronischer Zufuhr von Saccharose, Fructose, Sorbit und Xylit. *Med. Ernaehr.* **13**, 193.

Förster, H., Hoos, I., and Lerche, D. (1972b). Hepatische Glykogensynthese, Untersuchunger *in vivo* und *in vitro. Hoppe Seyler's Z. Physiol. Chem.* **353**, 1514.

Förster, H., Meyer, E., and Ziege, M. (1972c). Hepatische Glykogensynthese in Abhängigkeit von der Blutglucose-konzentration bei narkotisierten Ratten. *Klin. Wochenschr.* **50**, 478.

Förster, H., Meyer, E., and Ziege, M. (1972d). The intestinal absorption of glucose with the simultaneous determination of the arterioportal glucose concentration difference. *Rev. Eur. Etud. Clin. Biol.* **17**, 1084.

Förster, H., Lerche, D., and Hoos, I. (1972e). Einfluss von Fructose, Sorbit und Xylit auf die Gluconeogenese und auf die Glykogenspeicherung. *F. Kongr. Deut. Diabetes Ges.* Abstr., p. 52. Nanheim 4.-6.5.

Förster, H., Mehnert, H., Haslbeck, M., and Büchele, H. *Z. Ernaehrungswiss.* (in press).

Gericke, C., Rauschenbach, P., Kupke, I., and Lamprecht, W. (1968). Biosynthese des Glyceridglycerins aus Fructose. *Hoppe-Seyler's Z. Physiol. Chem.* **349**, 1055.

Haydon, R. K. (1969). The antiketogenic effect of polyhydric alcohols in rat liver slices. *Biochim. Biophys Acta* **46**, 598.

Heinz, F., Lamprecht, W., and Kirch, J. (1968). Enzymes of fructose metabolism in human liver. *J. Clin. Invest* **47**, 1826.

Hollmann, S. (1961). "Nichtglykolytische Stoffwechselwege der Glucose." Thieme, Stuttgart.

Keller, U., and Froesch, E. R. (1972). Vergleichende Untersuchungen über den Stoffwechsel von Xylit, Sorbit und Fructose beim Menschen. *Schweiz. Med. Wochenschr.* **102**, 1017.

Kupke, I., and Lamprecht, W. (1967). Einbau von uniform markierter Fructose in Leberlipide. *Hoppe-Seyler's Z. Physiol. Chem.* **348**, 17.

Lang, K. (1971). Xylit, Stoffwechsel und klinische Verwendung. *Klin. Wochenschr.* **49**, 233.

Leuthardt, F., and Stuhlfauth, K. (1960). Biochemische, physiologische und klinische Probleme des Fructose-stoffwechsels. *Med. Grundlagenforsch.*, **3**, 413.

Lieber, C. S. (1967). Chronic alcoholic injury in experimental animals and man: biochemical pathways and nutritional factors. *Fed. Proc., Fed. Amer. Soc. Exp. Biol.* **26**, 1443.

Lusk, G. (1912). Phlorinzinglucosurie. *Ergeb. Physiol.* **12**, 315.

Mehnert, H., Summa, J. D., and Förster, H. (1964). Untersuchungen zum Xylitstoffwechsel bei gesunden, leberkranken und diabetischen Personen. *Klin. Wochenschr.* **42**, 382.

Mehnert, H., Förster, H., Geser, C. A., Haslbeck, M., and Dehmel, K. H. (1970). Clinical use of carbohydrates in parenteral nutrition. *In* "Parenteral Nutrition" (C. H. Meng and D. H. Law, eds.), p. 112. Thomas, Springfield, Illinois.

Perheentupa, J., and Raivio, K. (1967). Fructose induced hyperuricemia. *Lancet* **2**, 528.

Popper, H., and Schaffner, F. (1961). "Die Leber," p. 574. Thieme, Stuttgart.

Schumer, W. (1971). Adverse reactions following xylitol infusion. *Metab., Clin. Exp.* **20**, 345.

Thomas, D. W., Edwards, J. B., Gilligan, J. E., Lawrence, J. R., and Edwards, R. G. (1972). Complications following intravenous administration of solutions containing xylitol. *Med. J. Austr.* **1**, 1238.

CHAPTER 18

Metabolism of Lactose and Galactose

R. G. HANSEN

Lactose is the primary carbohydrate source for developing mammals, and in humans it constitutes 40% of the energy consumed during the nursing period. Why lactose is the unique carbohydrate of milk is unclear, however, especially since most individuals can meet their galactose need by biosynthesis from glucose. Whatever the rationale for galactose in milk, its occurrence in glycoproteins and lipids, particularly in nervous tissue, has suggested a specific role. The physical and organoleptic properties of galactose may also be significant evolutionary determinants.

A. Lactose Catabolism

1. Microorganisms

The utilization of lactose for energy or structural purposes is preceded by hydrolysis to the hexoses prior to absorption (Fig. 1). Since bacteria and yeast can also utilize lactose, the characteristics of the enzymes involved in the hydrolysis have resulted in fundamental studies with microorganisms. The potential for improving the properties and acceptability of lactose in foods by hydrolyzing to the monosaccharides has given impetus to efforts to concentrate the enzyme from microbial sources. Sub-

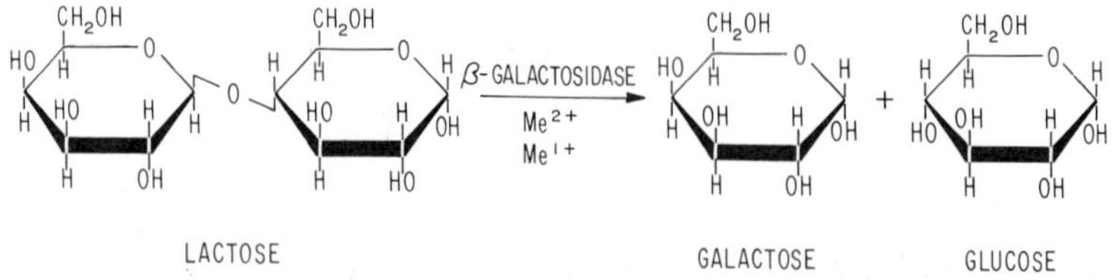

Fig. 1 Utilization of lactose for energy or structural purposes.

stantial progress has been made toward isolating and characterizing the enzyme from lactose-fermenting microorganisms (Wendorff *et al.*, 1971; Strom *et al.*, 1971; Hill and Huber 1971). In microorganisms, β-galactosidases are induced by galactose and by other sugars with a configuration related to galactose.

2. Mammalian Digestion

In mammals, the rate-limiting step in the digestive process appears to be the hydrolysis of the disaccharide into absorbable monosaccharides. Most of the disaccharidase activity is localized in the brush border fraction of the mucosal cells of the small intestine (Miller and Crane, 1961).

Since disaccharides, including lactose, are hydrolyzed by enzymes in the brush border membrane of the small intestine, digestive disturbances that alter the physical structure of the membranes decrease their capacity to metabolize these sugars. Lactase is more sensitive to such disturbances than are the other disaccharidases (Plotkin and Isselbacher, 1964); in fact, it is the first disaccharidase to be removed by digestive disturbance (Littman and Hammond, 1965). This suggests that it may be located more superficially in the microvilli.

Lactose may be the last disaccharidase to be restored when normal digestive function is being reestablished. Because undigested lactose cannot be absorbed, an excess quantity of fluid enters the bowel lumen to dilute the sugar, motility is increased, and the subject may develop abdominal cramps, bloating, or diarrhea (Gudmand-Hoyer *et al.*, 1970; Mc-Michael *et al.*, 1965). In milk-drinking young children, gastroenteritis may produce secondary complications involving faulty disaccharide absorption and a subsequent unavailability of sugar.

a. Lactase Deficiencies

Whether lactase is induced or constitutive has not been clearly established. In populations that traditionally depend on milk as a significant source of energy, most adults retain an ability to hydrolyze lactose. When lactose is not a significant component of the adult diet, however, the capacity to hydrolyze lactose seems to decline over time (Simoons, 1970). Apparently limited quantities of milk (up to a pint per day) and milk products are well tolerated even by some who are deficient in the enzyme. In some lactase-deficient people, however, a pint or more of milk can induce an unusual fermentation that leads to diarrhea, flatulence, or cramping abdominal pain. By contrast, children who have isomaltase–sucrase deficiency are relatively asymptomatic and can tolerate a normal dietary intake of sucrose.

When the synthetic disaccharide lactulose (galactose plus fructose) is consumed, many enzyme-deficient patients can tolerate over 50 gm (equivalent to two pints of milk) daily with very little change in the stool weight (Dahlqvist and Gryboski, 1965). Thus, gastrointestinal symptoms following the consumption of limited quantities of milk should be minimal in most lactase-deficient adults.

Lactose intolerance is common in many nonmilk consuming populations and has been well documented in Thailand (Flatz *et al.*, 1969; Keusch *et al.*, 1969a), based upon measurements taken after an oral dose of 1 gm of lactose per kilogram of body weight following an overnight fast. The ability to metabolize lactose, obviously present during infancy, disappears between the ages of one and four in most individuals in northern Thailand. These same individuals can metabolize sucrose normally; therefore, unspecific damage to intestinal mucosa is probably not the problem. Some of the people surveyed were dairy workers who consumed limited quantities of milk. These, too, were lactose-intolerant as adults. Genetic differences were also evident, however, since a few individuals retained the capacity to hydrolyze lactose as adults.

Evidence that lactase levels respond to an altered lactose intake is questionable (Gilat, 1971). Neither exposure to extra lactose (Cuatrecasas *et al.*, 1965; Keusch *et al.*, 1969b) nor removal of lactose from the diet for periods of 40–50 days altered lactase levels in intestinal tissue in adults. Perhaps of more significance, a group of 10 galactosemic children, ages 7–17 years, who had carefully avoided lactose-containing materials since early infancy, tested normal in lactose tolerance tests, suggesting that their lactase levels had not decreased during their long periods of lactose abstinence. The one exception was a 15-year-old black who was determined by biopsy to be lactase-deficient (Kogut *et al.*, 1967).

In populations that consume only limited quantities of milk as adults, a decreased capacity to metabolize lactose reflects a general, gradual, adaptive decline in enzymatic activity (Bolin and Davis, 1970). Relative to lactose intolerance in Australian-born Chinese, Bolin and Davis concluded that an environmental rather than a genetic factor was operative, citing a continuous intake of lactose in the more tolerant subjects. This was based on the incidence (56%) of lactose intolerance in the test group compared to an incidence of 95–100% in other Chinese residing in Australia and in indigenous Chinese in Singapore.

Another view holds that any adaptive response to lactose in humans is insignificant during the life-span. A genetic basis for adult lactase deficiency is indicated by the 70% of black Americans who are reported to be lactose intolerant, which duplicates the adult intolerance level of black Africans, where exposure to lactose is probably much less. Although the question is not really resolved in humans, adult lactase deficiency may be primarily under genetic control.

Intestinal lactase in adult humans may have occurred initially as a consequence of domestication of milk-producing livestock. In most western European adults and in those derived from Europe ethnically, lactase remains present throughout life; hence, the lactose in dairy foods is effectively digested. According to this concept, adaptation to the presence of lactose in the diet has required many generations over a period of several thousand years. Adult blacks in the United States have not developed intestinal lactase after exposure to lactose for 300 years.

The rarely occurring congenital lactose intolerance that is the result of a deficiency of lactase was named by Holzel (1968) as "alactasia." A limited number of infants having this defect have been documented, but the mode of inheritance is not clear. Another congenital lactose intolerance, associated with the inability of infants to hydrolyze lactose and the subsequent appearance of lactose in the urine, is probably a different and more complex syndrome. When it occurs, this disability generally seems to be secondary to mucosal damage associated with acute infectious diarrhea.

b. Human Galactosidases

Human intestinal tissue seems to have three β-galactosidases: (1) an enzyme specific for lactose with a maximum activity at pH 6.0 located in the brush border, (2) a β-galactosidase with a pH optimum of 4.5 in the lysosomes, and (3) a β-galactosidase in the cytoplasm with an optimum at 6.0 which is not active for lactose (Gray and Santiago, 1969). It has been postulated that the β-galactosidase in brush border is a precursor of the specific lactase, but this remains unsubstantiated.

By electrophoresis on cellulose acetate strips, human β-galactosidases have also been resolved into three components (Fluharty *et al.*, 1971). A fluorescent aglycone of galactose was used to locate the separated

$$GLC \longrightarrow GLC-6-P \rightleftharpoons GLC-1-P \rightleftharpoons UDP-GLC \rightleftharpoons UDP-GAL$$

$$UDP-GAL + GLC \longrightarrow GAL-GLC \ (LACTOSE) + UDP$$

Fig. 2 Biosynthesis of lactose.

protein. This procedure holds promise for further defining the function of the three distinct galactosidases.

B. Lactose Biosynthesis

It was early concluded (Gander *et al.*, 1957) that UDP-galactose was the donor and glucose 1-phosphate the acceptor for the synthesis of lactose. The product of the reaction, the phosphate ester of lactose, was postulated to have a role in its excretion by the glandular tissue. Using ^{14}C isotopes in the lactating cow and isolating the postulated intermediates and products, it was concluded that a major pathway for lactose synthesis was with UDP-galactose as the donor, and free glucose (not glucose 1-phosphate) as the acceptor (Fig. 2) (Hansen *et al.*, 1962). Tissue extracts subsequently were found to catalyze the synthesis of lactose from UDP-galactose and glucose, but contrary to expectation, with very low yield (Watkins and Hassic, 1962). Thus, in an unconvincing manner, the isotope studies in the whole animal were confirmed. A substantial advance was made when the lactose synthetase was resolved into two components.

1. Lactose Synthetase

The synthetase was shown to be a mixture of A protein fractionated from mammary glands, and α-lactalbumin, a protein normally found in milk (Brodbeck and Ebner, 1966; Brodbeck *et al.*, 1967). In the absence of α-lactalbumin, the A protein will catalyze galactosyl transfer to *N*-acetylglucosamine (Brew *et al.*, 1968), but glucose is not a good acceptor, having a high apparent Michaelis constant of 1 M (Fig. 3). In the presence of α-lactalbumin, the A protein effectively catalyzes the syn-

$$UDP-GAL + N-ACETYLGLUCOSAMINE-O-R \xrightarrow{\text{A PROTEIN}} GAL-N-ACETYLGLUCOSAMINE-O-R + UDP$$

$$UDP-GAL + GLC \xrightarrow[\substack{\alpha-LACTALBUMIN}]{\text{A PROTEIN}} GAL-GLC \ (LACTOSE) + UDP$$

Fig. 3 Galactosyltransferase.

thesis of lactose, decreasing the K_m about 1000-fold (Morrison and Ebner, 1971). Thus, α-lactalbumin is a specifier protein for the synthesis of lactose. The net effect of α-lactalbumin is to convert the enzyme from a glycosyltransferase in the biosynthetic pathway of complex polysaccharides to an efficient system for lactose synthesis.

The mammary gland is unique in that it can produce α-lactalbumin, and in so doing it makes glucose an effective substrate for the enzyme. Further, a high degree of chemical homology has been demonstrated in the amino acid sequence of hen's egg white lysozyme and bovine α-lactalbumin (Brew *et al.*, 1967). The three-dimensional models of α-lactalbumin and lysozyme also show a high degree of similarity. Functionally, the biosynthesis of lactose involves the formation of a $\beta(1 \rightarrow 4)$ glycosidic linkage, and such a linkage is hydrolyzed in the lysozyme reaction. It has thus been speculated that the two proteins have related evolutionary origins.

The principal function of the galactosyltransferase from tissues other than lactating mammary tissues is to transfer galactose to an appropriate carbohydrate side chain of glycoproteins and lipids. On this basis it is expected that the transferase would be widely distributed (Fitzgerald *et al.*, 1971).

As determined by disc gel electrophoresis, α-lactalbumins isolated from milk of pigs, sheep, and goats have two electrophoretically distinct forms, both of which are active in the lactose synthetase reaction and appear to be charged isomers resulting from slight variations in amino acid composition (Schmidt and Ebner, 1972). Genetic variance in bovine α-lactalbumin has been observed in electrophoretic separation; hence, it is possible that the two forms observed in the pig, goat, and sheep represent genetic variance.

The A protein in bovine skim milk has been purified of contaminating proteins. It is a single chain glycoprotein of molecular weight 44,000 (Clarke *et al.*, 1971; Klee and Klee, 1972). In the presence of one of the substrates, the A protein and α-lactalbumin form a stable complex that contains one molecule of each protein.

C. Galactose Metabolism

Galactose consumed in excess of developmental needs is metabolized for energy purposes by pathways that have been identified only during the last two decades.

The liver appears to be the primary site of the metabolism of galactose; however, other tissues (including the red cell) have this capacity;

hence, they can conveniently indicate the metabolic capability of the individual. The fundamental studies in microorganisms defined the major substrates and enzymes associated with the various pathways of metabolism of galactose.

1. Ancillary Pathways

The primary phosphorylation of galactose is at carbon-1 (Fig. 4) (Kosterlitz, 1943). Whether galactose is phosphorylated at carbon-6 by an enzyme in mammalian tissues is not clear. A claim (Inouye *et al.*, 1962) that Gal-6-P occurs in red cells has not been substantiated. Galactose 6-phosphate may also occur from the reaction of a phosphohexomutase from Gal-1-P, the primary product of phosphorylation (Fig. 4). Indications are that when Gal-1-P is not present Gal-6-P is not formed (Dahlqvist, 1971).

In humans the primary pathway of metabolism of galactose appears to be as the nucleoside diphosphate sugar (Caputo *et al.*, 1948; Leloir, 1951). The reductive and oxidative pathways of galactose metabolism in humans become significant when this major pathway is blocked (Fig. 4). When galactose accumulates, the presence of aldose reductase and reduced pyridine nucleotide promotes the formation of the hexitol, which

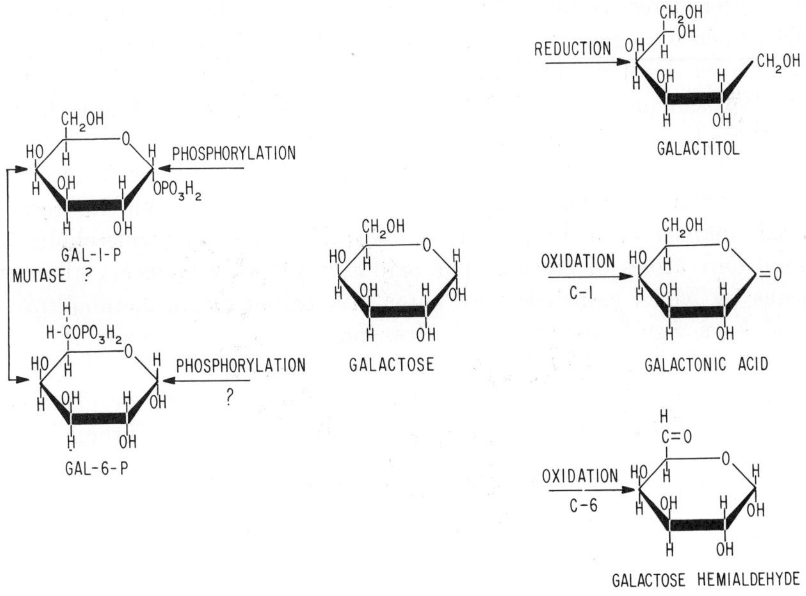

Fig. 4 Metabolism of galactose.

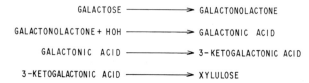

Fig. 5   Glucose oxidized to D-fructose; galacitol not oxidized.

may accumulate as a dead end product within the cell (Hers, 1960; Clements *et al.*, 1969). By contrast, the corresponding hexitol derived from glucose is specifically oxidized to D-fructose which can be metabolized (Fig. 5).

Intracellular accumulation of galactitol leads to osmotic swelling in the peripheral nerves of rats (Gabbay and Snider, 1972). A motor nerve conduction defect results, which is reversible on withdrawal of galactose. The presence of galactitol in peripheral nerves of human galactosemics has not yet been demonstrated as it has in brain (Wells *et al.*, 1965), lens (Gitzelmann *et al.*, 1967), and urine (Wells *et al.*, 1964). In the lens this galactitol accumulation leads to opacity.

Segal (Cuatrecasas and Segal, 1966) has hypothesized an oxidative pathway of galactose metabolism in the human (Fig. 6). Xylulose formed in this manner could be phosphorylated and converted to metabolically useful derivatives of glucose. The oxidative pathway offers an alternate means by which galactose may enter into carbohydrate metabolism. It could be a significant process in transferase-deficient humans who can tolerate and metabolize galactose in limited quantity.

Kozak and Wells (1971) concluded that an oxidative or glycolytic pathway of galactose metabolism does not function in the brain. Major

GALACTOSE ⟶ GALACTONOLACTONE

GALACTONOLACTONE + HOH ⟶ GALACTONIC ACID

GALACTONIC ACID ⟶ 3-KETOGALACTONIC ACID

3-KETOGALACTONIC ACID ⟶ XYLULOSE

Fig. 6   Hypothesized oxidative pathway of galactose metabolism in the human.

portions of [^{14}C]galactose administered are recovered as galactose and galactitol.

In contrast, galactonic acid has been isolated from the urine of a galactosemia patient given an oral dose of galactose (Bergren *et al.*, 1972). Indeed, about one-fifth of the ingested dose of galactose is excreted as galactonate by the galactosemic. Lesser amounts are found in the normal individual similarly treated.

2. Leloir Pathway and Galactosemia

The Leloir pathway constitutes the principal means of galactose utilization by humans.

$$\text{Gal} + \text{ATP} \xrightarrow{\text{kinase}} \text{Gal-1-P} + \text{ADP}$$

$$\text{Gal-1-P} + \text{UDP-Glc} \underset{}{\overset{\text{transferase}}{\rightleftharpoons}} \text{UDP-Gal} + \text{Glc-1-P}$$

$$\text{UDP-Gal} \underset{}{\overset{\text{epimerase}}{\rightleftharpoons}} \text{UDP-Glc}$$

$$\overline{\text{Gal} + \text{ATP} \rightleftharpoons \text{Glc-1-P} + \text{ADP}}$$

Discovering and understanding the human metabolic disorders in the metabolism of galactose came after identification of the pathways of conversion to glucose or other metabolizable products. Persons with galactosemia are deficient in transferase activity (Isselbacker *et al.*, 1956), which explains why tissues accumulate galactose, galactitol, and Gal-1-P. Despite their almost total lack of transferase activity, however, galactosemics can assimilate some galactose.

Since the absence of activity for the Gal-1-P-uridyltransferase is the basis for diagnosing galactosemia, the development of quantitative methods for the measurement of the transferase in erythrocytes allowed precise definition of interrelationships in families of patients with the disease (Donnell *et al.*, 1960; Schwarz *et al.*, 1961).

a. Genetics

Direct measurements of the enzyme in erythrocytes confirmed that all tested galactosemics had little or no transferase activity. Further, the parents, some siblings, and some other relatives of the tested patients had, on the average, one-half the normal level of enzyme. Both of the parents, as well as a paternal and maternal grandparent, must therefore be carriers, or genetic heterozygotes, before the disorder will be expressed in the offspring. This finding clearly establishes galactosemia as an autosomal-recessive disease.

In tested galactosemic families, the siblings demonstrate the expected Mendelian ratio of one galactosemic:two heterozygotes:one normal. At least one of the maternal and paternal grandparents of the patient and other relatives have been heterozygotes in the frequency predicted from the mode of inheritance.

Given an accurate estimate of the number of heteroyzgotes in the population of the United States (Hansen *et al.*, 1964), standard genetic calculations indicate about one galactosemic infant per 20,000 live births. In Great Britain, surveys of hospital records indicate that about one infant in 70,000 live births is diagnosed as a galactosemic (Schwarz *et al.*, 1961). Chemical considerations of the frequency of the heterozygote in the United States and in Denmark indicate that the disease may occur as often as once in 20,000 live births. Both clinical and chemical assessments of the expected disease rates are complex and of uncertain accuracy, however, and recent discoveries of apparent polymorphs (Beutler *et al.*, 1966) have further complicated the situation.

b. Other Variants

A so-called Duarte variant of Gal-1-P-uridyltransferase has been defined by refinement of the chemical assay for the transferase in blood cells. Family relationships in a limited sample have been given for the Duarte mutant. The heterozygote has about two-thirds the normal value, and the homozygote has one-third the normal value of the transferase. On the basis of the red cell value for the uridyltransferase, individuals have been identified who have both the classic and the Duarte structural defects. In the classic galactosemic, the defective protein is revealed by immunological procedures but is inactive according to chemical assays (Hansen, 1969; Tedesco and Mellman, 1971). Stuctural alteration is assumed to prevent the protein from performing its normal catalytic function.

Other transferase variants have been identified. Cuatrecasas and Segal (1966) showed that a so-called Negro variant may have 10% of the normal transferase activity based on assays of uridyltransferase in liver and intestinal tissues, even though the red cells have no activity. This is difficult to explain morphologically; hence, a more rigorous characterization of the mutant is needed before a suitable interpretation can be derived.

c. Treatment

The prescribed treatment for galactosemia is immediate removal from the diet of foods that contain lactose and other oligosaccharides incorporating galactose. The principal source of lactose is milk and products made from milk. It is obviously difficult, therefore, to achieve a diet entirely free of galactose. Hydrolyzed casein and soybean protein are standard sources of protein for the galactosemic. Since casein is prepared from milk, a day's supply of the hydrolyzed product could contain from a fraction of 1 gm to 3 or 4 gm of galactose. Soybean products obviously will also contain some galactose in the form of oligosaccharides. However, galactose in this form does not appear to be readily digested (Donnell *et al.*, 1969). Since the limits of tolerance for galactose are difficult to define precisely in individual patients, the best prescription is to reduce intake as much as is practicable.

d. Galactose Tolerance

The belief that maturity may bring some adaptive modifications of the capacity to metabolize galactose is probably not valid, since the amplitude and shape of the galactose tolerance curve does not change significantly with advancing age (Donnell *et al.*, 1969). No objective evidence supports the conclusion that the galactosemic somehow acquires an adaptive capacity to utilize galactose. Rather, milk consumption, and thus galactose intake, tends to decrease with maturity, both in total amount and in proportion to body size. Consequently, the tissues are exposed to less galactose and its metabolic derivatives. The apparently increased tolerance may therefore merely reflect only a decreased intake of the sugar.

e. Transferase

The unusual finding of a fresh mutation in human transferase has been reported (Mellman *et al.*, 1970). Parents of a child with galactosemia were heterozygotes as expected. Both maternal grandparents, however, were normal. Tests for nonpaternity failed to provide an alternate explanation.

Glucose-1-P, uridine mono-, di-, and triphosphates and nucleoside diphosphate–glucose derivatives inhibit the transferase (Segal and Rogers, 1971). From inhibitor constants and calculated intracellular levels of these compounds, it has been deduced that the uridine derivatives and Glc-1-P could have a physiologically regulatory function.

f. Toxicity of Galactose 1-Phosphate

Symptoms, other than eye cataracts, that are observed in patients with transferase deficiency may result from Gal-1-P in high concentration

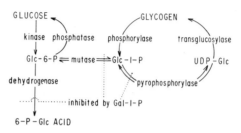

Fig. 7 Mutase, dehydrogenase, and pyrophosphorylase reactions.

within the cell. In humans suffering from lack of transferase, blood glucose levels are subnormal, indicating a disturbed carbohydrate metabolism. Three key reactions involving glucose have been reported to be affected by Gal-1-P: the mutase, dehydrogenase, and pyrophosphorylase reactions shown in Fig. 7 (Komrower, 1961; Oliver, 1961). The pyrophosphorylase functions in glycogen storage, the mutase is a factor in glycogen formation and utilization, and the dehydrogenase is essential to a major pathway of glucose utilization. In each case the evidence is based on kinetic studies of isolated enzymes, but not from human tissues; it is therefore somewhat presumptive. About a 50-fold excess of Gal-1-P over glucose phosphate is required for significant inhibition of the isolated reactions.

g. Uridine Diphosphate Galactose C-4 Epimerase

The epimerase-catalyzed reaction is the primary site of glucose formation from galactose when this sugar is being metabolized for energy. Conversely, when galactose is needed for structural purposes, it is derived from glucose at the nucleoside diphosphate hexose level via the same reaction (Fig. 8).

The UDP Gal-C-4 epimerase reaction requires DPN^+ as a cofactor. When the enzyme is fractionated from animal sources, the DPN^+ usually dissociates from the protein; *Escherchia coli* and yeast enzymes, however, retain DPN^+ more tenaciously. Reduced pyridine nucleotide strongly inhibits the animal enzymes. The specific hydrogen atoms involved are removed from one hexose substrate and replaced on the other in the opposite steric mode at carbon-4. This requires substantial change in conformation of the hexoses during the reaction with retention of the carbon-bound hydrogens.

There have been attempts to provide a molecular mechanism for the epimerase reaction and to specifically pinpoint the hydrogens involved

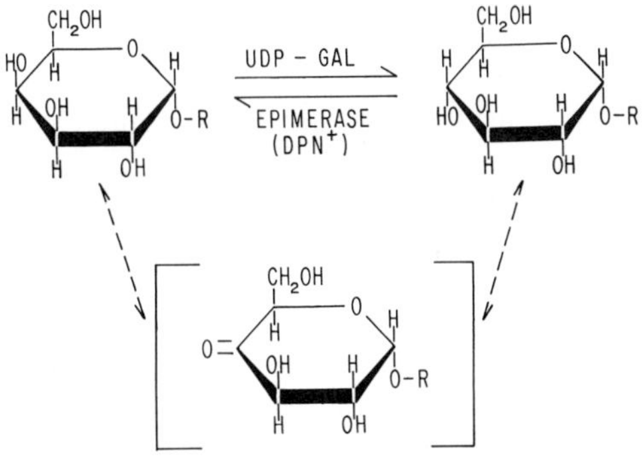

Fig. 8 Uridine Diphosphate Galactose C-4 Epimerase (R stands for uridine diphosphate).

in the interconversion. A "model" system wherein free hexose and UMP will reduce enzyme-bound DPN⁺ has been examined (Bertland *et al.*, 1971). When enzyme-bound DPN⁺ is reduced in a reaction containing epimerase, UMP, and hexose, the relative rates of reaction with glucose, glucose-3-*d*, glucose-4-*d* and glucose-5-*d* are 1, 0.3, 1, and 1, respectively, with enzyme (Davis *et al.*, 1971). The isotope effect with glucose-3-*d*, but not with glucose-4-*d* is suggested as evidence for a role of a 3-keto-hexose intermediate in the epimerase reaction. This isotope effect in the model reaction may be spurious as observed, and it awaits further elaboration. A keto intermediate at carbon-4 is the most plausible possibility in the epimerase reaction (Fig. 8) (Nelsestuen *et al.*, 1971).

It has been demonstrated that ethanol inhibits the clearance of galactose from the blood (Isselbacher and Krane, 1961). The increase in $NADH_2$ resulting from ethanol oxidation apparently inhibits the UDP Gal-C-4 epimerase (Forsander, 1972). The addition of NAD⁺ to a liver preparation partially prevents the effect of ethanol on $NADH_2$.

Bacteria, and presumably mammals, can carry out the transformations from hexose to hexuronic acid to pentose at the nucleoside diphosphate sugar level (Fig. 9) (Fan and Feingold, 1972). For polymers containing galactose or products derived from galactose such as galacturonic acid or pentose, these products are formed from the nucleoside diphosphate sugars that have been "activated" for glycosylation reactions.

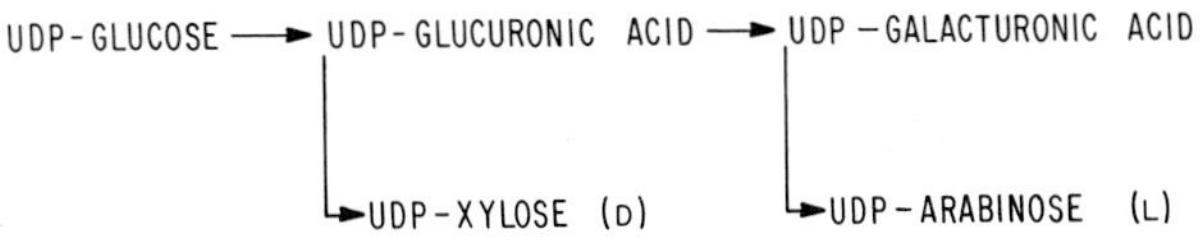

Fig. 9 Transformations from hexose to hexuronic acid to pentose.

3. Pyrophosphorylase Pathway

While the pyrophosphorylases appear to be primarily a means for the

$$\text{Hexose-1-P} + \text{NTP} \xrightleftharpoons{\text{pyrophosphorylase}} \text{NDP-hexose} + \text{PP}_i$$

biosynthesis of nucleoside diphosphate sugar intermediates, secondarily they may offer an alternative route of galactose metabolism.

In the absence of the primary (Leloir) pathway of metabolism via the transferase enzyme, this alternate means of galactose metabolism may be significant.

$$
\begin{aligned}
\text{Gal} + \text{ATP} &\xrightarrow{\text{kinase}} \text{Gal-1-P} + \text{ADP} \\
\text{Gal-1-P} + \text{UTP} &\xrightleftharpoons{\text{pyrophosphorylase}} \text{UDP-Gal} + \text{PP}_i \\
\text{UDP-Gal} &\xrightleftharpoons{\text{epimerase}} \text{UDP-Glc} \\
\underline{\text{UDP-Glc} + \text{PP}_i} &\xrightleftharpoons{\text{pyrophosphorylase}} \underline{\text{Glc-1-P} + \text{UTP}} \\
\text{ATP} + \text{Gal} &\longrightarrow \text{Glc-1-P} + \text{ADP}
\end{aligned}
$$

On the basis of partial purification of enyzmes extracted from animal livers, this alternative to the transferase pathway was proposed (Issel-bacher, 1958). The kinase and epimerase reactions also occur as in the pathway involving transferase, but the two "pyrophosphorylase" enzymes are unique to the above sequence. The pyrophosphorylase for UDP-Glc is widely distributed in nature and has been identified in a variety of tissues. The unique component of the proposed alternative pathway is the UDP-Gal pyrophosphorylase. Determinations of whether or not this enzyme is present in the liver are complicated by the non-specificity of the pyrophosphorylase for UDP-Glc (Albrecht *et al.*, 1966; Knop and Hansen, 1970). In crystalline form, the UDP-Glc pyrophosphorylase will catalyze the reaction with UDP-Gal at more than 10% of the rate of that of the principal substrate, thus complicating the interpretation of its metabolic function in the whole organism. Since there appears to be chemical (Ting and Hansen, 1968) and immunological (R. Gitzelmann and R. G. Hansen, unpublished manuscript, 1972) identity, one protein probably catalyzes both pyrophosphorylase reactions.

The UDP-Glc pyrophosphorylase constitutes almost 0.5% of the extractable protein from liver; hence, the capacity to metabolize galactose by this route could be appreciable. Regardless of the mechanism, this percentage substantiates the possibility of an alternate pathway for galactose disposition, especially when the primary Leloir pathway is blocked.

That the pyrophosphorylase pathway is largely responsible for galactose metabolism in the red cell from the galactosemic is indicated by the following observation. An antibody has been prepared from crystalline UDP-Glc pyrophosphorylase from human liver (R. Gitzelmann and R. G. Hansen, unpublished manuscript, 1972). This antibody, when incubated with extracts of galactosemic red cells, specifically precipitates the enzyme responsible for metabolism of galactose; hence, in the galactosemic, without the Leloir pathway, some galactose can be converted to glucose by his erythrocytes (and probably other tissues) via the pyrophosphorylase pathway.

The pyrophosphorylase pathway of metabolism has the potential not only to utilize galactose but also the sugar needed for normal human development is formed as follows:

$$\text{Glc-1-P} + \text{UTP} \overset{\text{pyrophosphorylase}}{\rightleftharpoons} \text{UDPGlc} + \text{PP}_i$$

$$\text{UDP-Glc} \overset{\text{epimerase}}{\rightleftharpoons} \text{UDP-Gal}$$

In galactosemics the further possibility exists that an abundance of pyrophosphorylase gives rise to Gal-1-P formation:

$$\text{UDP-Gal} + \text{PP}_i \overset{\text{pyrophosphorylase}}{\rightleftharpoons} \text{UTP} + \text{Gal-1-P}$$

4. Other Metabolic Defects in Galactose Metabolism

Defects in catalysis in all three steps in the Leloir pathway have now been described in humans. Following the discovery of the transferase defect, a number of kinaseless families were reported, and recently a child with an apparent defect in the epimerase step was described. In patients with a defect in the kinase (Gitzelmann, 1967), galactose and galactitol but not Gal-1-P accumulate in the tissues. The other toxic manifestations of transferaseless galactosemia do not occur in persons lacking the kinase, but cataracts develop in patients with either defect. Thus, galactitol might be the causative agent for cataract formation in both the kinase and transferase defects.

In humans, an epimerase deficit in red cells has been reported (Gitzelmann, 1972). If the liver and other organs are also epimerase deficient because of the function of the reaction in metabolism, the galactose intake would require careful balance. The dietary problem would be to provide adequate amounts of galactose to meet the requirements for this sugar and derived products but insufficient to provide toxic quantities of intermediates. In the absence of epimerase, presumably the kinase and pyrophosphorylase reactions would provide a biosynthetic route for UDP-Gal.

D. Galactose Biosynthesis in the Mammary Gland

It is generally assumed that galactose is formed from UDP-Glc in the mammary gland, catalyzed by the C-4 epimerase. In addition to defining, a primary pathway of lactose formation with UDP-Gal and free glucose being the immediate precursors, the isotope studies revealed the interesting possibility of an unusual mode of galactose synthesis (Hansen *et al.*, 1962). When [1,3-^{14}C]labeled glycerol was injected into lactating mammary glands, the patterns of labeling of glucose and galactose differed (Table I). The galactose portion of lactose contains more ^{14}C than does glucose. Further, carbons 4 and 6 of the galactose particularly reflect the injected substrate. UDP-Gal generally contained more label than did UDP-Glc, even though the latter is thought to be the normal biosynthetic precursor (Table II).

Two alternative explanations have been considered: (1) There are discrete centers of metabolic activity within the cell and glycerol (or glyceraldehyde) is intercepted differentially at these centers, and (2) UDP-

TABLE I

[14]C in Nucleotides from Mammary Tissue after Injection of [1,3-^{14}C]Glycerol

Nucleotide-Carbohydrate	Total cpm/ μmole hexose	cpm/μmole (C-4 = 100)					
		C-1	C-2	C-3	C-4	C-5	C-6
UDP-Glc	11,950	14.5	7.1	7.4	100	0.7	109.0
UDP-Gal	12,220	12.5	9.9	11.5	100	0.5	122
Lactose (glucose)	46.8	107.0	21.5	94.6	100	17.1	96.5
(galactose)	910.0	11.3	6.6	9.6	100	1.3	115.0

TABLE II

[14]C[a] in Nucleotides from Mammary Tissue after Injection of [1,3-^{14}C]Glycerol

Nucleotide	Experiment No.					
	1	2	3	4	5	6
UDP-Glc	11,950	10,260	10,830	11,373	30,264	13,190
UDP-Gal	12,220	10,856	11,380	11,966	38,213	17,600

[a] Counts per minute per micromole of hexose.

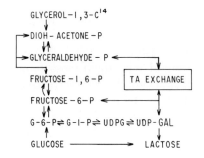

Fig. 10 Incorporation of glycerol into lactose.

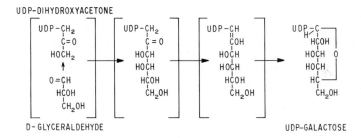

Fig. 11 UDP-dihydroxyacetone.

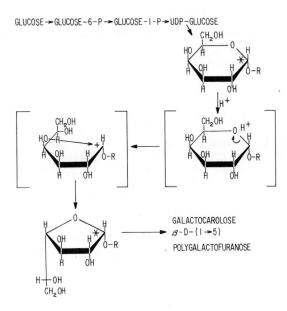

Fig. 12 Conversion of galactopyranose to galactofuranose (R stands for uridine diphosphate).

hexose is labeled by direct incorporation of triose (Fig. 10). The earlier but unconfirmed report of UDP-dihydroxyacetone offered one possible intermediate (Fig. 11) (Smith *et al.*, 1961). More recently (Trejo *et al.*, 1971), in the conversion of galactopyranose to galactofuranose at the nucleotide level an open chain hexose derivative has been postulated which could incorporate triose by a transaldolase-like exchange (Fig. 12).

References

Albrecht, G. J., Bass, S. T., Seifert, L. L., and Hansen, R. G. (1966). Crystallization and properties of uridine diphosphate glucose pyrophosphorylase from liver. *J. Biol. Chem.* **241**, 2968.

Bergren, W. R., Ng, W. G., and Donnell, G. N. (1972). Galactonic acid in galactosemia: Identification in the urine. *Science* **176**, 683–684.

Bertland, A. U., II, Seyama, Y., and Kalckar, H. M. (1971). Concerted reduction of yeast uridine diphosphate galactose 4-epimerase. *Biochemistry* **10**, 1545–1551.

Beutler, E., Baluda, M. C., Sturgeon, P., and Day, R. W. (1966). The genetics of galactose-1-phosphate uridyl transferase deficiency. *J. Lab. Clin. Med.* **68**, 646–658.

Bolin, T. D., and Davis, A. E. (1970). Lactose intolerance in Australian-born Chinese. *Australas. Ann. Med.* **19**, 40.

Brew, K., Vanaman, T. C., and Hill, R. L. (1967). Comparison of the amino acid sequence of bovine α-lactalbumin and hen's egg-white lysozyme. *J. Biol. Chem.* **242**, 3747–3749.

Brew, K., Vanaman, T. C., and Hill, R. L. (1968). The role of α-lactalbumin and the A protein in lactose synthetase: A unique mechanism for the control of a biological reaction. *Proc. Nat. Acad. Sci. U.S.* **59**, 491.

Brodbeck, U., and Ebner, K. E. (1966). Resolution of a soluble lactose synthetase into two protein components and solubilization of microsomal lactose synthetase. *J. Biol. Chem.* **241**, 762.

Brodbeck, U., Denton, W. L., Tanahashi, N., and Ebner, K. E. (1967). The isolation and identification of the B protein of lactose synthetase as α-lactalbumin. *J. Biol. Chem.* **242**, 1391–1397.

Caputo, R., Leloir, L. F., and Trucco, R. E. (1948). Lactase and lactose fermentation in *S. fragilis*. *Enzymologia*, **12**, 350.

Clarke, J. T. R., Wolfe, L. S., and Perlin, A. S. (1971). Evidence for a terminal α-D-galactopyranosyl residue in galactosylgalactosylglucosylceramide from human kidney. *J. Biol. Chem.* **246**, 5563–5569.

Clements, R., Weaver, J., and Winegrad, A. (1969). The distribution of polyol:NADP oxidoreductase in mammalian tissues. *Biochem. Biophys. Res. Commun.* **37**, 347–353.

Cuatrecasas, P., and Segal, S. (1966). Galactose conversion to D-xylulose: An alternate route of galactose metabolism. *Science* **153**, 549–550.

Cuatrecasas, P., Lockwood, D. H., and Caldwell, J. R. (1965). Lactase deficiency in the adult. *Lancet* **1**, 14–18.

Dahlqvist, A. (1971). A fluorometric method for the assay of galactose-1-phosphate in red blood cells. *J. Lab. Clin. Med.* **78**, 931–938.

Dahlqvist, A., and Gryboski, J. D. (1965). Inability of the human small-intestinal lactase to hydrolyze lactulose. *Biochim. Biophys. Acta* **110**, 635.

Davis, L., and Glaser, L. (1971). On the mechanism of the UDP-D-glucose-4′-epimerase evidence for a 3-keto-hexose intermediate. *Biochem. Biophys. Res. Commun.* **43**, 1429–1435.

Donnell, G. N., Bergren, W. R., Bretthauer, R. K., and Hansen, R. G. (1960). The enzymatic expression of heterozygosity in families of children with galactosemia. *Pediatrics* **25**, 572–581.

Donnell, G. N., Koch, R., and Bergren, W. R. (1969). Observations on results of management of galactosemic patients. *In* "Galactosemia" (D. Y. Y. Hsia, ed.), p. 247. Thomas, Springfield, Illinois.

Fan, D.-F., and Feingold, D. S. (1972). Biosynthesis of uridine diphosphate D-xylose. V. UDP-D-glucuronate and UDP-D-galacturonate carboxy-lyase of *Ampullariella digitata*. *Arch. Biochem. Biophys.* **148**, 576–580.

Fitzgerald, D. K., McKenzie, L., and Ebner, K. E. (1971). Galactosyl transferase activity in a variety of sources. *Biochim. Biophys. Acta* **235**, 425–428.

Flatz, G., Saegudom, C., and Sanguanbhokhai, T. (1969). Lactose intolerance in Thailand. *Nature (London)* **221**, 758–759.

Fluharty, A. L., Lassila, E. L., Porter, M. T., and Kihara, H. (1971). The electrophoretic separation of human beta-galactosidases on cellulose acetate. *Biochem. Med.* **5**, 158–164.

Forsander, O. A. (1972). Influence of ethanol on the redox state of the liver. *Biochim. Biophys. Acta* **268**, 253–256.

Gabbay, K. H., and Snider, J. J. (1972). Nerve conduction defect in galactose-fed rats. *Diabetes* **21**, 295–300.

Gander, J. E., Petersen, W. E., and Boyer, P. D. (1957). On the enzymic synthesis of lactose-1-PO₄. *Arch. Biochem. Biophys.* **69**, 85–99.

Gilat, T. (1971). Lactase—an adaptable enzyme? *Gastroenterology* **60**, 346–347.

Gitzelmann, R. (1967). Hereditary galactokinase deficiency, a newly recognized cause of juvenile cataracts. *Pediat. Res.* **1**, 14.

Gitzelmann, R. (1972). Deficiency of uridine diphosphate galactose 4-epimerase in blood cells of an apparently healthy infant. *Helv. Paediat. Acta* **27**, 125–130.

Gitzelmann, R., Curtius, H. C., and Schneller, I. (1967). Galactitol and galactose-1-phosphate in the lens of a galactosemia infant. *Exp. Eye Res.* **6**, 1–3.

Gray, G. M., and Santiago, N. A. (1969). Intestinal beta-galactosidases. I. Separation and characterization of three enzymes in normal human intestine. *J. Clin. Invest.* **48**, 716–726.

Gudmand-Høyer, E., Dahlqvist, A., and Jarnum, S. (1970). The clinical significance of lactose malabsorption. *Amer. J. Gastroenterol.* **53**, 460.

Hansen, R. G. (1969). Hereditary galactosemia. *J. Amer. Med. Ass.* **208**, 2077–2082.

Hansen, R. G., Wood, H. G., Peeters, G. J., Jacobson, B., and Wilken, J. (1962). Lactose synthesis. VI. Labeling of lactose precursors by glycerol-1,3-C¹⁴ and glucose-2-C¹⁴. *J. Biol. Chem.* **237**, 1034–1039.

Hansen, R. G., Bretthauer, R. K., Mayes, J., and Nordin, J. H. (1964). Estimation of frequency of occurrence of galactosemia in the population. *Proc. Soc. Exp. Biol. Med.* **115**, 560–563.

Hers, H. G. (1960). L'aldose-reductase. *Biochim. Biophys. Acta* **37**, 120.

Hill, J. A., and Huber, R. E. (1971). Effects of various concentrations of Na⁺ and Mg²⁺ on the activity of beta-galactosidase. *Biochim. Biophys. Acta* **250**, 530–537.

Holzel, A. (1968). Defects of sugar absorption. *Proc. Roy. Soc. Med.* **61**, 1095.

Inouye, T., Tannenbaum, M., Hsia, D. Y.-Y. (1962). Identification of galactose-6-phosphate in galactosemic erythrocytes. *Nature (London)* **193**, 67–68.

Isselbacher, K. J. (1958). A mammalian uridine-diphosphate galactose pyrophosphorylase. *J. Biol. Chem.* **232**, 429–444.

Isselbacher, K. J., and Krane, S. M. (1961). Studies on the mechanism of the inhibition of galactose oxidation by ethanol. *J. Biol. Chem.* **236**, 2394.

Isselbacher, K. J., Anderson, E. P., Kurahashi, K., and Kalckar, H. M. (1956). Congenital galactosemia, a single enzymatic block in galactose metabolism. *Science* **123**, 635.

Keusch, G. T., Troncale, F. J., Thavaramara, B., Prinyanont, P., Anderson, P. R., and Bhamarapravathi, N. (1969a). Lactose deficiency in Thailand: Effect of prolonged lactose feeding. *Amer. J. Clin. Nutr.* **22**, 638.

Keusch, G. T., Troncale, F. J., Miller, L. H., Promadhat, V., and Anderson, P. R. (1969b). Acquired lactose malabsorption in Thai children. *Pediatrics* **43**, 540–545.

Klee, W. A., and Klee, C. B. (1972). The interaction of α-lactalbumin and the A protein of lactose synthetase. *J. Biol. Chem.* **247**, 2336–2344.

Knop, J. K., and Hansen, R. G. (1970). Uridine diphosphate glucose pyrophosphorylase. IV. Crystallization and properties of the enzyme from human liver. *J. Biol. Chem.* **245**, 2499–2504.

Kogut, M. D., Donnell, G. N., and Shaw, K. N. (1967). Studies of lactose absorption in patients with galactosemia. *J. Pediat.* **71**, 75–81.

Komrower, G. M. (1961). Galactosemia. *Cereb. Palsy Bull.* **3**, 117–126.

Kosterlitz, H. W. (1943). The structure of the galactose-phosphate present in the liver during galactose assimilation. *Biochem. J.* **37**, 318.

Kozak, L. P., and Wells, W. W. (1971). Studies on the metabolic determinants of D-galactose-induced neurotoxicity in the chick. *J. Neurochem.* **18**, 2217–2228.

Leloir, L. F. (1951). The enzymatic transformation of uridine diphosphate glucose into a galactose derivative. *Arch. Biochem. Biophys.* **33**, 186–190.

Littman, A., and Hammond, J. B. (1965). Diarrhea in adults caused by deficiency in intestinal disaccharides. *Gastroenterology* **48**, 237–249.

McMichael, H. B., Webb, J., and Dawson, A. M. (1965). Lactase deficiency in adults. A cause of functional diarrhea. *Lancet* **1**, 717–720.

Mellman, W. J., Allen, F. H., Jr., Baker, L., and Tedesco, T. A. (1970). Direct evidence of mutation at the locus for galactose-1-phosphate uridyl transferase. *Pediatrics* **45**, 672–676.

Miller, D., and Crane, R. K. (1961). The digestive function of the epithelium of the small intestine. II. Localization of disaccharide hydrolysis in the isolated brush border portion of intestinal epithelial cells. *Biochim. Biophys. Acta* **52**, 293.

Morrison, J. F., and Ebner, K. E. (1971). Studies on galactosyltransferase. Kinetic effects of α-lactalbumin with N-acetylglucosamine and glucose as galactosyl group acceptors. *J. Biol. Chem.* **246**, 3992–3998.

Nelsestuen, G. L., and Kirkwood, S. (1971). The mechanism of action of the enzyme uridine diphosphoglucose 4-epimerase. Proof of an oxidation-reduction mechanism with direct transfer of hydrogen between substrate and the B-position of the enzyme-bound pyridine nucleotide. *J. Biol. Chem.* **246**, 7533–7543.

Oliver, I. T. (1961). Inhibitor studies on uridine diphosphoglucose pyrophosphorylase. *Biochim. Biophys. Acta* **52**, 75–81.

Plotkin, G. R., and Isselbacher, K. J. (1964). Secondary disaccharidase deficiency in adult celiac disease (non-tropical sprue) and other malabsorptive states. *N. Engl. J. Med.* **271**, 1033–1037.

Schmidt, D. V., and Ebner, K. E. (1972). Multiple forms of pig, sheep, and goat α-lactalbumin. *Biochim. Biophys. Acta* **263**, 714–720.

Schwarz, V., Wells, A. R., Holzel, A., and Komrower, G. M. (1961). A study of the genetics of galactosemia. *Ann. Hum. Gene.* **25,** 179.

Segal, S., and Rogers, S. (1971). Nucleotide inhibition of mammalian liver galactose-1-phosphate uridylytransferase. *Biochim. Biophys. Acta* **250,** 351–360.

Simoons, F. J. (1970). Primary adult lactose intolerance and the milking habit: A problem in biologic and cultural interrelations. II. A cultural historical hypothesis. *Amer. J. Dig. Dis.* **15,** 695–710.

Smith, E. E. B., Galloway, B., and Mills, G. T. (1961). Uridine diphosphate dihydroxyacetone. *Biochem. Biophys. Res. Commun.* **5,** 148–151.

Strom, R., Attardi, D. G., Forsén, S., Turini, P., Celada, F., and Antonini, E. (1971). The activation of beta-galactosidase by divalent and monovalent cations. *Eur. J. Biochem.* **23,** 118–124.

Tedesco, T. A., and Mellman, W. J. (1971). Galactosemia: Evidence for a structural gene mutation. *Science* **172,** 727–728.

Ting, W. K., and Hansen, R. G. (1968). Uridine diphosphate galactose pyrophosphorylase from calf liver. *Proc. Soc. Exp. Biol. Med.* **127,** 960–962.

Trejo, A. G., Haddock, J. W., Chittenden, G. J. F., and Baddiley, J. (1971). The biosynthesis of galactofuranosyl residues in galactocarolose. *Biochem. J.* **122,** 49–57.

Watkins, W. M., and Hassic, W. Z. (1962). The synthesis of lactose by particulate enzyme preparations from guinea pig and bovine mammary glands. *J. Biol. Chem.* **237,** 1432–1440.

Wells, W. W., Pittman, T. A., and Egan, T. J. (1964). The isolation and identification of galactitol from the urine of patients with galactosemia. *J. Biol. Chem.* **239,** 3192–3195.

Wells, W. W., Pittman, T. A., Wells, H. J., and Egan, T. J. (1965). The isolation and identification of galactitol from the brains of galactosemia patients, *J. Biol. Chem.* **240,** 1002–1004.

Wendorff, W. L., and Amundson, C. H. (1971). Characterization of beta-galactosidase from *Saccharomyces fragilis. J. Milk Food Technol.* **34,** 300–306.

CHAPTER 19

The Effects on Metabolism of Maltose and Higher Saccharides

IAN MACDONALD

Not very long ago it would have been ludicrous to even consider a contribution on the metabolism of maltose and higher saccharides since until recently the teaching was that all carbohydrates are broken down to glucose and metabolized as such. This is now known not to be so and furthermore it seems that the gut, although part of the "milieu externe," may play an important role in the regulation of activities in the "milieu interne," particularly in relation to carbohydrate metabolism (McIntyre et al., 1964). Since there is an enormous literature on the metabolism of glucose (the monosaccharide of maltose and higher polysaccharides), only those aspects of metabolism where a difference may exist between glucose and maltose and the higher saccharides will be considered.

A. Insoluble Higher Saccharides

Insoluble higher saccharides are largely composed of the starches whose chemical composition is very varied and whose metabolic effects can be divided into those which can be broken down by the gut amylases and those which cannot. A point sometimes overlooked is that cooking, by destroying the covering of the starch granule, converts a relatively undigestible starch to one that is rapidly absorbable. In man the blood glucose levels after the ingestion of cooked starch are identical with those after glucose (Sun and Shay, 1961), whereas after ingesting raw cornstarch the blood glucose level has a much lower profile (Crossley, 1966).

Since starches are insoluble they cannot have any direct effect on metabolism until they are converted to a soluble form. They may however have indirect consequences via their effects on intestinal microflora (Peterson *et al.*, 1953) or they may consist of molecules that cannot be hydrolyzed at all in the gut and affect metabolism only indirectly by their physical properties.

This latter group of higher saccharides contains such substances as cellulose, pectin, and pentosans, and there exists a recent rash of nonscientific literature on the metabolic benefits to man of consuming these nondigestible compounds, variously known as "fiber" or "roughage." Further consideration of these particular higher saccharides should be delayed until the results of controlled experiments become available.

B. Soluble Higher Saccharides and Maltose

Though the amount of starch consumed is extensive, less so in the developed than in the developing parts of the world, the remainder of this review will be concerned with the soluble higher saccharides because (a) the insoluble saccharides must be hydrolyzed to the soluble forms if absorption is to take place and (b) there seems to be a tendency by the food manufacturer to move away from the use of sucrose and toward the use of soluble higher saccharides.

Maltose, per se, is not consumed in large quantities, perhaps partly because of the difficulty of obtaining it pure, and partly because it may not have the properties needed by the food chemist. However, there is a distinct trend to a greater intake of "glucose syrup" (i.e., "corn syrup," "liquid glucose") which is a partial hydrolysate of starch (Wood, 1964). It is obvious that in a partial hydrolysis the stage when the breakdown process is halted can be varied and as far as the partial hydrolysis of starch is concerned, a measure of the extent of the breakdown is given in a unit known as a dextrose equivalent. This is the reducing power calculated as dextrose (glucose) and expressed as a dry weight basis.

$$\text{Dextrose equivalent} = \frac{\text{reducing sugar expressed as dextrose}}{\text{total solids in syrup}}$$

The composition of the glucose syrup as was laid down by the British Pharmaceutical Codex (1963) and the Pharmacopeia of the United States

TABLE I

The Composition of the "Standard" Partial Hydrolysate of Maize Starch

Composition of glucose syrup (% W/W)			
Glucose	14.9	Hexasaccharides	5.1
Maltose	11.0	Heptasaccharides	4.3
Trisaccharides	9.1	Octa- and higher saccharides	18.5
Tetrasaccharides	7.7	Mineral matter	0.3
Pentasaccharides	6.5	Water	22.6
	Dextrose equivalent = 43		

of America (1960) is seen in Table I. However, specifying the dextrose equivalent does not characterize a glucose syrup since that depends— among other things—on whether the hydrolysis of starch is carried out using acid or an enzyme.

The properties of maltose and its higher polysaccharides that commend them are that unlike glucose they have a lower osmotic pressure (and this can be varied depending on the dextrose equivalent) and therefore do not cause gut distention with water due to osmosis and consequently are not associated with abdominal discomfort, anorexia, and nausea as would be induced by strong glucose solutions of equal concentration. Maltose and the higher polysaccharides are less sweet, and it is also perhaps of more than passing interest that the production of glucose syrup is not expensive.

C. Effects on Alimentary Canal

1. Mouth

All dietary carbohydrates predispose to dental caries, but sucrose seems to be the most potent in this respect (Konig and Grenby, 1965). The experimental evidence in animals suggests that glucose syrup given in 20% solution produces less caries than a solution of sucrose of similar strength. On the other hand, spray-dried glucose syrup substituted for sucrose in the diet of rats did not reduce the level of fissure caries (Grenby, 1972). A trial of the effects on the teeth of glucose syrup instead of sucrose in fruit juice concentrates given to preschool children would show if this was an effective measure in reducing caries.

A recent trial in adults consuming a third of a pound of candies per day for 3 days showed that when glucose syrup replaced sucrose the extent of dental plaque was significantly less with the glucose syrup containing candies (Grenby and Bull, 1973).

The extent to which maltose and its polysaccharides stimulate the taste buds depends on the number of the smaller molecules present in the carbohydrate mixture, but one of the advantages claimed for glucose syrup is that it is less sweet than glucose alone and therefore more acceptable. This can be important in patients who need a high carbohydrate intake. The extent of the sweetness can be varied depending on the degree of hydrolysis achieved.

2. Absorption

The details of the absorption of disaccharides have been discussed elsewhere in this volume, but it is perhaps worth stressing those features that are especially relevant to maltose and glucose syrup. The blood glucose levels in the first few minutes after ingestion tend to be higher after glucose syrup than after an equivalent amount of glucose (Dodds *et al.*, 1959). This implies that gastric emptying and intestinal absorption are not delayed with glucose syrup, and in fact many physicians prefer giving glucose syrup for glucose tolerance tests. After comparable doses of glucose and maltose the blood sugar levels are similar (Matthews *et al.*, 1968).

In the first few weeks of life the infant has a relative deficiency of pancreatic amylase (Hadorn *et al.*, 1968) and, unlike adults, the blood glucose curve after cooked starch rises only slightly (Husband *et al.*, 1970). It is possible that glucose syrup, but not maltose, would produce a blood glucose level that was intermediate between that of glucose and of starch, and this may be of value in the treatment of hypoglycemia and in preventing the reactive hypoglycemia which follows giving glucose (Husband *et al.*, 1970).

After gastric and intestinal surgery, disaccharides may be absorbed as such (Gryboski *et al.*, 1963), and in view of the suggestion that maltose may be metabolized when given intravenously (*vide infra*) it could be advantageous to give maltose in preference to sucrose or lactose postoperatively in these patients.

It has been found that when a glucose syrup beverage is taken with a high fat meal the ensuing rise in the concentration of plasma triglyceride is significantly lower than when a simple carbonated drink is taken (Green *et al.*, 1971). This could result from the release by raised blood glucose levels of insulin, which then lowers plasma triglyceride concentration, perhaps by enhancing the activity of lipoprotein lipase (Perry and Corbett, 1964).

In a study lasting 14 weeks the sucrose in the diet of 19 men was replaced by glucose syrup with an artificial sweetner, and it was found that a standard glucose tolerance test at the end of this time resulted in a blood glucose curve that was less flat than the control curve (Fry, 1972).

Thus, the substitution of maltose and/or its higher polysaccharides for glucose or sucrose could, by its different behavior in the gut, affect the metabolism of the consumer, not so much by interfering with metabolic pathways directly, but by affecting the psyche or the speed of absorption under special circumstances.

D. Effects on Lipid Metabolism

Dietary carbohydrates are accepted by all as being readily converted in the body into fat, and carbohydrate-induced elevation of serum triglyceride has been described (Ahrens *et al.*, 1961; Fredrickson *et al.*, 1967). Apart from the increase in triglyceride laid down in the fat depots and the raised level of endogenous triglyceride in serum, even the liver does not escape the conversion of carbohydrate to triglyceride with accumulation of the latter (Benedict and Lee, 1937).

1. Liver

Diets containing a high proportion of carbohydrate with an inadequate protein intake in children lead to accumulation of lipid in the liver in the clinical condition known as kwashiorkor. With a similar diet, the Strasbourg goose lays down fat in its liver. The question then arises as to whether all dietary carbohydrates are equal in this respect. This question is not easy to answer in man, but in experimental animals sucrose is associated with more liver lipid than is glucose syrup, whereas a diet with raw cornstarch as the sole carbohydrate barely elevates the liver triglyceride (Macdonald, 1962).

In a recent experiment, female rats were given one of three mixtures of maltose polysaccharides. The mixtures contained either high molecular weight polysaccharides (dextrose equivalent 13) or the low molecular weight portion (dextrose equivalent 59) of glucose syrup or pure glucose (dextrose equivalent 100). The authors reported that as the molecular weight of the carbohydrate increased so did the weight, cholesterol, and free fatty acids of liver (Birch and Etheridge, 1973).

2. Serum

The influence of dietary carbohydrates on the level of triglyceride in the fasting serum of man is well documented, and there is quite a lot of evidence in man and experimental animals to show that not all dietary carbohydrates have the same quantitative effect on the serum concentration of endogenous triglyceride.

As will be discussed elsewhere, the dietary carbohydrate that is associated with an elevation of liver and serum triglyceride is fructose, either as such, or in the sucrose molecule. Among the glucose-containing polysaccharides in the diet there is some evidence to suggest that they do not all have the same effect on serum lipid concentration, even though these differences are small compared with those produced by fructose. If molecular weight is used as a ranking order parameter (starch, through glucose syrup and maltose to glucose) then the lower the molecular weight the greater the reduction in the concentration of both triglyceride and phospholipid of fasting serum of young men after 5 days on a high carbohydrate regimen (Macdonald, 1965).

In an acute study on fasting male baboons, glucose syrup tended to produce a greater fall in serum cholesterol and triglyceride levels than did glucose, but these differences, especially in the triglyceride level, were very small compared with those produced by an acute load of sucrose when the triglyceride level rose (Macdonald and Roberts, 1967).

In an experiment described earlier, in which 19 men lived for 14 weeks on a diet which contained less than 3 gm of sucrose per day, it was found that as soon as the men transferred to the diet where glucose syrup replaced the normal sucrose intake, the level of triglyceride in fasting serum fell. This fall was much more striking in 5 of the men who, before the experiment diet started, had raised levels of fasting serum triglyceride (Roberts, 1973). Thus, it seems possible that the exchange of sucrose for glucose syrup may have at least a temporary effect in lowering serum triglyceride when it is elevated, although it was noticed that on return to the normal sucrose-containing diet there was an overswing and marked elevation of the serum triglyceride level in all subjects. These changes may well have taken place had glucose been used instead of glucose syrup, but it is doubtful if the palatibility would have been so acceptable.

3. Adipose Tissue

Despite the statement made over 50 years ago that in man fructose "shows a tendency or preference to change into fat in the body" (Higgins, 1916) the assumption is still made and generally accepted that the source of energy in the diet is of no consequence, assuming it is absorbed. There are, however, a few experiments in animals, which should obviously be confirmed, that hint that to a small extent this may be an invalid assumption. In rats given a diet containing over 80% of the energy as carbohydrate it was found that after 26 weeks the percentage of fat in the carcass was less with glucose or glucose syrup than with sucrose or fructose (Allen and Leahy, 1966). In experiments using baboons in isoenergetic diets the weight gain at the end of 26 weeks was greater with fructose and sucrose than with glucose syrup in the diet (Brook and Noel, 1969). More recently, hypo-joule intakes for 11 days were found to lead in young men to a greater weight loss when glucose syrup was the dietary carbohydrate (86% energy intake) than when replaced by a comparable amount of sucrose. With young women the reverse was the case (Macdonald and Taylor, 1973).

E. As Energy Source

As a source of ready energy there is one report that shows that after the ingestion of [^{14}C]glucose in the rat the $^{14}CO_2$ produced for the following 80 min is considerably less than after the ingestion of [^{14}C]maltotriose (Martin and Young, 1967), thus suggesting that oral maltotriose is metabolized more rapidly than is oral glucose.

A preliminary report suggests that the ingestion of about 60 gm of glucose syrup before fairly severe exercise not only caused a rise in the blood glucose levels during exercise, but concurrent with this there was an improvement in the performance of the exercise in terms of increased work done and decreased heart rate (Green and Thomas, 1972). The improved effects may have been because carbohydrate was ingested, rather than the nature of the carbohydrate source, apart from its acceptability by trained athletes.

F. Metabolism of Maltose Given Intravenously

Under normal circumstances the carbohydrates presented to the liver and other metabolic sites are monosaccharides, and it was not considered of much value to present disaccharides to the general metabolic pool since no means of converting the disaccharides to monosaccharides was known. In fact, evidence exists to show that in those pathological conditions where disaccharide absorption does take place the sugar is excreted in the urine (Gryboski *et al.*, 1963). However, in 1967, it was reported that when 10 gm of maltose is given intravenously to man (and to rats) very little is excreted in the urine in the ensuing 24 hr, in contrast to infusions of lactose and sucrose (Weser and Sleisenger, 1967). Similarly, in patients with celiac disease the maltose output in the urine is not significantly different from controls after oral ingestion, but after ingesting lactose and sucrose the amount of the disaccharide in the 5-hr urine sample showed a fourfold increase over controls. Further investigations into the metabolism of infused maltose in the rat showed that circulating glucose and maltose are similarly metabolized, and that maltose metabolism probably procedes via hydrolysis to glucose (Young and Weser, 1970). More recent findings indicate that maltose stimulates insulin release without first being converted to glucose and that the insulin could enhance the entry of maltose into cells (Young and Weser, 1971).

The clinical significance of these findings on the metabolism of intravenous maltose is obvious. The osmotic inconvenience of monosaccharides in intravenous therapy would be reduced in that with maltose twice as much energy could be given per osmole. Furthermore, the later peak of $^{14}CO_2$ specific activity after maltose than glucose (Young and Weser, 1971) could mean a slower and therefore useful paying out of energy.

G. Clinical Value of Maltose and Higher Polysaccharides

1. The soluble polysaccharides are beyond doubt much more palatable and are less likely to give rise to nausea, hypoglycemia, etc., than glucose and are more easily broken down in the gut than some starches.

2. Glucose syrups have an assured place in the therapy of chronic kidney disease where fluid and protein intake must be restricted and the energy needs made up by fat or carbohydrate.

3. Glucose polymers are relatively noncariogenic when compared with sucrose.

4. There is evidence to suggest that glucose and its polymers may be less likely to induce a raised serum triglyceride than is sucrose.

5. For fattening the thin.

The disadvantages of glucose syrups are obviously those which apply to dietary carbohydrates in general, namely, diabetes mellitus, obesity, and carbohydrate-induced lipidemia. Specific clinical disadvantages of glucose syrups are that excess may give rise to diarrhea in normal infants, and it has been said that patients on long-term diets containing glucose syrup may get fungal infections of the mouth.

H. Conclusions

There are effects, on metabolism, of maltose and higher saccharides that are different from those of the constituent monosaccharide, glucose. Many of these differences are not marked, but nevertheless exist. The significance of these differences and especially the extra-metabolic influences on metabolism, which are readily demonstrated by maltose and the higher saccharides, are fields where further knowledge would not only be interesting but also clinically useful.

References

Ahrens, E. H., Hirsch, S., Oette, K., Farquhar, J. W., and Stein, Y. (1961). Carbohydrate-induced and fat-induced lipemia. *Trans. Ass. Amer. Physicians* **74**, 134–146.
Allen, R. J. L., and Leahy, J. S. (1966). Some effects of dietary dextrose, fructose, liquid glucose and sucrose in the adult male rat. *Brit. J. Nutr.* **20**, 339–347.
Benedict, F. G., and Lee, R. C. (1937). "Lipogenesis in the Animal Body, with Special Reference to the Physiology of the Goose." Carnegie Institution, Washington, D.C.
Birch, G. G., and Etheridge, I. J. (1973). Short-term effects of feeding rats with glucose syrup fractions and dextrose. *Brit. J. Nutr.* **29**, 87–93.
British Pharmaceutical Codex. (1963). Pharmaceutical Society, London.
Brook, M., and Noel, P. (1969). Influence of dietary liquid glucose, sucrose and fructose on body fat formation. *Nature (London)* **222**, 562–563.
Crossley, J. N. (1966). A relationship between carbohydrate tolerance and serum lipid concentration in healthy young men. *Proc. Nutr. Soc.* **25**, i–ii.
Dodds, C., Fairweather, F. A., Miller, A. L., and Rose, C. F. M. (1959). Blood-sugar response of normal adults to dextrose, sucrose, and liquid glucose. *Lancet* **1**, 485–488.
Fredrickson, D. S., Levy, R. I., and Lees, R. S. (1967). Fat transport in lipoproteins—an integrated approach to mechanisms and disorders. *N. Engl. J. Med.* **276**, 34, 94, 148, 215, and 273.
Fry, A. J. (1972). The effect of a "sucrose-free" diet on oral glucose tolerance in man. *Nutr. Metabol.* **14**, 313–323.
Green, L. F., and Thomas, V. (1972). Some effects of glucose syrup ingestion during vigorous exercises of differing intensities and duration. *Proc. Nutr. Soc.* **31**, 5A–6A.
Green, L. F., Dale, T. L. C., Ford, M. A., and Bagley, R. (1971). The effect of a glucose syrup drink on plasma triglyceride concentrations after a high fat meal and a low fat meal. *Proc. Nutr. Soc.* **30**, 92A.
Grenby, T. H. (1972). The effect of glucose syrup on dental caries in the rat. *Caries Res.* **6**, 52–59.
Grenby, T. H., and Bull, J. (1973). Changes in the dental plaque after eating sweets containing starch hydrolysates instead of sucrose. *Proc. Nutr. Soc.* **32**, 39A–40A.
Gryboski, J. D, Thayer, W. R., Gryboski, W. A., Gabrielson, I. W., and Spiro,

H. M. (1963). A defect in disaccharide metabolism after gastrojejunostomy. *N. Engl. J. Med.* **268**, 78–80.

Hadorn, B., Zoppi, G., Shmerling, D. H., Prader, A., McIntyre, I., and Anderson, C. M. (1968). Quantitative assessment of exocrine pancreatic functions in infants and children. *J. Pediat.* **73**, 39–50.

Higgins, H. L. (1916). The rapidity with which alcohol and some sugars may serve as nutrient. *Amer. J. Physiol.* **41**, 258–265.

Husband, J., Husband, P., and Mallinson, C. N. (1970). Gastric emptying of starch meals in the new-born. *Lancet* **2**, 290–292.

Konig, K. G., and Grenby, T. H. (1965). The effect of wheat grain fractions and sucrose mixtures on rat caries developing in two strains of rats maintained on different regimes and calculated by two different methods. *Arch. Oral. Biol.* **10**, 143–153.

Macdonald, I. (1962). Some influences of dietary carbohydrate on liver and depot lipids. *J. Physiol. (London)* **162**, 334–344.

Macdonald, I. (1965). The effects of various dietary carbohydrates on the serum lipids during a five-day regimen. *Clin. Sci.* **29**, 193–197.

Macdonald, I., and Roberts, J. B. (1967). The serum lipid response of baboons to various carbohydrate meals. *Metab., Clin. Res.* **16**, 572–579.

Macdonald, I., and Taylor, J. (1973). The effect of high carbohydrate: low energy diets on body weight in man. *Proc. Nutr. Soc.* **32**, 36A–37A.

McIntyre, N., Holdsworth, C. D., and Turner, D. S. (1964). New interpretation of oral glucose tolerance. *Lancet* **2**, 20–21.

Martin, A. F. M., and Young, F. G. (1967). Specific activity of carbon dioxide expired by rats after oral sucrose and other sugars. *Nature (London)* **215**, 885–886.

Matthews, D. M., Craft, I. L., and Crampton, R. F. (1968). Intestinal absorption of saccharides and peptides. *Lancet* **2**, 49.

Perry, W. F., and Corbett, B. N. (1964). Changes in plasma triglyceride concentration following the intravenous administration of glucose. *Can. J. Physiol. Pharmacol.* **42**, 353–356.

Peterson, G. E., Dick, E. C., and Johansson, K. R. (1953). Influence of dietary aureomycin and carbohydrate on growth, intestinal microflora and vitamin B_{12} synthesis of the rat. *J. Nutr.* **51**, 171–189.

Pharmacopeia of the United States of America. (1960). 16th Revision. U.S.P. XVI. U.S. Pharmacopeial Convention Inc., Easton, Pennsylvania.

Roberts, A. M. (1973). Effects of a sucrose-free diet on the serum-lipid levels of men in Antartica. *Lancet,* **i**, 1201–1204.

Sun, D. C. H., and Shay, H. (1961). An evaluation of the starch tolerance test in pancreatic insufficiency. *Gastroenterology* **40**, 379–82.

Weser, E., and Sleisenger, M. H. (1967). Metabolism of circulating disaccharides in man and the rat. *J. Clin. Invest.* **46**, 499–505.

Wood, F. (1964). Glucose syrups in food manufacture. *Proc. Int. Food Ind. Congr., 1964* pp. 153–158.

Young, J. M., and Weser, E. (1970). Effect of insulin on the metabolism of circulating maltose. *Endocrinology* **86**, 426–429.

Young, J. M., and Weser, E. (1971). The metabolism of circulating maltose in man. *J. Clin. Invest.* **50**, 986–991.

CHAPTER 20

Involvement of the Raffinose Family of Oligosaccharides in Flatulence

E. CRISTOFARO, F. MOTTU, AND
J. J. WUHRMANN

When reviewing the literature on the oligosaccharides of the raffinose family, it appears that these sugars have only aroused a moderate amount of interest in the field of chemistry and biochemistry from the time of their discovery at the beginning of this century until about 10 years ago when a group of German scientists, led by Taeufel, began to show some interest in their metabolism owing to their presence in rather large amounts in leguminous seeds known to be a staple food in many countries.

About 5 years ago, these sugars were first suspected of playing a part in flatus formation which, for a long time, had been associated with the ingestion of different leguminous seeds. The increased interest in the transformation of new protein sources into food systems, such as soybeans where these sugars are present, has incited several research groups to look more closely into their possible effect on flatulence.

Flatulence itself is a very old problem. Some authors at the beginning of this century discovered that even a slight increase in pressure, from

313

30 to 60 mm Hg, in rectal gas, may lead to various vegetative signs of discomfort, such as headache, dizziness, slight mental confusion, reduced ability to concentrate, and slight retinal edema with reduced accommodation, as mentioned by Askevold (1956).

There are other different causes of flatulence, as indicated by Alvarez (1942) and Debray and Veyne (1966) in their papers. Both described the mechanism by which gas goes in and out of the bowels. In contrast to swallowed air which is usually passed through the bowels easily, rapidly, and painlessly, gas resulting from the ingestion of food to which the patient is allergically sensitive, or which contains nonmetabolizable substances such as some carbohydrates, for example, seems to frequently remain trapped for hours in segments of the bowel which are tonically and painfully contracted.

Presumably, the first authors who attempted to quantitate the production of flatus in man after the ingestion of specific foods, and who were able to show that soybeans considerably increased gas production as well as its carbon dioxide percentage, were Blair *et al.* in 1947. From that time on, research was undertaken which could help to identify the offensive agents in order to be able to remove them from foods.

Steggerda asserted in 1961 that from his experiments on humans with beans:

> There is something in the bean that can alter the normal physiology of the gastrointestinal tract, so that flatulence production is markedly stimulated.

Further work has made it possible to establish the implication of the raffinose family of sugars in flatulence.

A. The Oligosaccharides of the Raffinose Family and Their Presence in Food Raw Materials

1. Development and Occurrence

The raffinose family of oligosaccharides is constituted of sugars related to raffinose by the fact of having one or more α-D-galactopyranosyl groups in their structure. These D-galactosyl groups are found in nature joined to sugars such as D-glucose, to sucrose, to certain polysaccharides, and to a few nonsugars such as glycerol and inositol (French, 1954).

Figure 1 shows the structural interrelationship of three important members of this family, where the basic unit is sucrose and α-galactose units are bound to glucose. Next to sucrose itself, *raffinose* is probably the most abundant oligosaccharide of the plant world.

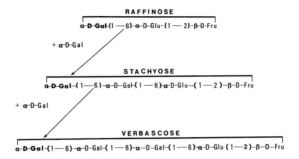

Fig. 1 Structural relationships of the raffinose oligosaccharides.

As stated by French (1954), *stachyose* enjoys the distinction of being the oldest and best known sugar of its class, the classic tetrasaccharide. It was isolated and first described by von Planta and Schulze in 1890. The original source of stachyose and the best source to date is the juice from the rhizomes of *Stachys tuberifera*. This plant has long been used in the Orient, Japan in particular, as a vegetable; it was widely grown in France and known as *"crosnes du Japon"* and in Anglo-Saxon countries as "Japanese artichokes." It contains up to 14% of stachyose, which means 73% of the solids of the tubers. *Verbascose*, the pentasaccharide, was first isolated in 1910 by Bourquelot and Bridel from the roots of mullein (*Verbascum thapsus*), an inedible plant. Among the common food raw materials one group of substances contains important amounts of the above mentioned sugars, the mature dry *leguminous seeds*.

Table I gives the analysis of carbohydrates of different types of beans made by Hardinge *et al.* (1965) and Cristofaro. The table shows that soybeans have the highest levels of raffinose (1.9%) and stachyose (5.2%), while chick-peas and pigeon peas have the highest levels of verbascose (4.2%). Kawamura and collaborators published the results of their analysis on soybeans in Japanese journals in 1966 and 1967. The levels found are in agreement with those presented in Table I.

2. Evolution in the Leguminous Seeds

It is known that the level of the oligosaccharides of the raffinose family, as well as starch, with the exception of soybeans, increases during maturation of the seeds, while sucrose decreases. This was demonstrated for peas by Shallenberger and Moyer (1961), and we have analyzed commercial frozen green peas without finding any trace of oligosaccharides. Tanusi (1972) found similar results for soybeans and remarked that in the cotyledon the oligosaccharides are gradually detected during growth

TABLE I

Oligosaccharides in Leguminous Seeds

Seeds	Raffinose		Stachyose		Verbascose	
	Edible portion (%)[a]	Dry matter (%)[b]	Edible portion (%)[a]	Dry matter (%)[b]	Edible portion (%)[a]	Dry matter (%)[b]
Beans, black mung, dry	0.5	—	1.8	—	3.7	—
Beans, green mung, dry	0.8	—	2.5	—	3.8	—
Chick peas	1.0	1.1	2.5	2.5	4.2	—
Cow peas	0.4	0.4	2.0	4.8	3.1	0.5
Field beans	0.5	0.2	2.1	1.2	3.6	4.0
Horse gram	0.7	—	2.0	—	3.1	—
Lentils	0.6	0.9	2.2	2.7	3.0	1.4
Lima beans, canned	—	—	0.2	—	—	—
Peas, green, dry	—	0.6	—	1.9	—	2.2
Peas, yellow, dry	—	0.3	—	1.7	—	2.2
Pigeon peas	1.1	—	2.7	—	4.1	—
Soybeans, dry	1.9	0.8	5.2	5.4	—	—

[a] Hardinge *et al.* (1965).
[b] Cristofaro, (unpublished, 1971).

in the sequence sucrose → raffinose → stachyose → verbascose. Gomyo and Nakamura (1966) were able to synthetize raffinose from uridine diphosphate galactose and sucrose using solutions of immature soybean enzymes. The germination of the leguminous seeds induces, on the contrary, a rapid decrease in the oligosaccharides followed by their disappearance after 4 days. Many papers were published by Japanese authors on this subject, Kawamura and Hurnich; in particular (1966).

Table II gives the data for soybeans found by Taeufel *et al.* (1960) and by Cristofaro showing that for stachyose the decrease is very rapid during the first 3 days of germination. Pazur *et al.* (1962) detected the presence of an α-galactosidase among the enzymes of germinating soybeans and demonstrated that the galactose produced during enzymatic hydrolysis was utilized. For the analysis of the sugars contained in beans an aqueous or alcoholic extract is made, clarified, then chromatographed. The oligosaccharides thus separated can be identified and their amounts measured by colorimetry, using for an example the method described by Dubois *et al.* (1956). A gas chromatography method was recently published by Delente and Ladenburg (1972), which permits results to be obtained in a relatively short time for defatted soybean meal extracts.

TABLE II

Changes in Oligosaccharides during Germination of Soybeans

Seeds	Raffinose		Stachyose		Verbascose	
	(gm)[a,b]	(%)[c,d]	(gm)[a,b]	(%)[c,d]	(gm)[a,b]	(%)[c,d]
Starting beans	0.64	1.24	2.90	5.04	Tr.[e]	Tr.[e]
Soaked beans (12 hr)	0.57	—	2.62	—	Tr.[e]	—
Soaked beans (24 hr)	0.55	—	2.17	—	Tr.[e]	—
Germination 1 day	0.51	0.64	1.64	4.02		Tr.[e]
Germination 2 days	0.26	0.13	1.18	3.40		
Germination 3 days	Tr.[e]	Tr.[e]	0.41	2.10		
Germination 4 days	Tr.[e]		0.18	1.06		
Germination 5 days	—		—	Tr.[e]		
Germination 6 days	Tr.[e]		Tr.[e]			

[a] Taeufel *et al.* (1960).
[b] Dry matter of 400 Chinese beans.
[c] Cristofaro (unpublished, 1972).
[d] Dry matter, Canadian beans.
[e] Here, Tr. stands for traces.

B. The Raffinose Family of Sugars in the Alimentary Canal

1. *Metabolism*

It is well known that the disaccharide lactose is metabolized by normal human subjects since its β-glycosidic linkage is split by lactase. There is evidence in the literature (Kosterlitz, 1937) that galactose is converted to glucose in the liver by phosphorylating steps similar to those described for fructose. Similarly, the metabolism of the members of the raffinose family of sugars containing α-galactose units would require the presence of an α-galactosidase. However, such an enzyme does not seem to be present in man and in other mammals.

Semenza and Auricchio (1962) and Prader *et al.* (1963) have extensively investigated the disaccharidases and oligosaccharidases of the human intestinal flora and mucosa. In particular, Gitzelmann and Auricchio (1965), in the case of a galactosemic child, did not find any α-galactosidase activity in homogenates of human small intestinal mucosa. They proved that only trace amounts of raffinose and stachyose were excreted in infants' urine when the latter were fed with soybean powder. These findings were reviewed recently by Semenza (1968).

Taeufel *et al.* in 1964–1965 published their first works on human and animal intestinal absorption and degradation of pure α-galactose containing oligosaccharides. They showed in particular (1965a) that raffinose, stachyose, and verbascose are able to pass intact only in very small amounts, less than 1% of the dose administered, through the intestinal wall and can be found in urine as shown in Table III. This passage is less for oligosaccharides of increasing molecular weights. These authors did not observe, on the other hand, that raffinose is hydrolyzed by the acid gastric juice, and they found that in human feces there is a low α-galactosidase activity and a considerable β-fructosidase activity, decomposing raffinose to fructose and melibiose.

They concluded that because of the absence of the necessary enzymes the galactosaccharides of the raffinose family passing into the bloodstream cannot be utilized by the organism. In 1967, they obtained analogous results with more extensive experimentation on rats. They obtained data for the oligosaccharide excretion in urine, as given in Table IV, which follows the same pattern as in man, showing again that the passages of the oligosaccharides through the intestinal wall decrease with an increase in the molecular weight.

When studying the intestinal behavior of different oligosaccharides, including raffinose, in contact with intestinal mucosa of rats, pigs, and humans, they showed for raffinose in particular that there is no mucosal enzymatic decomposition (Rutloff *et al.*, 1967). Finally, Krause *et al.* (1968) published further results concerning the intestinal absorption of sugars of the raffinose family, as given in Fig. 2, where it is evident that there is a rapid decrease of the absorption with increasing molecular weights.

TABLE III

Galacto-oligosaccharides Excretion in Human Urine[a]

Sugar (per os[b] 4 gm)	No. of subjects	Average excretion during 24 hr (mg)	Average amount excreted (%)
Melibiose	4	22.0 ± 1.8	0.60 ± 0.06
Raffinose	10	14.6 ± 5.1	0.39 ± 0.13
Stachyose	8	4.5 ± 1.5	0.18 ± 0.06
Verbascose	8	2.1 ± 1.0	0.08 ± 0.04

[a] From Taeufel *et al.* (1965).
[b] Per os, by mouth.

TABLE IV

Galacto-Oligosaccharides Excretion
in Rat Urine[a]

Sugar (per os 100 mg)	Average excretion during 24 hr (mg)
Melibiose	2.2 ± 0.2
Raffinose	4.9 ± 0.4
Stachyose	0.9 ± 0.3
Verbascose	0.5 ± 0.2

[a] From Taeufel *et al.* (1967).

2. Utilization by Intestinal Bacteria

All the researchers mentioned above belonged to the same team operating in Germany. This group gave a very important contribution to the study of the metabolism of the sugars of the raffinose family. In view of the fact that only very small amounts of raffinose in particular go

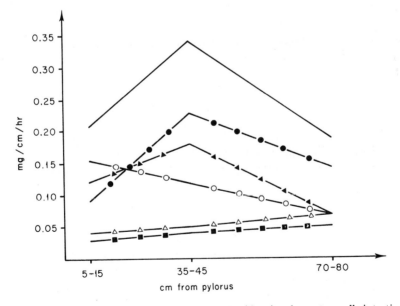

Fig. 2 In vitro absorption rates of oligosaccharides in the rat small intestine: (——) glucose, (●) sucrose, (▲) raffinose, (○) lactose, (△) stachyose, and (■) verbascose. From Krause *et al.* (1968).

into the bloodstream and are excreted by urine, they concluded that this sugar during digestion remains within the lumen and gets into distal parts where it is *microbially metabolized.*

After the hypothesis regarding the responsibility of intestinal microflora in the utilization of oligosaccharides was made, different authors started working on the subject. In 1966, Calloway *et al.* worked *in vitro* with ileal and colonic dejecta from humans. They made the interesting observation that organisms from both intestinal segments are able to utilize stachyose. This could confirm the suggestion that the intestinal flora may contain an α-1,6,-galactosidase. A gram-positive, nonmotile microaerophile organism that utilizes stachyose was isolated from the pig intestine.

In 1967–1968, Steggerda, performing experiments both in man and animals and *in vitro,* investigated the type of flora specifically responsible for flatus with its high concentration of carbon dioxide following the ingestion of bean products. He found that this bacterial flora is the gram-positive, spore-forming anaerobic bacterial type, known as the *clostridia group* which usually inhabits the small intestine and colon of man and animals.

On the other hand, the clostridia group of anaerobic bacteria has been described by Breed *et al.* (1957) as giving a stormy, fermenting response upon exposure to mono- and oligosaccharides. Steggerda (1968) also mentioned that *in vitro* the pH of the media in which the organisms are reacting definitely becomes more acid when gas is produced. These results were confirmed by Richards *et al.* (1968) who proved that the bacteria must be *Clostridium perfringens.*

In 1969, Rockland *et al.* demonstrated that the growth of clostridia is stimulated by a substrate containing beans. Hydrogen and carbon dioxide, the major constituents in flatus gases, were also found to be the primary gases collected. In 1970, Kurtzman and Halbrook isolated a carbohydrate-rich fraction from beans and found that a certain portion of the flatulent beans must not be absorbed by the gut and must be metabolized by *C. perfringens.*

In the same years, Rackis *et al.* (1967, 1970b) confirmed that clostridia can use oligosaccharides of the raffinose family. As shown in Table V, these sugars added to cultures isolated from dog colons can develop important amounts of gas with percentages of carbon dioxide and hydrogen as high as those given by monosaccharides, glucose in particular. They deduced from their experiments that the gas producing factor is closely associated with the oligosaccharides of the leguminous seeds. Having seen that glucose produces gas at a faster rate than either fructose or galactose, they suggested that a breakdown of the oligosaccharides to monosac-

TABLE V

Anaerobic Gas Production with Carbohydrates—Anaerobic Cultures Isolated from Dog Colon Biopsies[a,b]

Substrate	Total gas produced (ml)				Gas composition (%)	
	1 hr	6 hr	12 hr	24 hr	CO_2	H_2
Monosaccharides						
Glucose	0	23	38	51	37	62
Fructose	0	10	20	37	41	58
Galactose	0	8	20	36	38	61
Oligosaccharides						
Maltose	0	19	38	43	38	61
Sucrose	0	19	27	30	32	68
Raffinose	0	9	25	30	35	65
Stachyose	0	10	27	31	37	63
Control	0	1.5	1.5	2.0		

[a] From Rackis *et al.* (1970b). (Reprinted from Food Technology/Journal of Food Science, vol. 15, pp. 634–639, 1970. Copyright © by Institute of Food Technologists.)

[b] One to ten milliliter of thioglycollate anaerobic bacteria medium, 0.1 gm of substrate added.

charides occurs before the gas producing mechanism takes place. It is thus understandable that a longer delay must occur before the onset of gas production with such oligosaccharides as stachyose and raffinose, particularly if glucose is the preferred substrate.

3. Trials on Animals

The tests *in vitro* have shown that the oligosaccharides of the raffinose family can be utilized by the bacteria present in the intestine. Trials on animals were undertaken in order to better understand the nature and the mechanism of gas production. Different points had to be cleared up such as the site in the alimentary canal where the gas is formed, the modifications of the intestinal environment as a result of the composition of the gas, and the nature of the flatus-producing factors and the possible implication of the oligosaccharides of the raffinose family.

Experiments were first made with leguminous seeds or their extracts, then extended to pure sugars of the raffinose family. Hedin and Adachi in 1962 developed a technique for the measurement of gas in the digestive tract of rats in order to compare the effect of a dehydrated red bean diet to a reference casein diet. As shown in Fig. 3, they found that gas is formed mainly in the intestine 2–8 hr after feeding. They also established in the same experimentation that a red bean diet provokes an increase in the volume of gas and in the percentages of carbon dioxide and hydrogen. They remarked that from the second to the eleventh day of feeding, there was a gradual decrease in the hydrogen production, whereas the percentage of carbon dioxide rose to 90 and the total production remained constant. The daily peak occurred approximately 4 hr after feeding.

Richards and Steggerda (1966) and Steggerda (1968) investigated in dogs the relationship between the colon and various areas of the small intestine as gas production areas following the introduction of different diets into surgically prepared intestinal segments. Table VI gives some of their results, showing that the highest volume of gas and percentage of carbon dioxide are found in the colon. They also showed in these experiments that the volume of gas is increased by a navy bean homogenate, and indicated that contrary to some current beliefs, bacterial action in the duodenum, jejunum, and ileum of the dog may add significantly to the total intestinal gas volume.

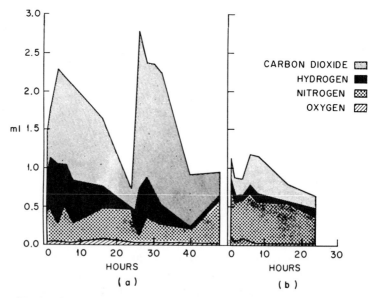

Fig. 3 Total volume and composition of gases found in the intestines of rats at given intervals after feeding. (a) Dehydrated red bean diet and (b) casein diet. From Hedin and Adachi (1962).

TABLE VI

Factors Influencing Volume and Composition of Gas Production in Various
Areas of the Intestine of the Dog[a]

Group 5 dogs	Treatment	Intestinal segments	Mean gas vol (ml/3 hr)	Mean gas composition			
				% CO_2	% O_2	% N_2	% H_2
1	Methyl cellulose	Duodenum	0.00				
		Jejunum	1.50	9.30	13.90	76.80	.00
		Ileum	1.50	9.30	13.90	76.80	.00
		Colon	1.50	9.30	13.90	76.80	.00
2	Navy bean homogenate	Duodenum	5.70	21.26	6.47	40.37	33.90
		Jejunum	4.90	21.00	4.66	51.50	27.88
		Ileum	15.00	21.67	6.00	54.65	27.74
		Colon	31.90	34.26	3.80	28.81	33.16

[a] From Steggerda (1968).

In 1968, Richards *et al.* confirmed previous results by feeding diets containing beans directly into segments of the intestine of the dog. They concluded that some constituent in the bean must interact with certain anaerobic bacteria inhabiting the small and large intestine and suggested that the flatus factor responsible can be related to oligosaccharides.

Because of the implication of oligosaccharides in flatus formation and of the presence in leguminous seeds of important amounts of members of the raffinose family, it was decided to study the effect on rats of soybean oligosaccharides and of pure single oligosaccharides (Cristofaro *et al.*, 1970). In the initial experiment, we were interested in testing the effect on rats of crude oligosaccharide fractions from soybeans. Two groups of diets were prepared, one based on soybean milk, the other on casein; each group had three levels of oligosaccharides, i.e., 0.45% raffinose and 0.85% stachyose for the low level, 0.45% raffinose and 2.05% stachyose for the medium, and 0.90% raffinose and 4.05% stachyose for the high level. The results obtained, presented in Fig. 4, show that in each group of diets there is a marked increase in the volume of the intestinal gas, containing a high percentage of carbon dioxide, when the oligosaccharides were at medium and high levels, i.e., when stachyose was present at 2 and 4% levels.

In a second trial, we tested three diets based on casein with an addition of pure stachyose at medium level (2.05%), of pure raffinose (3.55%), and of lactose (2.4%) calculated to give the same amount of galactose as the stachyose diet. Soybean milk and casein with oligosaccharide diets at

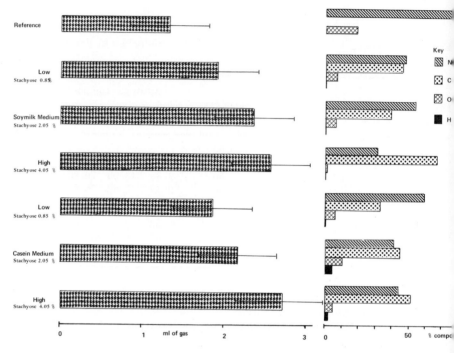

Fig. 4 Trial on rats: total gas volumes and composition.

medium levels are included for a repetition. Lactose was incorporated to see if there was a difference between α- and β-galactose–glucose bonds. The results obtained, given in Fig. 5, demonstrated the prevailing influence of stachyose on flatus formation. The diet containing this sugar produced a highly significant increase in the volume of gas. Raffinose and lactose induced an insignificant increase in flatus. Soybean and casein with soybean carbohydrates confirmed with a lower significance the results of the first trial.

It was considered necessary to check the effect of verbascose in order to ascertain whether one more galactose unit in the molecule could produce a greater volume of gas (Cristofaro *et al.*, 1972). Four diets containing either stachyose or verbascose (2.5% of the dry matter) and differing in the type of protein used were studied by assays on rats. Figure 6 shows that the volume of intestinal gas is markedly influenced by verbascose as well as by stachyose. The difference between these two oligosaccharides is slight. Thus, the presence of one more α-galactose unit in the molecule has apparently no marked effect, contrary to what had been previously observed for stachyose versus raffinose. Carbon dioxide is always present in very high amounts accompanied by some hydrogen.

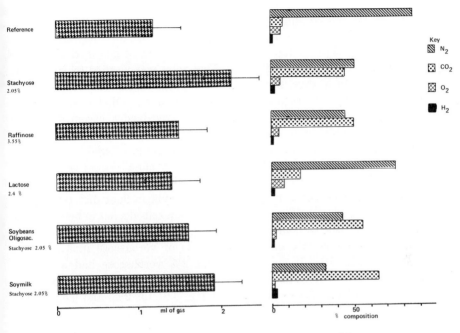

Fig. 5 Trial on rats: total gas volumes and composition.

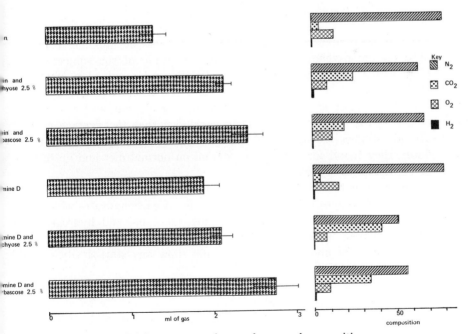

Fig. 6 Trial on rats: total gas volumes and composition.

Soy protein isolate appears to have no effect per se on flatulence. The high mean volume of gas measured for the reference diet may result from a certain dispersion of the results. The formation of gas in the intestine of the rats takes place in the cecum. This point was verified by subjecting digesting rats to X-rays. Between the fourth and eighth hour of digestion, gas bubbles could be seen in rats having received a diet containing defatted soybean flour, while in those having received a casein diet there were no gas bubbles as shown in Fig. 7.

The presence of carbon dioxide in the intestinal gas found by Steggerda (1968) suggested that the pH of the alimentary canal should be measured during digestion. Figure 8 shows that a marked pH decrease (7.2–6.4) takes place in the cecum of rats fed with a soybean flour diet (stachyose, 2% of dry matter). Figure 9 shows a marked decrease between the fourth and eighteenth hour after the meal.

Finally, the influence of different levels of verbascose in a diet was measured (Fig. 10). It was shown that 2% verbascose had a marked effect. The mean pH values of eight measurements were 6.98 for the reference diet, 6.76 for the diet containing 1%, 6.61 for 2%, and 6.52 for 3% verbascose.

4. Trials on Humans

Trials on humans are more difficult than those on animals. They depend on the psychological and physical attitude of the subject. Their interpretation is therefore more complicated and can lead to imprecise conclusions.

In 1947, Blair *et al.* studied in humans the volume and composition of gas in the digestive tract. They concluded that the average content of male subjects is about one liter. During a day's collection of rectal flatus, they found for one subject 200 ml on normal diet and up to 2600 ml on diets with soybeans.

Askevold in 1956 declared that one of the factors responsible for flatus volume and composition is the fermentation of carbohydrates which produces large quantities of carbon dioxide, methane, and hydrogen. His trials on humans with a large choice of diets, comprising green dried peas soaked for 24 hr and then boiled, did not show any statistically reliable result.

Steggerda, in 1961, when working on bean diets, made a striking observation on the average flatus production. While on the basal diet gas volume is 16 ml/hr with 10–12% carbon dioxide, on a bean diet the average volume changes to 190 ml/hr with above 50% carbon dioxide.

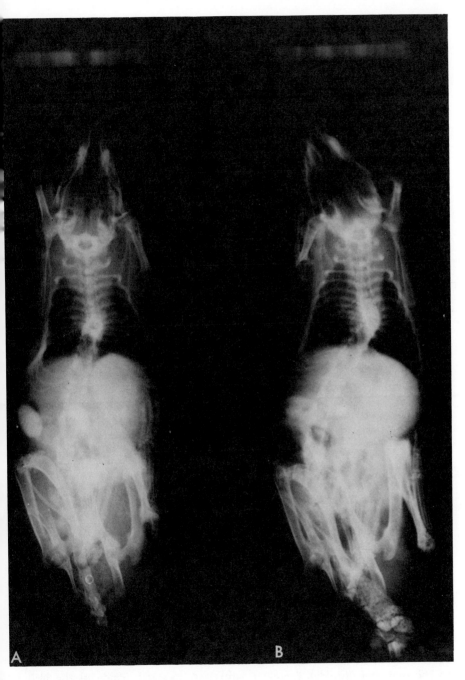

Fig. 7 Radiograph of digesting rats: (A) Casein and (B) defatted soy flour. Diets contain 20% protein.

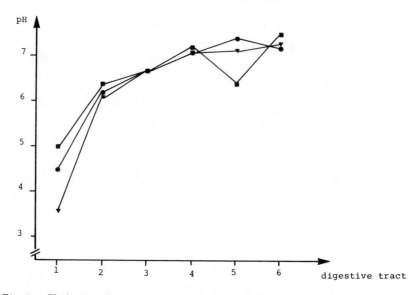

Fig. 8 pH during digestion. Diets containing 20% protein: (●) casein, (▼) meat and (■) soybean flour. Key: 1, stomach; 2, duodenum; 3, small intestine first third; 4, small intestine second third; 5, cecum; and 6, colon.

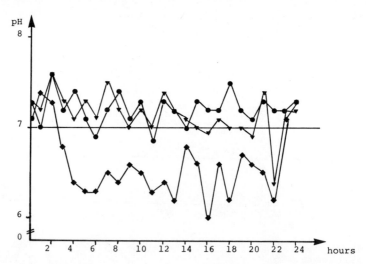

Fig. 9 Cecum pH during digestion. Diets containing 20% protein: (●) casein, (▼) meat, and (♦) soybean flour.

But only in 1966 did he deduce from the results obtained (see Table VII) that the flatulent factor in soybeans does not reside in the hull, fat, or protein fraction but is in some way associated with the low mo-

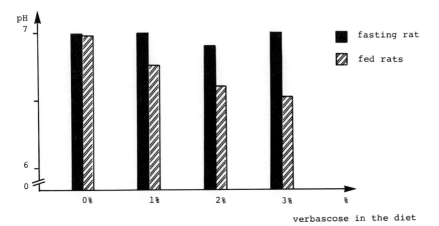

Fig. 10   Mean cecum pH during digestion.

lecular fraction of the carbohydrates present in soybeans in agreement with Rackis *et al.* (1967, 1970). He stated, for the first time, that this fraction contains relatively large amounts of *raffinose* and *stachyose*. His

TABLE VII

Effects of Soy Products on Flatus in Man[a]

		Flatus volume (ml/hr)	
Product[c]	Daily intake (gm)	Average	Range
Full-fat soybean flour[b]	146	30	0–75
Defatted soybean flour[b]	146	71	0–290
Soy protein concentrate[b]	146	36	0–98
Soy proteinate[b]	146	2	0–20
Water-insoluble residue[b,d]	146	13	0–30
Whey solids[e]	48	300[f]	
80% ethanol extractives[e]	27	240	220–260
Navy bean meal[b]	146	179	5–465
Basal diet[b]	146	13	0–28

[a] From Rackis *et al.* (1970b).
[b] Steggerda *et al.* (1966).
[c] All products were toasted with live steam at 100°C for 40 min.
[d] Fed at a level three times higher than that present in the defatted soybean flour diet.
[e] Amount equal to that present in 146 gm of defatted soybean flour.
[f] One subject, otherwise four subjects per test.

results were fairly well summarized in a paper he published in 1968 where he concluded that low molecular weight carbohydrates are responsible for flatus.

Textured soybean flours were given to humans by Werner *et al.* (1971) who noticed cases of flatus complaint (5 out of 6 patients) when the amount ingested was 50 gm or more of dry material, in contradiction with Kies and Fox (1971) whose subjects did not complain of flatulence when they received 25 and 50 gm. In 1972, Murphy *et al.* published their results on human subjects for chemical and physical fractions of beans, *Phaseolus vulgaris*, and for raffinose and stachyose added to a nonflatulent diet at levels higher than those in beans. They found that the fractions rich in carbohydrates were producing the highest amounts of carbon dioxide, but 4–7 hr after ingestion, raffinose and stachyose did not give higher amounts of carbon dioxide than the control diet. Hickey *et al.* (1972) studied the flatulent effect of commercial wheat cereals and milling fractions in voluntary subjects, and found that the most flatulent of all was the product containing the highest amounts of raffinose and stachyose.

There is some contradiction among the authors having made trials on humans with regard to the implication of raffinose and stachyose, where the results for these sugars did not show the same effect on humans as on rats. Presumably one of the reasons for this is that the time needed for the complete utilization of these sugars by the intestinal bacteria can be very long, as indicated by Rackis *et al.* (1970b).

C. The Raffinose Family of Sugars in Processed Foods and Their Elimination

As discussed above, the raffinose family of sugars is present in food raw materials such as mature dried peas and beans, known as staple foods in many countries. In soybeans, which are one of today's promising new source of proteins, stachyose is present in a rather high amount (Table I) as well as in full-fat flours; in defatted flours the oligosaccharides account for 15% of the total carbohydrates. However, these sugars are only present in trace amounts in soy protein concentrate and isolate. Recent technological developments have shown that soybeans and soybean flours are quite suitable for the preparation of protein beverages and textured proteins called meat analogs. In these cases, flatus formation in relation with oligosaccharides of the raffinose family is one of the problem areas as mentioned by Wolf and Cowan (1971) in their recent monograph.

1. Flatus Inhibition

Different authors have demonstrated *in vitro* and *in vivo* that the addition of antibiotics or of bacteriostats such as Vioform or Mexaform to diets containing leguminous seeds inhibited flatus formation (Steggerda, 1968; Richards and Steggerda, 1966). The addition of such substances to foods for human consumption is obviously unacceptable.

Rackis *et al.* (1970b) showed *in vitro* that in soybeans there is an inhibiting factor for gas production. This factor is associated with certain phenolic compounds inhibiting the activity of the intestinal anaerobic bacteria. It is therefore a bacteriostatic and not a bactericidal compound.

Arai *et al.* (1966) identified isoflavones in soybeans and in small amounts phenolic acids such as syringic, ferulic, and chlorogenic acids in decreasing order. These substances were tested *in vitro* by Rackis *et al.* (1970b) who found that isoflavones had no effect, while syringic acid had an effect greater than that of the other phenolic acids.

2. Heat Treatment in the Presence of Water

Processing of leguminous seeds, soybeans in particular, can be accompanied by a lowering of the oligosaccharides present, especially when the technological steps include treatments in the presence of water. It is known that stachyose and verbascose are fairly well solubilized in water. Kawamura (1968) found that autoclaving of defatted soybean flakes reduces the sugars. Yau-Lai Lo *et al.* (1968) soaked whole soybeans for the preparation of soybean milk. They found that after 24 hr of soaking, 3.5% of stachyose was present in the extracted substance.

On the basis of these results, we may consider that soaking and/or blanching of leguminous seeds might well have an influence on the oligosaccharides content. These points are worthy of further investigation. Extrusion cooking of flours, particularly defatted soybean flour, for the preparation of textured vegetable protein, takes place at around 150°C and in the presence of 20–40% water. However, this treatment does not have any influence on the oligosaccharides level of the material used.

3. Enzymatic Treatments

Enzymatic treatments could be applied either with a preparation rich in α-galactosidase or with microorganisms capable of utilizing the oligosaccharides of the raffinose family.

First of all, mention should be made of the germinating beans (Table II). A patent was granted in the United States in 1968 (Okumura and

Wilkinson, 1968) for the preparation of soybean milk from sprouted beans; it is evident that the oligosaccharides residual level is a function of the duration of the sprouting process.

Different species of microorganisms, bacteria, and fungi are able to synthesize α-galactosidase. In 1963, Yu-Teh Li *et al.*, having deduced that this type of enzyme was less known, studied its formation and behavior in the presence of raffinose and stachyose. Suzuki *et al.*, in 1966 and 1969, published their work on the preparation and mode of action of α-galactosidase from two types of microorganisms. One of these publications led to obtaining a French patent in 1968 (Agence Ind. Sci. et Tech., 1968) which corresponded to two Japanese applications. This enzymatic hydrolysis is applied to raffinose in molasses of the sugar industry (Yamane, 1971).

In 1972, the Rohm & Haas Company applied for a patent in relation with an enzyme able to degrade oligosaccharides of the raffinose family. We have checked the ability of this enzyme to split galactose-containing oligosaccharides in soybean milk but, similarly to Calloway *et al.* (1971), found the final product unpalatable. In 1969, Lyons *et al.* made a survey of different strains of actinomycetales for the specific purpose of decomposing flatus-forming oligosaccharides.

An enzyme preparation from *Aspergillus saitoi*, possessing both α-galactosidase and invertase activity, was studied on soybean milk by Sugimoto and van Buren (1970). They found that at the natural pH of soybean milk the addition of a small amount of the enzyme preparation resulted in complete hydrolysis of the galacto-oligosaccharides. The milk obtained was organoleptically acceptable because proteases were absent from the enzyme preparation. Two other types of degradation of the oligosaccharides resulting from microorganisms, applied in the preparation of traditional Oriental foods, may be mentioned. They are tempeh, an Indonesian food, and soy sauce. For tempeh, Shallenberger *et al.* (1967) found that these sugars disappeared after about 72 hr of fermentation, and Calloway *et al.* (1971) have shown that the product was essentially nonflatulent. Our results were analogous to those of these investigators. Soy sauce sugars were studied by Yoshino (1951), who was unable to detect any oligosaccharides in the finished product.

D. Conclusions

The studies made to date for the purpose of elucidating the metabolism of the sugars of the raffinose family support the fact that these carbohydrates are not digested because the human alimentary canal does not produce the enzyme, α-galactosidase, and they are not resorbed by the intestinal wall because their molecules are too large. They therefore come in contact with bacteria which inhabit the lower parts of the intestine and which are able to utilize them with the subsequent formation of flatus, characterized by the presence of high amounts of carbon dioxide.

Trials on animals have shown that the molecular weight of these oligosaccharides has an influence on flatus formation. Stachyose, the tetraholoside, and verbascose, the pentaholoside, have marked effects, while raffinose, the triholoside, has an insignificant effect. On the other hand, it has been shown in rats that flatus is accompanied by a lowering of the pH, at least by 0.5 unit. This modification of the intestinal environment is considered to be important and might affect the metabolism of other substances.

E. Research Needed

The tests on human subjects have so far not given analogous results to those obtained on animals. Further research is definitely needed in this field in order to establish the levels of oligosaccharides that start to produce flatulence in humans. These findings will be important for the acceptability of some new types of soy products.

It is interesting to note that flatulence has never been registered as a complaint among the populations of Middle Eastern and Asian countries, the first being traditionally accustomed to consuming chick-peas, and the second soybeans in considerable amounts. Many hypotheses could be advanced to provide an explanation for what may be the result of special local cooking habits, a particular constitution of the intestinal flora or of the enzymatic system, or merely a life-long adaptation to the food.

It is clear that stachyose and verbascose are involved in flatulence provoked by the consumption of leguminous seeds, but it should be noted that they may not be the sole responsible factors. However, it is suggested that the oligosaccharides of the raffinose family be further investigated, in particular with regard to the determination of their levels and the elucidation of their effects in foods for human consumption.

References

Agence Ind. Sci. et Tech. (1968). Procédés de décomposition du raffinose en galactose et en sucrose et nouveaux produits ainsi obtenus. French Patent 1,555,370.

Alvarez, W. C. (1942). What causes flatulence? *J. Amer. Med. Ass.* **120,** 21–25.

Arai, S., Suzuki, H., Fujimaki, M., and Sukurai, Y. (1966). Studies on flavor compounds in soybean. Part II. Phenolic acids in defatted soybean flour. *Agr. Biol. Chem.* **30,** 364–369.

Askevold, F. (1956). Investigation on the influence of diet on the quality and composition of intestinal gas in humans. *Scand. J. Clin. Lab. Invest.* **8,** 87–94.

Blair, H. A., Dern, R. J., Bates, P. L. (1947). The measurement of volume of gas in the digestive tract. *Amer. J. Physiol.* **149**, 688–707.

Bourquelot, E., and Bridel, M. (1910). New sugar verbascose obtained from mullein root. *C. R. Acad. Sci.* **151**, 760–762.

Breed, R. S., Murray, E. G., and Smith, N. R. (1957). Bergey's Manual of Determinative Bacteriology 7th edit., pp. 638 and 666–667, Williams & Wilkins, Baltimore, Maryland.

Calloway, D. H., Colasito, D. J., and Mathews, R. D. (1966). Gases produced by human intestinal microflora. *Nature (London)* **212**, 1238–1239.

Calloway, D. H., Hickey, C. A., and Murphy, E. L. (1971). Reduction of intestinal gas forming properties of legumes by traditional and experimental food processing methods. *J. Food Sci.* **36**, 251–255.

Cristofaro, E., Mottu, F., and Wuhrmann, J. J. (1970). Study of the effect of stachyose and raffinose on the flatulence activity of soymilk. *3rd Int. Congress Food Sci. and Tech.* (unpublished).

Cristofaro, E., Mottu, F., and Wuhrmann, J. J. (1972). Study of the effect of flatulence of leguminous seeds oligosaccharides. *57th Meet. Amer. Ass. Cereal Chem. 3rd Int. Congress Food Sci. and Tech.* (unpublished).

Debray, P., and Veyne, S. (1966). Les ballonnements abdominaux, physiopathologie, etiologie, traitement. *Aliment. Vie.* **54**, 260–269.

Delente, J., and Ladenburg, K. (1972). Quantitative determination of the oligosaccharides in defatted soybean meal by gas-liquid chromatography. *J. Food Sci.* **37**, 372–374.

Dubois, M., Gilles, K. A., Hamilton, J. K., Rebers, P. A., and Smith, F. (1956). Colorimetric method for determination of sugars and related substances. *Anal. Chem.* **28**, 350–356.

French, D. (1954). The raffinose family of oligosaccharides. *Advan. Carbohyd. Chem.* **9**, 149–184.

Gitzelmann, R., Auricchio, S. (1965). The handling of soya alpha-galactosides by a normal and a galactosemic child. *Pediatrics* **36**, 231–235.

Gomyo, T., Nakamura, N. (1966). Biosynthesis of raffinose from uridine diphosphate galactose and sucrose by an enzyme preparation of immature soybeans. *Agr. Biol. Chem.* **30**, 425–427.

Hardinge, M. G., Swarner, J. B., and Crooks, H. (1965). Carbohydrates in foods. *J. Amer. Diet. Ass.* **46**, 197–204.

Hedin, P. A., and Adachi, R. A. (1962). Effect of diet and time of feeding on gastrointestinal gas production in rats. *J. Nutr.* **77**, 229–235.

Hickey, C. A., Murphy, E. L., and Calloway, D. H. (1972). Intestinal gas production following ingestion of commercial wheat cereals and milling fractions. *Cereal Chem.* **49**, 276–283.

Kawamura, S. (1967). Quantitative paper chromatography of sugars of the cotyledon, hull and hypocotyl of soybeans of selected varieties. *Kagowa Daigaku Nogakubu Gakujutsu Hokuku* **18**, 117–131.

Kawamura, S. (1968). Changes of sugars and decrease in available lysine on autoclaving defatted soybean flakes. *J. Jap. Soc. Food Nutr.* **20**, 478–481.

Kawamura, S., and Hurnichi, A. (1966). Changes in soybean carbohydrates during growth and germination. II Changes of sugars during germination. *Kagowa Daigaku Nojakuku Galenjutsu Hokoku* **17**, 99–103.

Kawamura, S., and Minoru, T. (1966). Sugars of the cotyledon, hull, and hypocotyl of soybeans. *Eiyo To Shokuryo* **19**, 268–275.

Kies, C., and Fox, H. M. (1971). Comparison of the protein nutritional value of TVP, methionine enriched TVP and beef at two levels of intake for human adults. *J. Food Sci.* **36**, 841–845.

Kosterlitz, H. W. (1937). The presence of a galactose-phosphate in the livers of rabbits assimilating galactose. *Biochem. J.* **31**, 2217–2234.

Krause, W., Taeufel, K., Ruttloff, H., and Maune R. (1968). Zur enzymatischen Spaltung und Resorption von di und höheren Oligosacchariden im Intestinaltrakt von Tier und Mensch. *Ernaehrungsforschung* **13,** 161–169.

Kurtzman, R. H., and Halbrook, W. U. (1970). Polysaccharide from dry navy beans, *Phaseolus vulgaris:* its isolation and stimulation of *Clostridium per-fringens. Appl. Microbiol.* **20,** 715–719.

Lyons, A. J., Jr., Pridham, T. G., and Hesseltine, C. W. (1969). Survey of some actinomycetales for α-galactosidase activity. *Appl. Microbiol.* **18,** 579–583.

Murphy, E. C., Horsley, H., and Burr, H. K. (1972). Fractionation of dry bean extracts which increase carbon dioxide egestion in human flatus. *J. Agr. Food Chem.* **20,** 813–817.

Okumura, G. K., and Wilkinson, J. E. (1968). Process of producing soy milk from sprouted soybeans. U.S. Patent 3,399,997.

Pazur, J. H., Shadaksharaswamy, M., and Meidell, G. E. (1962). The metabolism of oligosaccharides in germinating soybeans, *Glycine max. Arch. Biochem. Biophys.* **99,** 78–85.

Prader, A., Semenza, G., and Auricchio, S. (1963). Intestinale Absorption und Malabsorption der Disaccharide. *Schweiz. Med. Wochenschr.* **93,** 1272–1279.

Rackis, J. J., Sessa, D. J., and Honig, D. H. (1967). Isolation and characterization of flavor and flatulence factors in soybean meal. *Proc. Int. Conf. "Soybean Protein Foods," 1966* ARS-71-35, pp. 100–109.

Rackis, J. J., Honig, H. D., Sessa, D. J., and Steggerda, F. R. (1970a). Flavor and flatulence factors in soybean protein products. *J. Agr. Food Chem.* **18,** 977–982.

Rackis, J. J., Sessa, D. J., Steggerda, F. R., Shimuzu, J., Anderson, J., and Pearl, S. L. (1970b). Soybean factor relating to gas production by intestinal bacteria. *J. Food Sci.* **35,** 634–639.

Richards, E. A., and Steggerda, F. R. (1966). Production and inhibition of gas in various regions in the intestine of the dog. *Proc. Soc. Exp. Biol. Med.* **122,** 573–576.

Richards, E. A., Steggerda, F. R., and Murata, A. (1968). Relationship of bean substitutes and certain intestinal bacteria to gas production in the dog. *Gastroenterology* **55,** 502–509.

Rockland, L. B., Gardiner, B. L., and Pieczarka, D. (1969). Stimulation of gas production and growth of *Clostridium perfringens* type A by legumes. *J. Food Sci.* **34,** 411–414.

Rohm & Haas Company. (1972). Process for rendering innocuous flatulence-producing saccharides. U.S. Patent 3,632,346.

Ruttloff, H., Taeufel, A., Krause, W., Haenel, H., and Taeufel, K. (1967). Die intestinal-enzymatische Spaltung von Galakto-Oligosacchariden in Darm von Tier und Mensch mit besonderer Berücksichtigung von *Lactobacillus bifidus.* II. Zum intestinalen Verhalten der Lactulose. *Nahrung* **11,** 39–46.

Semenza, G. (1968). Intestinal oligosaccharidases and disaccharidases. *In* "Handbook of Physiology" Amer. Physiol. Soc., J. Field, ed.), Sect. 6, Vol. V, pp. 2543–2566. Williams & Wilkins, Baltimore, Maryland.

Semenza, G., and Auricchio, S. (1962). Chromatographic separation of human intestinal disaccharidases. *Biochim. Biophys. Acta* **65,** 173–175.

Shallenberger, R. S., and Moyer, J. C. (1961). Relation between changes in glucose, fructose, galactose, sucrose and stachyose, and the formation of starch in peas. *J. Agr. Food Chem.* **9,** 137–140.

Shallenberger, R. S., Hand, D. B., and Steinkraus, K. H. (1967). Changes in sucrose, raffinose and stachyose during tempeh fermentation. *U.S., Dept. Agr., Agr. Res. Serv.* ARS-74-41, 68–71.

Steggerda, F. R. (1961). Relation between consumption of dry beans and flatulence in human subjects. *Dry Bean Res. Conf.* pp. 14–15. (Edited by Western Regional Research Laboratory, Albany, California.)

Steggerda, F. R. (1967). Physiological effects of soybean fractions. *Proc. Int. Conf, "Soybean Protein Foods," 1966* ARS-71-35, pp. 94–99.

Steggerda, F. R. (1968). Gastrointestinal gas following food consumption. *Ann. N.Y. Acad. Sci.* **150,** 57–66.

Steggerda, F. R., and Dimmick, J. F. (1966). Effect of bean diets on concentration of carbon dioxide in flatus. *Amer. J. Clin. Nutr.* **19,** 120–124.

Steggerda, F. R., Richards, E. A., and Rackis, J. J. (1966). Effects of various soybean products on flatulence in the adult man. *Proc. Soc. Exp. Biol. Med.* **121,** 1235–1239.

Sugimoto, H., and van Buren, J. P. (1970). Removal of oligosaccharides from milk. *J. Food Sci.* **35,** 655–660.

Suzuki, H., Ozawa, Y., and Tanabe, O. (1966). Studies on the decomposition of raffinose by α-galactosidase of actinomycetes. *Agr. Biol. Chem.* **30,** 1039–1046.

Suzuki, H., Ozawa, Y., Dota, H., and Yoshida, H. (1969). Studies on the decomposition of raffinose by α-galactosidase of mold. *Agr. Biol. Chem.* **33,** 501–513.

Taeufel, A., Krause, W., Ruttloff, H., and Taeufel, K. (1964). Intestinale Resorption und Ausscheidung von Raffinose. *Nahrung* **8,** 107–108.

Taeufel, K., Steinbach, K. J., and Vogel, E. (1960). Mono- und Oligosaccharide einiger Leguminosensamen sowie ihr Verhalten bei Lagerung und Keimung. *Z. Lebensm. Unters. Forsch.* **112,** 31–40.

Taeufel, K., Ruttloff, H., Krause, W., Taeufel, A., and Vetter, K. (1965 a). Zum intestinalen Verhalten von Galakto-Oligosacchariden beim Menschen. *Klin. Wochenschr.* **43,** 268–272.

Taeufel, K., Noack, R., Ruttloff, H., and Krause, W. (1965 b). Zum biochemisch-intestinalen Verhalten der Kohlenhydrate bei Mensch und Tier. Teil I and II. *Med. Ernaehr.* **6,** 253–256 and 289–293.

Taeufel, K., Krause, W., Ruttloff, H., and Maune, R. (1967). Zur intestinalen Spaltung von Oligosacchariden. *Z. Gesamte Exp. Med. Einschl. Exp. Chir.* **144,** 54–66.

Tanusi, S. (1972). Changes of carbohydrate contents of the soybean seed during growth (cotyledon, hull and hypocotyl). *J. Jap. Soc. Food Nutr.* **25,** 89–93.

von Planta, A., and Schulze, E. (1890). Ueber ein neues krystallisierbares Kohlenhydrat. *Ber. Deut. Chem. Ges.* **23,** 1692–1699.

Werner, I., Abrahamsson, L., and Hambraeus, L. (1971). Clinical and metabolic studies on textured soya protein. Dept. Intern. Med., University of Uppsala Sweden (unpublished).

Wolf, M. J., and Cowan, J. C. (1971). Soybeans as a food source. *CRC Food Tech.* **2,** 81–158.

Yamane, T. (1971). The decomposition of raffinose by α-galactosidase. *Sucr. Belge* **90,** 345–348.

Yau-Lai Lo, W., Steinkraus, K. H., Hand, D. B., Hackler, L. R., and Wilkens, W. F. (1968). Soaking soybeans before extraction as it affects chemical composition and yield of soy-milk. *Food Technol. (Chicago)* **22,** 138–140.

Yoshino, H. (1951). Studies on soy-sauce by paper-partition chromatography. II. Sugars in soy-sauce. *Nippon Jozo Kyokai Zasshi* **46,** 7–13.

Yu-Teh Li, Su-Chen C. Li, and Sheltar, M. R. (1963). α-Galactosidase from *Diplococcus pneumoniae*. *Arch. Biochem. Biophys.* **103,** 436–442.

Disorders Related to Sugar Metabolism: Metabolic Abnormalities

CHAPTER 21

Inborn Errors of Carbohydrate Metabolism

H. G. HERS

The main inborn errors of carbohydrate metabolism recognized up to now have been classified in Table I according to the metabolic process involved. Enzymatic deficiencies that concern the digestion of carbohydrates and their absorption were discussed by Herman (Chapter 11, this volume). Pentosuria and the various forms of galactosemia and of fructosuria were covered by Touster (Chapter 15), Hansen, (Chapter 18) and Hue (Chapter 22). Therefore, this chapter will treat only the deficiencies that concern gluconeogenesis and storage disorders. In the latter group, glycogen storage diseases and mucopolysaccharidosis include a large number of subgroups, each of them caused by the specific defect of one enzyme.

A. Inborn Errors of Gluconeogenesis

1. Importance of Gluconeogenesis

The mechanisms of gluconeogenesis and its control have been recently reviewed by Exton (1972). The liver is the major site of gluconeogenesis with the kidney becoming an important site during starvation and acido-

TABLE I

Inborn Errors of Carbohydrate Metabolism

	Disease	Enzyme affected
Digestion	Sucrose intolerance	Maltase–sucrase (Burgess *et al.*, 1964)
	Lactose intolerance	Lactase (Holzel *et al.*, 1959)
Absorption	Glucose–galactose intolerance	Glucose–galactose carrier (Eggermont and Loeb, 1966)
Interconversion of sugars	Galactosemia	Gal-1-P-uridylltransferase (Isselbacher *et al.*, 1956)
		Galactokinase (Gitzelmann, 1965)
	Benign fructosuria	Fructokinase (Schapira *et al.*, 1962)
	Fructose intolerance	Liver aldolase (Hers and Joassin, 1961)
	Pentosuria	L-Xylulose reductase (Wang and van Eys, 1970)
Gluconeogenesis	Hexosediphosphatase deficiency	Hexosediphosphatase (Baker and Winegrad, 1970)
	Type I glycogenosis	Glucose-6-phosphatase (Cori and Cori, 1952)
Storage disorders	Glycogenoses	See Fig. 1
	Mucopolysaccharidoses	See Table II

sis. Gluconeogenesis performs several functions. It provides glucose to the body during starvation; it allows the free utilization of lactate and glycerol produced in small amounts under basal conditions and in increased amounts during exercise or heightened lipolytic activity; it also provides NH_3 in the kidney to counteract acidosis; and, finally, it allows the metabolism of some amino acids absorbed from the alimentary tract or released during protein breakdown in muscle. The main enzymes proper to gluconeogenesis are those involved in the conversion of pyruvate to oxalacetate and phosphoenol pyruvate, hexosediphosphatase, and glucose-6-phosphatase. Pathological conditions are known in which one of the two last enzymes is deficient.

2. Hexosediphosphatase Deficiency

Hexosediphosphatase catalyzes the irreversible hydrolysis of fructose diphosphate into fructose 6-phosphate. The enzyme first purified from rabbit liver by Gomori (1943) is an alkaline phosphatase, completely inactive at physiological pH. It was recognized by Hers and Kusaka (1953) that the native enzyme present in fresh liver displays an important activity at pH 7; this neutral activity is easily lost upon purification of the enzyme. Hexosediphosphatase is strongly inhibited by AMP at physiological concentration (Taketa and Pogell, 1963). The enzyme has been extensively purified, and its properties have been reviewed (Pontremoli and Horecker, 1970).

Hexosediphosphatase deficiency was described for the first time in 1970 by Baker and Winegrad in a 5-year-old girl who had frequent episodes of hypoglycemia and severe metabolic acidosis. Other cases have been more recently reported (Baerlocher *et al.*, 1971; Hülsmann and Fernandes, 1971; Pagliara *et al.*, 1972).

In hexosediphosphatase deficiency, there is a failure to maintain glucose homeostasis during prolonged fasting. Lactic acid and alanine, two important substrates of gluconeogenesis, accumulate in blood and there is a metabolic acidosis. Other features of the clinical symptomatology include hepatomegaly with fatty changes of the liver and muscular hypotonia.

The mechanism by which hypoglycemia arises after a fasting of some 20 hr is obviously related to the fact that, at that time, liver glycogen stores are depleted and blood glucose can only be provided by gluconeogenesis. An observation more difficult to explain is that the patients are to some extent sensitive to fructose or glycerol; these substrates produce a hypoglycemic and hypophosphatemic response together with hyperlactacidemia, metabolic acidosis, and hyperuricemia. These changes are presumably in relation to the inability of the patients to form glucose from fructose or glycerol, which accumulate in the liver in the form of fructose 1-phosphate or glycerophosphate before being converted into lactic acid.

3. Glucose-6-phosphatase Deficiency (Type I Glycogenosis)

Type I glycogenosis has been known for 20 years to result from a deficiency of glucose-6-phosphatase (Cori and Cori, 1952). This enzyme is required not only for the conversion of glycogen to glucose but also for gluconeogenesis. It is bound to the endoplasmic reticulum of the liver and kidney, which are also the tissues affected in type I glycogenosis, whereas the muscle is entirely normal. The clinical features include an important hepatomegaly, lactaciduria, and lactic acidosis sometimes complicated by hyperuricemia. A severe hypoglycemia is observed upon fasting and may cause the death of the patients.

Several functional tests may be used for the diagnosis of type I glycogenosis. One of them is the measurement of the conversion of either galactose (Schwartz *et al.*, 1957) or fructose (Hers, 1959) into glucose, which does not occur in the absence of glucose-6-phosphatase. Another test is based on the determination of the turnover rate of [2-³H]glucose as compared to that of [¹⁴C]glucose. The half-life of the [2-³H]glucose is indeed normally shorter than that of [¹⁴C]glucose because of the loss of [³H] through the Cori cycle or through the futile recycling of glucose into glucose 6-phosphate and fructose 6-phosphate. Since glucose-6-phosphatase is required for this recycling, the difference in turnover rate is not observed in patients with type I glycogenosis (Hue and Hers, 1972; Van Hoof *et al.*, 1972).

B. Glycogen Storage Diseases

1. Introduction

Glycogen is the polysaccharide present in nearly all animal cells but particularly abundant in liver and muscle. It is made of a very large number of glucosyl units assembled by a α-glucosidic linkage either to carbon-4 or to carbon-6 of the preceding unit. The mechanisms of glycogen synthesis and glycogen degradation are summarized in Fig. 1 (for a review, see Ryman and Whelan, 1971). The main steps of glycogen synthesis are the formation of the α-1,4 linkages by transfer of a glucosyl unit from UDPG through the action of glycogen synthetase (reaction 4) and the formation of the 1,6 linkages by the 1,4 $\rightarrow$ 1,6 transfer of an oligosaccharide through the action of the branching enzyme (reaction 5). There are two mechanisms for glycogen degradation. One is phosphorolytic and involves the transfer of a glucosyl unit from the nonreducing end of the polysaccharide to inorganic phosphate by phosphorylase (reaction 6) and the hydrolysis of the 1,6 linkages by amylo-1,6-glucosidase (reaction 7). The other mechanism is hydrolytic and lysosomal. It requires only one enzyme, the acid α-glucosidase, which can completely hydrolyze glycogen into glucose (reaction 12).

The glycogen storage diseases include a series of inherited pathological conditions in which there is a large excess of glycogen in tissues mostly in liver and muscle (for reviews, see Hers, 1964a; Howell, 1972). In the following section the various types of glycogenosis are described and classified according to the main metabolic pathway involved.

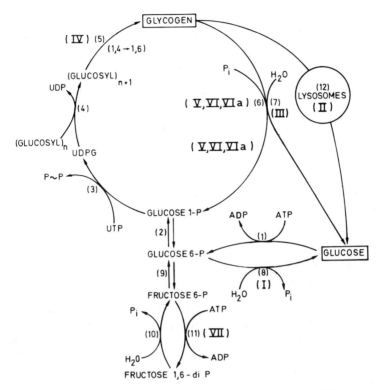

Fig. 1 The metabolism of glycogen. Enzymes: (1) glucokinase, (2) phosphoglucomutase, (3) UDPG-pyrophosphorylase, (4) glycogen synthetase, (5) branching enzyme, (6) phosphorylase, (7) amylo-1,6-glucosidase, (8) glucose-6-phosphatase (liver and kidney), (9) hexosephosphate isomerase, (10) fructosediphosphatase, (11) phosphofructokinase, (12) acid α-glucosidase. Roman numerals refer to the type of glycogen storage disease.

2. *Inborn Errors of Glycogen Synthesis*

a. *Type IV Glycogenosis*

Type IV glycogenosis described by Andersen (1956) is characterized by the accumulation in tissues of a glycogen with very long outer chains, similar to amylopectin in this respect (Illingworth and Cori, 1952). This disorder is sometimes called "amylpectinosis." A deficiency of branching enzymes has been demonstrated by Brown and Brown (1966) in one patient. It must be pointed out, however, that this abnormality does not give an entirely satisfactory explanation for the disorder since the glycogen which accumulates in patients is branched to some extent, although less branched than normally. Apparently the amylopectin-like glycogen is very poorly soluble and difficult to degrade by phosphorylase.

Type IV glycogenosis is an extremely rare disorder; in this laboratory, in which several hundred cases of glycogen storage diseases have been investigated during the last 15 years, only one was of type IV.

b. Other Disorders

A deficiency of glycogen synthetase has been claimed to be the cause of congenital familial hypoglycemia. The deficiency was only observed once in one needle biopsy (Lewis *ét al.*, 1963), and as discussed elsewhere (Hers, 1964b), the claim was not adequately supported. It has never been confirmed.

3. Inborn Errors of the Phosphorolytic Degradation of Glycogen

The enzymes to be considered here are not only phosphorylase and amylo-1,6-glucosidase, which are directly involved in glycogen breakdown, but also glucose-6-phosphatase and the glycolytic enzymes which are responsible for further metabolism of the hexose phosphates. The deficiency of glucose-6-phosphatase has already been described above (Section A,3).

a. Inborn Errors in the Phosphorylase System

(1) THE PHOSPHORYLASE SYSTEM.   There are two forms of phosphorylase in tissues, one of them, called *a*, is fully active; the other called *b*, is generally inactive in physiological conditions. The two forms are interconvertible by phosphorylation and dephosphorylation, under the action of specific kinases and phosphatases, as represented in Fig. 2. The kinase system is stimulated by various hormones including epinephrine and glucagon in the liver, epinephrine alone in the muscle, and this stimulation is mediated by 3',5'-cyclic AMP. In the liver, the phosphorylase phosphatase is also greatly stimulated by glucose.

Phosphorylase is a different protein in muscle and liver, and it is therefore expected that different mutations would affect the enzyme in each of these tissues.

(2) GLYCOGENOSIS TYPE V OR MCARDLE DISEASE.   In this disorder (McArdle, 1951) there is a complete deficiency of muscle phosphorylase

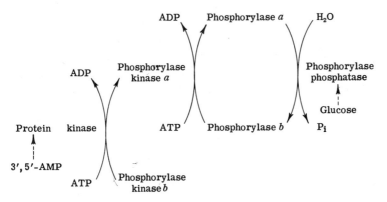

Fig. 2 Enzymatic interconversion of the two forms of phosphorylase by phosphorylation and dephosphorylation. The cyclic AMP-dependent protein kinase acts as a phosphorylase kinase kinase.

(Mommaerts *et al.*, 1959; Larner and Villar-Palasi, 1959; Schmid *et al.*, 1959), whereas phosphorylase kinase and phosphorylase phosphatase are normally active. The disease may be unnoticed during childhood and manifest itself in the young adult mostly by muscular pain, weakness, and stiffness following slight exertion. In contrast to the normal procedure, the blood lactate falls during exercise. The evolution of the disorder is progressive. Weakness and wasting of individual muscle groups appear with increasing severity. This loss of muscle mass is associated with a marked excretion of creatine in the urine.

(3) TYPE VI AND TYPE VIa GLYCOGENOSES. Type VI glycogenosis (Hers, 1959; Hers and Van Hoof, 1968) includes cases of the hepatomegalic type of glycogen storage diseases (Von Gierke's disease) which cannot be classified as type I, type III, or type IV because the activities of glucose-6-phosphatase, amylo-1,6-glucosidase, or branching enzymes are normal. In the liver of many of these patients, phosphorylase is poorly active. Liver phosphorylase activity is however very variable from one patient to another as it is also sometimes in normal liver, and there is not yet a clear demonstration that the primary defect is on the formation of phosphorylase itself. In one subgroup of the disorder, called VIa, it has been demonstrated by Huijing (1967) that phosphorylase kinase is deficient. This deficiency was initially demonstrated in leukocytes but has now been observed also in liver and in erythrocytes in this laboratory. As indicated in Fig. 3 erythrocytes seem to be the ideal material to demonstrate the defect. When a hemolysate of normal subjects is incubated in the presence of glycogen, a spontaneous activation of phosphorylase

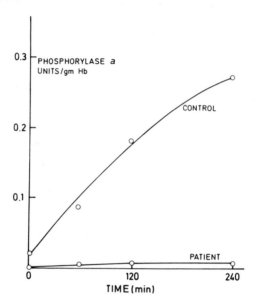

Fig. 3 Activation of phosphorylase in a hemolysate. A 20% lysate of human red cells was incubated at 20° in the presence of 0.1% glycogen and 2×10^{-6} M cyclic AMP. (Unpublished observation of B. Lederer.)

occurs; this activation does not occur with the blood of the patients. All patients analyzed to date are male, and the disease is recognized to be sex-linked (Huijing and Fernandes, 1969).

b. Deficiencies of Amylo-1,6-glucosidase (Type III Glycogenosis)

Amylo-1,6-glucosidase catalyzes the complex reaction by which a phosphorylase limit dextrin is converted into glycogen (Fig. 4). The limit dextrin has short outer chains, and the first step catalyzed by amylo-1,6-glucosidase is a transglucosylation of a maltotriose unit from a side chain to a main chain. In the second step, glucosyl residues located by a 1,6 linkage are taken out by hydrolysis (Walker and Whelan, 1960). The activity of amylo-1,6-glucosidase can be measured by a number of methods (Hers *et al.*, 1967).

In type III glycogenosis there is a deficiency of amylo-1,6-glucosidase in practically all tissues of the body (Illingworth *et al.*, 1956). The polysaccharide which accumulates in the tissues has the structure of a phosphorylase limit dextrin, in which the outer chains are much shorter than in glycogen. This is why the disorder is sometimes called limit dextrinosis. The patients suffer from hepatomegaly and hypoglycemia, and the clinical picture is not very different from that in type I or in type VI glyco-

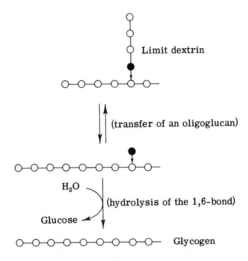

Fig. 4 The mechanism of action of amylo-1,6-glucosidase: (◯) Glucose residues in α-1,4 linkage; (●) residues in α-1,6 linkage; (—) α-1,4-linkage; and (↓) α-1,6 linkage.

gen storage diseases. There exist several subgroups of the disorder. In type IIIa, the amylo-1,6-glucosidase is completely inactive, whatever the method used for its detection. In other subgroups an important residual activity can be measured in liver or in muscle by only certain methods (Van Hoof and Hers, 1967). These subgroups presumably result from allelic mutations of the same gene. The unequal distribution of the residual activities in liver and muscle is poorly understood.

c. Deficiency of Phosphofructokinase (Type VII Glycogenosis)

Phosphofructokinase catalyzes the irreversible phosphorylation of fructose 6-phosphate into fructose diphosphate at the expense of ATP. This reaction is considered as the limiting step in glycolysis in many tissues. Deficiency of this enzyme was described by Tarui *et al.* (1965) in the muscle of three sibs. In these patients, ischemic exercise caused no rise in venous lactate level. Their muscles contained 2–4% glycogen and also a marked excess of glucose 6-phosphate and fructose 6-phosphate.

4. Inborn Error of the Hydrolytic Degradation of Glycogen

Type II glycogenosis resulting from the absence of acid α-glucosidase is studied in Section C,3.

C. Inborn Lysosomal Diseases

The inborn lysosomal diseases are the various disorders caused by the congenital defect of one of the acid hydrolases normally present in lysosomes (Hers, 1965). Type II glycogenosis was the first lysosomal disease to be clearly recognized. Its study can be used as a model for the understanding of other storage disorders that are consecutive to the primary defect of one lysosomal enzyme.

1. The Concept of Lysosomes

Lysosomes (lytic bodies) are cytoplasmic particles that contain a great number of hydrolytic, digestive enzymes (for a review, see de Duve and Wattiaux, 1966). These enzymes are separated from the surrounding cytoplasm by a lipoprotein membrane which prevents them from acting on external substrates. All of them have an acid optimal pH. More than forty hydrolytic enzymes have been localized in lysosomes; it is very likely that additional ones will be recognized in the future.

The lytic action of all lysosomal enzymes points clearly to the fact that these particles make up an intracellular digestive system. All substances which penetrate into the cell by endocytosis (phagocytosis and pinocytosis) are initially isolated inside vacuoles called phagosomes, which later fuse with lysosomes (see Fig. 5). They are then destroyed by the digestive enzymes. When indigestible or poorly digestible substances like polyvinylpyrrolidone, Triton X-100, or dextran are injected into rats, they accumulate in lysosomes which become greatly enlarged (see Fig. 6b). Another function of lysosomes is the digestion of the cytoplasm itself. In the process of autophagy, a small area of the cytoplasm is surrounded by membranes, thereby being sequestered. The vacuoles so formed fuse with lysosomes giving birth to "autophagic vacuoles" also called "autophagosomes" (see Fig. 6a). The macromolecules that are isolated in these vacuoles are digested almost completely by the lysosomal enzymes. This is also the fate of all cellular structures such as mitochondria, endoplasmic reticulum, and glycogen. This polysaccharide is converted into glucose by the lysosomal acid α-glucosidase. During these processes, the material to be digested and the enzyme involved never come in contact with the cytoplasm from which they are separated by the lysosomal membrane. The small molecules that are formed during this digestion pass through the lysosomal membrane and are reutilized by the cell. The lysosomal digestion is schematically represented in Fig. 5.

Under the electron microscope, lysosomes normally appear as "dense

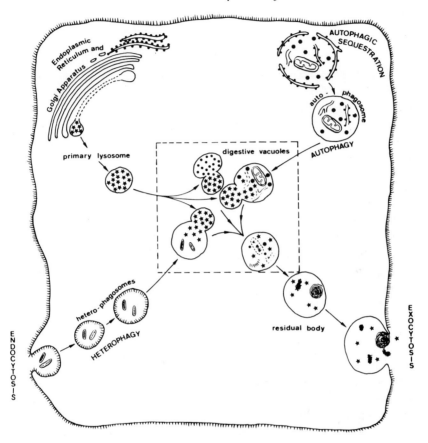

Fig. 5 Schematic representation of the lysosomal system.

bodies," the diameter of which is usually about 0.5 μ (see Fig. 6a). The dense bodies exhibit a polymorphic internal structure that is believed to be made mostly of undigestible residues left behind in lysosomes. These structures are the hallmark of the "residual bodies." Residues can be extruded from certain cells by "exocytosis" (the reverse of endocytosis), but in no case have they been observed free in the cytoplasm.

2. The Concept of Inborn Lysosomal Diseases

Inborn lysosomal diseases are inborn errors of metabolism in which the metabolic block is in the intralysosomal digestion. The existence of numerous enzymes with a similar digestive function inside lysosomes presented nature with an opportunity for a large number of genetic protein alterations, the morphological and functional consequences of which

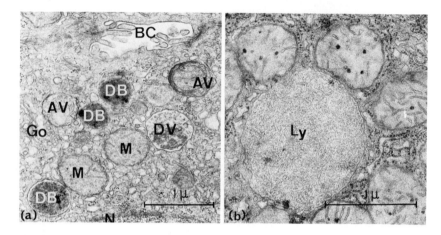

Fig. 6 Several aspects of lysosomes in rat hepatocytes. (a) In the liver of a normal rat one recognizes a part of the nucleus (N), two mitochondria (M), the Golgi apparatus (Go), a biliary canaliculus (BC), three dense bodies (DB), one digestive vacuole (DV), and two autophagic vacuoles (AV), one of them containing a mitochondrion. (b) In the liver of a rat which has received an injection of dextran (1 mg/gm of body weight) 2 days before a greatly enlarged lysosome (Ly) is seen, surrounded by several mitochondria. Photograph from F. Van Hoof.

can show many similarities. These consequences could be predicted from our knowledge of physiology of lysosomes. The main characteristics of inborn lysosomal diseases are as follows (Hers, 1965).

1. All lysosomal diseases are storage diseases characterized by the deposition, inside the vacuolar system, of all substances that normally would have been degraded by the missing enzyme. Under the electron microscope, lysosomes appear greatly enlarged by various substances.

2. The stored material is generally heterogeneous. This is because digestive enzymes, like the acid hydrolases of lysosomes, are not highly specific for one substrate but only for one type of linkage; for instance, β-galactosidase (its absence is responsible for G_{M1}-gangliosidosis, also called pseudo-Hurler's disease) is specific for the β-galactosidic linkages and liberates a galactose residue either from a preexisting oligo- or polysaccharide, or from a glycolipid. In the absence of this enzyme, one must expect a mixed accumulation of polysaccharides and glycolipids to occur. This characteristic is of great importance since most lipidoses and mucopolysaccharidoses are in fact lipomucopolysaccharidoses in which there is a simultaneous accumulation of mucopolysaccharides and glycolipids in many tissues.

3. Like other hereditary disorders, the inborn lysosomal diseases affect

most tissues of the body. The electron microscope shows abnormal structures in lysosomes even in tissues that are slightly affected clinically.

4. These diseases are progressive because, at birth, the lysosomes have not yet had time to accumulate a large amount of material. In Pompe's disease (see below), most of the patients die within the first year of life. In other lysosomal disorders, the progressivity is prolonged over several years or even several decades.

3. Type II Glycogenosis

Type II glycogenosis, also called Pompe's disease, is the most serious form of glycogenosis; most of the affected children die from muscular weakness when they are about 6 months old. There is an enormous accumulation of glycogen in nearly all tissues. This deposition contrasts with the ability of the patients to mobilize normally their glycogen under the action of glucagon or epinephrine. Accordingly, all enzymes which participate in the phosphorolytic degradation of glycogen are present in normal amounts in these patients. This is why the pathogeny of the disease remained mysterious for a long time.

In 1963, it was shown that human liver and muscle contain an α-glucosidase that can hydrolyze glycogen into glucose and which is most active at pH 4. This α-glucosidase is bound to lysosomes (Lejeune *et al.*, 1963) and is absent from the tissues of patients affected with type II glycogenosis (Hers, 1963). It is highly probable that the role of this glucosidase is to hydrolyze glycogen which would have penetrated into the lysosomes. Since there is no glycogen in blood plasma, where it would be rapidly destroyed by α-amylase, one must assume that glycogen has not penetrated into lysosomes by phagocytosis but through autophagy. It is expected that, being deprived of α-glucosidase, the patients affected with type II glycogenosis will accumulate glycogen inside the vacuolar system derived from the lysosomes. In agreement with this interpretation, it has been shown with the electron microscope that there are actually two localizations for glycogen in the liver of patients with Pompe's disease. One part of the polysaccharide is in the cytoplasm as normally and in a normal amount, whereas the other is concentrated inside vacuoles clearly separated from the rest of the cytoplasm by a membrane (see Fig. 7). This second part is the glycogen present in excess in the tissue. No dense bodies, the normal form of hepatic lysosomes, can be seen. We have many reasons to believe that the vacuoles filled with glycogen are enlarged lysosomes (Baudhuin *et al.*, 1964).

Similar pictures have been seen in numerous types of cells of patients with type II glycogenosis but were never observed in normal tissues.

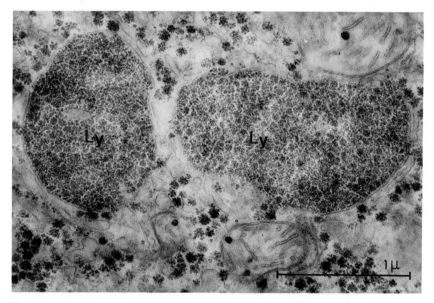

Fig. 7 Part of a hepatocyte of a patient with type II glycogenosis. Two glycogen-filled vacuoles (Ly) limited by a single membrane can be recognized. Glycogen (dark spots) is also freely dispersed in the cytoplasm. From Baudhuin *et al.* (1964).

Their presence in amniotic cells allows the prenatal diagnosis of the disease (Hug *et al.*, 1970).

It is not clear at the present time how this enlargement of the lysosomal system causes the muscular degeneration characteristic of the disease. It is possible that the enlarged lysosomes are particularly fragile and that under the effect of muscular contraction their membrane ruptures allowing the lysosomal cathepsins to reach the cytoplasm.

4. Mucopolysaccharidoses

The clinical syndrome known as Hurler's syndrome or gargoylism is characterized by numerous skeletal deformities, corneal opacities, visceromegaly, mental retardation, and urinary excretion of mucopolysaccharides. Several subgroups of the disorder have been recognized mostly on the basis of the chemistry of the mucopolysaccharide which is excreted, or on the recognition of a precise enzymatic defect. It is also remarkable that the disorder was initially classified as a lipidosis on the basis of histochemical evidence of lipid storage in the brain.

The first indication that the Hurler's syndrome could be considered as a lysosomal disorder came in 1964 from the ultrastructural study of

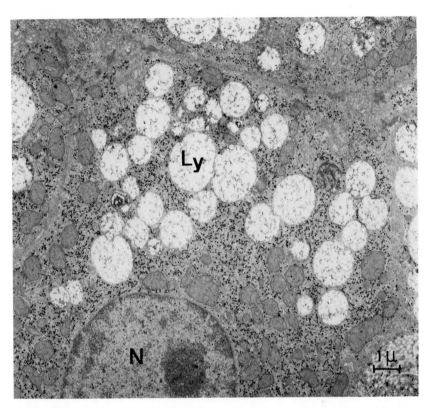

Fig. 8 Hepatocytes of a patient with Hurler's disease. The lysosomes (Ly) are dilated and appear as clear vacuoles. Photograph from F. Van Hoof.

the liver of patients (Van Hoof and Hers, 1964). As shown in Fig. 8, this study revealed the presence of huge vacuoles that could be considered as enlarged lysosomes. These structures are indeed similar to those obtained experimentally in the rat by administration of dextran or of polyvinylpyrrolidone or other compounds which are known to accumulate inside lysosomes. Furthermore, no normal dense bodies could be seen in this liver. The publication of these results makes it readily clear that the mucopolysaccharides present in excess in tissue of the patients were actually located inside lysosomes. Since mucopolysaccharides are made by fibroblasts and also found in the urine, it appeared that the excess of these compounds in many tissues presumably resulted from the endocytosis of circulating mucopolysaccharides. The primary cause of the disease could be either an excess of synthesis by the fibroblasts or a defect of degradation by other tissues. It is now a general agreement that the second interpretation is correct and that the disorder is the result of the

TABLE II

Deficiencies of a Lysosomal Acid Hydrolase in Mucopolysaccharidoses

Disease	Material stored	Missing enzyme	References
Type I (Hurler)	Dermatan sulfate + heparan sulfate	α-L-Iduronidase	Matalon and Dorfman (1972) Bach *et al.* (1972)
Type V (Scheie)	Dermatan sulfate	α-L-Iduronidase (allelic with type I)	Bach *et al.* (1972) McKusick *et al.* (1972)
β-Glucuronidase deficiency A	?	β-D-Glucuronidase N-Sulfatase	Neufeld and Cantz (1973) Kresse and Neufeld (1972)
Type III (Sanfilippo) B	Heparan sulfate	N-Acetyl-α-glucosaminidase	O'Brien (1972) von Figura and Kresse (1972)
Gₘ₁-gangliosidosis	Gₘ₁-ganglioside + keratan sulfate	β-D-Galactosidase	Van Hoof and Hers (1968) Okada and O'Brien (1968) Dacremont and Kint (1968)
Fucosidosis	Fucosides	α-L-Fucosidase	Van Hoof and Hers (1968)
Mannosidosis	Mannosides	α-D-Mannosidase	Öckermann (1967), Carroll *et al.* (1972)
Mucosulfatidosis	Mucopolysaccharides + sulfatides	Arylsulfatases A + B + C	Austin (1965)

specific defect of one acid hydrolase which is involved in the degradation of mucopolysaccharides in the lysosomes. Since mucopolysaccharides are very complex structures which contain many sugar derivatives, a large number of hydrolytic enzymes are required for their complete degradation, each of them being possibly affected by a mutation. At the present time, as many as eight different defects, listed in Table II, have been recognized as the primary cause of the same number of subgroups of gargoylism. In the brain of the patient, one usually finds "zebra bodies" in which lamellar structures indicate the presence of glycolipids.

5. Other Storage Diseases

Besides type II glycogenosis and the mucopolysaccharidoses, the following diseases have been clearly attributed to a deficit of one lysosomal glycosidase. Since the depot material is a glycolipid, these disorders are usually not classified as inborn errors of carbohydrate metabolism. In each case, the investigation with the electron microscope has shown the vacuolar localization of the deposit.

Gaucher's disease is the result of the deficiency of β-glucosidase (Brady *et al.*, 1965; Patrick, 1965). The deposit is made of glucocerebrosides.

In Fabry's disease the α-galactosidase is inactive (Kint, 1970). A trihexoside ceramide accumulates.

Tay-Sachs disease is the result of a deficiency of hexosaminidase, of which two isoenzymes are known. In the most frequent type of the disease, the brain is primarily affected and only one of the two isoenzymes is deficient (Okada and O'Brien, 1968). In a much rarer form, the two isoenzymes are inactive and the condition is extended to many tissues. Various gangliosides with a terminal *N*-acetylgalactosamine group are found in excess (Sandhoff *et al.*, 1968).

More complete information on the role of lysosomes in storage diseases as well as appropriate references will be found in a recent book (Hers and Van Hoof, 1973).

References

Andersen, D. H. (1956). Familial cirrhosis of the liver with storage of abnormal glycogen. *Lab. Invest.* **5**, 11–20.

Austin, J. H. (1965). Mental retardation. Metachromatic leucodystrophy. *In* "Medical Aspects of Mental Retardation" (C. H. Carter, ed.), pp. 768–813. Thomas, Springfield, Illinois.

Bach, G., Friedman, R., Weissman, B., and Neufeld, E. F. (1972). The defect in the Hurler and Scheie syndromes: Deficiency of α-L-iduronidase. *Proc. Nat. Acad. Sci. U.S.* **69**, 2048–2051.

354   H. G. Hers

Baerlocher, K., Gitzelmann, R., Nüssli, R., and Dumermuth, G. (1971). Infantile lactic acidosis due to hereditary fructose 1,6-diphosphatase deficiency. *Helv. Paediat. Acta* **26,** 489–506.

Baker, L., and Winegrad. A. I. (1970). Fasting hypoglycaemia and metabolic acidosis associated with deficiency of hepatic fructose 1,6-diphosphatase activity. *Lancet* **1,** 13–16.

Baudhuin, P., Hers, H. G., and Loeb, H. (1964). An electron microscopic and biochemical study of type II glycogenosis. *Lab. Invest.* **13,** 1139–1152.

Brady, R. O., Kanfer, J. N., and Shapiro, D. (1965). Metabolism of glucocerebrosides. II. Evidence of an enzymatic deficiency in Gaucher's disease. *Biochem. Biophys. Res. Commun.* **18,** 221–225.

Brown, B. I., and Brown, D. H. (1966). Lack of an α-1,4 glucan: α-1,4-glucan-6-glucosyltransferase in a case of type IV glycogenosis. *Proc. Nat. Acad. Sci. U.S.* **56,** 725–729.

Burgess, E. A., Levin, B., Mahalanabis, D., and Tonge, R. E. (1964). Hereditary sucrose intolerance: Levels of sucrose activity in jejunal mucosa. *Arch. Dis. Childhood* **39,** 431–433.

Carroll, M., Dance, N., Masson, P. K., Robinson, D., and Winchester, B. G. (1972). Human mannosidosis. The enzymic defect. *Biochem. Biophys. Res. Commun.* **49,** 579–583.

Cori, G. T., and Cori, C. F. (1952). Glucose-6-phosphatase of the liver in glycogen storage disease. *J. Biol. Chem.* **199,** 661–667.

Dacremont, G., and Kint, J. A. (1968). G_{M1}-ganglioside accumulation and β-galactosidase deficiency in a case of G_{M1}-gangliosidosis (Landing disease). *Clin. Chim. Acta* **21,** 421–425.

de Duve, C., and Wattiaux, R. (1966). Functions of lysosomes. *Annu. Rev. Physiol.* **28,** 435–492.

Eggermont, E., and Loeb, H. (1966). Glucose-galactose intolerance. *Lancet* **2,** 343–344.

Exton, J. H. (1972). Gluconeogenesis. *Metabolism.* **21,** 945–990.

Gitzelmann, R. (1965). Deficiency of erythrocyte galactokinase in a patient with galactose diabetes. *Lancet* **2,** 670–671.

Gomori, G. (1943). Hexosediphosphatase. *J. Biol. Chem.* **148,** 139–149.

Hers, H. G. (1959). Etudes enzymatiques sur fragments hépatiques. Application à la classification des glycogénoses. *Rev. Int. Hepatol.* **9,** 35–55.

Hers, H. G. (1963). α-Glucosidase deficiency in generalized glycogen storage disease (Pompe's disease). *Biochem. J.* **86,** 11–16.

Hers, H. G. (1964a). Glycogen storage disease. *Advan. Metab. Disord.* **1,** 1–44 and 335–336.

Hers, H. G. (1964b). Comments. *In* "Control of Glycogen Metabolism" (W. J. Whelan, ed.), p. 386. Little, Brown, Boston, Massachusetts.

Hers, H. G. (1965). Inborn lysosomal diseases. *Gastroenterology* **48,** 625–633.

Hers, H. G., and Joassin, G. (1961). Anomalie de l'aldolase hépatique dans l'intolérance au fructose. *Enzyme Biol. Clin.* **1,** 4–14.

Hers, H. G., and Kusaka, T. (1953). Le métabolisme du fructose-1-phosphate dans le foie. *Biochim. Biophys. Acta* **11,** 427–437.

Hers, H. G., and Van Hoof, F. (1968). Glycogen-storage diseases: Type II and type VI glycogenosis. *In* "Carbohydrate Metabolism and its Disorders" (F. Dickens, P. J. Randle, and W. J. Whelan, eds.), Vol. 2, pp. 151–168. Academic Press, New York.

Hers, H. G, and Van Hoof, F. (1973). "Lysosomes and Storage Diseases." Academic Press, New York.

Hers, H. G., Verhue, W., and Van Hoof, (1967). The determination of amylo-1,6-glucosidase. *Eur. J. Biochem.* **2**, 257–264.

Holzel, A., Schwarz, V., and Sutcliffe, K. W. (1959). Defective lactose absorption causing malnutrition in infancy. *Lancet* **1**, 1126–1128.

Howell, R. R. (1972). The glycogen storage disease. *In* "The Metabolic Basis of Inherited Disease" (J. B. Stanbury, J. B. Wyngaarden, and D. S. Fredrickson, eds.), 3rd ed., pp. 149–173. McGraw-Hill, New York.

Hue, L., and Hers, H. G. (1972). The turnover of 2-³H glucose: An assay for diagnosis of type I glycogenosis. *In* "Biochemistry of the Glycosidic Linkage; An Integrated View" (R. Piras and H. G. Pontis, eds.), Vol. 2, pp. 681–686. Academic Press, New York.

Hug, G., Schubert, W. K., and Soukup, S. (1970). Prenatal diagnosis of type II glycogenosis. *Lancet* **1**, 1002.

Huijing, F. (1967). Phosphorylase kinase in leukocytes of normal subjects and patients with glycogen storage disease. *Biochim. Biophys. Acta* **148**, 601–603.

Huijing, F., and Fernandes, J. (1969). X-chromosomal inheritance of liver glycogenosis with phosphorylase kinase deficiency. *Amer. J. Hum. Genet.* **21**, 275–284.

Hülsmann, W. C., and Fernandes, J. (1971). A child with lactacidemia and fructose-diphosphatase deficiency in the liver. *Pedia. Res.* **5**, 633–637.

Illingworth, B., and Cori, G. T. (1952). Structure of glycogens and amylopectins. III. Normal and abnormal human glycogen. *J. Biol. Chem.* **199**, 653–660.

Illingworth, B., Cori, G. T., and Cori, C. F. (1956). Amylo-1,6-glucosidase in muscle tissue in generalized glycogen storage disease. *J. Biol. Chem.* **218**, 123–129.

Isselbacher, K., Anderson, E. P., Kurahashi, K., and Kalckar, H. M. (1956). Congenital galactosemia, a single enzymatic block in galactose metabolism. *Science* **123**, 635–636.

Kint, J. A. (1970). Fabry's disease: Alpha-galactosidase deficiency. *Science* **167**, 1268–1269.

Kresse, H., and Neufeld, E. F. (1972). The Sanfilippo A corrective factor. *J. Biol. Chem.* **247**, 2164–2170.

Larner, J., and Villar-Palasi, C. (1959). Enzymes in a glycogen storage myopathy. *Proc. Nat. Acad. Sci. U.S.* **45**, 1234.

Lejeune, N., Thinès-Sempoux, D., and Hers, H. G. (1963). Tissue fractionation studies. 16. Intracellular distribution and properties of α-glucosidases in rat liver. *Biochem. J.* **86**, 16–21.

Lewis, G. M., Spencer-Peet, J., and Stewart, K. M. (1963). Infantile hypoglycemia due to inherited deficiency of glycogen synthetase in liver. *Arch. Dis. Childhood* **38**, 40–58.

McArdle, B. (1951). Myopathy due to a defect in muscle glycogen breakdown. *Clin. Sci.* **10**, 13–35.

McKusick, V. A., Howell, R. R., Hussels, I. E., Neufeld, E. F., and Stevenson, R. E. (1972). Allelism, non-allelism and genetic compounds among the mucopolysaccharidoses. *Lancet* **1**, 993–996.

Matalon, R., and Dorfman, A. (1972). Hurler's syndrome, and α-L-iduronidase deficiency. *Biochem. Biophys. Res. Commun.* **47**, 959–964.

Mommaerts, W. F. H. M., Illingworth, B., Parson, C. M., Guillory, R. J., and Seraydarian, K. (1959). A functional disorder of muscle associated with the absence of phosphorylase. *Proc. Nat. Acad. Sci. U.S.* **45**, 791–797.

Neufeld, E. F., and Cantz, M. (1973). The mucopolysaccharidoses studied in cell culture. *In* "Lysosomes and Storage Diseases" (H. G. Hers and F. Van Hoof, eds.). Academic Press, New York.

O'Brien, J. S. (1972). Sanfilippo syndrome: Profound deficiency of α-acetylglucosaminidase activity in organs and skin fibroblasts from type B patients. *Proc. Nat. Acad. Sci. U.S.* **69**, 1720–1722.

Öckermann, P. A. (1967). A generalized storage disorder resembling Hurler's syndrome. *Lancet* **2**, 239–241.

Okada, S., and O'Brien, J. S. (1968). Generalized gangliosidosis: β-galactosidase deficiency. *Science* **160**, 1002–1004.

Pagliara, A. S., Karl, I. E., Keating, J. P., Brown, B. I., and Kipnis, D. M. (1972). Hepatic fructose-1,6-diphosphatase deficiency. A cause of lactic acidosis and hypoglycemia in infancy. *J. Clin. Invest.* **51**, 2115–2123.

Patrick, A. D. (1965). A deficiency of glucocerebrosidase in Gaucher's disease. *Biochem. J.* **97**, 17–18.

Pontremoli, S., and Horecker, B. L. (1970). Fructose-1,6-diphosphatase from rabbit liver. *Curr. Top. Cell. Regul.* **2**, 173–199.

Ryman, B. E., and Whelan, W. J. (1971). New aspects of glycogen metabolism. *Advan. Enzymol.* **34**, 285–443.

Sandhoff, K., Andreae, U., and Jatzkewitz, H. (1968). Deficient hexosaminidase activity in an exceptional case of Tay-Sachs disease with additional storage of kidney globoside in visceral organs. *Life Sci.* **7**, 283–288.

Schapira, F., Schapira, G., and Dreyfus, J. C. (1962). La lésion enzymatique de la fructosurie bénigne. *Enzyme Biol. Clin.* **1**, 170–175.

Schmid, R., Robbins, P. W., and Traut, R. R. (1959). Glycogen synthesis in muscle lacking phosphorylase. *Proc. Nat. Acad. Sci. U.S.* **45**, 1236–1240.

Schwartz, R., Ashmore, J., and Renold, A. E. (1957). Galactose tolerance in glycogen storage disease. *Pediatrics* **19**, 585–595.

Taketa, K., and Pogell, B. M. (1963). Reversible inactivation and inhibition of liver fructose-1,6-diphosphatase by adenosine nucleotides. *Biochem. Biophys. Res. Commun.* **12**, 229–235.

Tarui, S., Okuno, G., Ikura, Y., Takana, T., Suda, M., and Nishika, M. (1965). Phosphofructokinase deficiency in skeletal muscle. An new type of glycogenosis. *Biochem. Biophys. Res. Commun.* **19**, 517.

Van Hoof, F., and Hers, H. G. (1964). L'ultrastructure des cellules hépatiques dans la maladie de Hurler (gargoylisme). *C. R. Acad. Sci.* **259**, 1281–1283.

Van Hoof, F., and Hers, H. G. (1967). The subgroups of type III glycogenosis. *Eur. J. Biochem.* **2**, 265–270.

Van Hoof, F., and Hers, H. G. (1968). The abnormalities of lysosomal enzymes in mucopolysaccharidoses. *Eur. J. Biochem.* **7**, 34–44.

Van Hoof, F., Hue, L., de Barsy, T., Jacquemin, P., Devos, P., and Hers, H. G. (1972). Glycogen storage diseases. *Biochimie* **54**, 745–751.

von Figura, K., and Kresse, H. (1972). The Sanfilippo B corrective factor: A *N*-acetyl-α-D-glucosaminidase. *Biochem. Biophys. Res. Commun.* **48**, 262–269.

Walker, G. J., and Whelan, W. J. (1960). The mechanism of carbohydrase action. 8. Structures of the muscle-phosphorylase limit dextrins of glycogen and amylopectin. *Biochem. J.* **76**, 264–268.

Wang, Y. M., and van Eys, J. (1970). The enzymatic defect in essential pentosuria. *N. Engl. J. Med.* **282**, 892–896.

CHAPTER 22

The Metabolism and Toxic Effects of Fructose

L. HUE

The first part of this chapter is devoted to the description of the reactions involved in the metabolic pathway of fructose in the liver. Since three different pathways have been proposed for the metabolism of D-glyceraldehyde, particular attention is given to the discussion of the metabolism of this triose which is formed from fructose in the liver. The second part deals with the toxicity of fructose in relation to the hypoglycemia of hereditary fructose intolerance. This fructose-induced hypoglycemia is characterized by a lack of response to glucagon. It is shown that this unresponsiveness to glucagon is not satisfactorily explained by the known toxic effects of fructose on the level of adenine nucleotides in the liver but might result from an inhibition of phosphorylase a activity by fructose 1-phosphate.

A. Metabolism of Fructose

The utilization of fructose by peripheral tissues *in vivo* seems to be negligible (Wick *et al.*, 1953; Weichselbaum *et al.*, 1953) because glucose, which is always present in blood, prevents either the uptake of fructose or the phosphorylation of fructose by hexokinase (Mackler and Guest,

1953; Nakada, 1956); adipose tissue, however, could be an exception (Froesch and Ginsberg, 1962).

Fructose is metabolized mainly in liver (Levine and Huddlestun, 1947; Mendeloff and Weichselbaum, 1953), kidney (Reinecke, 1944), and intestinal mucosa (Bollman and Mann, 1931). These tissues possess a number of enzymes which are responsible for a specialized metabolic pathway of fructose. It has been shown that the hepatocyte is freely permeable to fructose (Cahill *et al.*, 1958).

1. Metabolic Pathway of Fructose in the Liver

a. Phosphorylation of Fructose

Liver hexokinase can catalyze the phosphorylation of glucose, fructose, and mannose, but the enzyme has 20 times more affinity for glucose than for fructose (Sols and Crane, 1954). Glucose therefore prevents the phosphorylation of fructose into fructose 6-phosphate. Furthermore, glucokinase which is also present in the liver, is inactive on fructose (Vinuela *et al.*, 1963).

Fructose is phosphorylated into fructose 1-phosphate by fructokinase (Leuthardt and Testa, 1951; Cori *et al.*, 1951; Hers, 1952a). The enzyme has recently been purified to approximately 1000-fold from rat liver (Adelman *et al.*, 1967; Sanchez *et al.*, 1971a). The phosphoryl donor is the Mg-ATP complex and the enzyme has an absolute requirement for K^+ (Hers, 1952b; Sanchez *et al.*, 1971b). It is strongly inhibited by ADP, one of its reaction products (Parks *et al.*, 1957), and this inhibition is partially reversed by K^+ (Sanchez *et al.*, 1971b). Fructokinase is not specific for fructose since it also catalyzes the phosphorylation of L-sorbose, D-tagatose, D-xylulose, and L-galactoheptulose and the affinity for the ketoses depends on the concentration of potassium (Adelman *et al.*, 1967; Sanchez *et al.*, 1971a).

b. Splitting of Fructose 1-Phosphate

Fructose 1-phosphate is split by liver aldolase or aldolase B (Hers and Kusaka, 1953; Leuthardt *et al.*, 1953). The same enzyme catalyzes the splitting of fructose 1-phosphate and the condensation of two triose phosphates into fructose diphosphate. The maximal activity of the liver enzyme on fructose diphosphate is the same as on fructose 1-phosphate; this is expressed by the cleavage ratio equal to one (Hers and Jacques, 1953). On the contrary, the muscle and the brain enzymes have a fructose diphosphate:fructose 1-phosphate ratio equal to 50 and 10, respectively. Furthermore, the affinity of liver aldolase for fructose 1-phosphate is 10 or 30 times greater than that of brain or muscle aldolase (Penhoet *et al.*, 1969a,b). The liver aldolase is therefore well adapted to play its dual role in glycolysis and in fructose metabolism.

The cleavage of the fructose molecule into two C-3 compounds *in vivo* has been demonstrated with the use of isotopes. It has indeed been observed that liver glycogen formed in rats from either [1-¹⁴C]fructose (Hers, 1955) or [6-¹⁴C]fructose (Landau and Merlevede, 1963; Rauschenbach and Lamprecht, 1964) was labeled both in C-1 and C-6. This result implies that D-glyceraldehyde, which is formed after splitting of fructose 1-phosphate, is transformed in a triose phosphate.

c. *Metabolism of* D-*Glyceraldehyde*

There are three mechanisms by which D-glyceraldehyde, which is formed in the liver upon the splitting of fructose 1-phosphate, can be converted into a triose phosphate (Fig. 1).

(1) D-Glyceraldehyde can be directly phosphorylated into D-glyceraldehyde 3-phosphate by triokinase. This enzyme also phosphorylates dihydroxyacetone (Hers and Kusaka, 1953). Triokinase has been partially purified from guinea pig liver (Hers, 1962), beef liver (Heinz and Lamprecht, 1961) and rat liver (Frandsen and Grunnet, 1971). The rat liver

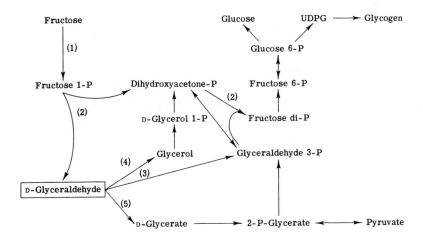

Fig. 1 The metabolism of fructose and the glyceraldehyde crossroads in the liver. (1) fructokinase, (2) liver aldolase, (3) triokinase, (4) alcohol dehydrogenase or aldose reductase, and (5) aldehyde dehydrogenase.

enzyme has a K_m for D-glyceraldehyde and dihydroxyacetone equal to about $10^{-5}\ M$ (Sillero *et al.*, 1969; Frandsen and Grunnet, 1971).

(2) D-Glyceraldehyde can also be reduced to glycerol by NADH under the action of alcohol dehydrogenase (Wolf and Leuthardt, 1953) or by NADPH in a reaction catalyzed by an aldose reductase (Hers, 1960). Glycerol would then be phosphorylated into D-glycerol 1-phosphate (Bublitz and Kennedy, 1954a) which is then oxidized into dihydroxyacetone phosphate. As discussed previously (Hers, 1957) the formation of [1,6-^{14}C]glycogen from [6-^{14}C]fructose allows one to discard the participation of this pathway, which would result in the formation of [3,4-^{14}C]glycogen. This labeling of carbon-3 and -4 is expected from the stereospecificity of glycerol kinase (Bublitz and Kennedy, 1954b).

(3) As a third possibility it has been proposed that D-glyceraldehyde, after oxidation to glyceric acid (Leuthardt *et al.*, 1953; Lamprecht and Heinz, 1958) and phosphorylation (Holzer and Holldorf, 1957; Ichihara and Greenberg, 1957; Lamprecht *et al.*, 1959) reaches glycolysis at the level of 2-phosphoglycerate. The main argument in favor of this hypothesis is the accumulation of D-glycerate in the liver after a fructose load (Kattermann and Holzer, 1961). Rauschenbach and Lamprecht (1964) have even claimed that triokinase has no functional significance and that most of the D-glyceraldehyde produced from fructose and converted into glycogen is oxidized to D-glycerate. This conclusion was based on their observation that there is in glycogen slightly less ^{14}C in C-1 originating from [6-^{14}C]fructose than ^{14}C in C-6 originating from [1-^{14}C]fructose.

On the basis of the kinetic properties of the various enzymes that act on D-glyceraldehyde, Sillero *et al.* (1969) recently came to the conclusion that triokinase is the major enzyme involved. Furthermore, triokinase is one of the three enzymes which are specialized in the metabolism of fructose. These enzymes, fructokinase, aldolase B, and triokinase, are found in tissues known to metabolize fructose, namely, liver, kidney, and to some extent intestinal mucosa (Kranhold *et al.*, 1969; Heinz, 1968a). Finally, it has been reported that a fructose diet significantly increased the enzyme activities of various enzymes and that triokinase was the most affected one (Heinz, 1968b; Veneziale, 1972). The latter results are, however, in contradiction with the data presented by Sillero *et al.* (1969).

The problem of the glyceraldehyde crossroads was investigated by measuring the conversion of [4-^{3}H,6-^{14}C]fructose into glycogen (Hue and Hers, 1972). The fate of ^{3}H and ^{14}C during the conversion of [4-^{3}H,6-^{14}C]fructose into fructose diphosphate is shown in Fig. 2. After splitting of fructose 1-phosphate, D-glyceraldehyde is labeled on C-1 with ^{3}H and on C-3 with ^{14}C. If D-[1-^{3}H,3-^{14}C]glyceraldehyde is oxidized into

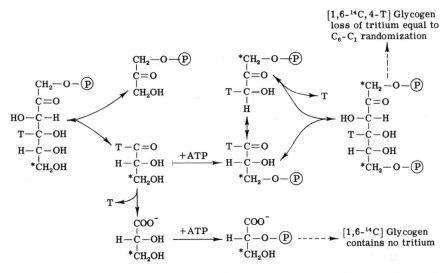

Fig. 2 Labilization of tritium (T) and randomization of carbon in the conversion of [4-³H,6-¹⁴C]fructose into fructose diphosphate.

D-glycerate, ³H will obviously be lost and glycogen will contain no tritium. On the contrary, if D-[1-³H,3-¹⁴C]glyceraldehyde is phosphorylated, the label will stay on C-1 after phosphorylation and will still be on dihydroxyacetone phosphate after isomerization. Triosephosphate isomerase is able to distinguish between the two carbinol hydrogens of dihydroxyacetone phosphate, and the hydrogen coming from the aldehydohydrogen of glyceraldehyde 3-phosphate is not labilized during the isomerization whereas the other one is exchanged with a proton (Rieder and Rose, 1959). However, during the condensation of the two triose phosphates the tritium present in dihydroxyacetone phosphate will be lost (Rieder and Rose, 1959), and this loss of tritium will be equal to the degree of C-6 to C-1 randomization. Accordingly, after the injection of [4-³H,6-¹⁴C]fructose, the loss of ³H relative to ¹⁴C found in glycogen and proper to fructose metabolism was equal to the C-6 to C-1 randomization. This is in agreement with the hypothesis that D-glyceraldehyde is phosphorylated by triokinase. This conclusion is further supported by the fact that during the *in vitro* oxidation of tritiated D-glyceraldehyde there is an isotopic discrimination against the tritiated glyceraldehyde. This yields a progressive accumulation of [1-³H]glyceraldehyde relative to [3-¹⁴C]glyceraldehyde. If this oxidation were operative *in vivo*, there would also be an increase of the ³H:¹⁴C ratio in glyceraldehyde from the liver. The fact that this increase was not found to occur *in vivo* is against the participation of the oxidative pathway in the metabolism of D-glyceraldehyde by the liver. The experimental data are therefore in good agreement with the hypothesis that D-glyceraldehyde formed from fructose is phosphorylated by triokinase prior to its conversion to glycogen.

2. Inborn Errors of the Fructose Metabolism

The subject was recently reviewed by Froesch (1972).

Essential fructosuria is characterized by the absence of fructokinase in the liver (Schapira *et al.*, 1961) and presumably also in kidney and intestine. As a consequence, the fructose ingested accumulates in blood and is excreted as such in urine. This observation is indirect evidence for the preponderant role played by fructokinase in the metabolism of fructose.

The second disorder is known as hereditary fructose intolerance. The administration of fructose to patients is followed by a tremendous fall in blood glucose level and this hypoglycemia is further characterized by its unresponsiveness to glucagon; this hormone is known to induce a hyperglycemia by a stimulation of the glycogen breakdown in the liver. Biochemically the disorder is characterized by a modification of the properties of liver aldolase (Hers and Joassin, 1961). The activity of this enzyme is greatly depressed on both substrates, but this reduction predominantly affects the splitting of fructose 1-phosphate. Immunological studies have shown that the liver of the patients contains a protein sharing the immunological properties of purified liver aldolase (Nordmann *et al.*, 1968). Since liver aldolase of the patients is now unable to metabolize fructose 1-phosphate, one can assume that the toxic effects of fructose are related to the accumulation of fructose 1-phosphate in their liver. The consequences of such an accumulation are best studied in experimental models which consist in animals receiving large amounts of fructose intravenously (see Section B).

B. Toxicity of Fructose for the Normal Liver

When large doses of fructose are injected intravenously to the animals, the concentration of fructose 1-phosphate in the liver can reach 10–15 mM within a few minutes after the injection (Günther *et al.*, 1967; Heinz and Junghänel, 1969; Burch *et al.*, 1969) and the level depends on the dose injected (Burch *et al.*, 1970). As a consequence of this accumulation of fructose 1-phosphate, the intracellular concentrations of inorganic phosphate and ATP fall to about one-third to one-fourth the normal value (Mäenpää *et al.,* 1968) and this effect, also, is dose-dependent (Raivio *et al.,* 1969). Furthermore, the total amount of adenine nucleotides decreases and AMP is degraded into IMP which accumulates (Woods, *et al.,* 1970) and is further metabolized to uric acid and allantoin in rat (Mäenpää *et al.,* 1968). The same effect of large doses of fructose on liver adenine nucleotides has also been observed in man by Bode and co-workers (1971), and an increased level of uric acid in blood after fructose infusion has been demonstrated by Perheentupa and Raivio (1967). As a consequence of this depletion of adenine nucleotides, the synthesis of protein in the liver is disturbed (Mäenpää *et al.,* 1968; Mäenpää, 1972). Some ultrastructural changes have been reported to occur in the liver after fructose (Goldblatt *et al.,* 1970; Phillips *et al.,* 1970).

The clinical importance of the fructose-induced depletion of adenine nucleotides in the liver, as well as of the fructose-induced hyperuricemia, remains to be investigated. Since fructose is not utilized as such by muscle, heart, or brain but is in great part converted to glucose by the liver, there is no rationale for the clinical use of large doses of fructose injected intravenously. Considering the potential toxicity of fructose for the liver, this practice should be completely abandoned.

Knowing the effect of fructose in the normal liver one can now try to understand the toxicity of fructose in hereditary fructose intolerance.

C. Toxicity of Fructose in Hereditary Fructose Intolerance

The administration of fructose to patients with hereditary fructose intolerance is followed by a fall in the concentration of glucose and of inorganic phosphate in the blood. As said above, the hypoglycemia is unresponsive to glucagon, whereas the disappearance of inorganic phosphate indicated that the hepatic levels of phosphate and ATP are low. The mechanism by which the administration of fructose prevents the glycogenolytic action of glucagon has been studied (Van den Berghe *et al.,* 1973). The mode of action of glucagon is shown in Fig. 3. ATP is involved at three levels: the production of cyclic AMP from ATP by adenyl cyclase, the phosphorylation of phosphorylase kinase, and the phosphorylation of phosphorylase. The high concentration of fructose 1-phosphate in the liver is another factor which might interfere with one of these enzymes.

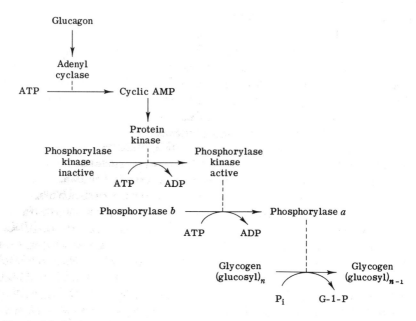

Fig. 3 Mechanism of the glycogenolytic action of glucagon by cyclic AMP. The scheme has been demonstrated to occur in muscle (Krebs, 1972) and presumably operates also in liver.

1. The Activity of Adenyl Cyclase

Since the K_m of liver adenyl cyclase for ATP (Pohl *et al.*, 1971) is of the same order of magnitude as the actual concentration of ATP in the liver, a substance which decreases this concentration will also affect the activity of adenyl cyclase. Cyclic AMP was measured in the liver 3 min after the injection of glucagon in animals which had been previously treated in order to decrease their ATP level (Fig. 4). Glycerol and fructose induce a fall of ATP by trapping inorganic phosphate (Burch *et al.*, 1970) and ethionine by sequestering the adenosyl residue (Farber *et al.*, 1964). In control animals cyclic AMP concentration rose from the basal value of 0.8 μM to about 60 μM after glucagon. This increase was markedly influenced by the treatment with fructose, ethionine, or glycerol, and a good correlation was found between the level of ATP and the amount of cyclic AMP produced under the action of glucagon. In human subjects, cyclic AMP was measured in urine. As shown in Table I, after the administration of glucagon the excretion of cyclic AMP was as a mean 10-fold increased. In control children, this effect of glucagon was not modified by the previous administration of a small dose of fructose (250 mg/kg), whereas in fructose intolerant patients it was greatly

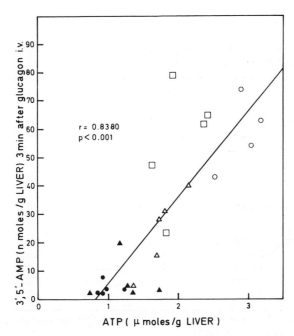

Fig. 4 Correlation between the hepatic concentration of ATP and the level of cyclic AMP obtained after the administration of glucagon: (○) control, (□) glycerol, (△) ethionine, (●) fructose, and (▲) ethionine plus fructose. Glycerol and fructose (5 mg/gm of body weight) were injected intraperitoneally 20 min before glucagon (0.1 μg/gm of body weight). Ethionine (1 mg/gm of body weight) was given intraperitoneally 4 hr before glucagon injection. After Van den Berghe *et al.* (1973).

diminished although usually not completely suppressed by that treatment.

2. The Response to Cyclic AMP

The administration of fructose to patients with fructose intolerance not only causes a marked decrease in the activity of their liver adenyl cyclase but also completely abolishes the glycogenolytic action of cyclic AMP itself (Fig. 5).

As indicated in Fig. 3, several steps in the glycogenolytic effect of cyclic AMP could be altered by the decreased ATP level or be inhibited by fructose 1-phosphate. In contrast with glucagon-stimulated adenyl cyclase, protein kinase (Krebs, 1972) and phosphorylase kinase (Krebs *et al.*, 1964) have a K_m for ATP which is at least one order of magnitude below the actual concentration of ATP in the liver. It is thus probable that

TABLE I

The Effect of a Fructose Load on the Stimulation of the Urinary Excretion of Cyclic AMP by Glucagon in Human Subjects[a]

		Without fructose		After fructose	
Subject	Age	Before glucagon	After glucagon	Before glucagon	After glucagon
Controls					
VP	3 weeks	—	—	14.7	68.6
VDH	8 weeks	8.7	91.0	11.3	143.1
TA	18 months	11.8	69.3	16.6	81.6
VL	13 yr	4.1	26.3	4.6	33.9
Hereditary fructose intolerant					
PN	7 weeks	16.3	41.5	16.8	16.0
TB	4 months	8.1	236.9	12.6	24.7
LJ	5 months	10.2	80.1	12.2	14.4
PT	3 yr	6.3	69.5	7.8	20.5

[a] Results are expressed as nanomoles of cyclic AMP per milligram of creatinine. After Van den Berghe *et al.* (1973).

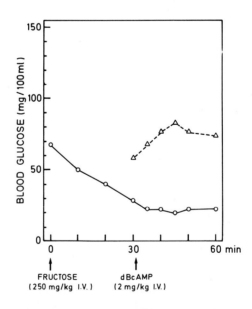

Fig. 5 Effect of dibutyryl cyclic AMP (dBcAMP) on glycemia in a patient with hereditary fructose intolerance. The test was performed in the absence of fructose (△) and 30 min after a fructose load (○). After Van den Berghe *et al.* (1973).

the activity of these two enzymes is not affected by the change in concentration of ATP. Furthermore, the activities of protein kinase and phosphorylase kinase were not found to be inhibited by fructose 1-phosphate at the concentration present in the liver after a load of fructose. Finally, although the administration of fructose to mice provokes an important reduction of the capacity of the liver to form cyclic AMP, the level obtained after the injection of glucagon is still large enough to cause an activation of liver phosphorylase, and this normal activation is further evidence that no abnormal behavior has to be found at the level of the two kinases.

The fact that in patients with hereditary fructose intolerance blood glucose can be raised by the administration of galactose during a fructose-induced hypoglycemia (Cornblath *et al.*, 1963) indicates that the conversion of glucose 1-phosphate to glucose 6-phosphate and free glucose is not impaired. The only remaining steps which could be affected by fructose are those catalyzed by phosphorylase *a* and amylo-1,6-glucosidase. The activity of the latter was not affected by the presence of fructose 1-phosphate.

3. *Activity of Phosphorylase* a

The K_m of phosphorylase *a* for phosphate is of the same order of magnitude as the actual phosphate concentration (Maddaiah and Madsen, 1966). The decreased phosphate concentration in the liver after the administration of fructose (Mäenpää *et al.*, 1968) will therefore affect the activity of phosphorylase *a*. Furthermore, it was found that fructose 1-phosphate is a competitive inhibitor of liver phosphorylase *a*. Figure 6 shows that the activity of phosphorylase *a* can be reduced to about

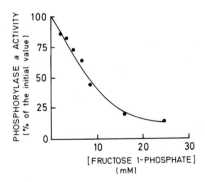

Fig. 6 Influence of fructose 1-phosphate on the activity of purified dog liver phosphorylase *a*. The phosphorolytic activity was measured in the presence of 1.5 m*M* inorganic phosphate and increasing amounts of fructose 1-phosphate.

25% of its initial value by the addition of 10–15 mM fructose 1-phosphate, which is the concentration found in the liver of mice after the injection of high amounts of fructose. This inhibition by fructose 1-phosphate has already been observed by Maddaiah and Madsen (1966).

D. Summary

In summary, fructose greatly interferes with the production of cyclic AMP by adenyl cyclase. This effect depends on the decreased level of ATP in the liver. However, the level of cyclic AMP is still large enough to activate glycogenolysis in mice and, on the other hand, the administration of cyclic AMP to patients intolerant to fructose does not correct the fructose-induced hypoglycemia. The lack of response to glucagon in the patients is best explained by a decreased activity of liver phosphorylase a as a result of the fall in the concentration of inorganic phosphate and of the inhibition by fructose 1-phosphate which accumulates.

References

Adelman, R. C., Ballard, F. J., and Weinhouse, S. (1967). Purification and properties of rat liver fructokinase. *J. Biol. Chem.* **242,** 3360–3365.

Bode, C., Schumacker, H., Goebell, H., Zelder, O., and Pelzel, H. (1971). Fructose induced depletion of liver adenine nucleotides in man. *Horm. Metab. Res.* **3,** 289–290.

Bollman, J. L., and Mann, F. C. (1931). The physiology of the liver; the utilization of fructose following complete removal of liver. *Amer. J. Physiol.* **96,** 683–695.

Bublitz, C., and Kennedy, E. P. (1954a). Synthesis of phosphatides in isolated mitochondria. The enzymatic phosphorylation of glycerol. *J. Biol. Chem.* **211,** 951–961.

Bublitz, C., and Kennedy, E. P. (1954b). A note on the asymmetrical metabolism of glycerol. *J. Biol. Chem.* **211,** 963–967.

Burch, H. B., Max, P., Chyu, K., and Lowry, O. H. (1969). Metabolic intermediates in liver of rats given large amounts of fructose or dihydroxyacetone. *Biochem. Biophys. Res. Commun.* **34,** 619–626.

Burch, H. B., Lowry, O. H., Meinhardt, L., Max, P., Jr., and Chyu, K. (1970). Effects of fructose, dihydroxyacetone, glycerol and glucose on metabolites and related compounds in liver and kidney. *J. Biol. Chem.* **245,** 2092–2102.

Cahill, G. F., Jr., Ashmore, J., Earle, A. S., and Zottu, S. (1958). Glucose penetration into liver. *Amer. J. Physiol.* **192,** 491–496.

Cori, G. T., Ochoa, S., Slein, M. W., and Cori, C. F. (1951). The metabolism of fructose in liver. Isolation of fructose 1-phosphate and inorganic pyrophosphate. *Biochim. Biophys. Acta* **7,** 304–317.

Cornblath, M., Rosenthal, I. M., Reisner, S. H., Wybregt, S. H., and Crane, R. K. (1963). Hereditary fructose intolerance. *N. Engl. J. Med.* **269,** 1271–1278.

Farber, E., Shull, K. H., Villa-Trevino, S., Lombardi, B., and Thomas, M. (1964). Biochemical pathology of acute hepatic adenosine triphosphate deficiency. *Nature (London)* **203,** 34–40.

Frandsen, E. K., and Grunnet, N. (1971). Kinetic properties of triokinase from rat liver. *Eur. J. Biochem.* **23,** 588–592.

Froesch, E. R. (1972). Essential fructosuria and hereditary fructose intolerance. *In* "The Metabolic Basis of Inherited Disease" (J. B. Stanbury, J. B. Wyngaarden, and D. S. Frederickson, eds.), 3rd ed., pp. 131–148. McGraw-Hill, New York.

Froesch, E. R., and Ginsberg, J. L. (1962). Fructose metabolism of adipose tissue. I. Comparison of fructose and glucose metabolism in epididymal adipose tissue of normal rats. *J. Biol. Chem.* **237,** 3317–3324.

Goldblatt, P. J., Witschi, H., Friedman, M. A., Sullivan, R. J., and Shull, K. H. (1970). Some structural and functional consequences of hepatic adenosine triphosphate deficiency induced by intraperitoneal fructose administration. *Lab. Invest.* **23,** 378–385.

Günther, M. A., Sillero, A., and Sols, A. (1967). Fructokinase assay with a specific spectrophotometric method using 1-phosphofructokinase. *Enzymol. Biol. Clin.* **8,** 341–352.

Heinz, F. (1968a). Messung der Enzymaktivitäten in der Dünndarmmucosa der Ratte. *Hoppe-Seyler's Z. Physiol. Chem.* **349,** 339–344.

Heinz, F. (1968b). Enzyme des Fructosestoffwechsels: Änderungen von Enzymeaktivitäten in Leber und Niere der Ratte bei fructose und glucose reicher Ernährung. *Hoppe Seyler's Z. Physiol. Chem.* **349,** 399–404.

Heinz, F., and Junghänel, J. (1969). Metabolitmuster in Rattenleber nach Fructoseapplikation. *Hoppe Seyler's Z. Physiol. Chem.* **350,** 859–866.

Heinz, F., and Lamprecht, W. (1961). Anreicherung und Charakterisierung einer Triosekinase aus Leber. Zur Biochemie des Fructosestoffwechsels. III. *Hoppe Seyler's Z. Physiol. Chem.* **324,** 88–100.

Hers, H. G. (1952a). La fructokinase du foie. *Biochim. Biophys. Acta* **8,** 416–423.

Hers, H. G. (1952b). Rôle du magnésium et du potassium dans la réaction fructokinasique. *Biochim. Biophys. Acta* **8,** 424–430.

Hers, H. G. (1955). The conversion of fructose 1-C¹⁴ and sorbitol 1-C¹⁴ to liver and muscle glycogen in the rat. *J. Biol. Chem.* **214,** 373–381.

Hers, H. G. (1957). "Le métabolisme du fructose." Editions Arscia, Bruxelles.

Hers, H. G. (1960). L'aldose réductase. *Biochim. Biophys. Acta* **37,** 120–126.

Hers, H. G. (1962). Triokinase. *In* "Methods in Enzymology" (S. P. Colowick and N. O. Kaplan, eds.), Vol. 5, pp. 362–364. Academic Press, New York.

Hers, H. G., and Jacques, P. J. (1953). Hétérogénéité des aldolases. *Arch. Int. Physiol.* **61,** 260–261.

Hers, H. G., and Joassin, G. (1961). Anomalie de l'aldolase hépatique dans l'intolérance au fructose. *Enzymol. Biol. Clin.* **1,** 4–14.

Hers, H. G., and Kusaka, T. (1953). Le métabolisme du fructose-1-phosphate dans le foie. *Biochim. Biophys. Acta* **11,** 427–437.

Holzer, H., and Holldorf, A. (1957). Anreicherung, Charakterisierung und biologische Bedeutung einer D-Glycerat-kinase aus Rattenleber. *Biochem. Z.* **329,** 283–291.

Hue, L., and Hers, H. G. (1972). The conversion of [4-³H]fructose and of [4-³H]glucose to liver glycogen in the mouse. An investigation of the glyceraldehyde crossroads. *Eur. J. Biochem.* **29,** 268–275.

Ichihara, A., and Greenberg, D. M. (1957). Studies on the purification and properties of D-glyceric acid kinase of liver. *J. Biol. Chem.* **225,** 949–958.

Kattermann, R., and Holzer, H. (1961). D-Glycerat beim Fructoseabbau in der Leber. *Biochem. Z.* **334,** 218–226.

Kranhold, J. F., Loh, D., and Morris, R. C., Jr. (1969). Renal fructose-metabolizing enzymes: Significance in hereditary fructose intolerance. *Science* **165**, 402–403.

Krebs, E. G. (1972). Protein kinases. *Curr. Top. Cell. Regul.* **5**, 99–133.

Krebs, E. G., Love, D. S., Bratvold, G. E., Trayser, K. A., Meyer, W. L., and Fischer, E. H. (1964). Purification and properties of rabbit skeletal muscle phosphorylase *b* kinase. *Biochemistry* **3**, 1022–1033.

Lamprecht, W., and Heinz, F. (1958). Isolierung von Glycerinaldehyd-Dehydrogenase aus Rattenleber. Zur Biochemie des Fructosestoffwechsels. *Z. Naturforsch.* *B* **13**, 464–465.

Lamprecht, W., Diamantstein, T., Heinz, F., and Balde, P. (1959). Phosphorylierung von D-Glycerinsäure zu 2-Phospho-D-glycerinsäure mit Glyceratkinase in Leber. I. Zur Biochemie des Fructosestoffwechsels. II. *Hoppe Seyler's Z. Physiol. Chem.* **316**, 97–112.

Landau, B. R., and Merlevede. W. (1963). Initial reactions in the metabolism of D- and L-glyceraldehyde by rat liver. *J. Biol. Chem.* **238**, 861–867.

Leuthardt, F., and Testa, E. (1951). Die Phosphorylierung der Fructose in der Leber. II. Mitteilung. *Helv. Chim. Acta* **34**, 931–938.

Leuthardt, F., Testa, E., and Wolf, H. P. (1953). Der enzymatische Abbau der Fructose 1-phosphats in der Leber. III. Mitteilung über den Stoffwechsel der Fructose in der Leber. *Helv. Chim. Acta* **36**, 227–251.

Levine, R., and Huddlestun, B. (1947). The comparative action of insulin on the disposal of intravenous fructose and glucose. *Fed. Proc., Fed. Amer. Soc. Exp. Biol.* **6**, 151.

Mackler, B., and Guest, G. M. (1953). Effects of insulin and glucose on utilization of fructose by isolated rat diaphragm. *Proc. Soc. Exp. Biol. Med.* **83**, 327–329.

Maddaiah, V. T., and Madsen, N. B. (1966). Kinetics of purified liver phosphorylase. *J. Biol. Chem.* **241**, 3873–3881.

Mäenpää, P. H. (1972). Fructose and liver protein synthesis. *Acta. Med. Scand., Suppl.* **542**, 115–118.

Mäenpää, P. H., Raivio, K. O., and Kekomäki, M. P. (1968). Liver adenine nucleotides: Fructose-induced depletion and its effect on protein synthesis. *Science* **161**, 1253-1254.

Mendeloff, A. I., and Weichselbaum, T. E. (1953). Role of the human liver in the assimilation of intravenously administered fructose. *Metab., Clin. Exp.* **2**, 450–458.

Nakada, H. I. (1956). The metabolism of fructose by isolated rat diaphragms. *J. Biol. Chem.* **219**, 319–326.

Nordmann, Y., Schapira, F., and Dreyfus, J. C. (1968). A structurally modified liver aldolase in fructose intolerance: Immunological and kinetic evidence. *Biochem. Biophys. Res. Commun.* **31**, 884–889.

Parks, R. E., Ben-Gershom, E., and Lardy, H. A. (1957). Liver fructokinase. *J. Biol. Chem.* **227**, 231–242.

Penhoet, E. E., Kochman, M., and Rutter, W. J. (1969a). Isolation of fructose diphosphate aldolases A, B and C. *Biochemistry* **8**, 4391–4395.

Penhoet, E. E., Kochman, M., and Rutter, W. J. (1969b). Molecular and catalytic properties of aldolase C. *Biochemistry* **8**, 4396–4402.

Perheentupa, J., and Raivio, K. (1967). Fructose-induced hyperuricaemia. *Lancet* **2**, 528–531.

Phillips, M. J., Hetenyi, G., Jr., and Adachi, F. (1970). Ultrastructural hepatocellular alterations induced by *in vivo* fructose infusion. *Lab. Invest.* **22**, 370–379.

Pohl, S. L., Birnbaumer, L., and Rodbell, M. (1971). The glucagon sensitive adenyl cyclase system in plasma membranes of rat liver. *J. Biol. Chem.* **246**, 1849–1856.

Raivio, K. O., Kekomäki, M. P., and Mäenpää, P. H. (1969). Depletion of liver adenine nucleotides induced by D-fructose, dose-dependence and specificity of the fructose effect. *Biochem. Pharmacol.* **18**, 2615–2624.

Rauschenbach, P., and Lamprecht, W. (1964). Einbau von ¹⁴C-markierter Glucose und Fructose in Leberglykogen. Zum Fructose-Stoffwechsel in der Leber. *Hoppe Seyler's Z. Physiol. Chem.* **339**, 277–292.

Reinecke, R. M. (1944). The kidney as a locus of fructose metabolism. *Amer. J. Physiol.* **141**, 669–676.

Rieder, S. V., and Rose, I. A. (1959). The mechanism of triosephosphate isomerase reaction. *J. Biol. Chem.* **234**, 1007–1010.

Sanchez, J. J., Gonzalez, N. S., and Pontis, H. G. (1971a). Fructokinase from rat liver. I. Purification and properties. *Biochim. Biophys. Acta* **227**, 67–78.

Sanchez, J. J., Gonzalez, N. S., and Pontis, H. G. (1971b). Fructokinase from rat liver. II. The role of K⁺ on the enzyme activity. *Biochim. Biophys. Acta* **227**, 79–85.

Schapira, F., Schapira, G., and Dreyfus, J. C. (1961). La lésion enzymatique de la fructosurie bénigne. *Enzymol. Biol. Clin.* **1**, 170–175.

Sillero, M. A. G., Sillero, A., and Sols, A. (1969). Enzymes involved in fructose metabolism in liver and the glyceraldehyde metabolic crossroads. *Eur. J. Biochem.* **10**, 345–350.

Sols, A., and Crane, R. K. (1954). Substrate specificity of brain hexokinase. *J. Biol. Chem.* **210**, 581–595.

Van den Berghe, G., Hue, L., and Hers, H. G. (1973). Effect of the administration of fructose on the glycogenolytic action of glucagon. An investigation of the pathogeny of hereditary fructose intolerance. *Biochem. J.* **134**, 637–645.

Veneziale, C. M. (1972). Regulation of D-triokinase and NAD-linked glycerol-dehydrogenase activities in rat liver. *Eur. J. Biochem.* **31**, 59–62.

Vinuela, E., Salas, M., and Sols, A. (1963). Glucokinase and hexokinase in liver in relation to glycogen synthesis. *J. Biol. Chem.* **238**, PC1175–PC1177.

Weichselbaum, T. E., Margraf, H. W., and Elman, R. (1953). Metabolism of intravenously infused fructose in man. *Metab., Clin. Exp.* **2**, 434–449.

Wick, A. N., Sherill, J. W., and Drury, D. R. (1953). The metabolism of fructose by the extrahepatic tissues. *Diabetes* **2**, 465–468.

Wolf, H. P., and Leuthardt, F. (1953). Über die Glycerindehydrase der Leber. *Helv. Chim. Acta* **36**, 1463–1467.

Woods, H. F., Eggleston, L. V., and Krebs, H. A. (1970). The cause of hepatic accumulation of fructose 1-phosphate on fructose loading. *Biochem. J.* **119**, 501–510.

Disorders Related to Sugar Metabolism: Galactose-Containing Sugars and the Eye

CHAPTER 23

Cataractogenic Effects of Lactose and Galactose

JIN H. KINOSHITA

An observation that a diet enriched with large amounts of lactose or galactose fed to rats induces cataracts has been a subject of considerable interest. First of all, this experimental cataract is the easiest type to reproduce in animals. Secondly, many investigators believe that studies on experimental galactose cataract may shed light on cataracts in human galactosemia. It is, therefore, understandable why so many studies have been undertaken ever since the initial observation was made by Mitchell and Dodge (1935).

This review is not intended to be an exhaustive survey of this subject but is limited to the new concepts and developments in the understanding of the cataractogenic effects of galactose.

A. Lactose and Galactose Effects on the Ocular Lens

In the middle thirties there was considerable interest in examining the various dietary factors that play a role in the development of cataracts. It was observed that a diet containing high levels of lactose was effective in producing opacities in the lens. On the other hand, sugars such as maltose, starch, sucrose, or dextrose had no similar deleterious effects on the lens. Apparently, there was something unusual about lactose that induced cataracts. These earlier observations were made by Mitchell and Dodge (1935) and by Yudkin and Arnold (1935). Further studies revealed that it was the galactose component of lactose that was the cataractogenic agent. Galactose was clearly much more effective than lactose; for example, young rats fed a diet rich in galactose induced a cataract in about a 2-week period, while lactose feeding required anywhere from about 8–12 weeks for cataracts to appear. Thus, it was definitely established that galactose was the cataractogenic sugar that seemed most effective in producing cataracts in young rats.

When it was later shown that galactosemic patients developed cataracts, there was a renewed interest in the fact that galactose cataracts can be produced in the rats. The possibility existed that insights into the mechanism of cataract development might be gained by studying the experimental galactose cataracts.

The morphological changes that occur during the development of the cataract in rats fed galactose appear consistently (Kinoshita, *et al.*, 1968a; Sipple, 1966). Vacuoles first develop in the equatorial region of the lens after the young rats are maintained on the galactose diet for 3 days. Later these vacuoles appear to spread to the anterior surface. The vacuoles continue to develop for about a 2-week period and then suddenly the lens nucleus becomes opaque. Much later the entire lens becomes opaque. The time required for the appearance of the initial vacuoles and the dense nuclear opacity is surprisingly reproducible from animal to animal.

The cataract induced by lactose feeding takes longer to develop. The course appears somewhat different in that sometimes a dense opacity is observed in the nucleus and at other times the opacity occurs first in the cortex. The difference is due to the age of the animals. This is illustrated by the fact that the galactose feeding in the older, in contrast to younger, animals results in cortical opacities. Even when the experiments are begun with young rats, it takes so long to develop cataracts by lactose feeding, and the time of appearance of the opacity is so variable, the age factor becomes an important consideration.

Rats appear better able to tolerate high levels of galactose than lactose. Diets as high as 70% sugar are employed in these experiments (Mitchell and Dodge, 1935; Yudkin and Arnold, 1935). Since galactose is much more effective, usually a diet composed of 50–25% of this sugar is sufficient to induce cataracts. Although some inhibition of growth in the galactose-fed animal is observed, generally the animal appears in better health than is observed in lactose feeding.

B. Mechanism of Cataract Development

For many years the mechanism of the cataractogenic effects of galactose has been under close scrutiny. It was first thought that galactose or a galactose derivative in some way interfered with the glucose metabolism in the lens (Kinoshita, 1962). Since glycolysis is the primary source of energy in the lens, any interference of this pathway would result in a shortage of ATP which could lead to cataract production. When the story of human galactosemia was unfolded revealing clearly that the galactose-1-phosphate uridyltransferase was the enzyme deficient in patients with this disease (Kalckar *et al.*, 1956), investigators looked into the possibility that galactose 1-phosphate was the agent that interfered with glucose metabolism in the lens. However, no clear-cut evidence could be obtained to implicate this phosphate intermediate as having an inhibitory effect on lens glucose metabolism.

In any cataract, many biochemical parameters are affected, and it has been most difficult to single out the primary factor that initiates the cataractogenic events. The procedure followed in these studies is to determine what changes are first observed in the development of cataracts. Attempts were then made to relate other changes to the initial change.

A major change to occur that can be observed early in the course of cataract development is the marked drop in glutathione and amino acids (Reddy, 1965; Sippel, 1966; Kinoshita *et al.*, 1969). In the lens, the glutathione is normally present in unusually high amounts. In galactose cataract there is a precipitous drop in the glutathione content and concomitantly there is lowering in the level of free amino acids. It appears that changes in the level of amino acids and glutathione are somehow closely related to the factor that initiates the cataract.

The clue to the primary mechanism in the cataract development in the galactose-fed rat was provided by histopathologists (Friedenwald and Rytel, 1955). They found that the earliest change observable is the accumulation of fluid within the lens fibers. Thus, the development of hydropic lens fibers was the first microscopic change detectable in galactose cataract. The question raised by this observation was what caused these lens fibers to swell. The obvious answer was that the swelling had to result either from an accumulation of electrolytes or of abnormal metabolites. At the very early stages in the galactose cataract there appears to be no net increase in electrolytes. Consequently, the osmotic swelling would appear to result from accumulation of an abnormal metabolite.

C. Galactitol Formation in the Lens

The identification of the metabolite of galactose that could produce the osmotic change was provided by the observation made by van Heyningen (1959). She found the presence of a sugar alcohol which she identified as galactitol (dulcitol) in the lens of a galactose-fed rat. Subsequent experiments revealed the presence of aldose reductase, the enzyme that reduced aldoses to sugar alcohols in a reaction involving NADPH (van Heyningen, 1959; Hayman and Kinoshita, 1965). Aldose reductase in conjunction with polyol dehydrogenase constitutes the sorbitol pathway, a mechanism previously studied in the seminal vesicles by Hers (1960). This pathway was shown to function in the lens as well. Instead of glucose, galactose and other aldoses also serve as substrate for aldose reductase (Fig. 1).

There are other properties of galactitol which make it a good candidate for being responsible for the osmotic change (Kinoshita, 1965). Galactitol is a sugar alcohol, and it has been known for some time that sugar alcohols in general do not penetrate biological membranes very readily. Furthermore, galactitol is not a substrate for polyol dehydrogenase; thus, unlike sorbitol (Fig. 1), it is not further metabolized. Since it is not readily lost from the lens fibers by diffusion nor is it further metabolized, galactitol synthesis may lead to sufficient accumulation to produce osmotic swelling.

A series of experiments were thus conducted to demonstrate that in the lenses of rats fed galactose the accumulation of galactitol does lead to lens swelling (Kinoshita, 1965). It can be seen in Fig. 2 that there is indeed an early accumulation of galactitol that is accompanied by an increase in lens hydration. As the diet is continued the level of dulcitol

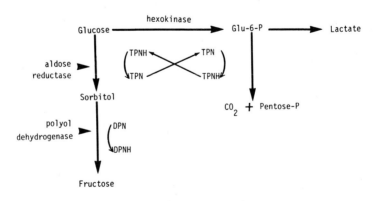

Fig. 1 Sorbitol pathway.

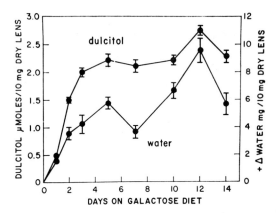

Fig. 2 Increases in galactitol and in water content in the lens of galactose-fed rats. From Kinoshita *et al.* (1962).

is increased and appears to level off. This increase in sugar alcohol content is paralleled by an increase in lens hydration. In this experiment the concentration of galactitol at day 5 when the vacuoles are first visible is about 100 mM.

The experiment relating the galactitol retention to osmotic change was also duplicated *in vitro* (Kinoshita, 1965; Sippel, 1966). The lens *in situ* is in a state of tissue culture; thus, by removing the lens and placing it in an artificial environment, one can actually duplicate the same sequence of events that occurs in the experimental animal. This is accomplished by incubating the lens in a medium rich in galactose. Under these *in vitro* conditions, one can more precisely demonstrate the relationship of the accumulation of sugar alcohol and the increase in lens hydration that was demonstrated *in vivo*. These findings support the contention that galactitol formed within the lens fibers accumulates to a high enough level to cause osmotic swelling and probably accounts for the hydropic lens fibers observed histologically.

D. Secondary Changes

1. Amino Acid Loss

The fact that the retention of galactitol can account for the osmotic change which, in turn, may be related to the appearance of hydropic lens fibers, suggests that the primary event in cataractogenesis is the formation of galactitol. All other changes that occur must somehow be related either to the accumulation of galactitol or the concomitant increase in lens hydration. If this thesis is correct, the drop in levels of amino acids and glutathione, changes observed early in the course of cataract development, must occur as a result of the primary event. To demonstrate this experimentally, the lens was incubated in a high galactose medium, but it was prevented from swelling despite galactitol accumulation by progressively increasing the tonicity of the medium during the 48-hr incubation period (Kinoshita *et al.*, 1969). This was accomplished by adding sugar alcohol to the medium during the course of the incubation. Under these conditions, when the galactose-exposed lens retained its normal volume, it was shown that the glutathione level and the amino acid concentration were maintained close to normal levels (Table I).

TABLE I

Amino Acid Content of Galactose-Exposed Lens in Osmotically Compensated Medium[a]

Amino acid	% Amino acid in galactose medium	% Amino acid in osmotically compensated medium
Glutathione	50	95
Aspartic acid	38	68
Glutamic acid	20	76
Glycine	20	75
Alanine	13	50
Valine	42	100
Leucine	20	80
Isoleucine	32	89
Phenylalanine	17	79
Tyrosine	18	88
Serine	33	113
Threonine	45	88
Proline	15	45
Methionine	31	118
Lysine	72	88
Histidine	22	75
Arginine	75	95
Taurine	6	102

[a] The free amino acid levels in the lens incubated for 2 days in a galactose-containing medium are compared with those in a control medium. The exact description of experiment is described by Kinoshita *et al.* (1969).

2. Electrolyte Changes

The other change to occur during the development of the galactose cataract is an alteration in electrolyte distribution (Kinoshita and Merola, 1964; Kinoshita, 1965). This change is observed later in the course of cataract development than the drop in amino acids, but it is a phenomenon which becomes important during the late stages of the galactose cataract. First of all, there is a slow decrease in the level of potassium and an increase in sodium initially, but these changes become magnified as the cataract progresses; for example, when the vacuoles first become visible, the initial vacuolar stage, there is no change in cation levels, but in the late vacuolar stage, a period of 12 days on the galactose diet, there is a noticeable increase in sodium and decrease in potassium (Fig. 3). At the nuclear cataract stage, a marked increase in sodium and chloride accompanied by a large increase in hydration is observed.

In the initial vacuolar stage, the increase in hydration is the result of the movement of water into the lens to osmotically compensate for the accumulation of galactitol. This is the reason the sodium and potassium concentrations appear lower, based on a water basis, but remain unchanged on a dry weight basis (Figs. 3 and 4). In the late vacuolar stage, the galactitol level is still substantial and, in addition, there is a slight

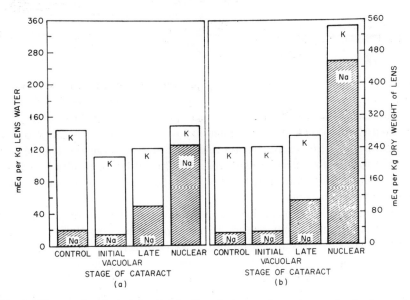

Fig. 3 Changes in electrolyte distribution during various stages of galactose cataracts. (a) Lens water basis and (b) dry weight basis. From Kinoshita (1965, p. 794).

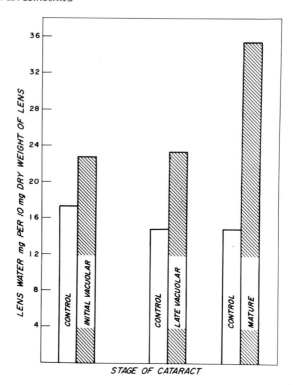

Fig. 4 Changes in lens hydration during various stages of galactose cataracts. From Kinoshita and Merola (1964).

increase in total cation as a result of a rise in sodium. Thus, at this stage the presence of galactitol and increase in sodium ions both contribute to the osmotic change (Figs. 3 and 4). In the nuclear cataract stage marked changes in the lens permeability occur. No longer is the lens able to retain galactitol or K^+. It appears that at this stage the lens is permeable to all substances except proteins. This situation results in a Donnan swelling, as exemplified by the tremendous increases in Na^+ and Cl^- and in hydration (Figs. 3 and 4).

The changes in permeability to electrolytes as the cataract progresses have been studied using tracers (Thoft and Kinoshita, 1968). In these experiments increases in cation fluxes were observed even in the initial vacuolar stage. It appears that the osmotic swelling as a result of the presence of sugar alcohol increases permeability to K^+ and Na^+; the lens at this stage is able to compensate by accelerating the cation pump so that the actual levels of cations do not change. As the cataract progresses, however, the lens becomes more leaky and a stage is reached where the cation pump can no longer compensate for the increase in permeability. It is at this point that swelling develops according to the Donnan principle.

In the studies *in vitro*, these changes to electrolytes can be prevented when the lens exposed to galactose is incubated in an osmotically compensated medium (Kinoshita *et al.*, 1968a). As in the experiments dealing with loss in amino acids and glutathione, if the galactose-exposed lens is kept from swelling the alteration in electrolyte distribution is maintained within normal values.

A similar observation was made by Broekhuyse (1968) on inositol levels in the lens. This compound, also present in substantial levels, appears to leak out rapidly from a lens exposed to galactose. This change could also be avoided by maintaining the normal state of hydration during prolonged incubation in the presence of galactose by use of an osmotically compensated medium.

All these observations are summarized in Fig. 4. The mechanism that triggers the cataractous process is the conversion of galactose to galactitol by aldose reductase. Galactitol accumulates to unusually high levels because it does not readily leak out of the lens nor is it removed by metabolism. Thus, a hypertonic condition is created which is corrected by the movement of water into the lens fibers. The resulting swelling has deleterious effects. It causes the hydrops to form which disintegrates to form vacuoles in some areas of the lens. It increases permeability to amino acids lowering its concentration to interfere with protein synthesis. It reduces inositol levels. It increases permeability to cations so that K^+ concentration begins to decrease and Na^+ increases. The lens is able to compensate for the leakiness to cations initially by increasing its cation pump activity. However, as the cataract progresses and the increase in permeability continues, the cation pump can no longer effectively exclude Na^+. These conditions lead finally to the nuclear cataract stage marked by a pronounced increase in Na^+ and Cl^-. A complete loss in selective permeability occurs, leading to a Donnan swelling.

E. Aldose Reductase Inhibitors

Since it is obvious that aldose reductase plays a central role in the pathogenesis of this type of cataract, the approach to preventing cataract would be to seek an appropriate inhibitor of the enzyme (Jedziniak and Kinoshita, 1971). Through a series of experiments the structural properties required of a compound to be an effective aldose reductase inhibitor were revealed. One important characteristic is that the compound must have an acid group, and the second requirement is that it must possess a hydrophobic region. One compound with the necessary requirements is tetramethyleneglutaric acid (TMG), a carboxylic acid with carboxylic tetramethylene group as the hydrophobic region. This compound was found to be an effective inhibitor of lens aldose reductase.

The effectiveness of TMG on the galactose-exposed lens is shown in Fig. 5. The lens exposed to galactose-containing medium for a 3-day period results in progressive increases in galactitol and in water content (Kinoshita *et al.*, 1968b). After the second day a ring of opacities in the form of vacuoles appears at the lens periphery. These opacities become quite prominent by the end of the third day. No change in clarity or transparency is observed in the lens in the control medium. In the medium containing galactose and TMG, the level of galactitol is strikingly reduced, as is the lens hydration. Furthermore, the lens remains perfectly transparent and is indistinguishable from the lens in the control medium. Moreover, the presence of TMG prevented the changes in levels of glutathione and amino acids from occurring. Thus, the early changes that occur in galactose cataract are greatly minimized by an aldose reductase inhibitor (Fig. 6). Unfortunately, TMG has been ineffective *in vivo*. However, a further search for other inhibitors has been undertaken by Dvornik (1970) of the Ayerst Laboratories and promising inhibitors have been uncovered. It would thus appear that further advances in preventing this kind of cataract should be forthcoming.

F. Research Needs

As illustrated in this review, considerable information is available regarding the initial factors and early stages of the galactose cataracts in rats. However, more information is needed to understand the factors involved in the later stages of the cataract; for example, the mechanism of formation of the dense nuclear opacity remains obscure. In addition, rats appear particularly susceptible to the galactose effects on lens. The

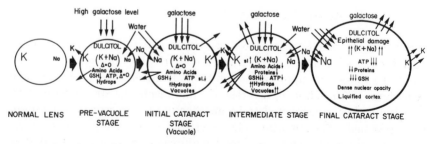

Fig. 5 Summary of events during development of galactose cataract.

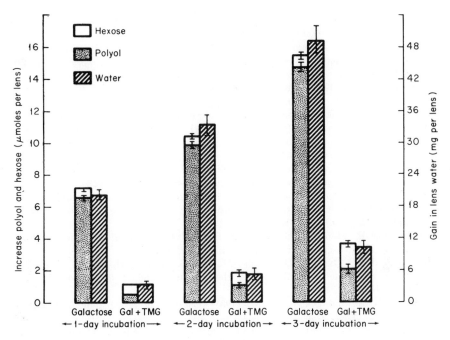

Fig. 6 The effect of an aldose reductase inhibitor on the lens exposed to galactose. [Reproduced from Kinoshita *et al.* (1958b). *Biochim. Biophys. Acta,* 158(1968) p. 474 (Fig. 2).]

question which remains unanswered is why it is more difficult to induce cataract with galactose feeding in other animals than it is in rats.

There has been some evidence that suggests that sugar alcohol accumulation is responsible for diabetic and xylose cataracts as it is in galactose cataract. However, more definitive information is needed to link clearly the primary role of aldose reductase in these other two forms of sugar cataracts.

References

Broekhuyse, R. M. (1968). Changes in myo-inositol permeability in the lens due to cataractous condition. *Biochim. Biophys. Acta* **163,** 269–273.

Dvornik, D. (1970). Aldose reductase inhibitors. *Gordon Res. Conf. Proc., Conf. Med. Chem.,* p. 61.

Friedenwald, J. S., and Rytel, D. (1955). Contributions to the histopathology of cataract. *AMA Arch. Ophthalmol.* **53,** 825–831.

Hayman, S., and Kinoshita, J. H. (1965). Isolation and properties of lens aldose reductase. *J. Biol. Chem.* **240,** 877–882.

Hers, H. G. (1960). Aldose reductase. *Biochim. Biophys. Acta* **37,** 120–126.

Jedziniak, J., and Kinoshita, J. H. (1971). Activators and inhibitors of lens aldose reductase. *Invest. Ophthalmol.* **10,** 357–366.

Kalckar, H. M., Anderson, E. P., and Isselbacher, K. J. (1956). Galactosemia, a congenital defect in a nucleotide transferase. *Biochim. Biophys. Acta* **20,** 262–268.

Kinoshita, J. H. (1962). Annual review, physiology chemistry of the eye. *Arch. Ophthalmol.* **68,** 554–570.

Kinoshita, J. H. (1965). Cataracts in galactosemia. *Invest. Ophthalmol.* **4,** 786–799.

Kinoshita, J. H., Barber, W. G., Merola, L. O., and Tung, B. (1969). Changes in the levels of free amino acids and myoinositol in the galactose-exposed lens. *Invest. Ophthalmol.* **8,** 625–632.

Kinoshita, J. H., and Merola, L. O. (1964). Hydration of the lens during the development of galactose cataract. *Invest. Ophthalmol.* **3,** 577–584.

Kinoshita, J. H., Merola, L. O., Satolv, K., and Dikmak, E. (1962). Osmotic changes caused by the accumulation of dulcitol in the lenses of rats fed with galactose. *Nature (London)* **194,** 1085.

Kinoshita, J. H., Merola, L. O., and Tung, B. (1968a). Changes in cation permeability in the galactose-exposed rabbit lens. *Exp. Eye Res.* **1,** 80–90.

Kinoshita, J. H., Dvornik, D., Kraml, M., and Gabbay, K. H. (1968b). The effect of aldose reductase inhibitor on the galactose-exposed lens. *Biochim. Biophys. Acta* **158,** 472–474.

Mitchell, H. S., and Dodge, W. M. (1935). Cataract in rats fed on high lactose rations. *J. Nutr.* **9,** 39–49.

Reddy, D. V. N. (1965). Amino acid transport in the lens in relation to sugar cataracts. *Invest. Ophthalmol.* **4,** 700–708.

Sippel, T. O. (1966). Changes in the water, protein and gluthathione contents of the lens in the course of galactose cataract development in rats. *Invest. Ophthalmol.* **5,** 568–575.

Thoft, R. A., and Kinoshita, J. H. (1968). The rate of potassium exchange of galactosemic rat lenses. *In* "Biochemistry of the Eye," pp. 383–387. Karger, Basel.

van Heyningen, R. (1959). Formation of polyols by the lens of the rat with sugar cataract. *Nature (London)* **184,** 194.

Yudkin, A. M., and Arnold, C. H. (1935). Cataracts produced in albino rats on ration containing a high proportion of lactose or galactose. *Arch. Ophthalmol.* **14,** 960–966.

Disorders Related to Sugar Metabolism: Obesity, Cardiovascular Disease and Hypertriglyceridemia

CHAPTER 24

Obesity

GOTTHARD SCHETTLER AND
GUENTER SCHLIERF

Obesity, a syndrome which may have a variety of causes, is characterized by an absolute or relative increase in body fat mass. For clinical purposes, body weights and heights, in most cases, are sufficient indicators of obesity. More accurate determination of body fat requires measurements such as skinfold thickness and lean body mass.

This review deals with the subject of human obesity in a rather sketchy manner by pointing out areas of particular interest and present research activity and refers to the role of carbohydrates whenever indicated. In so doing it is recognized that a great amount of very important work in this field will not be covered.

Obesity has been with man for many centuries (Fig. 1). In many societies it was associated with affluence and wealth as it still is in some populations. During the last decades it has become apparent that obesity clearly reduces live expectancy and increases morbidity and mortality from various causes (Fig. 2). Obesity is very common: It has been estimated that in Western countries up to one-third of the adult population exhibits various degrees of obesity. In many instances, overweight begins in childhood. According to a survey in West German children performed by Maaser and Droese (1971), 23–27% of 6–10-year-old children were overnourished.

389

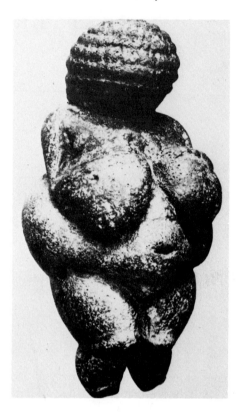

Fig. 1 Venus of Willendorf; approximately 30,000 B.C. Museum of Natural History, Vienna.

A. Pathogenesis

Excess body fat accumulates whenever energy intake from food exceeds energy expenditure (Fig. 3). Obesity therefore results from a positive energy balance which in turn may result from a number of causes.

During the last century there has been a general trend toward *increased calorie intake*. In Germany, this increase was from about 2600 to 3000 cal/person/day (Holtmeier, 1969). It is at least partly the result of increased availability of foods with high caloric density, i.e., high proportions of sugars and fat. Thus, for example, the sugar and fat consumption in Germany has increased from 49–90 to 90–130 gm/person/day since 1900 (Ernährungsbericht, 1969). This general trend is modified by individual factors such as family eating habits, which in conjunction with heredity are responsible for the high familial occurrence of obesity

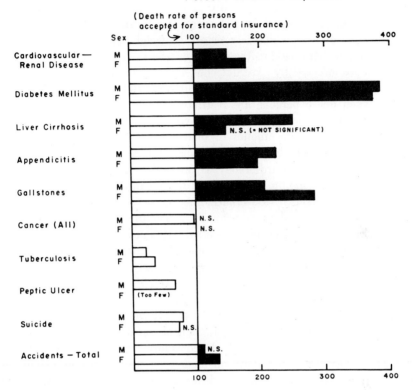

DEATHS —
Percent actual of expected

Fig. 2 Effects of obesity on susceptibility to various diseases. Black bars represent increased susceptibility in overweight individuals. [Reproduced from Marks, H. H.: Influence of obesity on morbidity and mortality. *Bull. N.Y. Acad. Med.* **36,** 296–312, 1960.]

Fig. 3 Schematic presentation of factors affecting energy balance (adapted from Irsigler teaching slide).

(Mayer, 1966). Possibly, overeating at a very early age is considerably more deleterious than later in life since it may result in an increased number of fat cells which may be irreversible and this forms the basis of a kind of obesity which is resistant to weight reduction (Hirsch and Han, 1969; Knittle, 1971). According to recent evidence the obese person is also more affected by external cues with regard to regulation of food intake than his lean contemporary (Stunkard, 1968). Hunger ratings run higher when obese persons are restricted to the intake of a rather unpleasant formula diet (Wooley, 1971), volume rather than caloric density tends to determine the amount of food eaten (Hashim and van Itallie, 1965), and the clock rather than intrinsic hunger mechanisms regulates eating behavior (Stunkard, 1968). If, as it is generally assumed, energy balance is mainly maintained by way of food intake, this regulation, by whatever mechanisms it takes place, must be set at an extremely sensitive level to keep body weight constant in a lifetime during which approximately 20,000 kg of food are consumed. Obviously, overeating is not always gluttony since small daily amounts of additional calories will, over the years, add up to increase of body fat. Thus, one excess teaspoon of sugar equal to 20 cal/day may lead to a weight increase by 1 kg during the course of one year.

In addition to the general increase in calorie intake, there has been a *decrease of energy expenditure* through physical activity, both at work and during off-work hours. Table I shows how the percentage of persons engaged in heavy physical work has decreased and the percentage of persons doing light work has increased during the last century in Germany. Even with comparable work loads, there are great differences in energy

TABLE I

Calorie Requirements and Proportions of Subjects with Heavy Occupational Work in 1850 and 1966

Work	Cal required/day	Percent of population	
		1850	1966
Light	2400	18	62
Heavy	3300	26	10
Very heavy	4000	16	1
Mean in 1850	3050		
Mean in 1966	2610		
	2950 = actual intake without alcohol		

expenditure between individuals (Mayer, 1966; Bloom and Eidex, 1967a). Frequently, the onset of obesity can be traced to the time when a previously physically active student takes an office job, a person is immobilized by a fracture, or by pregnancy. Obese people also tend to spend more hours in bed than lean controls (Bloom and Eidex, 1967b). Finally, modern clothing and heating, for people living in a cool climate, can save up to 200 cal/per day (Hötzel, 1970).

B. Metabolic Differences

In spite of many recognized conditions favoring the development of obesity, there has been a continuing search for some metabolic abnormality which by itself could be the cause or at least one cause of obesity. In fact, metabolic abnormalities such as summarized in Table II are quite frequently found in obese subjects. However, all metabolic aberrations found thus far are the result rather than the cause of obesity and can be reversed upon weight reduction.

This is particularly true for abnormalities of carbohydrate metabolism characterized by elevated fasting and postglucose plasma insulin levels and glucose intolerance. Hyperinsulinism while not *causing* obesity, nevertheless may aid in *maintaining* it by way of postprandial hypoglycemia and increased appetite. It is, therefore, useful to briefly consider present concepts of abnormal insulin secretion in obesity since modification of this particularly undesirable feature of the obese state may be quite useful for patient care. According to Bagdade *et al.* (1967), insulin resistance

TABLE II

Metabolic and Endocrine
Abnormalities in Obesity

Hyperinsulinism
Increased cortisol secretion
Disturbance of GH secretion
Increased dehydroepiandrosterone levels
Decreased glucose utilization
Reactive hypoglycemia
Increased liver glycogen
Increased levels of FFA
Decreased lipid mobilization with fasting
Increased cholesterol synthesis
Hypertriglyceridemia

---

results from an *increased body fat mass* and leads to elevated fasting as well as postglucose insulin levels. Glucose intolerance develops when insulin secretory reserve becomes exhausted, additional factors being heredity, age, and duration of obesity (Gries *et al.*, 1971). *In vitro* studies using isolated human adipose tissue cells have indeed shown insulin resistance of large adipose cells as compared to small ones (Gries *et al.*, 1971). There has been recent evidence (Drenick *et al.*, 1972; Grey and Kipnis, 1971) that preceding nutrient intake might also be an important determinant of insulin levels and insulin response in obesity. Thus, Drenick *et al.* were able to show that overfeeding after weight reduction quite early restores hyperinsulinemia and Grey and Kipnis reported that elevated basal insulin levels in obese subjects were found following high carbohydrate but not high fat diets (Fig. 4). These studies underline the necessity of restricting carbohydrates in obesity in order to restore insulin

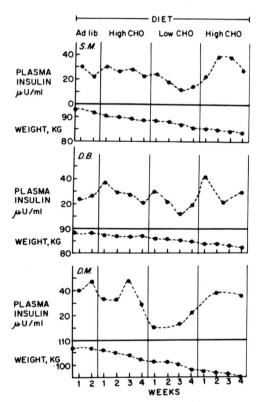

Fig. 4 Effect of high carbohydrate and low carbohydrate, hypocaloric diets on the basal plasma hyperinsulinemia of obese subjects. CHO Stands for carbohydrate (from Grey and Kipnis, 1971).

levels to normal and thus hopefully decrease appetite and fat deposition and, in the long run, prevent exhaustion of islet cell reserve.

C. Experimental Obesity in Man

A stimulating approach to the study of obesity has been provided by overfeeding experiments published in the United States as well as in Europe. Miller and Mumford, in 1967, reported on studies in altogether 16 normal subjects in whom they intended to produce overweight by overfeeding for periods up to 6 weeks. By comparing hypercaloric low and high protein diets they found that weight gain was less with the former as compared to the latter diet. With either diet, however, it remained considerably below the calculated weight gain (Fig. 5). In the United States, Sims and co-workers (1968), studying the effects of weight gain on vari-

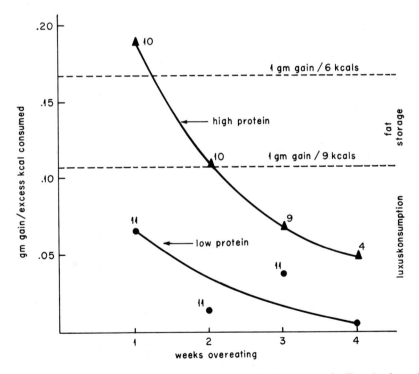

Fig. 5 Weekly gains in weight per excess kilocalorie consumed. Two horizontal lines represent the expected gains on the assumption that the gain is either entirely fat or 66% fat: values falling below the lines indicate "Luxuskonsumption." The numerals show the number of subjects on which the points are based (from Miller and Mumford, 1967).

ous metabolic parameters in prison volunteers, also published evidence that actual weight gain in some subjects was consistently less than expected. One of the men, for example, consumed at least 2000 excess calories per day for more than 100 days, but instead of the expected weight gain of about 2 kg/week, only showed a total gain of slightly more than 5 kg in more than 100 days. According to a French proverb: *"Obèse ne devient pas qui vent."*

In Germany, Kasper (1968), who set out to study maximal fat absorption in healthy young subjects, to his surprise could not detect any weight gain in some subjects consuming up to 6000 cal and 600 gm of fat per day (Fig. 6).

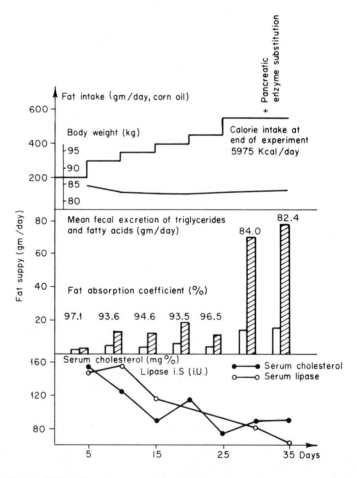

Fig. 6 Hyperalimentation in a healthy male subject, 23 years of age. Lack of weight gain with 5975 cal/day (from Kasper, 1968).

If, in the studies mentioned, all of the food was in fact consumed, absorption was normal, and physical activity not altered, excess calories must have been disposed by increased heat production. Indeed, there was evidence of an increased "thermic response" to food and exercise in the study of Miller and Mumford (1967), while neither basal metabolic rate nor thermic response to food intake in the resting state (specific dynamic action) was altered. While there was some indication that in their study not only calorie intake but also dietary composition affected weight gain during hyperalimentation, further evidence for this aspect has been presented by Irsigler (1969) in 12 obese subjects. The author found that oxygen consumption in exercising obese subjects was higher following fat and protein containing meals than following carbohydrate intake when calorie expenditure actually showed a decrease from the exercise base line.

In spite of shortcomings of the hyperalimentation studies, mainly because heat production was only estimated by indirect calorimetry and contradictory results by other authors (Newburgh, 1944), a tentative conclusion and working hypothesis could be formulated as follows:

1. "Normal subjects" when being overfed, have the ability to dispose of at least part of additional calories by increasing heat production following food intake under conditions of mild exercise. This "Luxuskonsumption" may be more marked with high fat intake.

2. In contrast, obese subjects may more efficiently conserve excess food calories, particularly with high carbohydrate diets (Irsigler, 1969).

While it is tempting to accept these theories to explain some facts about obesity which are quite familiar to the physician, obviously more work is needed in this area with rigidly controlled methods and conditions.

D. Therapeutic Aspects

Reduction of overweight by reducing body fat is only possible if negative calorie balance is introduced. Inspection of Fig. 3 reveals that this can be tried by either increasing physical activity or decreasing calorie intake.

Increasing physical activity, while definitely of value in prevention and long-term management of obesity, has only limited value for acute reduction of excess body weight. In order to lose 1 kg of adipose tissue, one would have to do more than 10 hr of very heavy physical work or exercise.

The most commonly used and effective approach for weight control
reduces daily caloric intake by limiting or eliminating those dietary constituents which contain elements of high caloric density such as fats and
sugars. The low calorie diet, therefore, is a low fat, low carbohydrate
diet in which sugars are eliminated. The latter action, as indicated
earlier, tends to decrease insulin requirements and insulin secretion and
thus may be of specific value in curbing appetite and decreasing
lipogenesis.

Although there have been studies suggesting greater rates of weight loss
with high fat as compared to high carbohydrate reducing diets (Kekwick
and Pawan, 1956; Kasper and Plock, 1970), controlled long-term studies
of this question do not appear to support this concept (Fig. 7) (Bortz
et al., 1968; Kinsell *et al.*, 1964). When the short-lived effect of water
and electrolyte loss is excluded, weight loss appears to occur independently of the dietary mixture used. Maximal rates of weight loss are
achieved by reducing calorie intake to zero. Studies in humans with total
fasting have been extended for periods up to 280 days without apparent
ill effects (Barnard *et al.*, 1969). Loss of lean body mass eventually limits
the duration of this kind of treatment. It is significantly less with con-

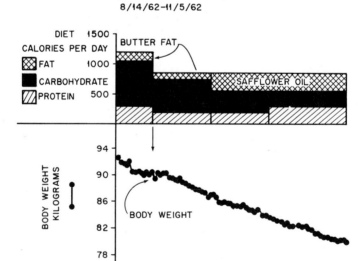

Fig. 7 The rate of weight loss is not appreciably altered by highly significant
changes in the relative amounts of fat and carbohydrate in the diet provided
the total caloric value remains unchanged. [Reproduced with permission from
Kinsell *et al.*, (1964) *Metabolism,* **13,** 195.]

ventional low calorie reducing diets where weight loss indeed mainly means loss of excess body fat (Sarett *et al.*, 1966).

E. Outlook and Research Needs

Obesity, prevalent at a high percentage in Western populations, increases morbidity and mortality from a variety of disorders. Prevention and treatment rests on avoidance of positive calorie balance and production of negative calorie balance, respectively. Reasons for elimination of sugars from reducing diets are insulin's properties to stimulate appetite and lipogenesis and decrease fat mobilization. Exhaustion of β-cell reserve may be delayed with low carbohydrate (sugar) intake. Treatment of obesity by low calorie diets in addition includes reduction of fat as a source of concentrated calories. Studies have been reported indicating that efficiency of energy utilization from diet may be subject to modification in certain individuals or conditions. These need to be confirmed with reliable methods.

Further work, with particular reference to the role of sugars, should include studies on the effect of various carbohydrates in infant formulas and the effectiveness of sugar alcohols which do not stimulate insulin release in reducing diets. Finally, medicine will have to enlist help from social and behavioral sciences if measures aimed at controlling overweight in populations are to be successfully instituted.

References

Bagdade, J. D., Bierman, E. L., and Porte, D., Jr. (1967). Significance of basal insulin levels in the evaluation of the insulin response to glucose in diabetic and nondiabetic subjects. *J. Clin. Invest.* **46,** 1349.

Barnard, D. L., Ford, J., Garnett, E. S., Mardell, R. J., and Whyman, A. E. (1969). Changes in body composition produced by prolonged total starvation and refeeding. *Metab., Clin. Exp.* **18,** 7.

Bloom, W. L., and Eidex, M. F. (1967a). Inactivity as a major factor in adult obesity. *Metab., Clin. Exp.* **16,** 679.

Bloom, W. L., and Eidex, M. F. (1967b). The comparison of energy expenditure in the obese and lean. *Metab., Clin. Exp.* **16,** 685.

Bortz, W. M., Howat, P., and Holmes, W. L. (1968). Fat, carbohydrate, salt and weight loss. Further studies. *Amer. J. Clin. Nutr.* **21,** 1291.

Deutsch. Ges. Ernähr. (1969). Ernährungs-bericht. Jos. Henrich Publ. Co., Frankfurt/M.

Drenick, E. J., Brickman, A. S., and Gold, E. M. (1972). Dissociation of the obesity-hyperinsulinism relationship following dietary restriction and hyperalimentation. *Amer. J. Clin. Nutr.* **25,** 746.

Ernährungsbericht (1969). Ed. Dtsch. Ges L. Ernährung. J. Henrich Publ. Co., Frankfurt.

Grey, N., and Kipnis, D. M. (1971). Effect of diet composition on the hyperinsulinemia of obesity. *N. Engl. J. Med.* **285**, 827.

Gries, F. A., Daweke, H., and Liebermeister, H. (1971). Pathophysiologie der Fettsucht mit besonderer Berücksichtigung der Störungen des Kohlenhydratstoffwechsels. *Monatsk. Aerztl. Fortbild.* **21**, 341.

Hashim, S. A., and van Itallie, T. B. (1965). Studies in normal and obese subjects using a monitored food dispensing device. *Ann. N.Y. Acad. Sci.* **131**, 654.

Hirsch, J. P., and Han, P. W. (1969). Cellularity of rat adipose tissue: Effects of growth, starvation and obesity. *J. Lipid Res.* **10**, 77.

Holtmeier, H. J. (1969). "Diät bei Übergewicht und gesunde Ernährung." Thieme, Stuttgart.

Hötzel, H. (1970). Ernährung und Übergewicht. *Deut. Aerztebl.* **67**, 2660.

Irsigler, K. (1969). Zum Energiehaushalt des menschlichen Organismus. *Wien. Klin. Wochenschr.* **81**, 845.

Kasper, J. (1968). Untersuchungen am Gesunden bei maximaller oraler Fettbelastung. *Med. und. Ernähr.* **9**, 193.

Kasper, H., and Plock, E. (1970). Das Verhalten des Körpergewichts unter fettreicher, kohlenhydratarmer Kost. *Verh. Deut. Ges. Inn. med.* **76**, 817.

Kekwick, A., and Pawan, G. L. S. (1956). Caloric intake in relation to body weight changes in the obese. *Lancet* **2**, 155.

Kinsell, L. W., Gunning, B., Michaels, G. D., Richardson, J., Cox, S. E., and Lemon, C. (1964). Calories do count. *Metab., Clin. Exp.* **13**, 195.

Knittle, J. L. (1971). Childhood obesity. *Bull. N.Y. Acad. Sci.* **47**, 579.

Maaser, R., and Droese, W. (1971). Overnutrition in West-German children. *Lancet* **2**, 545.

Marks, H. H. (1960). Influence of obesity on Morbidity and Mortality. *Bull. N.Y. Acad. Med.* **36**, 296–312.

Mayer, J. (1966). Some aspects of the problem of regulation of food intake and obesity. *N. Engl. J. Med.* **274**, 610, 662, and 722.

Miller, D. S., and Mumford, P. (1967). Gluttony, 1. An experimental study of overeating low—or high—protein diets. 2. Thermogenesis in overeating man. *Amer. J. Clin. Nutr.* **20**, 1212 and 1223.

Newburgh, L. H. (1944). Obesity. I. Energy metabolism. *Physiol. Rev.* **24**, 18.

Sarett, H., Longenecker, J. B., and Harkins, R. W. (1966). Weight loss and body composition. *J. Amer. Oil Chem. Soc.* **43**, 183.

Sims, E., A. H., Goldman, R. F., Gluck, C. M., Horton, E. S., Kelleher, P. C., and Rowe, D. W. (1968). Experimental obesity in man. *Trans. Ass. Amer. Physicians* **81**, 153.

Stunkard, A. J. (1968). Environment and obesity: Recent advances in our understanding of regulation of food intake in man. *Fed. Proc., Fed. Amer. Soc. Exp. Biol.* **27**, 1367.

Tepperman, J. (1962). "Metabolic and Endocrine Physiology." Yearbook Publ., Chicago, Illinois.

Wooley, O. W. (1971). Long-term food regulation in the obese and nonobese. *Psychosom. Med.* **33**, 436.

CHAPTER 25

Sugars in Cardiovascular Disease

FRANCISCO GRANDE

It is widely believed that the diet and, more specifically, the fat and cholesterol content of the diet, is one of the main environmental factors affecting the development of the atherosclerotic process and its cardiovascular complications. The emphasis on dietary fat and cholesterol has been historically determined by the early recognition of the presence of cholesterol and other lipids in the atheromatous plaque.

Animal experiments, epidemiological investigations in different populations, and follow-up studies have led to the concept that the influence of dietary fat and cholesterol on the development of atherosclerosis is related to the effect of these dietary components on the concentration of the serum lipids, particularly cholesterol.

Dietary carbohydrates, the main source of calories in the diets consumed by the great majority of the human race, have attracted less attention in this respect. During the last years, however, there has been growing interest in determining whether dietary carbohydrates have any influence on atherogenesis. In spite of the considerable literature already accumulated there seems to be little convincing evidence for a direct link between carbohydrate intake and development of atherosclerosis. On the other hand, some of the observations reported are of interest and have made a contribution to our understanding of the effects of the diet on the regulation of the serum lipid levels.

401

Rather than attempting an exhaustive review of the literature, a few of the main aspects of the problem will be discussed. Accordingly, this discussion will be concerned mainly with the following topics: (a) dietary carbohydrates and experimental atherosclerosis, (b) dietary carbohydrates and serum lipids in animals and man, and (c) sugar intake and coronary heart disease.

A. Dietary Carbohydrates and Production of Atherosclerosis in Experimental Animals

Early in this century, Anitschkow and other Russian authors demonstrated the production of atherosclerosis in the rabbit by feeding cholesterol dissolved in sunflower oil (Anitschkow, 1913, 1914; Anitschkow and Chalatow, 1913). Numerous studies reported during the 60 years elapsed since Anitschkow's original observation have demonstrated that it is possible to produce atherosclerotic lesions in a variety of animals fed with different experimental diets (Kritchevski, 1964). Most of the diets used were characterized by a high content of fat and cholesterol, but it has been demonstrated by various authors (Steiner and Dayton, 1956; Lambert *et al.*, 1958; Steiner *et al.*, 1959; Wigand, 1959; Vles and Kloeze, 1967; Kritchevski, *et al.*, 1968; Kritchevski and Tepper, 1968; Kritchevski, 1970) that high fat diets devoid of cholesterol also produce hypercholesterolemia and atherosclerosis. In the dog, however, the presence of cholesterol in the diet seems to be necessary to produce atherosclerotic lesions as indicated by a recent report (Lazzarini-Robertson *et al.*, 1972). It appears, therefore, that in a number of experimental animals it is possible to produce lesions resembling those found in human atherosclerosis by feeding diets which induce hypercholesterolemia (Tennent *et al.*, 1957; Stamler, 1966) whether or not containing added cholesterol (Tennent *et al.*, 1957; Wigand, 1959; Vles *et al.*, 1964; Vles and Kloeze, 1967, Kritchevski, 1970). The possible influence of the carbohydrate components of the diet on the development of the atherosclerotic process was, obviously, not given much consideration in the early studies. However, the observation that the hypercholesterolemic and atherogenic effects of certain fats in the rabbit, appear only when these fats are added to a semisynthetic diet, but not when they are added to green fodder or commercial chow, stimulated interest in the role of dietary components other than the added fat (Lambert *et al.*, 1958; Malmros and Wigand, 1959; Wigand, 1959). Kritchevski *et al.* (1968) have studied the atherogenic effect of cholesterol-free, semisynthetic diets containing 40% of their weight as carbohydrates, in rabbits. The results indicate that the hypercholesterolemic and atherogenic effects were greatest when the dietary carbohydrate was starch and smallest when it was glucose. These results are in contrast with previous observations by Portman *et al.* (1956) showing that rats fed diets containing cholesterol and cholic acid had higher serum cholesterol and β-lipoprotein concentration when the diet contained either sucrose, fructose, or glucose than when it contained starch. From these and other reports it may be concluded that the "atherogenic" effect of dietary carbohydrates can only be demonstrated under some specific dietary conditions and that it varies from one species to another. The relative atherogenic effect of the individual carbohydrates seems to be different in different species. In the rabbit, a diet containing 30% lactose and 0.35% cholesterol produced higher serum cholesterol and more extensive atherosclerotic lesions than a diet containing sucrose (Wells and Anderson, 1959).

Of particular interest in this connection are the studies recently reported by Lang and Barthel (1972) in males of three species of subhuman primates. The animals were fed, for periods of 16 months, diets containing (by weight) 10% fat, 0.5% cholesterol, and 66% of either sucrose or dextrin. All three species showed some elevation of the serum lipids when given the experimental diets, but the effects varied from species to species. Thus, in *Macaca mulatta* serum cholesterol and degree of coronary artery disease were greater with the dextrin than with the sucrose diet. The sucrose diet, on the other hand, increased the cholesterol content of the aorta more than the dextrin diet. In *Cebus albifrons*, the dextrin-fed animals had no higher serum cholesterol than those fed the sucrose diet but had significantly more intimal proliferation of the coronary arteries. Finally in *Macaca arctoides* no difference was detected between the sucrose and the dextrin diets with respect to their effects on the serum lipids or to the development of atherosclerotic lesions. The authors concluded from their observations: "The possibility that one class of carbohydrates may be more atherogenic than another is neither confirmed nor denied by the data." The differences between species observed in this study must serve as a reminder of the need for extreme caution in extrapolating results obtained in one species to another.

The experiments in animals show that dietary carbohydrates may have some influence modifying the effect of atherogenic diets, but it is doubted that they provide any evidence in favor of a specific atherogenic effect of a given carbohydrate. The experiments reported by Portman *et al.* (1961) are of interest in this connection. These authors studied the effect of long-term feeding of fat-free (and consequently carbohydrate-high) diets in Cebus monkeys. The conclusion of these studies was: "There is little support in the studies of the Cebus monkeys for the suggestion that essential fatty acid deficiency is responsible for abnormal cholesterol concentration or for the acceleration of the development of atherosclerosis." In other words, there is no evidence that the high carbohydrate diets used by these authors had any detectable atherogenic effect.

B. The Effect of Dietary Carbohydrates on the Serum Lipids

Because the effect of dietary fat and cholesterol on the development of the atherosclerotic process is generally believed to be mediated through the effect of these dietary components on the serum lipids, particularly serum cholesterol, numerous experiments have been performed in an effort to determine the influence of the carbohydrates in the diet on the serum lipid levels in man and animals. Some of the main results of these experiments will be briefly discussed below.

1. Effects of Sugars and Complex Carbohydrates on the Serum Lipids in Experimental Animals

a. Serum Cholesterol

Several reports in the literature indicate that the nature of the dietary carbohydrates influences the serum cholesterol levels in various animal species (Grande, 1967). In 1956 Portman et al., as already noted, observed that starch caused lower serum cholesterol than either sucrose, fructose, or glucose, in rats fed a diet with added cholesterol and cholic acid. On the other hand, Anderson (1969) reported lower serum and liver cholesterol in rats fed sucrose than in rats fed various forms of modified food starches. Rabbits and chickens fed diets with added cholesterol have been reported to have higher serum cholesterol levels when the diet contained sucrose than when it contained glucose (Grant and Fahrenbach, 1959). Rabbits fed diets containing 29% lactose and 0.35% cholesterol have higher serum cholesterol levels than rabbits fed a similar diet containing sucrose (Wells and Anderson, 1959). In the dog, however, exchange of starch for sucrose to the extent of 48% of the total calorie intake caused no significant change in serum cholesterol before or after removal of the thyroid gland (Schultz and Grande, 1968). Kritchevski et al. (1959) observed that germ-free chickens fed a high sucrose diet had higher serum cholesterol than chickens fed a high glucose diet.

These and other observations, while indicating differences between the dietary carbohydrates with respect to their effects on the level of serum cholesterol, cannot be duplicated in all animal species or in the same species under different experimental conditions. It seems therefore that instead of describing the results of the numerous individual reports it will be more useful to summarize the main results obtained in this particular line of research. The following facts seem to be reasonably well established at the present.

1. Different animal species and different strains of the same species respond differently to the same dietary manipulations (Portman *et al.*, 1956; Wells and Anderson, 1959; Taylor *et al.*, 1967; Schultz and Grande, 1968).

2. The serum cholesterol differences between the starch and the sucrose diets tend to disappear with time. Thus the difference observed by Portman *et al.* (1956) 3 weeks after the dietary exchange completely disappeared when the feeding was continued for 12–17 weeks (Fillios *et al.*, 1958).

3. According to a number of reports the serum cholesterol differences observed when feeding different carbohydrates are not manifested when diets without added cholesterol are used (Fillios *et al.*, 1958; Grant and Fahrenbach, 1959; Guggenheim *et al.*, 1960; Staub and Theissen, 1968; Bedö and Szigeti, 1967). Other authors, however, have found higher cholesterol with diets containing sucrose than with diets containing starch, in the absence of added cholesterol (Allen and Leahy, 1966).

4. Addition of antibiotics (or succinylsulfathiazole) to the diet causes the disappearance of the serum cholesterol differences seen when feeding different carbohydrates (Portman *et al.*, 1956; Grant and Fahrenbach, 1959; Portman and Stare, 1959). On the other hand, the hypercholesterolemic effect of lactose in the rabbit can be reproduced by adding antibiotics to the diet (Nelson *et al.*, 1953; Wells and Anderson, 1959; Wells *et al.*, 1962).

5. Germfree animals may respond differently than those conventionally reared with respect to the effect of a given dietary carbohydrate on serum cholesterol (Kritchevski *et al.*, 1959).

6. Modification of other components of the diet, for instance, the proteins, changes in some cases the effect of dietary carbohydrates on serum cholesterol (Farnell and Burns, 1962).

These facts illustrate the complexity of the problem. It seems clear that there is not a single pattern of response common to all the species examined and that the effects of the dietary carbohydrates on the serum cholesterol levels are influenced by the presence of dietary cholesterol, by modifications of the intestinal flora, and by changes in the composition of the diet.

b. Serum Triglycerides

Oral or intravenous administration of both glucose and sucrose to fasting rats produces a rapid decrease of the serum triglyceride level (Bragdon *et al.*, 1957; Baker *et al.*, 1968) which returns to the initial value within 2–3 hr. It has been suggested that this effect is the result, at least in part, of insulin release which causes a reduction in the liberation of FFA by the adipose tissue, thus decreasing the amount of fatty acids available for the formation of triglyceride by the liver (Nikkila, 1969).

In contrast with this acute effect, continuous feeding of various sugars causes hypertriglyceridemia which can be detected one day after starting the feeding. Numerous authors have demonstrated the elevation of serum triglyceride concentration in rats fed diets containing different sugars (Nath *et al.*, 1959; Bar-On and Stein, 1968; Kritchevski and Tepper, 1969; Hill, 1970; Naismith and Khan, 1970a,b) as compared to rats fed diets containing equivalent amounts of starch.

There are also differences among the various sugars with respect to the intensity of their hypertriglyceridemic effect. Nikkila and Ojala (1965, 1966) reported that fructose produces in the rat greater elevation of serum triglycerides than glucose, and this difference has been confirmed by other authors (Zakim *et al.*, 1967; Bar-On and Stein, 1968; Kritchevski and Tepper, 1969; Naismith and Khan, 1970a,b).

Macdonald and Roberts (1965) reported that diets containing sucrose produce higher levels of total serum lipids than diets containing either glucose or fructose, when given for 12 weeks to male rats. No such difference was observed when these three diets were given to female rats.

The elevation of serum triglycerides produced by diets containing sugars, in the rat, is influenced by the age of the animal. Chevalier *et al.* (1972) have reported that diets containing either fructose or sucrose produced elevations of serum triglycerides in mature rats but not in weanling rats.

Not all the animal species respond with elevations of serum triglycerides when sugars are substituted for starch in the diet. We have compared in dogs two diets containing respectively starch and sucrose in amounts corresponding to 48% of the total calorie intake. No significant differences in serum triglycerides were observed between the two dietary situations. However, when the experiment was repeated with the same dogs after removal of the thyroid gland, the serum triglyceride level was slightly, but significantly, higher with the sucrose diet (Schultz and Grande, 1968). Bar-On and Stein (1968) have shown that, as compared to glucose, fructose does not elevate serum triglycerides in the guinea pig.

The reasons for the differences between fructose and glucose in regard to their effects on the concentration of serum triglycerides in the rat have been actively investigated by various authors. It is believed that this difference is related to differences in the metabolism of these two sugars. From measurements of respiratory quotient it was concluded many years ago that fructose has a greater tendency to be converted into lipids than glucose (Higgins, 1916). More recently, it has been reported that, when fed to rats in equicaloric amounts, sucrose has greater fattening effects than glucose (Feyder, 1935). In his review on the control of triglyceride kinetics by carbohydrate metabolism and insulin, Nikkila (1969) concluded that the exact cause of the different effects of fructose and glucose is unknown, but lists the following facts as possible responsible factors: (1) Fructose causes greater formation of hepatic fatty acids and glycerol than glucose, (2) The inhibition of FFA release from the adipose tissue by fructose is smaller than that produced by glucose, and (3) fructose is less efficient than glucose in inducing lipoprotein lipase activity in the adipose tissue. These differences support the view that the greater hypertriglyceridemic effect of fructose may result from a combination of increased formation and decreased removal rate of circulating triglycerides.

The species differences in regard to the hypertriglyceridemic effect of fructose seem to be related to the form in which these sugars are absorbed. Fructose is absorbed in the rat as such, because of the low activity of glucose-6-phosphatase in the intestine of this animal. In the guinea pig fructose is absorbed as glucose because the intestine of this animal, in contrast to that of rat and man, is very efficient in converting fructose to glucose (Kiyasu and Chaikoff, 1957; Ginsburg and Hers, 1960). As already noted fructose and glucose have similar effects on the serum triglycerides in the guinea pig.

2. Effects of Sugars and Complex Carbohydrates on the Serum Lipids in Man

In metabolically normal men eating usual Western diets, an increase of the carbohydrate content of the diet, with corresponding reduction of its fat content, causes a decrease of serum cholesterol concentration. This fact was well documented more than 20 years ago when the rice–fruit diet was introduced for the treatment of hypertension (Kempner, 1948; Schwartz and Merlis, 1948; Chapman et al., 1950; Starke, 1950; Watkin et al., 1950; Hatch et al., 1955). Some of the authors noted also that, in addition to the decrease in serum cholesterol, the rice–fruit diet produced an elevation of the serum "neutral fat" in part of the patients

(Watkin *et al.*, 1950; Hatch *et al.*, 1955). Similar observations were reported by other workers (Ahrens *et al.*, 1957; Kuo and Carson, 1959; Brown and Page, 1960).

Antonis and Bersohn (1961) showed that changing from a low to a high carbohydrate diet caused a rise in fasting triglyceride concentration regardless of the type of fat in the preceding regime. The difference in carbohydrate content between the two diets corresponded to 25% of the total calorie intake. The average concentration of serum triglycerides reached a maximum of about double the starting value 3–5 weeks after the dietary change. After 32 weeks on the high carbohydrate diet most of the subjects had triglyceride levels similar to those at the beginning. There was a good deal of variability in the individual responses; some of the subjects showed only moderate increases in triglyceride concentration, while others developed gross hyperlipemia.

In the same year Ahrens *et al.* (1961) published their classic paper demonstrating that subjects who had elevated triglyceride levels when eating their usual mixed diets showed considerable elevations of plasma cholesterol, phospholipids, and triglycerides when given a high carbohydrate diet. Ahrens *et al.* (1961) introduced the concept of "carbohydrate-induced lipemia" and postulated that this is a common phenomenon "especially in areas of the world distinguished by caloric abundance and obesity." This paper stimulated considerable interest in the study of the effect of dietary carbohydrates on the serum lipids of man and should be considered as an important milestone in the development of our knowledge about the effect of the various dietary components on the regulation of the serum lipid levels in man.

The preceding data indicate that dietary carbohydrates have different effects on the various lipid fractions and that the effects observed in normolipemic individuals differ from those observed in certain hyperlipemic patients. It will therefore be convenient to consider separately the effect of the various dietary carbohydrates in normolipemic individuals and in hyperlipemic patients.

a. The Effect of Dietary Carbohydrates on the Serum Lipids of Normolipemic Subjects

(1) SERUM CHOLESTEROL. It can be stated, in general terms, that the mixed carbohydrates of usual human diets have an effect on the serum cholesterol levels of metabolically normal individuals which is intermediate between that of the saturated and the polyunsaturated glycerides. Isocaloric substitution of mixed dietary carbohydrates for saturated glycerides causes a decrease of serum cholesterol, whereas substitution

of mixed carbohydrates for polyunsaturated glycerides causes an elevation (Keys *et al.*, 1965). No significant change in serum cholesterol level is observed when the mixed dietary carbohydrates are exchanged for monoene fatty acid glycerides or by glycerides having twice the amount of polyunsaturated as saturated fatty acids (Keys *et al.*, 1958; Grande *et al.*, 1972). Most of the serum cholesterol changes observed in controlled dietary experiments by isocaloric exchange of fats and carbohydrates are consistent with the view that these changes mainly result from the changes in the fatty acid composition of the diet and that the mixed carbohydrates of the diet are practically neutral in their effects on serum cholesterol levels (Ahrens *et al.*, 1957; Hegsted *et al.*, 1965; Keys *et al.*, 1965).

The serum cholesterol differences observed between populations subsisting on different diets, however, are not always entirely accounted for by the differences in dietary fatty acid composition. For this reason, we have been interested in examining whether dietary carbohydrates of different origin might have a different effect on man's serum lipids, which can account for the differences in serum cholesterol levels not explained by the differences in fat and cholesterol intake.

In the first experiment (Keys *et al.*, 1960) two diets differing in the proportion of calories provided by carbohydrates from various food sources were compared. One of the diets derived 17% of its calories from sucrose and milk sugar, whereas in the other diet the same proportion of calories was supplied by carbohydrates from fresh fruits, vegetables, and legumes. The comparisons were made with two levels of dietary fat (16 and 31% of the total calories, respectively). There was a small difference in protein content corresponding to 4% of the total calorie intake, and a difference in cholesterol content of some 40 mg/1000 kcal between the two diets. The diets were fed for periods of 6 weeks. The results demonstrated that independent of the fat intake the diet rich in sucrose and milk sugar produced mean serum cholesterol levels higher by 18 mg/100 ml than the diet rich in carbohydrates from fruits, vegetables, and legumes. It was concluded from this experiment that sucrose and milk sugar tend to produce higher serum cholesterol levels than equal calories of carbohydrates contained in fruits, vegetables, and legumes.

Based on the idea that sucrose produces higher serum cholesterol levels than complex carbohydrates (mainly starch) (Antar and Ohlson, 1965; Hodges and Krehl, 1965), many other authors have made dietary experiments involving the exchange of sucrose and starch. A summary of some of these experiments is presented in Table I.

As shown in the table the sucrose-containing diets produced higher serum cholesterol levels than the diets containing isocaloric amounts of

TABLE I

Effect on Serum Cholesterol of Exchanging Sucrose and Complex Carbohydrates from Various Sources (mainly starch) in the Diet (Normolipemic Individuals)[a]

Source of starch	Exchange % of total cal	No. of subjects	Duration of dietary periods	Serum cholesterol mg/100 ml			p	Remarks	Ref.
				Sucrose	Starch	Sucrose minus starch[b]			
1. Fruits, vegetables, and legumes	17	28	6 weeks	180	162	18 ± 2.2	<0.01	Two levels of fat	Keys et al. (1960)
2. Rice	19	6	25 days	202	196	6 ± 8.5	>0.05		Irwin et al. (1964)
3. Bread and potatoes	17	12	3 weeks	214	210	4 ± 5.7	>0.05		Grande et al. (1965)
4. Leguminous seeds	17	12	3 weeks	221	202	19 ± 5.2	<0.01		Grande et al. (1965)
5. Cereals and potatoes	23	18	4 weeks	247	236	11 ± 3.1	<0.01	Coconut oil	McGandy et al. (1966)
6. Cereals and potatoes	23	18	4 weeks	207	197	10 ± 4.2	<0.05	Olive oil	McGandy et al. (1966)
7. Cereals and potatoes	23	18	4 weeks	172	164	8 ± 3.4	<0.05	Safflower oil	McGandy et al. (1966)
8. Bread	35	15	2 weeks	213	200	13 ± 8.1	>0.05		Groen et al. (1966)
9. Bread	35	15	4-5 weeks	227	203	24 ± 7.0	<0.01		Groen et al. (1966)
10. Wheat starch	40	10	30 days	153	144	9		Young women	Klugh and Irwin (1966)
11. Cereals and potatoes	32	9	4 weeks	234	231	3 ± 4.6	>0.05	11 measurements, 1 woman	Dunnigan et al. (1970)
12. Wheatflour	16	12	2 weeks	194	187	7 ± 4.4	>0.05		Anderson et al. (1972)
13. Mixed vegetables	16	12	2 weeks	194	172	22 ± 3.8	<0.01		Anderson et al. (1972)
14. Wheat flour	16	12	2 weeks	155	157	-2			J. T. Anderson, unpublished
15. Chick-peas	16	12	2 weeks	155	157	-2			J. T. Anderson, unpublished
16. String beans and peas	16	12	2 weeks	155	156	-1			J. T. Anderson, unpublished

[a] All subjects were males with the exception of experiment in line 10 and 1 woman in line 11.
[b] Mean ± SE.

complex carbohydrates from various sources in a number of experiments but not in all of them. Other workers found no difference in serum cholesterol levels between sucrose and starch (Akinyanju *et al.*, 1968; Szanto and Yudkin, 1969). Furthermore, the effects observed in some of the experiments were not reproduced when the. experiments were repeated under comparable conditions. Thus, the- difference between the sucrose diet and the diet containing complex carbohydrates from leguminous seeds was not observed when the experiment was done using either chick-peas or string beans and peas as the source of carbohydrates. The reasons for these discrepancies are not readily apparent. Factors such as the magnitude of the dietary exchange, the duration of the dietary periods, and the age and sex of the subjects should be considered in this respect.

In the experiment by Klugh and Irwin (1966, line 10 in Table I) the exchange between sucrose and wheat starch corresponded to 40% of the calorie intake, but the cholesterol difference is close to that observed in one of the experiments with wheat flour (line 12) in which the dietary exchange corresponded only to 16% of the total calorie intake. It must be noted that the experiment reported by Klugh an Irwin (1966) was done with young women whereas all other experiments in Table I were performed with men, except that reported by Dunnigan *et al.* with 1 woman and 8 men.

The experiment by Groen *et al.* (1966, lines 8 and 9) is of particular interest with regard to the importance of the duration of the dietary periods. These authors compared sucrose and starch (bread) at the level of 35% of the total calorie intake. The present author's calculation of their data indicates that the difference between the two diets was not significant 2 weeks after the dietary exchange, but it became significant when the means of the cholesterol levels at 4 and 5 weeks after the dietary change were compared.

In connection with the question of the duration of the dietary periods the experiment reported by Mathur *et al.* (1968) should be considered here. In this experiment 30 men were fed a diet containing 156 gm/day of butterfat (about 50% of the total calorie intake). When the men changed from their usual diet to the butter diet the serum cholesterol rose from a mean of 123 to 206 mg/100 ml. The men were then given a diet containing the same amount of butterfat, but containing chick peas in place of wheat flour. This corresponded to the exchange of chick peas for wheat flour carbohydrates at the level of 24% of the total calorie intake. The important point of this experiment is that there was practically no change in serum cholesterol concentration 4 weeks after the dietary exchange. Later on the cholesterol level decreased to a lowest level of 160 mg/100 ml 24 weeks after the dietary change. It appears, therefore, that the cholesterol lowering effect of chick peas, relative to wheat flour, takes considerable time to develop. The failure to demonstrate any cholesterol lowering effect with chick peas might have resulted from the short duration of the experimental periods.

Other experiments reported in the literature have compared sucrose and starch at higher intake levels than those presented in Table I.

The experiments by Macdonald (1966c) involve a very drastic dietary exchange achieved by feeding daily 500 gm of either raw corn (maize), starch, or sucrose. In one of these experiments (Macdonald and Braithwaite, 1964) the subjects were 7 males 21–41 years old. No change in serum cholesterol was observed when these subjects changed from a "free choice diet" to the sucrose diet, but a decrease of serum cholesterol was noted when the starch diet was substituted for the free choice diet indicating that starch produces lower serum cholesterol than sucrose. The effect of the dietary exchange was described as the slope of the regression line for serum cholesterol concentration over a period of 25 days of observation. The authors estimated that the starch diet decreased serum cholesterol at the rate of 2.3 mg/100 ml per day. As noted (Grande, 1967), this may not be the best way of estimating the effect of the dietary change, but in any case the serum cholesterol change was greater than those changes reported in Table I indicating that these large dietary exchanges of starch for sucrose produce marked changes in serum cholesterol concentration.

Another interesting aspect of Macdonald's work refers to the influence of sex on the effect of dietary carbohydrates (Macdonald and Nowakowska, 1964; Macdonald, 1965a, 1966a). As compared to starch, the cholesterol-raising effect of sucrose observed in men was not observed in young women. Recalculation of Macdonald's data (Grande, 1967) demonstrates no significant serum cholesterol differences between sucrose and starch in these young women. In a group of 6 postmenopausal women, however, sucrose produced mean serum cholesterol higher by 30 mg/100 ml (SE ± 11.0) than starch. This difference was significant at the probability level of p = 0.04. It appears, therefore, that postmenopausal women, like men, have lower serum cholesterol when eating starch than when eating sucrose. It should be noted, however, that the postmenopausal women lost weight during the experiment and that the serum cholesterol decrease was correlated with the weight loss (Macdonald, 1966a).

Macdonald's experiments show that in male individuals large dietary exchanges of sucrose and starch cause significant serum cholesterol changes, sucrose being hypercholesterolemic relative to starch. The question is whether these large dietary contrasts occur among people eating diets usually consumed by human populations. The amounts of sucrose used by Macdonald *et al.* are twice as great as the largest daily intakes of sucrose reported in the literature for any population (Lopez *et al.*, 1966). Accordingly, it is difficult to believe that the differences in serum cholesterol among populations can be simply explained by differences in sugar and starch intake.

With more moderate exchanges, such as those presented in Table I, the greatest short-term changes in serum cholesterol were those described by Groen *et al.* (1966) when exchanging sucrose and bread, and in some of the present author's experiments with leguminous seeds (Grande *et al.*, 1965). Whether the apparent hypocholesterolemic effect, relative to sucrose, observed in these experiments really results from starch, other "complex" carbohydrates, or other components of the natural foods used is, of course, an open question. It must be noted that comparisons of starches extracted from different cereals (wheat, corn, and rice) and from potatoes showed no significant differences in serum cholesterol concentration when these starches were introduced into the daily diet of male subjects in the amount of 175 gm (26–27% of the daily calorie intake) (Grande, 1967). In view of the present author's failure (and that of others) to demonstrate a significant difference between sucrose and starch in a number of experiments it is difficult to understand the results reported by Pleshkov (1964). This author noted a mean increase of serum cholesterol concentration of 19 mg/100 ml (SE ± 8.9, p < 0.05) in 30 atherosclerotic patients (25 men and 5 women) who were given for 10 days a daily supplement of 50 gm of sucrose in addition to their usual diet. As previously noted (Grande, 1967), only two single serum cholesterol values (before and after sugar administration) were reported in this publication.

There are not many comparisons of the effects of different sugars on the serum cholesterol levels in metabolically normal individuals. In 1963, experimental results were reported in which glucose, sucrose, and lactose were compared (Anderson *et al.*, 1963). The dietary exchanges involved were 233 gm daily of either sucrose, anhydrous glucose, and a mixture of 104 gm of lactose and 129 gm of anhydrous glucose. The comparisons were made on 12 men whose average calorie intake was about 3000 kcal/day. The dietary substitution corresponded to 31% of the total calorie intake. No significant differences in serum cholesterol were found in these experiments as shown in Table II which gives the mean serum cholesterol levels at the end of 3 weeks for the various diets.

It is clear that in normolipemic individual sucrose did not produce significantly higher serum cholesterol levels than either glucose or a mixture of lactose and glucose.

Similar negative results have been reported by Shamma'a and

TABLE II

Serum Cholesterol Levels (mg/100 ml) of 12 Men[a]

Diet			
Glucose	Sucrose	Lactose and glucose	Mean difference
180 ± 6.5	185 ± 7.9		5 ± 3.5
177 ± 9.7		178 ± 11.2	1 ± 3.4
	179 ± 11.1	174 ± 10.1	5 ± 4.5

[a] At the end of 3 weeks of eating diets containing either glucose, sucrose, or lactose and glucose in amounts corresponding to 31% of the total caloric intake. Means and SE (Anderson *et al.*, 1963).

Al-Khalidi (1963) on three students fed a formula diet containing 50% of the calories as either glucose, sucrose, lactose or galactose. The experimental periods were 16–25 days long. The authors found no differences in serum cholesterol concentration between the four sugars used. Wells and Anderson (1962) reported a decrease of serum cholesterol in 4 young men when they changed from a diet containing 160 gm of lactose per day to a diet containing an equivalent amount of sucrose. Average "normal" (presumably when eating the usual diet) serum cholesterol was 185 mg/100 ml. Serum cholesterol rose to 231 mg/100 ml after eating the lactose diet for 5 weeks and decreased to 162 mg/100 ml 3 weeks after changing from the lactose diet to the diet containing sucrose. No switchback of the diets was made in this experiment.

McGandy *et al.* (1966) have reported that a high "sugar-lactose" diet containing 111 mg of lactose and 111 gm of sucrose daily caused significantly higher serum cholesterol than a "high-sugar" diet containing 222 gm of sucrose when the dietary fat was coconut oil. No difference between the two diets was found, however, when the dietary fat was either olive oil or safflower oil. The possibility that the effects of dietary sugars on serum cholesterol are influenced by the type of dietary fat is suggested by these results.

Macdonald (1966b) has reported direct comparisons of the effects of fructose and glucose on serum cholesterol concentration in man and in pre- and postmenopausal women receiving a fat-free diet. No significant differences in serum cholesterol concentration between the two sugars were found in any of the groups.

In evaluating the significance of the experiments testing the effects of different sugars on serum cholesterol concentration it is important to keep in mind that these effects are profoundly influenced by the composition of the rest of the diet. This is strikingly illustrated by the results reported by Winitz *et al.* (1964). These authors fed 18 subjects with a "chemical diet" containing amino acids, vitamins, and minerals. The only source of fat was 2 gm daily of ethyl linoleate. About 90% of the caloric value of the diet was provided by glucose (555 gm/liter of aqueous dietary solution). A marked decrease of serum cholesterol was observed when the subjects changed from their usual diet to the chemical diet. The mean decrease was 67 mg/100 ml, 4 weeks after the change of diet. When the chemical diet was replaced by a similar diet containing 75% of glucose and 25% of sucrose instead of glucose alone, the serum cholesterol levels rose on the average 48 mg/100 ml in 3 weeks. The diet containing glucose alone was given again and the serum cholesterol concentration, once more, decreased. It is evident in this experiment that giving sucrose in exchange for a part of the glucose, caused an elevation of serum cholesterol concentration, but it is obvious that such serum cholesterol change was produced in the presence of a diet unlike any natural diet spontaneously consumed by man. This experiment is indeed of great importance in showing that the effects on the serum lipids of exchanging dietary carbohydrates may be drastically affected by the nature of the diet. It is clear that under the conditions of this experiment sucrose has a cholesterol-raising effect as compared to glucose, but it is also clear that such an effect cannot be demonstrated when sucrose and glucose are exchanged in the presence of usual diets made up of natural foodstuffs.

Because some populations characterized by low serum cholesterol levels and low incidence of coronary heart disease subsist on diets which are rich in undigestible carbohydrates, it has been suggested that some of the components of the fraction of the dietary carbohydrates usually called "fiber" or "unavailable carbohydrates" may have hypocholesterolemic effects (Higginson and Pepler, 1954; Walker and Arvidson, 1954; Bersohn *et al.*, 1956; Hardinge *et al.*, 1958; Walker, 1961; Trowell, 1972). Laboratory experiments have shown, however, that daily supplements of 15 gm of cellulose do not affect serum cholesterol concentration in man (Keys *et al.*, 1961) and this result has been confirmed by others (Prather, 1964). On the other hand, the same amount of pectin added to the daily diet caused after 3 weeks an average decrease in serum cholesterol concentration of the order of 5%, which was statistically significant. This effect of pectin has been confirmed by a number of studies in man and animals, and it has been shown that pectin prevents atherosclerosis in cholesterol-fed animals (Wells and Ershoff, 1961; Fisher *et al.*, 1964, 1965, 1966b; Fausch and Anderson, 1965; Palmer and Dixon, 1966). The effect of pectin in animals depends on the presence of dietary cholesterol (Fausch and

Anderson, 1965; Fisher *et al.*, 1965, 1966a), it is probably the result of a decrease in the absorption of cholesterol and bile acids (Hyun *et al.*, 1963; Leveille and Sauberlich, 1966), and it is not modified by antibiotics or by succinylsulfathiazole. Although the hypocholesterolemic effect of pectin seems to be well established, it is doubtful that it can play an important role in maintaining low serum cholesterol in human populations. The amount of pectin needed to obtain the effect observed in experiments seems to be higher than existing in usual human diets (Keys *et al.*, 1961).

Other complex carbohydrates like some gums have been reported to possess a cholesterol-lowering effect in man (Fahrenbach *et al.*, 1965). Polysaccharides having a hypocholesterolemic effect in rats fed a high fat diet containing cholesterol have been isolated from unmilled rice and from black gram peas (*Phaseolus mungo*) (Devi and Kurup, 1972; Vigayagopalan and Kurup, 1972).

The evidence which has been reviewed seems to indicate that, although under certain circumstances it is possible to demonstrate difference between different dietary carbohydrates with respect to their effects on serum cholesterol concentration, these effects are unlikely to be very important within the range of natural diets usually eaten by man.

In one of his publications, Macdonald (1966b) concluded: "The serum cholesterol level is very sensitive to the amount and type of lipid in the diet, and if it is affected by the amount and type of dietary carbohydrate per se, then this influence is small compared with that of the dietary lipid." In the same vein, McGandy *et al.* (1966) have stated: "The carbohydrate effects on serum cholesterol are of much smaller order of magnitude than the fat effects and manipulation of the source of dietary carbohydrates would appear to add little to the dietary regulation of blood cholesterol." More recently, these authors have stated: "We conclude that the practical significance of differences in dietary carbohydrate is minimal in comparison to those related to dietary fat and cholesterol" (McGandy *et al.*, 1967).

These statements very aptly summarize the present status of our knowledge in regard to the effect of dietary carbohydrates on serum cholesterol concentration in normolipemic individuals.

(2) SERUM TRIGLYCERIDES. As shown by Antonis and Bersohn (1961), when normolipemic individuals change from a high fat diet (40% of fat calories) to a high carbohydrate diet (15% of fat calories) the fasting serum triglyceride level rises. The triglyceride elevation was demonstrated already a few days after the dietary change, reached a maximum

in 3–5 weeks, and slowly disappeared. By the end of 32 weeks almost all of the subjects had returned to their initial values.

The transient nature of this phenomenon is well documented by the fact that individuals habitually subsisting on high carbohydrate diets do not have high fasting serum triglycerides (Antonis and Bersohn, 1960; Scott *et al.*, 1963, 1964; Hodges and Krehl, 1965). The elevation of serum triglycerides produced by increasing the carbohydrate content of the diet with simultaneous reduction of the fat content has been confirmed by numerous authors (Anderson, 1967). The effect, however, seems to be less pronounced in young women, as shown by the experiments reported by Beveridge *et al.* (1964). These authors changed a group of university students from their self-selected diets to a high carbohydrate formula diet. The carbohydrate was 90% dextrimaltose and 10% sucrose. The serum triglyceride level of 11 men showed a twofold increase 8–12 days after the dietary change and an increase of 50% after 16 days. In contrast 14 women showed a nonsignificant increase in serum triglycerides. Anderson (1967) has reviewed some of the experiments performed in the laboratory (Grande *et al.*, 1961; Anderson *et al.*, 1963). In one of the experiments 16 middle-aged men were fed alternately for periods of 3 weeks each of four diets deriving 28% of the total calories respectively from corn oil; a mixture of coconut oil and olive oil (58:42); a mixture of beef fat and safflower oil (94:6); and a mixture of sucrose, glucose, and sugars from fruit juices (66:17:17). The corresponding fasting levels of serum triglycerides when eating these diets were (means ± SE): 86 ± 12, 101 ± 11, 125 ± 17, and 173 ± 23. The serum triglyceride levels when the subjects ate the diet containing the carbohydrate mixture were significantly higher than those found with any of the other diets.

The magnitude of the triglyceride response appears to be related to that of the dietary exchange, but there is considerable variability among individuals with no evidence of abnormal lipid metabolism and with fasting serum triglyceride levels within the "normal" range when eating their usual diets (Anderson, 1967).

We have recently reported a study of the effects on serum triglyceride levels of exchanging fat and carbohydrates in a group of 38 middle-aged mentally retarded men (Grande *et al.*, 1972).

The fats exchanged were mixtures of natural oils so prepared as to have twice the amount of polyunsaturated as saturated fatty acids, and produced identical serum cholesterol levels, regardless of the amount fed, as well as identical serum triglycerides when fed at the same caloric level. When these fats were replaced by carbohydrates (mainly sucrose) in an amount corresponding to 25% of the total calorie intake the serum tri-

glyceride concentration increased. The mean increase was 75 mg/100 ml
(p < 0.01) at the end of 4 weeks. Replacement of these fats for carbohydrate, at the level of 17% of the total calorie intake, caused a mean elevation of serum triglycerides of 33 mg/100 ml (p < 0.01). The substitution
of carbohydrate for fat produced, therefore, a marked increase in serum
triglyceride which was roughly proportional to the increase in the carbohydrate content of the diet. There was a slight tendency for the increase
in serum triglycerides to be greater in those men who had initially the
higher triglyceride levels. However, when the 10 men with the highest
and the 10 men with the lowest serum triglycerides were compared it
was found that the mean elevations of serum triglycerides produced when
increasing the carbohydrate content of the diet by 25% of the total calorie
intake (and corresponding decrease in the fat content) for the two groups
were not significantly different from each other.

There has been considerable interest in studying the effect on serum
triglycerides of different kinds of dietary carbohydrates. Particular attention has been given by various workers to the comparison of the so-called
complex carbohydrates (mainly starch) and sugars, particularly sucrose.
Although many of the authors believe that sugars, and more specifically
sucrose, are hypertriglyceridemic relative to starch (Antar and Ohlson,
1965; Hodges and Krehl, 1965), there is considerable disagreement among
the different reports. The limited number of subjects, the variability of
the individual responses, and the differences in the extent and duration
of the dietary exchanges make the evaluation of the results very difficult.
This difficulty is well illustrated by the observations of H. B. Brown
quoted in Anderson's review (1967). In an exchange of simple and complex carbohydrates corresponding to 50% of the total calorie intake for
a 5-day period, Brown noted that the serum triglyceride level of 17 normal young men was on the average 75 mg/100 ml higher when eating
the diet rich in simple carbohydrates than when eating the complex carbohydrate diet. However, in a similar experiment with half as great an
exchange of simple and complex carbohydrates over a period of 18 days,
she found no difference in serum triglycerides. In addition, Brown noticed
considerable individual differences and advised caution in the evaluation
of the triglyceride responses.

The experiments by Macdonald (1965a,b, 1966a,b,c; Macdonald and
Braithwaite, 1964), as already mentioned in the discussion on serum
cholesterol, involved the exchange of large amounts of starch and sucrose
(500 gm/day or about 73% of the energy intake) in the presence of very
limited amounts of dietary fat. When the subjects changed from their
customary diets to the high sucrose diet there was an increase in plasma
triglycerides, but as noted by Anderson (1967) the increases were small

compared to those reported by other workers using more limited dietary exchanges. No increase of serum triglycerides was observed when the subjects changed from their usual diet to the starch diet. From this observation it was concluded that sucrose produces higher serum triglycerides than starch.

Not all the authors have been able to confirm these results (Stare, 1967), and it has been suggested (Lees, 1965) that poor absorption of the raw starch used in the experiments might have been responsible for Macdonald's results. Lees (1965) fed 7 young men and 1 woman purified diets deriving 10% of the calories from proteins and 90% from either cooked wheat and rice starch or sucrose. The sucrose diet caused an elevation of serum triglycerides of 88 mg/100 ml above the control level. The starch diet, however, caused an elevation of 109 mg/100 ml.

Klugh and Irwin (1966) compared cooked wheat starch and sucrose in two groups of young women (5 women per group). Each group ate one of the diets for a period of 30 days and the diets were reversed for another 30 days. The serum triglycerides, calculated by Anderson (1967) from the total serum fatty acids figures reported by the authors, were lower by 60 mg/100 during the sucrose periods. Kuo (1965; Kuo *et al.*, 1967) was unable to produce elevation of serum triglycerides in young men by doubling their daily intake of sucrose.

Comparing diets containing mixtures with different proportions of starch and sucrose, Nestel *et al.* (1970) found slightly higher serum triglycerides in normal young men when eating the diet rich in sucrose than when eating the diet rich in starch. Large differences were observed by these authors when similar dietary exchanges were tested in hyperlipidemic patients.

In our laboratory a comparison was made in two groups of young male university students (12 men per group) of diets containing either sucrose or starch from different sources (wheat flour, vegetables, and leguminous seeds). The carbohydrate supplements corresponded to 16% of the total calorie intake and were incorporated into a basic diet of the usual American type. Total calorie intake was individually adjusted and body weight was maintained constant throughout the experiments. The duration of the dietary periods was 2 weeks. No differences in serum triglyceride concentration between sucrose and starch were detected in these comparisons (Anderson *et al.*, 1972). These results seem to indicate that the hypertriglyceridemic effect of sucrose relative to starch cannot be demonstrated when the dietary exchanges are kept within limits compatible with usual food habits for a short time.

There is only limited information about the effect of different sugars on serum triglyceride concentration in normal individuals. In our experiments (Anderson *et al.*, 1963) we compared the effects of glucose, sucrose, and a mixture of lactose sucrose fed in amounts corresponding to 900 kcal/day (31% of the total calorie intake) in three groups, each of 8 middle aged men. The sugars were incorporated into a lowfat basic diet made up of usual foodstuffs. The duration of the dietary periods was 3 weeks. The three sugar diets with a fat content corresponding to 12% of the total calorie intake caused a striking elevation of serum triglycerides as compared with the control diet, which had a fat content equivalent to 35% of the total calorie intake, but the differences between the individual sugars were less evident. As compared with glucose, sucrose produced serum triglycerides higher by 32 mg/100 ml, and this difference is significant at the level of p = 0.04. The lactose–glucose diet caused higher serum triglycerides than the glucose diet, but the difference did not reach statistical significance. Although this experiment gives some indication of the hypertriglyceridemic effect of sucrose as compared to glucose, the limited number of subjects and the low fat content of the diet demand some caution in interpreting these results.

Macdonald reported (1966b) a study in 5 young men comparing the effects of mixtures of glucose and starch, fructose and starch, and fructose and glucose, which were fed at the dose of 7.5 gm/kg/day together with 50 gm/day of calcium caseinate and a vitamin preparation. The mixtures containing fructose produced significantly higher triglyceride levels than either the control diet or the glucose–starch diet. Similar, but perhaps more pronounced effects, were obtained in postmenopausal but not in premenopausal women.

In the recent experiments by Nestel *et al.* (1970) glucose and fructose raised triglycerides to a similar extent. The subjects ate diets made with usual foods.

b. Effect of Dietary Carbohydrates on the Serum Lipids of Hyperlipemic Patients

It is well established that, as reported by Ahrens *et al.* (1961), dietary carbohydrates have marked effects on the serum lipid levels of individuals whose serum is visibly lipemic when they are eating their customary diets. When these individuals change from a diet of usual fat content to a low fat, high carbohydrate diet, their serum lipids, particularly the triglycerides, show a remarkable increase which has been well documented in the literature. Since others in this book will discuss the effects of dietary carbohydrates on the serum lipids in the various types of hyperlipemia, only a brief review of the more pertinent reports will be given **here.**

That hyperlipemic patients respond to dietary manipulations differently than normolipemic individuals had been noted by, among others, Ahrens *et al.* (1957), Kuo and Carson (1959), and Brown and Page (1960). Following the publication of Ahrens *et al.* (1961) a number of reports have appeared confirming the hyperlipidemic effect of dietary carbohydrates in hyperlipemic subjects.

Farquhar *et al.* (1966) reported observations in 15 subjects who were fed alternately high fat and high carbohydrate formula diets. The difference between the two diets consisted on the exchange of corn oil for dextrin in amounts corresponding to 68% of the total calorie intake. Both diets contained the same amount of protein (15% of the total calories) and lactose (17% of the total calories) contained in skimmed-milk powder. It follows that the low fat diet was actually a fat-free diet. The diets were fed alternately; the high fat diet was given for periods of 2–10 weeks and the high carbohydrate diet for periods of 2–7 weeks. The average level of plasma triglycerides during the high fat period (mean of 4–16 samples obtained twice weekly) was 305 mg/100 ml. That for the first 4 weeks of the high carbohydrate period was 624 mg/100 ml. All the patients, but one who had a higher triglyceride concentration when eating the high fat diet than when eating the carbohydrate diet, had some increase in serum triglycerides when changing from the high fat to the high carbohydrate diet. There were considerable differences among the subjects in the triglyceride response to the high carbohydrate diet. Seven of the patients had serum triglycerides below 150 mg/100 ml when eating the high fat diet. The authors noted: "Carbohydrate-induced lipemia is not an all or none phenomenon, but rather that the majority of patients will increase their triglyceride production and concentration to a variable extent as a consequence of the high carbohydrate diet."

Some observations indicate that in hyperlipemic individuals sucrose has a greater effect elevating plasma triglycerides than starch. Kuo and Basset (1965) studied 5 middle-aged atherosclerotic patients who had fasting serum triglyceride levels above 350 mg/100 ml when eating their self-selected diets. These diets were estimated to supply about 2100 kcal daily and contained 60–70 gm of fat, 100–120 gm of protein, and 240–290 gm of carbohydrates. The carbohydrates of the self-selected diet were a mixture of starch and sucrose and it is stated that the amount of sucrose was 180–205 gm/day, corresponding approximately to 70–75% of the total carbohydrate intake. The subjects were fed alternately for periods of 4–6 weeks diets comparable to the self-selected diet but containing carbohydrates mainly as either sucrose or starch. The "sugar" diet contained 80–85% of the total carbohydrates as sucrose, whereas the "starch" diet contained no other sugars than those provided by a small glass of fruit juice. As compared with the levels observed when eating the self-selected diet, serum cholesterol decreased slightly in 3 subjects and rose in 2 when they were given the sucrose diet and decreased in the 5 subjects when the starch diet was given. Serum triglycerides rose markedly with the sugar diet and decreased with the starch diet. In other words, the sucrose diet gave higher and the starch diet much lower triglycerides than the self-selected diet. The data show a marked difference between sucrose and starch regarding their effects on serum triglycerides. The substitution of sugar for starch at the level of 40–45% of the total calorie intake produced a very marked elevation of the serum triglycerides, in contrast to the findings in normolipemic individuals, which have been discussed, and with observations in normal individuals by the same authors (Kuo, 1965). It is remarkable that the change from the self-selected diet to the sugar diet produced large increases of serum triglycerides in some of the patients because, according to the data given by the authors, 70–75% of the total carbohydrate content of the self-selected diet was sucrose. Thus, it would appear that increase in the sucrose content of the diet corresponding only to some 5% of the total calorie intake, with corresponding decrease of the starch intake, had a marked hypertriglyceridemic effect in some of these patients.

Substitution of sucrose for starch caused a more than threefold increase of plasma triglycerides in 2 hypercholesterolemic boys who had fasting triglycerides of 62 and 77 mg/100 ml when eating the self-selected diet. In contrast with the 5 hyperlipemic subjects these two boys had higher serum triglycerides when eating the starch diet than when eating the self-selected diet. The serum triglyceride differences were 33 and 49 mg/100 ml, respectively. In these 2 subjects the serum cholesterol levels were lower with both the sucrose and the starch diets than with the self-selected diet. This result is difficult to understand since the total amounts of fat and carbohydrate were supposed to be practically the same in the three diets.

Kaufmann *et al.* (1966) studied 6 patients (4 hypertriglyceridemic classified as "carbohydrate inducible," 1 hypertriglyceridemic or "mixed type," and 1 hypercholesterolemic). They were given diets of low fat content (3–12% of the total calorie intake) and high in either starch or sucrose. Calculation of the data given by the authors indicates that the exchange of sugar for starch corresponded on the average to 62% of the total calorie intake. The diets were fed for periods of 10–40 days. The serum triglyceride levels when the subjects were eating the starch diets were somewhat lower than those observed with the self-selected diets, at variance with the common idea that high carbohydrate diets elevate the serum triglycerides. The sucrose diet caused higher serum triglycerides than the starch diet in all the subjects. In a more recent paper (1967) these authors noted no difference in serum triglycerides when sucrose and starch (300 gm/day) were exchanged in the diet of 2 hyperlipemic individuals who had fasting serum triglycerides below 1000/100 ml when maintained with an ad libitum diet and in 2 normal subjects. Although the individual differences are considerable, these reports clearly support the view that replacement of starch by sucrose does indeed produce an elevation of serum triglycerides in hyperlipemic individuals.

In a more recent experiment, Nestel *et al.* (1970) also noted that the effect of sucrose on triglyceride concentration was greater than that of starch. These authors compared glucose and fructose and found that these two monosaccharides have the same effect on serum triglycerides in both normal and hyperlipemic subjects. On the other hand, Kaufmann *et al.* (1967) concluded: ". . . where significant differences in serum triglyceride level were found, fructose produced higher levels than starch, sucrose or glucose, and glucose higher levels than starch."

A series of experiments by Antar and his colleagues (Birchwood *et al.,* 1970; Little *et al.,* 1970; Antar *et al.,* 1970) suggests that the effect on serum triglycerides of exchanging sucrose and starch in the diet of hyperlipidemic patients may be affected by the nature of the fat in the diet. When the substitution was made in the presence of diets containing polyunsaturated fat and of low cholesterol content, there was little difference in serum triglyceride concentration. However, the substitution of sucrose for starch at the level of 40% of the total calorie intake produced significant elevations of serum cholesterol, phospholipids, and triglycerides in all the subjects when they were eating a diet containing saturated fat (17.5% of total calories) and cholesterol (310 mg/1000 kcal).

On the other hand, Hodges *et al.* (1967) noted in normolipemic subjects that the serum triglyceride differences between the sugar and the starch diets were greater when the exchanges were made in the presence of low fat diets.

Porte *et al.* (1966) noted no difference in serum triglyceride concentration between starch and glucose diets in two lipemic patients. These authors observed that calorie restriction causes marked decrease in serum triglycerides and advised caution in interpreting the results of exchanging carbohydrates in the diet, which, in their opinion, may be related to changing calorie balance more than to the type of carbohydrates.

C. Sugar Intake and Coronary Heart Disease

Yudkin (1957, 1963, 1964, 1966, 1967, 1969) has proposed that the consumption of sucrose is an important factor in the etiology of coronary heart disease, and this view has attracted attention in the scientific as well as in the lay literature.

Some of the limitations of this view have been discussed in McGandy's *et al.* review (1967). More recently, two extensive critical reviews of Yudkin's theory have been published (Keys, 1971; Walker, 1971). The analyses of current evidence presented by the two reviewers coincide in showing that Yudkin's view is not supported by the information at hand. One of the reviewers (Walker, 1971) concludes: "although evidence is incomplete, such evidence as is available does not significantly incriminate sugar." The other (Keys, 1971) concludes: "The theory is not supported by acceptable clinical, epidemiological, theoretical or experimental evidence."

Since these two reviewers have made an exhaustive analysis of Yudkin's theory it seems that very little can be added to this discussion. Accordingly, only a few brief comments will be made on the main points of the argument. Readers desiring more complete information should consult the original reviews. Yudkin's theory is based on four main arguments which may be summarized as follows:

1. The differences among countries in mortality attributed to coronary heart disease are closely correlated with differences in the per capita consumption of sugar in these countries.

2. The increase in the consumption of sugar over time has been reflected in an increase in the incidence of coronary heart disease.

3. Men afflicted with coronary heart disease are characterized by an unusually high consumption of sugar.

4. Sucrose has certain metabolic effects such as induction of hyperinsulinism and elevation of the serum lipids which may influence the development of the atherosclerotic process.

When the average sugar consumption calculated from food disappearance data is compared with the figure for coronary mortality derived from the national vital statistics for different countries a correlation is observed. But as noted (Walker, 1971; Keys, 1971) these two sources of information do not give a true picture of either the incidence of coronary heart disease or the actual consumption of sugar. Furthermore, a number of exceptions have been noted. Thus, the per capita sugar consumption in Sweden is higher than that in Finland, but the age-specific coronary death rate in Sweden is lower than in Finland (Keys, 1971). Some countries not included in Yudkin's original data such as Cuba, Venezuela, Colombia, Costa Rica, and Honduras are known to have remarkably high consumption of sugar and low incidence of coronary heart disease (Keys, 1971; Walker, 1971; Lopez *et al.*, 1966).

Finally, the data of the population studies reported by Keys (1971) indicate that the correlation between sugar intake and incidence of coronary heart disease ($r = 0.78$) is a little lower than that between intake of saturated fatty acids and incidence of coronary heart disease ($r = 0.86$), and that, as noted by others, there is a high correlation between the contents of sucrose and saturated fatty acids in the diets ($r = 0.84$), which is adequate to explain the relationship between sugar intake and incidence of coronary heart disease.

Little *et al.* (1967), on the other hand, have noted a positive correlation between total fat intake and the serum levels of cholesterol, phospholipids, and S_f 20–400 lipoproteins, but no correlation with either total carbohydrates or sucrose intakes.

Time trends in food consumption and mortality are difficult to evaluate, but the data available do not support the view that sugar consumption and coronary mortality have grown hand in hand. Thus, as reported by Keys (1971) the figures from the U.S. Department of Agriculture show yearly per capita averages for sugar consumption of 51.7 kg in the 1920's and 50.2 kg in the 1960's. The United States' vital statistics, on the other hand, show a considerable increase in coronary death rate between these two periods. In his analysis of the time trends of mortality in relation with changes in the diet, Ashton (1965) noted correlations of 0.64 and 0.55, respectively, between death rates and fat consumption, and death rates and sugar consumption, but death rates showed a correlation of $r = 0.78$ with time. Ashton warned against direct causal interpretation of this type of correlation, noting that similar relationships have been demonstrated between coronary mortality and television and automobile licenses.

Yudkin and his colleagues (Yudkin and Roddy, 1964; Yudkin and Morland, 1967) have reported that coronary patients have higher sucrose intake than controls. In their first report (Yudkin and Roddy, 1964) three groups of age-matched men were studied: 20 survivors of a myocardial infarction, patients with peripheral vascular disease, and controls (patients who were in the hospital because of orthopedic problems). The average intakes of sucrose, determined by a questionnaire were, respectively, 132, 141, and 77 mg/day. As noted by McGandy *et al.* (1967) the two groups of cardiovascular patients had sugar intakes which are of the same order of magnitude as the average per capita consumption of sucrose in Great Britain (139 gm/day) previously reported by Yudkin (1964). Similar criticism has been expressed by others (Marr and Heady, 1964). It appears therefore that the difference reported is a result of unusually low sugar intake of the control group rather than of the high sucrose intake

TABLE III

Sugar Consumption by Male Coronary Patients and Control Subjects

Author	Number of subjects		Sugar intake (gm/day)		p^a
	Patients	Controls	Patients	Controls	
Little *et al.* (1965)	86	84	47	65	<0.01
Papp *et al.* (1965)	20	20	121	117	>0.05
Begg *et al.* (1967)	63	33	39^b	55^b	<0.05
Paul *et al.* (1968)	66	85	116	96	—
Finegan *et al.* (1968)	100	50	66	69	>0.05
Burns-Cox *et al.* (1969)	80	160	100	97	>0.05
Howell and Wilson (1969)	170	1158	67	79	>0.05
Working Party Medical Research Council (1970)	150	275	122	113	>0.05
Gatti (1970)	47	31	57	45	>0.05

[a] Probability of chance occurrence of difference in sugar intake between patients and controls.

[b] Sugar intake in hot drinks alone.

by the patients. The second report presents similar data. The limitations of these studies have been discussed by Keys (1971) and by Walker (1971).

Other workers have been unable to confirm these results as shown by the data summarized in Table III and by the data of Keen and Rose (1964). The data by Paul *et al.* (1968) given in this table are of interest. They show a somewhat higher intake of sucrose for the patients than for the controls; but these authors have found that sugar consumption has a strong association with cigarette smoking. Logit analysis of the data demonstrated that the primary association is between cigarette smoking and coronary heart disease rather than between sugar intake and coronary heart disease. Similar results have been reported by Bennet *et al.* (1970) and by Elwood *et al.* (1970). In summary, then, the weight of evidence seems to be against any direct association between high sucrose intake and development of coronary heart disease.

The possibility that sugar has an influence on the blood lipids has been explored by numerous workers. As noted in the discussion of the effects of various carbohydrates the modest effects of sucrose observed in normolipemic individuals does not give support to the idea that sucrose influences the development of the atherosclerotic process because of its effects on the blood lipids. Yudkin and his associates have been unable to provide convincing evidence of elevation of serum cholesterol and phospholipids by sucrose (Akinyanju *et al.*, 1968; Szanto and Yudkin, 1969). The effect of sucrose on the blood lipids of hyperlipemic individuals that has been discussed is of interest, but the fact that these individuals are particularly sensitive to sucrose does not prove that previous high intake of sucrose is the cause of this particular sensitivity. As noted by Walker (1971): "a fundamental premise to this hypothesis is the validity of the belief that carefully controlled studies have shown that individuals who develop coronary heart disease have been consuming more sugar than those who do not."

Since the data available fail to support the view that coronary patients, as compared to controls, are characterized by habitual high consumption of sugar, it is impossible to attribute the sensitivity to sucrose observed in some of these patients to high intake of that sugar in their usual diets.

D. Final Comments—The Need for Future Research

As indicated by the analysis presented in the preceding section and by the more extensive critical reviews of Walker (1971) and Keys (1971), the evidence currently available does not support the view that high intake of sucrose is a major factor in the development of coronary heart disease. Yudkin (1971) has criticized the validity of the control groups used in various studies, but Walker (1971) concluded that the general validity of sucrose intake data obtained by dietary questionnaires such as those summarized in Table III is not in doubt. Whether more carefully conducted surveys comparing the sucrose intake of coronary patients and properly matched control subjects will change this picture is a moot question.

Walker (1971) has suggested another approach consisting in determining the rate of incidence of coronary heart disease among groups of population living in Western countries, and known to have widely different habitual levels of sucrose intake. In his opinion more intensive research with such particular groups might produce more trustworthy information. The difficulty is that the diets consumed by such population groups, in addition to their different levels of sucrose, are likely to differ also in other respects. Moreover, as noted by Lowenstein (1964), the influence of diet on coronary heart disease has to be examined in the context of the whole way of life of any group.

Most of the experimental evidence presented in the discussion fails to demonstrate any striking difference between sucrose and starch in regard to their effects on the serum lipids of normolipemic individuals. With the exception of a few experiments, and of experiments with extremely large dietary exchanges, it appears that in normal men eating diets made up of usual foodstuffs there is little difference between these two carbohydrates. This is in contrast with the remarkable difference observed when starch and sucrose were interchanged in the diet of hyperlipemic patients. In these individuals sucrose tends to produce much higher serum lipid levels, particularly triglycerides, than starch. It would appear therefore that some hyperlipemic individuals are particularly susceptible to sucrose, but the reason for this susceptibility is obscure. This is an area in which more research is obviously needed.

The differences between sucrose and starch, regarding their effects on the serum lipid levels, reported in hyperlipemic patients have been observed in experiments of a few weeks duration. Since the elevations of plasma triglycerides produced by high carbohydrate diets tend to disappear with time, it will be of interest to know whether such differences persist in these patients after longer periods of observation.

The relationship between insulin secretion and the elevation of plasma triglycerides caused by high carbohydrate diets has attracted considerable interest. Farquhar *et al.* (1966) demonstrated a good correlation between the plasma insulin levels and the magnitude of the triglyceride response when substituting dextrin for corn oil, and Reaven *et al.* (1967) reported that the degree of hypertriglyceridemia was highly correlated with the insulin response elicited by the ingestion of the high carbohydrate formula diet. These authors suggested that hypertriglyceridemia is secondary to exaggerated postprandial increases in plasma insulin concentration. This concept is of considerable interest for the present discussion because of the abnormally elevated insulin responses observed among coronary patients following the oral administration of glucose (Tzagournis *et al.*, 1967). On the other hand, Nikkila (1969) believed that the problem of the causal relationship between hyperinsulinism and hypertriglyceridemia is largely unsolved. Obviously it will be important to determine whether the different triglyceride responses to sucrose and starch observed in the hyperlipemic patients are related to differences in the insulin responses induced by these carbohydrates in such patients. In this connection it should be noted that Nestel *et al.* (1970) reported that serum triglyceride concentration and insulin responsiveness to glucose were always higher when the patients were eating the sucrose diet than when they were eating the starch diet.

Total calorie intake is an important consideration in the evaluation of the effect of dietary carbohydrates on the serum lipids in man. Nestel *et al.* (1970) reported that eucaloric carbohydrate-rich diets produced only modest triglyceride responses in normolipemic individuals whereas doubling the calorie intake led to marked hypertriglyceridemia. Conversely, a reduction of the calorie intake produces striking decreases of serum triglycerides in various forms of hypertriglyceridemia (Bierman and Porte, 1968; Nestel, 1966). As noted in the discussion some authors (Porte *et al.*, 1966) believed that the changes in serum triglycerides observed when exchanging dietary carbohydrates may be related to changes in energy balance more than to the type of carbohydrates.

Related to the problem of energy balance is the question of the influence of obesity on the serum triglyceride responses to dietary carbohydrates. Bierman and Porte (1968) observed a close association between obesity and the insulin and triglyceride responses to carbohydrate-rich diets, but there are conflicting results in the literature. Nestel *et al.* (1970) have suggested that whereas hyperinsulinemia is commonly found in hypertriglyceridemic and overweight individuals, hypertriglyceridemia may also occur in response to carbohydrate diets in the absence of high insulin levels. A recent paper by Goldrick *et al.* (1972) reported that overfeeding with glucose and fructose failed to elevate the serum triglycerides in two underweight young men.

The preceding comments are intended to illustrate a few of the factors which seem to influence the responses of the serum lipids to dietary carbohydrates. It is clear that much work needs to be done before we can have a proper understanding of the mechanisms by which dietary carbohydrates influence the concentration of the serum lipids in man.

References

Ahrens, E. H., Insull, W., Blomstrand, R., Hirsch, J., Tsaltas, T. T., and Peterson, M. L. (1957). Influence of dietary fats on serum lipid levels in man. *Lancet* **1**, 943–953.

Ahrens, E. H., Hirsch, J., Oette, K., Farquhar, J. W., and Stein, Y. (1961). Carbohydrate-induced and fat-induced lipemia. *Trans. Ass. Amer. Physicians* **74**, 134–146.

Akinyanju, P. A., Qureshi. R. U., Salter, A. J., and Yudkin, J. (1968). Effect of an "atherogenic" diet containing starch or sucrose on the blood lipids of young men. *Nature (London)* **218**, 975–977.

Allen, R. J. L., and Leahy, J. S. (1966). Some effects of dietary dextrose, fructose, liquid glucose, and sucrose in the adult male rat. *Brit. J. Nutr.* **20**, 339–347.

Anderson, J. T. (1967). Dietary carbohydrates and serum triglycerides. *Amer. J. Clin. Nutr.* **20**, 168–175.

Anderson, J. T., Grande, F., Matsumoto, Y., and Keys, A. (1963). Glucose, sucrose and lactose in the diet and blood lipids in man. *J. Nutr.* **79**, 349–359.

Anderson, J. T., Grande, F., Foster, N., and Keys, A. (1972). Different dietary carbohydrates and blood lipids in man. *Proc. Int. Congr. Nutr. 9th, 1972* p. 64.

Anderson, T. A. (1969). Effect of carbohydrate source on serum and hepatic cholesterol levels in the cholesterol-fed rat. *Proc. Soc. Exp. Biol. Med.* **130**, 884–887.

430   *Francisco Grande*

Anitschkow, N. (1913). Über die Veränderungen der Kaninchenaorta bei experimenteller cholesterinsteatose. *Beitr. Pathol. Anat. Allg. Pathol.* **56**, 379–404.

Anitschkow, N. (1914). Über die Atherosklerose der Aorta beim Kaninchen und über deren Entstehungsbedingungen. *Beitr. Pathol. Anat. Allg. Pathol.* **59**, 306–348.

Anitschkow, N., and Chalatow, S. (1913). Über experimentelle Cholesterinsteatose und ihre Bedeutung für die Entstehung einiger pathologischer Prozesse. *Zentralbl. Allg. Pathol. Pathol. Anat.* **24**, 1–9.

Antar, M. A., and Ohlson, M. A. (1965). Effect of simple and complex carbohydrates upon total lipids, nonphospholipids and different fractions of phospholipids of serum in young men and women. *J. Nutr.* **85**, 329–337.

Antar, M. A., Little, J. A., Lucas, C., Buckley, G. C., and Csima, A. (1970). Interrelationship between the kinds of dietary carbohydrate and fat in hyperlipoproteinemic patients. 3. Synergistic effect of sucrose and animal fat on serum lipids. *Atherosclerosis* **11**, 191–201.

Antonis, A., and Bersohn, I. (1960). Serum-triglyceride levels in South African Europeans and Bantu, and in ischemic heart disease. *Lancet* **1**, 998–1002.

Antonis, A., and Bersohn, I. (1961). The influence of diet on serum triglycerides. *Lancet* **1**, 3–9.

Ashton, W. L. (1965). Dietary fat and dietary sugar. *Lancet* **1**, 653–654.

Baker, N., Garfinkel, A. S., and Schotz, M. C. (1968). Hepatic triglyceride secretion in relation to lipogenesis and free fatty acid mobilization in fasted and glucose-refed rats. *J. Lipid Res.* **9**, 1–7.

Bar-On, H., and Stein, Y. (1968). Effect of glucose and fructose administration on lipid metabolism in the rat. *J. Nutr.* **94**, 95–105.

Bedö, M., and Szigeti, A. (1967). Die Wirkung der alimentären Kohlenhydrate auf der Stoffwechsel der Ratte. *Nahrung* **11**, 305–310.

Begg, T. B., Preston, S. R., and Healy, M. J. R. (1967). Dietary habits of patients with occlusive arterial disease. *Atti Conv. Int. Aspetti Diet. Infanzia Seneszenza, 5th, 1967* pp. 66–75.

Bennett, A. E., Doll, R., and Howell, R. W. (1970). Sugar consumption and cigarette smoking. *Lancet* **1**, 1011–1014.

Bersohn, I., Walker, A. R. P., and Higginson, J. (1956). Coronary heart disease and dietary fat. *S. Afr. Med. J.* **30**, 411–412.

Beveridge, J. M. R., Jagannathan, S. N., and Connell, W. F. (1964). The effect of the type and amount of dietary fat on the level of plasma triglycerides in human subjects in the postabsorptive state. *Can. J. Biochem. Physiol.* **42**, 999–1003.

Bierman, E. L., and Porte, D. (1968). Carbohydrate intolerance and lipemia. *Ann. Intern. Med.* **68**, 926–933.

Birchwood, B. L., Little, J. A., Antar, M. A., Lucas, C., Buckley, G. C., Csima, A., and Kallos, A. (1970). Interrelationship between the kinds of dietary carbohydrate and fat in hyperlipidemic patients. 2. Sucrose and starch with mixed saturated and polyunsaturated fats. *Atherosclerosis* **11**, 183–190.

Bragdon, J. H., Havel, R. J., and Gordon, R. S. (1957). Effects of carbohydrate feeding on serum lipids and lipoproteins in the rat. *Amer. J. Physiol.* **189**, 63–67.

Brown, H. B., and Page, I. H. (1960). Variable responses of hyperlipemic patients to altered food patterns. *J. Amer. Med. Ass.* **173**, 248–252.

Burns-Cox, C. J., Doll, R., and Ball, K. P. (1969). Sugar intake and coronary heart disease. *Brit. Heart J.* **31**, 485–490.

Chapman, C. B., Gibbons, T., and Henschel, A. (1950). The effect of the rice-fruit diet on the composition of the body. *New Engl. J. Med.* **243**, 899–905.

Chevalier, M., Wiley, J. H., and Leveille, G. A. (1972). The age-dependent response of serum triglycerides to dietary fructose. *Proc. Soc. Exp. Biol. Med.* **139**, 220–222.

Cohen, A. M., and Teitelbaum, A. (1968). Effects of glucose, fructose and starch on lipogenesis in rats. *Life Sci.* **7**, 23–29.

Devi, K. S., and Kurup, P. A. (1972). Hypolipidemic activity of *Phaseolus mungo* (black gram) in rats fed a high-fat high-cholesterol diet. Isolation of a protein and a polysaccharide fraction. *Atherosclerosis* **15**, 223–230.

Dunnigan, M. G., Fyfe, T., McKiddie, M. T., and Crosbie, S. M. (1970). The effects of isocaloric exchange of dietary starch and sucrose on glucose tolerance, plasma insulin and serum lipids in man. *Clin. Sci.* **38**, 1–9.

Elwood, P. C., Waters, W. E., Moore, S., and Sweetman, P. (1970). Sucrose consumption and heart disease in the community. *Lancet* **1**, 1014–1016.

Fahrenbach, M. J., Riccardi, B. A., Saunders, J. C., Laurie, I. N., and Heider, J. G. (1965). Comparative effect of Guar Gum and pectin on human serum cholesterol levels. *Circulation* **32**, Suppl. II, 11–12.

Farnell, D. R., and Burns, J. J. (1962). Dietary starch: Effect on cholesterol levels. *Metab., Clin. Exp.* **11**, 566–571.

Farquhar, J. W., Franck, A., Gross, R. C., and Reaven, G. M. (1966). Glucose, insulin and triglyceride responses to high and low carbohydrate diets in man. *J. Clin. Invest.* **45**, 1648–1656.

Fausch, H. D., and Anderson, T. A. (1965). Influence of citrus pectin feeding on lipid metabolism and body composition of swine. *J. Nutr.* **85**, 145–149.

Feyder, S. (1935). Fat formation from sucrose and glucose. *J. Nutr.* **9**, 457–468.

Fillios, L. C., Naito, C., Andrus, S. B., Portman, O. W., and Martin, R. S. (1958). Variations in cardiovascular sudanophilia with changes in the dietary levels of protein. *Amer. J. Physiol.* **194**, 275–279.

Finegan, A., Hickey, N., Maurer, B., and Muleahy, R. (1968). Diet and coronary heart disease: Dietary analysis on 100 male patients. *Amer. J. Clin. Nutr.* **21**, 143–148.

Fisher, H., Griminger, P., and Weiss, H. S. (1964). Avian atherosclerosis: Retardation by pectin. *Science* **146**, 1063–1064.

Fisher, H., Griminger, P., Sostman, E. R., and Brush, M. K. (1965). Dietary pectin and blood cholesterol. *J. Nutr.* **86**, 113 (Letter to the Editor).

Fisher, H., Vander Noot, G. W., McGrath, W. S., and Griminger, P. (1966a). Dietary pectin and plasma cholesterol in swine. *J. Atheroscler. Res.* **6**, 190–197.

Fisher, H., Soller, W. G., and Griminger, P. (1966b). The retardation by pectin of cholesterol-induced atherosclerosis in the fowl. *J. Atheroscler. Res.* **6**, 292–298.

Gatti, E. (1970). Primi resultati di una inchiesta sulla abitudine alimentari di paziente ospedalizatti per cardiopatia ischemica. *In* "Nutrition and Cardiovascular Disease" (F. Fidanza *et al.*, eds.), pp. 265–277. Morgagni, Rome.

Ginsburg, V., and Hers, H. G. (1960). On the conversion of fructose to glucose by guinea pig intestine. *Biochim. Biophys. Acta* **38**, 427–434.

Goldrick, R. B., Havenstein, N., Carroll, K. F., and Reardon, M. (1972). Effect of overfeeding on lipid and carbohydrate metabolism in lean young adults. *Metab., Clin. Exp.* **21**, 761–770.

Grande, F. (1967). Dietary carbohydrates and serum cholesterol. *Amer. J. Clin. Nutr.* **20**, 176–184.

Grande, F., Anderson, J. T., and Keys, A. (1961). The influence of chain length of the saturated fatty acids on their effect on serum cholesterol concentration in man. *J. Nutr.* **74**, 420–428.

Grande, F., Anderson, J. T., and Keys, A. (1965). Effects of carbohydrates of leguminous seeds, wheat and potatoes on serum cholesterol concentration in man. *J. Nutr.* **86**, 313–317.

Grande, F., Anderson, J. T., and Keys, A. (1972). Diets of different fatty acid composition producing identical serum cholesterol levels in man. *Amer. J. Clin. Nutr.* **25**, 53–60.

Grant, W. C., and Fahrenbach, M. J. (1959). Effect of dietary sucrose and glucose on plasma cholesterol in chicks and rabbits. *Proc. Soc. Exp. Biol. Med.* **100**, 250–252.

Groen, J. J., Balogh, M., Yaron, E., and Cohen, A. M. (1966). Effect of interchanging bread and sucrose as main source of carbohydrate in a low fat diet on the serum cholesterol levels of healthy volunteer subjects. *Amer. J. Clin. Nutr.* **19**, 46–58.

Guggenheim, K., Ilan, J., and Peretz, E. (1960). Effect of dietary carbohydrates and aureomycin on serum and liver cholesterol in rats. *J. Nutr.* **72**, 93–98.

Hardinge, M. G., Chambers, A. C., Crooks, H., and Stare, F. J. (1958). Nutritional studies of vegetarians. III. Dietary level of fiber. *Amer. J. Clin. Nutr.* **6**, 523–525.

Hatch, F. T., Abell, L. L., and Kendall, F. E. (1955). Effect of restriction of dietary fat and cholesterol upon serum lipids and lipoproteins in patients with hypertension. *Amer. J. Med.* **19**, 48–60.

Hegsted, D. M., McGandy, R. B., Myers, M. L., and Stare, F. J. (1965). Quantitative effects of dietary fat on serum cholesterol in man. *Amer. J. Clin. Nutr.* **17**, 281–295.

Higgins, H. L. (1916). The rapidity with which alcohol and some sugars may serve as nutriment. *Amer. J. Physiol.* **41**, 258–265.

Higginson, J., and Pepler, W. J. (1954). Fat intake, serum cholesterol concentration and atherosclerosis in South African Bantus. II. Atherosclerosis and coronary artery disease. *J. Clin. Invest.* **33**, 1366–1371.

Hill, P. (1970). Effect of fructose on rat liver lipids. *Lipids* **5**, 621–627.

Hodges, R. E., and Krehl, W. A. (1965). The role of carbohydrates in lipid metabolism. *Amer. J. Clin. Nutr.* **17**, 334–346.

Hodges, R. E., Krehl, W. A., Stone, D. B., and Lopez, A. (1967). Dietary carbohydrates and low cholesterol diet: Effects on serum lipids in man. *Amer. J. Clin. Nutr.* **20**, 198–208.

Howell, R. W., and Wilson, D. G. (1969). Dietary sugar and ischemic heart disease. *Brit. Med. J.* **3**, 145–148.

Hyun, S. A., Vahouny, G. V., and Treadwell, C. R. (1963). Effect of hypocholesterolemic agents on intestinal absorption. *Proc. Soc. Exp. Biol. Med.* **112**, 496–501.

Irwin, M. I., Taylor, D. D., and Feeley, R. M. (1964). Serum lipid levels, fat, nitrogen, and mineral metabolism of young men associated with kind of dietary carbohydrate. *J. Nutr.* **82**, 338–342.

Kaufmann, N. A., Poznanski, R., Blondheim, S. H., and Stein, Y. (1966). Changes in serum lipid levels of hyperlipemic patients following the feeding of starch, sucrose and glucose. *Amer. J. Clin. Nutr.* **18**, 261–269.

Kaufmann, N. A., Poznanski, R., Blondheim, S. H., and Stein, Y. (1967). Comparison of effects of fructose, sucrose glucose and starch on serum lipids in patients with hypertriglyceridemia and normal subjects. *Amer. J. Clin. Nutr.* **20**, 131–132.

Keen, H., and Rose, G. (1964). Dietary fat and dietary sugar. *Lancet* **2**, 362 (Letter to the editor).

Kempner, W. (1948). Treatment of hypertensive vascular disease with rice diet. *Amer. J. Med.* **4**, 545–577.

Keys, A. (1971). Sucrose in the diet and coronary heart disease. *Atherosclerosis* **14**, 193–202.

Keys, A., Anderson, J. T., and Grande, F. (1958). Effect on serum cholesterol in man of mono-ene fatty acid (oleic acid) in the diet. *Proc. Soc. Exp. Biol. Med.* **9**, 387–391.

Keys, A., Anderson, J. T., and Grande, F. (1960). Diet-type (fat constant) and blood lipids in man. *J. Nutr.* **70**, 257–266.

Keys, A., Grande, F., and Anderson, J. T. (1961). Fiber and pectin in the diet and serum cholesterol in man. *Proc. Soc. Exp. Biol. Med.* **106**, 555–558.

Keys, A., Anderson J. T., and Grande, F. (1965). Serum cholesterol response to changes in the diet. I. Iodine value versus 2S-P. *Metab., Clin. Exp.* **14**, 747–758.

Kiyasu, J. Y., and Chaikoff, I. L. (1957). On the manner of transport of absorbed fructose. *J. Biol. Chem.* **224**, 935–939.

Klugh, C. A., and Irwin, M. I. (1966). Serum levels of young women as related to source of dietary carbohydrate. *Fed. Proc., Fed. Amer. Soc. Exp. Biol.* **25**, 672 (abstr.)

Kritchevski, D. (1964). Experimental atherosclerosis. *In* "Lipid Pharmacology" (R. Paoletti, ed.), pp. 63–130. Academic Press, New York.

Kritchevski, D. (1970). Role of cholesterol vehicle in experimental atherosclerosis. *Amer. J. Clin. Nutr.* **23**, 1105–1110.

Kritchevski, D., and Tepper, S. A. (1968). Experimental atherosclerosis in rabbits fed cholesterol-free diets: Influence of chow components. *J. Atheroscler. Res.* **8**, 357–369.

Kritchevski, D., and Tepper, S. A. (1969). Influence of dietary carbohydrate on lipid metabolism in rats. *Med. Exp.* **19**, 329–341.

Kritchevski, D., Kolman, R. R., Guttmacher, R. M., and Forbes, M. (1959). Influence of dietary carbohydrate and protein on serum and liver cholesterol in germ-free chicken. *Arch. Biochem. Biophys.* **85**, 444–451.

Kritchevski, D., Sallata, P., and Tepper, S. A. (1968). Experimental atherosclerosis in rabbits fed cholesterol-free diets. Part 2. Influence of various carbohydrates. *J. Atheroscler. Res.* **8**, 697–703.

Kuo, P. T. (1965). Dietary sugar in production of hyperglyceridemia in patients with hyperlipemia and atherosclerosis. *Trans. Ass. Amer. Physicians* **78**, 97–116.

Kuo, P. T., and Bassett, D. R. (1965). Dietary sugar in the production of hyperglyceridemia. *Ann. Intern. Med.* **62**, 1199–1212.

Kuo, P. T., and Carson, J. C. (1959). Dietary fats and diurnal serum triglyceride levels in man. *J. Clin. Invest.* **38**, 1384–1393.

Kuo, P. T., Feng, L., Cohen, N. N., Fitts, W. T., and Miller, L. D. (1967). Dietary carbohydrates in hyperlipemia (hyperglyceridemia), hepatic and adipose tissue lipogenic activities. *Amer. J. Clin. Nutr.* **20**, 116–125.

Lambert, G. F., Miller, J. P., Olsen, R. T., and Frost, D. V. (1958). Hypercholesterolemia and atherosclerosis in rabbits by purified high fat rations devoid of cholesterol. *Proc. Soc. Exp. Biol. Med.* **97**, 544–549.

Lang, C. M., and Barthel, C. H. (1972). Effects of simple and complex carbohydrates on serum lipids and atherosclerosis in non-human primates. *Amer. J. Clin. Nutr.* **25**, 470–475.

Lazzarini-Robertson, A., Butkus, A., Ehrhart, L. A., and Lewis, L. A. (1972). Experi-

434 *Francisco Grande*

mental arteriosclerosis in dogs. Evaluation of anatomopathological findings. *Atherosclerosis* **15**, 307–325.

Lees, R. S. (1965). Plasma lipid response to two types of dietary carbohydrate. *Clin. Res.* **13**, 549 (abstr.).

Leveille, G. A., and Sauberlich, H. E. (1966). Mechanism of the cholesterol depressing effect of pectin in the cholesterol-fed rat. *J. Nutr.* **88**, 209–214.

Little, J. A., Shanoff, H. M., Csima, A., Redmond, S. E., and Yano, R. (1965). Diet and serum lipids in male survivors of myocardial infarction. *Lancet* **1**, 933–935.

Little, J. A., Shanoff, H. M., and Csima, A. (1967). Dietary carbohydrate and fat, serum lipoprotein and human atherosclerosis. *Amer. J. Clin. Nutr.* **20**, 133–138.

Little, J. A., Birchwood, B. L., Simmons, D. A., Antar, M. A., Kallos, A., Buckley, G. C., and Csima, A. (1970). Interrelationship between the kinds of dietary carbohydrate and fat in hyperlipoproteinemic patients. 1. Sucrose and starch with polyunsaturated fat. *Atherosclerosis* **11**, 173–181.

Lopez, A., Hodges, R. E., and Krehl, W. A. (1966). Some interesting relationships between dietary carbohydrates and serum cholesterol. *Amer. J. Clin. Nutr.* **18**, 149–153.

Lowenstein, F. W. (1964). Epidemiological investigations in relation to diet in groups who show little atherosclerosis and are almost free of coronary ischemic heart disease. *Amer. J. Clin. Nutr.* **15**, 175–186.

Macdonald, I. (1965a). The lipid response of young women to dietary carbohydrates. *Amer. J. Clin. Nutr.* **16**, 458–463.

Macdonald, I. (1965b). The effects of various dietary carbohydrates on the serum lipids during a five day regimen. *Clin. Sci.* **29**, 193–197.

Macdonald, I. (1966a). The lipid response of post-menopausal women to dietary carbohydrates. *Amer. J. Clin. Nutr.* **18**, 86–90.

Macdonald, I. (1966b). Influences of fructose and glucose on serum lipid levels in men and pre- and post-menopausal women. *Amer. J. Clin. Nutr.* **18**, 369–372.

Macdonald, I. (1966c). Lipid responses to dietary carbohydrates. *Advan. Lipid Res.* **4**, 39–67.

Macdonald, I., and Braithwaite, D. M. (1964). The influence of dietary carbohydrates on the lipid pattern in serum and in adipose tissue. *Clin. Sci.* **27**, 27–30.

Macdonald, I., and Nowakowska, A. (1964). Sex differences in the lipid response to dietary carbohydrate. *Proc. Nutr. Soc.* **23**, 33A–34A (abstr.).

Macdonald, I., and Roberts, J. B. (1965). Incorporation of various C¹⁴ dietary carbohydrates into serum and liver lipids. *Metab., Clin. Exp.* **14**, 991–999.

McGandy, R. B., Hegsted, D. M., Myers, M. L., and Stare, F. J. (1966). Dietary carbohydrates and serum cholesterol levels in man. *Amer. J. Clin. Nutr.* **18**, 237–242.

McGandy, R. B., Hegsted, D. M., and Stare, F. J. (1967). Dietary fats, carbohydrates and atherosclerotic vascular disease. *N. Engl. J. Med.* **277**, 186–192 and 241–247.

Malmros, H., and Wigand, G. (1959). Atherosclerosis and deficiency of essential fatty acids. *Lancet* **2**, 749–751.

Marr, J. W., and Heady, J. A. (1964). Levels of dietary sucrose in patients with occlusive atherosclerotic disease. *Lancet* **2**, 146 (Letter to the Editor).

Mathur, K. S., Kahn, M. A., and Sharma, R. D. (1968). Hypocholesterolemic effect of Bengal gram. A long-term study in man. *Brit. Med. J.* **1**, 30–31.

Naismith, D. J., and Khan, N. A. (1970a). Tissue lipogenesis in rats fed different carbohydrates. *Proc. Nutr. Soc.* **29**, 63A–64A (abstr.).

Naismith, D. J., and Khan, N. A. (1970b). Differences in the throughput of triglycerides in the plasma of rats fed various carbohydrates. *Proc. Nutr. Soc.* **29,** 64A–65A (abstr.).

Nath, N., Harper, A. E., and Elvehjem, C. A. (1959). Diet and cholesteremia. IV. Effects of carbohydrate and nicotinic acid. *Proc. Soc. Exp. Biol. Med.* **102,** 571–574.

Nelson, D., Ivy, A. C., Altschul, R., and Wilheim, R. (1953). Effect of aureomycin on experimental atherosclerosis and serum cholesterol. *AMA Arch. Pathol.* **56,** 262–267.

Nestel, P. J. (1966). Carbohydrate-induced hypertriglyceridemia and glucose utilization of ischemic heart disease. *Metab., Clin. Exp.* **15,** 787–795.

Nestel, P., Carroll, K. F., and Havenstein, N. 1970. Plasma triglyceride response to carbohydrates, fats and calorie intake. *Metab., Clin. Sci.* **19,** 1–18.

Nikkilä, E. A. (1969). Control of plasma and liver triglyceride kinetics by carbohydrate metabolism and insulin. *Advan. Lipid Res.* **7,** 63–134.

Nikkilä, E. A., and Ojala, K. (1965). Induction of hyperglyceridemia by fructose in the rat. *Life Sci.* **4,** 937–943.

Nikkilä, E. A., and Ojala, K. (1966). Acute effects of fructose and glucose on the concentration and removal rate of plasma triglycerides. *Life Sci.* **5,** 89–94.

Palmer, G. H., and Dixon, D. G. (1966). Effect of pectin dose on serum cholesterol levels. *Amer. J. Clin. Nutr.* **18,** 437–442.

Papp, O. A., Padilla, L., and Johnson, A. L. (1965). Dietary intake in patients with and without myocardial infarction. *Lancet* **2,** 259–261.

Paul, O., MacMillan, A., McKean, H., and Park, H. (1968). Sucrose intake and coronary heart disease. *Lancet* **2,** 1049–1051.

Pleshkov, A. M. (1964). Effect of long-term intake of easily absorbed carbohydrates (sugars) on blood lipid levels in patients with atherosclerosis. *Fed. Proc., Fed. Amer. Soc. Exp. Biol.* **23,** Part II, T334–336.

Porte, D., Bierman, E. L., and Bagdade, J. D. (1966). Substitution of dietary starch for dextrose in hyperlipemic subjects. *Proc. Soc. Exp. Biol. Med.* **123,** 814–816.

Portman, O. W., and Stare, F. J. (1959). Dietary regulation of serum cholesterol levels. *Physiol. Rev.* **39,** 407–442.

Portman, O. W., Lawry, E. Y., and Bruno, D. (1956). Effect of dietary carbohydrate on experimentally induced hypercholesteremia and hyperbetalipoproteinemia in rats. *Proc. Soc. Exp. Biol. Med.* **91,** 321–323.

Portman, O. W., Andrus, S. B., Pollard, D., and Bruno, D. (1961). Effects of long-term feeding of fat-free diets to Cebus monkeys. *J. Nutr.* **74,** 429–440.

Prather, E. S. (1964). Effect of cellulose on serum lipids in young women. *J. Amer. Dietet. Ass.* **45,** 230–233.

Reaven, G. M., Lerner, R. L., Stern, M. P., and Farquhar, J. W. (1967). Role of insulin in endogenous hypertriglyceridemia. *J. Clin. Invest.* **46,** 1756–1767.

Schultz, A. L., and Grande, F. (1968). Effects of starch and sucrose on the serum lipids of dogs before and after thyroidectomy. *J. Nutr.* **94,** 71–73.

Schwartz, W. B., and Merlis, J. K. (1948). Nitrogen balance studies on the Kempner rice diet. *J. Clin. Invest.* **27,** 406–411.

Scott, R. F., Likimani, J. C., Morrison, E. S., Thuku, J. J., and Thomas, W. A. (1963). Esterified serum fatty acids in subjects eating high and low cholesterol diets. *Amer. J. Clin. Nutr.* **13,** 82–91.

Scott, R. F., Lee, K. T., Kim, D. N., Morrison, E. S., and Goodale, F. (1964). Fatty acids of serum and adipose tissue in six groups eating natural diets containing 7 to 40 per cent fat. *Amer. J. Clin. Nutr.* **14,** 280–290.

436   *Francisco Grande*

Shamma'a, M., and Al-Khalidi, U. (1963). Dietary carbohydrates and serum cholesterol in man. *Amer. J. Clin. Nutr.* **13,** 194–196.

Stamler, J. (1966). Metabolism and atherosclerosis, a review of data and theories and discussion of controversial questions. *In* "Controversy in Internal Medicine" (F. J. Ingelfinger, A. S. Relman, and M. Finland, eds.), p. 27. Saunders, Philadelphia, Pennsylvania.

Stare, F. J. (1967). Dietary fats and carbohydrates, blood lipids and coronary heart disease. *Amer. J. Clin. Nutr.* **20,** 149–151.

Starke, H. (1950). Effect of the rice diet on the serum cholesterol fraction of 154 patients with hypertensive vascular disease. *Amer. J. Med.* **9,** 494–499.

Staub, H. W., and Theissen, R. (1968). Dietary carbohydrates and serum cholesterol in rats. *J. Nutr.* **95,** 633–638.

Steiner, A., and Dayton, S. (1956). Production of hyperlipemia and early atherosclerosis in rabbits by high vegetable fat diet. *Circ. Res.* **4,** 62–66.

Steiner, A., Varsos, A., and Samuel, P. (1959). Effect of saturated and unsaturated fats on the concentration of serum cholesterol and experimental atherosclerosis. *Circ. Res.* **7,** 448–453.

Szanto, S., and Yudkin, J. (1969). The effect of dietary sucrose on blood lipids, serum insulin and body weight in human volunteers. *Postgrad. Med. J.* **45,** 602–607.

Taylor, D. D., Conway, E. S., Schuster, E. M., and Adams, M. (1967). Influence of dietary carbohydrates on liver content and on serum lipids in relation to age and strain of rat. *J. Nutr.* **91,** 275–282.

Tennent, D. M., Siegel, H., Gunther, W. K., Ott, W. H., and Mushett, C. W. (1957). Lipid patterns and atherogenesis in cholesterol fed chickens. *Proc. Soc. Exp. Biol. Med.* **96,** 679–683.

Trowell, H. (1972). Ischemic heart disease and dietary fiber. *Amer. J. Clin. Nutr.* **25,** 926–932.

Tzagournis, M., Seidenstricker, J. F., and Hamwi, G. J. (1967). Serum insulin, carbohydrate and lipid abnormalities in patients with premature heart disease. *Ann. Intern. Med.* **67,** 42–47.

Vigayagopalan, P., and Kurup, P. A. (1972). Hypolipidemic activity of whole paddy in rats fed a high-fat-high-cholesterol diet. Isolation of an active fraction from the husk and bran. *Atherosclerosis* **15,** 215–222.

Vles, R. O., and Kloeze, J. (1967). Effect of feeding alternately maize oil and coconut oil on atherosclerosis in rabbits. *J. Atheroscler. Res.* **7,** 59–68.

Vles, R. O., Buller, J., Gottenlos, J. J., and Thomasson, H. J. (1964). Influence of type of dietary fat on cholesterol-induced atherosclerosis in the rabbit. *J. Atheroscler. Res.* **4,** 170–183.

Walker, A. R. P. (1961). Crude fiber, bowel motility and pattern of diet. *S. Afr. Med. J.* **35,** 114–115.

Walker, A. R. P. (1971). Sugar intake and coronary heart disease. *Atherosclerosis* **14,** 137–152.

Walker, A. R. P., and Arvidson, U. B. (1954). Fat intake, serum cholesterol concentration and atherosclerosis in the South African Bantu. I. Low fat intake and the age trend of serum cholesterol concentration in the South African Bantu. *J. Clin. Invest.* **33,** 1358–1365.

Watkin, D. M., Froeb, H. F., Hatch, F. T., and Gutman, A. B. (1950). Effects of diet in essential hypertension. II. Results with unmodified Kempner rice diet in fifty hospitalized patients. *Amer. J. Med.* **9,** 441–493.

Wells, A. F., and Ershoff, B. H. (1961). Beneficial effects of pectin in prevention of hypercholesterolemia and increase of liver cholesterol in cholesterol-fed rats. *J. Nutr.* **74**, 87–92.

Wells, W. W., and Anderson, S. C. (1959). The increased severity of atherosclerosis in rabbits on a lactose-containing diet. *J. Nutr.* **68**, 541–549.

Wells, W. W., and Anderson, S. C. (1962). The effect of dietary lactose on the serum cholesterol levels of human subjects. *Fed. Proc., Fed. Amer. Soc. Exp. Biol.* **21**, 100 (abstr.).

Wells, W. W., Quan-Ma, R., Cook, C. R., and Anderson, S. C. (1962). Lactose diets and cholesterol metabolism. II. Effect of dietary cholesterol, succinylsulfathiazole and mode of feeding on atherogenesis in the rabbit. *J. Nutr.* **76**, 41–47.

Wigand, G. (1959). Production of hypercholesterolemia and atherosclerosis in rabbits by feeding different fats without supplementary cholesterol. *Acta Med. Scand.* **166**, Suppl. 351.

Winitz, M., Graff, J., and Seedman, D. A. (1964). Effect of dietary carbohydrate on serum cholesterol levels. *Arch. Biochem. Biophys.* **108**, 576–579.

Working Party Medical Research Council. (1970). Dietary sugar intake in men with myocardial infarction. *Lancet* **2**, 1265–1271.

Yudkin, J. (1957). Diet and coronary thrombosis: Hypothesis and fact. *Lancet* **2**, 155–162.

Yudkin, J. (1963). Nutrition and palatability with special reference to obesity, myocardial infarction and other diseases of civilization. *Lancet* **1**, 1335–1338.

Yudkin, J. (1964). Dietary fat and dietary sugar in relation to ischemic heart disease and diabetes. *Lancet* **2**, 4–5.

Yudkin, J. (1966). Dietetic aspects of atherosclerosis. *Angiology* **17**, 127–133.

Yudkin, J. (1967). Evolutionary and historical changes in dietary carbohydrates. *Amer. J. Clin. Nutr.* **20**, 108–115.

Yudkin, J. (1969). Dietary sugar and coronary heart disease. *Nutr. News* **32**, 9–12.

Yudkin, J. (1971). Sugar consumption and myocardial infarction. *Lancet* **1**, 296–297.

Yudkin, J., and Morland, J. (1967). Sugar intake and myocardial infarction. *Amer. J. Clin. Nutr.* **20**, 503–506.

Yudkin, J., and Roddy, J. (1964). Levels of dietary sucrose in patients with occlusive atherosclerotic disease. *Lancet* **2**, 6–8.

Zakim, D., Pardini, R. S., Herman, R. H., and Sauberlich, H. E. (1967). Mechanism for the differential effects of high carbohydrate diets on lipogenesis in rat liver. *Biochim. Biophys. Acta* **144**, 242–251.

CHAPTER 26

Influence of Dietary Fructose and Sucrose on Serum Triglycerides in Hypertriglyceridemia and Diabetes

ESKO A. NIKKILÄ

High dietary intake of fructose or sucrose is accompanied by a marked elevation of serum triglyceride level in rats (Nikkilä and Ojala, 1965) and in man (Macdonald and Braithwaite, 1964; Kuo and Bassett, 1965, Macdonald, 1966; Kaufmann et al., 1966a). Both of these sugars also enhance the alimentary hyperlipemia (Nikkilä and Pelkonen, 1966; Mann et al., 1971) in contrast to glucose which reduces the serum triglyceride response to a fatty meal. The mechanisms of these effects are not clear, but a number of possible explanations have been suggested (Nikkilä, 1969). Thus, a part of the fructose molecule itself is directly converted

439

to hepatic and plasma lipids since, in contrast to glucose, it is mainly taken up by the liver and metabolized through triose phosphates which are precursors of glycerol 3-phosphate. The increase of the latter metabolite occurs in the liver after fructose (Wieland and Matchinsky, 1962; Bässler and Stein, 1967; Woods *et al.*, 1970), and this might stimulate the esterification of fatty acids and hepatic lipogenesis. The fatty acid synthesis in the liver is in fact increased by high fructose and high sucrose diets as compared to starch (Hill *et al.*, 1954; Bar-On and Stein, 1968; Bruckdorfer *et al.*, 1972). Fructose added to rat liver perfusate increases the secretion of VLDL-triglyceride into the medium (Topping and Mayes, 1970).

The hyperglyceridemic effect of fructose and sucrose in man has been demonstrated by using unphysiologically large quantities of these sugars. In these experiments the daily sugar intake has varied from 200 to 500 gm corresponding to 40–100% of total calorie consumption (Macdonald, 1966; Kaufmann *et al.*, 1966b; Kuo and Bassett, 1965; Antar *et al.*, 1970; Barter *et al.*, 1971). Results with more conventional intakes of sugars have been controversial. Dunnigan *et al.* (1970) did not find any change in serum lipid levels on exchanging starch and sucrose in normoglyceridemic subjects, whereas Little *et al.* (1970) reported an increase of plasma triglyceride level when 20% of calories was given as sucrose substituting for starch. Also, Mann *et al.* (1970) found significantly lower serum triglyceride levels in human volunteers on low sucrose diet as compared to moderate sucrose intake (85 gm/day). However, in this study the subjects lost weight during the low sucrose period, and this weakens the argument for a direct effect of sucrose.

Evidently then, knowledge of the possible hyperlipidemic effect of moderate dietary intake of simple sugars is still insufficient. It is also not known whether fructose is comparable to sucrose in this respect or whether it might show some advantages over the latter. This difference could be expected from the fact that insulin response to fructose is much less than it is to sucrose. Hyperinsulinism is believed to be an important factor contributing to development of hypertriglyceridemia.

Since fructose is now generally available in many countries it is of considerable clinical interest to know whether it can be used by patients whose sucrose intake is restricted, i.e., those with diabetes and hypertriglyceridemia. To answer this question a comparison was made of fructose- or sucrose-containing and sugar-free diets in a number of patients belonging to either of these categories or both. The study was made under strictly controlled ward conditions and, therefore, the results are not necessarily valid under ambulatory everyday life.

A. Primary Endogenous Hypertriglyceridemia without Diabetes

Five patients with serum triglyceride level above 200 mg/100 ml and with normal fasting blood glucose were studied during three successive dietary periods the length of which varied from 10 to 20 days. The first period was preceded by 1–2 weeks' hospitalization during which the amount of diet was adjusted to keep the weight constant, and the serum lipid levels reached a steady state. Under the three experimental periods the diet was isocaloric and contained 45% of calories as carbohydrate, 35% as fat (mainly saturated), and 20% as protein. The only constituent which was varied was the type of carbohydrate. During "starch" period a minimum of 80% of carbohydrate calories was derived from starch the rest coming from approximately 25 gm of lactose and 20 gm of sucrose. During "fructose" period 75–90 gm of fructose were added and an equivalent amount of starch was withdrawn. The fructose was not incorporated into the food but was added as a powder on deserts or in coffee and tea. During "sucrose" diet starch was substituted for 75–90 gm of sucrose. With the "endogenous" sucrose contained in the basic food the total sucrose consumption during this period was approximately 100 gm/day corresponding to 20–25% of total calorie supply. The sequence of the three periods varied in a random fashion.

During the study, weight was controlled every day and serum lipids were measured every second day. At the end of each dietary period the plasma triglyceride turnover was determined by [^{3}H]glycerol labeling technique (Farquhar *et al.*, 1965). Efforts were made to keep the weight constant by adjusting the calories on any systematic trend in weight. Even though this was not always successful the average weight changes remained small.

In 3 of the 5 subjects studied the average serum triglyceride level was higher during fructose diet as compared to starch. However, the mean value of the whole group during fructose period was not significantly different from the level measured during sugar-free diet. On the other hand, sucrose diet caused in all subjects an increase of serum triglyceride over the values recorded both during starch and fructose diets. The turnover study indicated that sucrose increased the production rate of plasma triglycerides (TG) without influencing the removal efficiency (Table I). The serum cholesterol concentration was not significantly influenced by the quality of carbohydrate,

This study supports the conclusion that sucrose increases serum triglyc-

TABLE I

Mean Plasma Triglyceride Concentration and Turnover Rate[a]

Diet	TG concentration[b] (mg/100 ml)	TG fractional turnover (h⁻¹)	TG turnover rate[c] (mg/hr/kg)
Starch	220 ± 45	0.180 ± 0.041	14.1 ± 2.1
Fructose	229 ± 36	0.172 ± 0.020	14.4 ± 2.3
Sucrose	270 ± 95[d]	0.179 ± 0.044	16.8 ± 4.5

[a] Five subjects with primary endogenous hypertriglyceridemia (Type IIB or IV) during three different diets.
[b] Each patient had 5–10 measurements during each period.
[c] Measured only at the end of each dietary period.
[d] $p < 0.01$. The figures are mean ± SD.

eride in subjects with primary hypertriglyceridemia even when the intake does not exceed the average level of consumption recorded in most countries (Walker, 1971). There are good reasons to suppose that under free dietary conditions this effect is not less, but can well be more, pronounced than during the strictly controlled ward conditions of the present study. Fructose given in equivalent amounts seems to be somewhat less hyperglyceridemic than sucrose, but the evidence for this is so far only suggestive. Kaufmann *et al.* (1966b), who have made the only previous study where sucrose and fructose were compared with each other (both administered in huge doses), found somewhat higher serum triglyceride levels during fructose as compared to sucrose feeding.

It is known that the influence of sucrose (and possibly also of fructose) on serum triglycerides is dependent on other dietary factors. The effect is attenuated by a diet rich in polyunsaturated fats (Macdonald, 1967; Birchwood *et al.*, 1970). Premenopausal women are more resistant to the sucrose induction of hyperglyceridemia than men (Macdonald, 1965) and estrogen treatment of male baboons reduces the rise of serum triglycerides induced by sucrose (Coltart and Macdonald, 1971). These effects might be related to the more efficient triglyceride removal mechanism which is present in young women as compared to men (Nikkilä and Kekki, 1971).

B. Untreated Adult-Onset Type of Diabetes with or without Hypertriglyceridemia

A comparative study of the effects of fructose and sucrose on serum triglyceride levels and on diabetic control was made in 6 untreated non-

ketotic diabetic subjects. The diets and other procedures were similar
to those described above, but triglyceride turnover study was omitted
and instead a daily glucose "curve" was recorded at least once during
each dietary period. Two patients refused to eat sucrose which indicates
the aversion of diabetics to sugar (educational or endogenous?).

The mean triglyceride levels for each patient are recorded in Table II.
It is seen that fructose increased the triglyceride significantly only in
2 subjects as compared to sugar-free (starch) diet, whereas in the 4 remaining patients the values were essentially similar during fructose and
starch periods. During sucrose 3 of the 4 subjects showed a rising trend
of serum triglyceride, but the difference from sugar-free period was rather
slight.

The fasting blood glucose values during each dietary period appear
in Table III. If compared with the sugar-free period the values were
higher during fructose diet in 4 of the 6 subjects and during sucrose diet
in only 1 of the 4 subjects. The difference is, however, unexpectedly small.
Similarly, the average daily blood glucose was increased relatively little
by feeding fructose or sucrose. In urinary glucose output a definite increase occurred during the fructose diet in 2 subjects.

It seems thus that the response of diabetics to moderate amounts of
fructose or sucrose in the diet is individually variable, some subjects showing a marked increase of both blood glucose and serum triglyceride levels
while other patients are remarkably resistant to the dietary sugars. This
is the situation in untreated stage, and it is to be expected that the influence is diminished when the diabetes is brought under adequate control.

TABLE II

Effect of Dietary Fructose and Sucrose on Plasma
Triglyceride in Untreated Adult-Onset Type
Diabetes[a]

Case	Starch	Fructose[b]	Sucrose[b]
K.H.	2.44	2.94	2.35
T.V.	3.40	4.39	—
V.N.	1.34	1.19	1.65
Kl.H.	1.77	1.65	2.08
J.G.	1.38	1.44	1.59
V.P.	1.67	2.03	—
Mean	2.00	2.27	—

[a] Mean of 5–8 determinations in each period, mM.
[b] Daily dose 80–100 gm substituting for starch.

TABLE III

Effect of Dietary Fructose and Sucrose on Blood Glucose in
Untreated Adult-Onset Type Diabetes[a]

Case	Starch	Fructose[b]	Sucrose[b]
K.H.	14.1 ± 1.1	13.6 ± 1.0	14.2 ± 1.3
T.V.	10.7 ± 0.8	12.0 ± 0.7	—
V.N.	6.9 ± 0.4	6.2 ± 0.3	5.7 ± 0.4
Kl.H.	14.4 ± 1.5	14.7 ± 0.7	14.1 ± 1.1
J.G.	13.4 ± 0.7	14.5 ± 0.6	15.2 ± 0.7
V.P.	9.5 ± 0.4	11.8 ± 0.9	—
Mean	11.3	12.1	12.3

[a] Mean ±SD of 6–10 values for each period, mM.
[b] Daily dose 80–100 gm substituting for starch.

It should be emphasized, again, that these results need not be valid under
ambulatory conditions.

C. Insulin-Treated Adult Diabetics

The literature on the effects of dietary fructose on diabetic control in
juvenile diabetes is extensive and it cannot be reviewed here. However,
only a few well-controlled studies on the subject have been carried out
(Hiller, 1955; Moorhouse and Kark, 1957), and these have shown that
dietary fructose might influence the diabetes to a favorable direction and
not have any adverse effects. A metabolic ward study was carried out
in 10 insulin-dependent diabetics, whose disease was relatively stable
with a constant maintenance dose of insulin but not always strictly controlled (Pelkonen *et al.*, 1972). These patients went through three different dietary periods, 10 days each. The first and third periods were similar
containing starch as the major carbohydrate. During the middle period 75
gm of fructose were added to the diet and a similar amount of starch
was reduced. The results are given in Tables IV and V which show that
during the fructose period both blood glucose and serum triglyceride were
slightly higher than during both control periods, but the differences were
not statistically significant.

On the basis of this and previous evidence it thus seems justified to
conclude that insulin-dependent diabetics can use a moderate amount
of fructose in their diet without any short-term deleterious effects on the
metabolic parameters of their disease. Similar or possibly even more favorable results have been obtained in a parallel study carried out at the

TABLE IV

Urinary Output of Glucose, Fasting Blood Glucose, and Body Weight[a]

	Starch I	Fructose	Starch II
Urinary glucose (gm/day)	17.3 ± 5.2	15.7 ± 4.5	12.9 ± 3.2
Fasting blood glucose (mmole/liter)	7.9 ± 0.6	8.6 ± 0.8	7.7 ± 0.7
Body weight (kg)	60.6 ± 2.5	59.8 ± 2.5	59.5 ± 2.4

[a] In 10 insulin-dependent patients during three successive dietary periods (mean ±S.E.).

Childrens Hospital of Helsinki (Åkerblom *et al.*, 1972). In this experiment diabetic children were given 1 gm of fructose per kilogram of body weight per day under home conditions and there was no evidence of impairment of their diabetic control.

D. *Influence of Fructose on Experimental Diabetes*

Streptozotocin produces a transient or permanent insulin deficiency diabetes, the severity of which is related to the dose of the drug. Using this experimental model, it seemed of interest to study the influence of fructose feeding on blood glucose and serum triglyceride. Diabetes of moderate degree was induced in rats with streptozotocin dose of 40 mg/kg and the animals were given free access to 10% fructose solution. After 2 weeks the fructose-fed and control diabetic rats did not differ from each other in regard to blood glucose or serum triglyceride values (Fig. 1). Only after 6 weeks had the animals on fructose developed a significantly higher

TABLE V

Fasting Levels of Plasma Cholesterol, Triglyceride, and Plasma FFA in Insulin-Dependent Diabetes[a]

	Starch I	Fructose	Starch II
Cholesterol (mmole/liter)	4.99 ± 0.12	5.10 ± 0.11	5.26 ± 0.15
Triglyceride (mmole/liter)	1.00 ± 0.06	1.22 ± 0.08	1.05 ± 0.14
FFA (mmole/liter)	0.61 ± 0.03	0.71 ± 0.04^{b}	0.60 ± 0.03

[a] Plasma cholesterol and triglyceride (mean ±S.E.) in 8 and plasma FFA (mean ±S.E.) in 10 insulin-dependent diabetic patients during three successive dietary periods.
[b] $p < 0.05$.

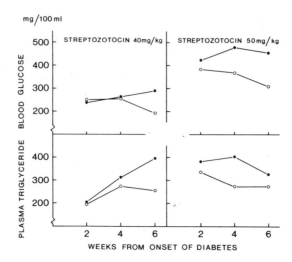

Fig. 1 Effect of fructose on serum triglyceride in streptozotocin-diabetic rats. Filled circles: rats receiving fructose. Open circles: rats fed with chow only.

hyperglycemia than the diabetic controls, and simultaneously the triglyceride level had also markedly increased. With 50 mg/kg of streptozotocin a severe diabetes was induced, and in these animals fructose feeding clearly aggravated the diabetic state. The triglyceridemia closely paralleled the changes of blood glucose, but fructose did not induce more hypertriglyceridemia in severely diabetic rats than in nondiabetic control animals (Fig. 1).

E. Conclusions

In daily amounts, which are close to average intake in different populations, sucrose increases serum triglyceride levels in subjects with primary hypertriglyceridemia. Fructose seems to be less active in this respect. Diabetics with either uncontrolled disease or treated with insulin can consume 75 gm of fructose daily without any evidence of adverse effects on metabolic signs of their disease.

References

Åkerblom, H. K., Siltanen, I., and Kallio, A.-K. (1972). Does dietary fructose affect the control of diabetes in children? *Symp. Clin. Metab. Aspects Fructose, Acta Med. Scand., Suppl. 542,* pp. 195–202.

Antar, M. A., Little, J. A., Lucas, C., Buckley, G. C., and Csima, A. (1970). Interrelationships between the kinds of dietary carbohydrate and fat in hyperlipoproteinemic patients. Part 3. Synergistic effect of sucrose and animal fat on serum lipids. *Atherosclerosis* **11,** 191–202.

Bar-On, H., and Stein, Y. (1968). Effect of glucose and fructose administration on lipid metabolism in the rat. *J. Nutr.* **94**, 95–105.

Barter, P. J., Carroll, K. F., and Nestel, P. J. (1971). Diurnal fluctuations in triglyceride, free fatty acids, and insulin during sucrose consumption and insulin infusion in man. *J. Clin. Invest.* **50**, 583–591.

Bässler, K. H., and Stein, G. (1967). Biochemische Grundlagen für Wirkungsunterschiede zwischen Sorbit und Fructose. *Hoppe-Seyler's Z. Physiol. Chem.* **348**, 533–539.

Birchwood, B. L., Little, J. A., Antar, M. A., Lucas, C., Buckley, G. C., Csima, A., and Kallos, A. (1970). Interrelationship between the kinds of dietary carbohydrate and fat in hyperlipoproteinemic patients. Part 2. Sucrose and starch with mixed saturated and polyunsaturated fats. *Atherosclerosis* **11**, 183–190.

Bruckdorfer, K. R., Khan, I. H., and Yudkin, J. (1972). Fatty acid synthetase activity in the liver and adipose tissue of rats fed with various carbohydrates. *Biochem. J.* **129**, 439–446.

Coltart, T. M., and Macdonald, I. (1971). Effect of sex hormones on fasting serum triglycerides in baboons given high sucrose diets. *Brit. J. Nutr.* **25**, 323–331.

Dunnigan, M. G., Fyfe, T., McKiddie, M. T., and Crosbie, S. M. (1970). The effects of isocaloric exchange of dietary starch and sucrose on glucose tolerance, plasma insulin, and serum lipids in man. *Clin. Sci.* **38**, 1–9.

Farquhar, J. W., Gross, R. C., Wagner, R. M., and Reaven, G. M. (1965). Validation of an incompletely coupled two-compartment non-recycling catenary model for turnover of liver and plasma triglyceride in man. *J. Lipid Res.* **6**, 119–134.

Hill, R., Baker, N., and Chaikoff, I. L. (1954). Altered metabolic patterns induced in the normal rat by feeding an adequate diet containing fructose as sole carbohydrate. *J. Biol. Chem.* **209**, 705–716.

Hiller, J. (1955). Die Laevuloseverwertung des acidotischen Diabetes mellitus. I. Mitteilung zur Analyse der Acetonurie und der Glucosurie. *Z. Klin. Med.* **153**, 388–396.

Kaufmann, N. A., Poznanski, R., Blondheim, S. H., and Stein, Y. (1966a). Changes in serum lipid levels of hyperlipemic patients following the feeding of starch, sucrose and glucose. *Amer. J. Clin. Nutr.* **18**, 261–269.

Kaufmann, N. A., Poznanski, R., Blondheim, S. H., and Stein, Y. (1966b). Effect of fructose, glucose, sucrose and starch on serum lipids in carbohydrate induced hypertriglyceridemia and in normal subjects. *Isr. J. Med. Sci.* **2**, 715–726.

Kuo, P. T., and Bassett, D. R. (1965). Dietary sugar in the production of hyperglyceridemia. *Ann. Intern. Med.* **62**, 1199–1212.

Little, J. A., Birchwood, B. L., Simmons, D. A., Antar, M. A., Kallio, A. K., Buckley, G. C., and Csima, A. (1970). Interrelationship between the kinds of dietary carbohydrate and fat in hyperlipoproteinemic patients. Part 1. Sucrose and starch with polyunsaturated fat. *Atherosclerosis* **11**, 173–182.

Macdonald, I. (1965). The lipid response of young women to dietary carbohydrates. *Amer. J. Clin. Nutr.* **16**, 458–463.

Macdonald, I. (1966). Influence of fructose and glucose on serum lipid levels in men and pre- and postmenopausal women. *Amer. J. Clin. Nutr.* **18**, 369–372.

Macdonald, I. (1967). Inter-relationship between the influences of dietary carbohydrates and fats on fasting serum lipids. *Amer. J. Clin. Nutr.* **20**, 345–351.

Macdonald, I., and Braithwaite, D. M. (1964). The influence of dietary carbohydrates on the lipid pattern in serum and in adipose tissue. *Clin. Sci.* **27**, 23–30.

Mann, J. I., Truswell, A. S., Hendricks, D. A., and Manning, F. (1970). Effects on serum-lipids in normal men of reducing dietary sucrose or starch for five months. *Lancet* **1**, 870–871.

Mann, J. I., Truswell, A. S., and Pimstone, B. L. (1971). The different effects of oral sucrose and glucose on alimentary lipaemia. *Clin. Sci.* **41**, 123–129.

Moorhouse, J. A., and Kark, R. M. (1957). Fructose and diabetes. *Amer. J. Med.* **23**, 46–58.

Nikkilä, E. A. (1969). Control of plasma and liver triglyceride kinetics by carbohydrate metabolism and insulin. *Advan. Lipid Res.* **7**, 63–134.

Nikkilä, E. A., and Kekki, M. (1971). Polymorphism of plasma triglyceride kinetics in normal human adult subjects. *Acta Med. Scand.* **190**, 49–59.

Nikkilä, E. A., and Ojala, K. (1965). Induction of hypertriglyceridemia by fructose in the rat. *Life Sci.* **4**, 937–943.

Nikkilä, E. A., and Pelkonen, R. (1966). Enhancement of alimentary hyperglyceridemia by fructose and glycerol in man. *Proc. Soc. Exp. Biol. Med.* **123**, 91–94.

Pelkonen, R., Aro, A., and Nikkilä, E. A. (1972). Metabolic effects of dietary fructose in insulin dependent diabetes of adults. *Symp. Clin. Metab. Aspects Fructose, Acta Med. Scand., Suppl. 542*, pp. 187–193.

Topping, D. L., and Mayes, P. A. (1970). Direct stimulation by insulin and fructose of very-low-density lipoprotein secretion by the perfused liver. *Biochem. J.* **119**, 48P.

Walker, A. R. P. (1971). Sugar intake and coronary heart disease. *Atherosclerosis* **14**, 137–152.

Wieland, O., and Matchinsky, F. (1962). Zur Natur den antiketogenen Wirkung von Glycerin und Fructose. *Life Sci.* **2**, 49–54.

Woods, H. F., Eggleston, L. V., and Krebs, H. A. (1970). The cause of hepatic accumulation of fructose 1-phosphate on fructose loading. *Biochem. J.* **119**, 501–510.

Disorders Related
to Sugar Metabolism:
Diabetes

CHAPTER 27

Pediatric Diabetes

MARVIN CORNBLATH

The traditional concepts of diabetes mellitus in infants and children require revision as the spectrum of pediatric diabetes expands. The multiplicity of presentations includes the impact on the fetus and the newborn of the diabetic mother, transient diabetes mellitus in the neonate, classic growth onset or "juvenile" diabetes, and maturity onset type in childhood. The clinical manifestations, diagnostic criteria, pathogenesis, management, and research needs are unique and specific for each. This presentation will be limited to transient diabetes that occurs in the neonate, maturity onset diabetes that has been recognized in children as young as 4 years of age, and classic insulin-deficient, growth onset juvenile diabetes that can occur at any age from birth through adolescence.

The incidence of diabetes mellitus in childhood is approximately one per two to three thousand children. Approximately 5% of all patients with diabetes mellitus are children less than 15 years of age. The incidence in males and females is equal (Paulsen and Colle, 1969; Kohrman and Weil, 1971; Renold et al., 1972).

A. Transient Diabetes Mellitus in the Neonate

Originally reported in 1852 by Kitselle (cited by Lawrence and Mc-Cance, 1931) this is perhaps the oldest type of diabetes mellitus known in the pediatric age group. This syndrome of probable multiple etiology also represents an example of one of the longest follow-ups reported in the literature. The first well-documented patient described in 1926 by Ramsey was reported to be a normal adult in 1953 (Arey, 1953). The sex distribution is equal. A positive family history for diabetes mellitus has been reported in approximately one-third of all patients. Recent interest in the secretion of insulin in the neonate has stimulated the reports of a number of these patients (Ferguson and Milner, 1970; Milner *et al.*, 1971; Schiff *et al.*, 1972; Pagliara *et al.*, 1973). The association of this syndrome with the newborn infant who is both underweight and undergrown for his period of gestation may provide new insights into the growth promoting activity of insulin on the fetus *in utero* (Cornblath and Schwartz, 1966; Gentz and Cornblath, 1969; Schiff *et al.*, 1972).

1. Clinical Manifestations

The age of onset can range from 4 days to 6 weeks with an average of 15 days (see Gentz and Cornblath, 1969). The clinical manifestations of this condition are characterized by weight loss, dehydration, and acidosis associated with polyuria and fever. Infants may lose as much as 30% of their body weight yet remain "open eyed and alert" (Hutchinson *et al.*, 1962). The occurrence of failure to thrive, sudden onset of dehydration and fever in spite of an adequate fluid intake and without vomiting or diarrhea are uniquely characteristic of neonatal diabetes mellitus.

2. Pathogenesis

Although originally considered to be the result of an associated infection or underlying intrauterine insult, the availability of the radioimmunoassay for insulin (Berson and Yalow, 1962) has indicated that the underlying difficulty may be an inappropriate insulin response to hyperglycemia (Gentz and Cornblath, 1969; Milner *et al.*, 1971; Schiff *et al.*, 1972). In the presence of markedly elevated blood glucose, the plasma insulin values remain normal or low. A study of siblings suggests that an inappropriate low insulin value precedes the onset of the hyperglycemia (Milner *et al.*, 1971). The insulin response to caffeine has been reported to be normal in one patient (Pagliara *et al.*, 1973) in contrast to the reduced response to glucose, arginine, and tolbutamide. The established and well-documented poor insulinogenic effect of glucose in both the term and premature neonate would suggest that transient diabetes reflects a further suppression of this normal mechanism. The transient nature of this syndrome is supported by the increase in insulin responsiveness to similar stimuli after recovery has occurred.

3. Management

Management of this condition involves restoring fluid and electrolyte losses and providing the amounts of insulin necessary to maintain metabolic homeostasis in the neonate. Fluid losses must be estimated and may approximate 10–30% of the body weight. These should be replaced within 12–24 hr in addition to the daily requirement of the neonate which is about 150 cm³/kg/24 hr. Replacement fluids should contain electrolytes in hypotonic concentrations as well as sources of free water. Insulin must be administered cautiously in quantities not to exceed 1–3 units/kg/day in divided doses. If the blood sugar is excessively high (>1200 mg/100 ml), a larger initial dose of 1–2 units/kg may be necessary. Maintaining blood glucose values between 150 and 200 mg/100 ml appears to provide optimal control in that glycosuria does not occur, and growth and development are normal, without the damaging consequences of iatrogenic hypoglycemia. Over-insulinization has been associated with profound hypoglycemic episodes (Gentz and Cornblath, 1969).

4. Prognosis

The number of follow-ups of these infants has not been adequate to predict their ultimate outcome. Some patients require no exogenous insulin while others may require therapy for several weeks or months. Although the majority have done well, approximately 10% of the infants have been mentally retarded. Whether this is the result of the hyperosmolar effects of the transient diabetes, of a preceding episode of hypoglycemia, or of iatrogenic hypoglycemia induced by excessive insulin dosage in an attempt to maintain normal blood glucoses remains unclear. Insufficiency of other endocrine functions has not been documented although investigations of growth hormone secretion, adrenal cortical activity, and thyroid function have been few to date. If appropriately treated, it would appear that the prognosis both for growth and development as well as for the abnormality in insulin secretion and metabolism may be excellent. Schiff *et al.* (1972) suggested that insulin therapy may have been important in inducing normal growth in one infant with transient diabetes who was underweight for his period of gestation.

B. Growth Onset—Juvenile Diabetes Mellitus

With peak ages of onset between 4 and 6 years and prepubertal (10–12 years), growth onset diabetes can occur in the neonate or in adult life.

Approximately 5% of all diabetics have the onset of their disease prior to 15 years of age (Danowski, 1957; White, 1965; Paulsen and Colle, 1969; Kohrman and Weil, 1971; Renold *et al.*, 1972).

1. Clinical Manifestations

Although previously reported to have "an acute onset often with rapid progression to diabetic coma" (White, 1960), it is noteworthy that this manner of presentation is no longer characteristic of the child with insulin-deficient diabetes. In a few, postprandial or late afternoon hypoglycemia (weakness, drowsiness, sweating, etc.) may be the earliest manifestation, followed in days, weeks, or months by the classic triad of polyuria, polydipsia, and polyphagia. Because of the diversity both in duration and in clinical manifestations in children, delays and errors in diagnosis may occur if diabetes mellitus is not suspected. Polyuria and polydipsia that occurs in 75% of the children (Danowski, 1957) have been confused with urinary tract infections. Nocturia, polyphagia, fatigue, and weight loss which occur in 40–50% of the children have been ascribed to emotional problems. Clinical manifestations of juvenile diabetes that are less frequent and often overlooked include abdominal pain, vulvitis, pruritis, recurring enuresis in a previously toilet trained child, visual defects, e.g., cataracts, and irritability. Once the disease is suspected, the diagnosis can be established easily by a routine urinalysis that reveals glycosuria and ketonuria and by a random blood sugar that is above 180 mg/100 ml.

2. Pathogenesis

At the time of overt clinical manifestation, the evidence would support that beta cell failure has occurred in the pancreas (Parker *et al.*, 1968). The exact etiology is unknown but the event may be precipitated by infection, a growth spurt, or other stresses. As a result of the absolute insulin deficiency which has been documented following a variety of insulin secreting stimuli such as glucose, tolbutamide, arginine, and glucagon (Parker *et al.*, 1968; Drash *et al.*, 1968; Cerasi and Luft, 1970), hyperglycemia ensues. In order to augment the peripheral utilization of glucose in the absence of insulin, cortisol, growth hormone, and glucagon levels (Unger, 1971) are elevated and gluconeogenesis is probably proceeding at a maximal rate. This results in a hyperglycemia exceeding the renal threshold and the subsequent glucosuria. Up to 40–50% of the caloric intake may be lost as urinary glucose. With each gram of glucose, there is an obligatory water loss of 10 ml plus associated losses in both extracellular and intracellular electrolytes, i.e., potassium, magnesium, and phosphate. The loss of water results in excessive thirst with the drinking of large quantities of water. In order to compensate for the loss of calories, about one-half of the patients will demonstrate polyphagia while the rest have a loss of appetite. In time, in both groups weight loss occurs. Continuing lipolysis and inability to maintain normal homeostasis even in the presence of hyperglycemia result in ketosis. With the development of ketosis, there are additional losses of cations since only about 25% of the ketone bodies can be excreted as free organic acids, about 25% neutralized by ammonia production in the kidney, and the remaining 50% result in urinary losses of sodium, potassium, magnesium, and even calcium. As this decompensation continues, carbonic acid accumulates with an increase production of CO_2 and water leading to hyperpnea and a reduction in pCO_2 as well as a progressive accumulation of hydrogen ion resulting in an uncompensated acidosis. Characteristic of ketoacidosis in children is vomiting and abdominal pain which may mimic an acute surgical abdomen, e.g., appendicitis. With vomiting and an inability to maintain fluid intake to compensate for the losses resulting from the polyuria and ketonuria, a rapid progression occurs to coma and even death if intervention with appropriate therapy does not occur.

The basic underlying etiology of the beta cell failure remains unknown. A number of hypotheses as summarized in Fig. 1 have been proposed (see review by Kohrman and Weil, 1971). These include exhaustion phenomena resulting from increased insulin secretion as a result of an indefinite period of pre-diabetes that may be characterized by peripheral resistance to insulin action or to the secretion of an inactive insulin

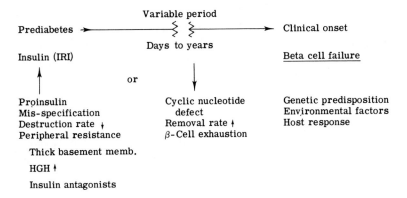

Fig. 1 A summary of various hypotheses of the pathogenesis of growth onset juvenile diabetes mellitus; modified from Kohrman and Weil (1971).

and/or proinsulin molecule (Melani *et al.*, 1970; Rubenstein and Steiner, 1970). The peripheral resistance to insulin action may also be the result of the early vascular changes characterized by the thickening of the basement membrane (Siperstein *et al.*, 1968; Siperstein, 1970) or increased secretion of growth hormone (Yde, 1969; Boden *et al.*, 1968; Hansen, 1970) or other insulin antagonists. None of these has been specifically identified or verified. However, beta cell failure may also represent a genetic predisposition or be the result of environmental factors such as refined sugars or trace elements either in excess or deficient amounts in the diet. Another hypothesis suggests that an immune response to the misspecified (O'Brien, 1970) or antigenically active physiologically inactive molecule results in insulin insufficiency. Again, the evidence is inconclusive and requires verification.

3. Management

a. Acute Episode of Diabetic Ketoacidosis

Basic principles in the treatment of the acute episode of diabetic ketoacidosis consist of the following guidelines: (1) the replacement of intracellular and extracellular water and electrolyte losses, (2) the reduction of ketone bodies, and (3) the correction of lowered pH. These are best accomplished by the provision of adequate amounts of insulin, electrolytes, and glucose and, in some instances, the treatment of the underlying infection which may have precipitated the episode of ketoacidosis. Rarely, if circulatory collapse has occurred, the circulation must be restored with a plasma expander prior to initiating therapy. Adequate amounts of insulin usually are 1–2 units/kg/dose at 4–6 hr intervals. Sufficient amounts of hypotonic fluid to compensate for fluid losses, which may be as great as 10% of the body weight, must be given to restore both extracellular and intracellular electrolyte losses. In view of the relatively small glucose reserve in the child even in the presence of hyperglycemia, glucose containing solutions as well as those with an excess of sodium compared to chloride are indicated. A number of regimens have been described in detail (Klein, 1971; Paulsen and Colle, 1969). It is important to emphasize that the treatment of children is different from that of adults and the early administration of glucose, the cautious use of insulin and the addition of potassium once urinary flow has been established are critical. The rapid correction of pH with sodium bicarbonate must be avoided to prevent precipitous falls in potassium levels, overshoot alkalosis and paradoxical intracerebral acidosis (Posner and Plum, 1967). Cerebral edema can result from sudden water shifts in the presence of excess sodium, falls in blood glucose, and can be fatal (Young and Bradley, 1967; Taubin and Matz, 1968).

b. Long-Term Management

The principal goal of therapy in each child is a normal life in every aspect: physical, psychological, and social. Criteria for adequate metabolic control are the absence of symptoms (polyuria, polydipsia, and nocturia) of ketonuria and of hypoglycemia. The vagrancies of appetite and the variability of activity and growth in children preclude demanding stricter metabolic criteria of control that may unduly compromise our principal goal. Recent reports (Kohrman and Ward, 1971) support our contention that the vascular and degenerative complications cannot be ascribed simply to hyperglycemia or degree of control (Larsson *et al.*, 1962; Knowles *et al.*, 1965; Deckert and Paulsen, 1968; Ahuja *et al.*, 1969; Adnitt and Taylor, 1970). Furthermore, once overt diabetes has been present for 2 years or longer, it is almost impossible to maintain normoglycemia and aglycosuria in the majority of children without producing frequent hypoglycemic reactions. In children, urine testing 2–4 times a day for sugar (Clinitest tablet) and ketones is a reliable measure of homeostasis and may only be necessary during periods of rapid changes in insulin requirement such as illness, growth, or emotional crisis. Obtaining frequent blood sugar values have been useless in our experience.

(1) DIET. The diet is unmeasured but not totally "free." The child is permitted to eat the same meals as his family and concentrated carbohydrates are discouraged for the entire family. Since carbohydrate, fat, and protein metabolism as well as caloric utilization may be influenced by exercise, infection, and emotional disturbances, hunger is a useful, and the usual, guide to nutritional needs. This is in addition to the estimated caloric need based on age, height, and body build. The role of diet in the management of diabetes mellitus is important in every patient if one is to attain the following goals in therapy:

1. As normal a metabolic state as possible—recognizing complete physiological and biochemical "normalcy" is impossible when the patient is receiving insulin injections.
2. Ideal body weight—obesity being undesirable in everyone and especially in the diabetic.
3. Normal weight gain and growth in the child.
4. Necessity of fitting the intake, amount, and distribution of food to the action of the injected insulin.

The data that polyunsaturated fats may be useful in minimizing long-term vascular complications in the diabetic are not adequate as yet to recommend the routine use of an unsaturated fat diet for all diabetic children.

(2) INSULIN. In the vast majority of diabetic children, one injection of an intermediate acting insulin has been sufficient for control. Following the initial episode, remissions of the diabetic state occur in about 30% of the children, especially in males (White, 1965). During this period, which may last days or months, the dose of insulin may drop precipitously. Gradually insulin requirements increase and are directly proportional to the duration of the disease and the age and size of the patient. Changes in dosage are based on the records of urinary excretion of glucose and acetone, growth patterns, and clinical manifestations.

(3) EMOTIONAL PROBLEMS. The degree of emotional adjustment to the disease has a direct positive correlation with adequacy of control (Baker *et al.*, 1969; Kohrman and Weil, 1971) and is dependent upon the personality of the child and the parent–child interaction. Of a group of 54 juvenile diabetics ranging in age from 18 months to 21 years of age, in a period of 6 years, 48 patients were hospitalized 76 times either at the time of original diagnosis for initial regulation or for an episode of ketoacidosis, whereas 6 patients were hospitalized 81 times. This latter frequency of hospitalizaton occurred in children with severe emotional maladjustments usually resulting from family discord, broken homes, extreme poverty, or a total lack of understanding of the diabetes. In children with severe emotional lability that can be translated into severe ketoacidosis and diabetic decompensation within a period of several hours, Baker *et al.* (1969) have reported beta adrenergic blockade as a useful therapeutic tool. Additional data are necessary before specific agents and dosages can be recommended.

The physician must be aware of the meaning of the disease to the parents and of the feelings of guilt or inadequacy it may engender. This can best be understood by gaining insight into the relationship of the child and family prior to the onset of the disease. Assistance should be given to both parents and child so that the disease does not become the focus of neurotic interaction. The services of a social worker plus psychiatric consultation are an integral part of therapy. Children over 9 years of age should be encouraged to assume much of the responsibilities for their own therapy. Patience, leniency in management, plus avoidance of severe criticism may minimize some of the problems (Kaye and Baker, 1965).

In summary, the child with the clinical syndrome of diabetes mellitus presents unique problems in management. No one criteria determines success or failure. Our goal of an emotionally stable, physically and socially integrated child precludes rigid management. In fact, when "nine per cent of . . . long-term patients were disturbed seriously enough to attempt suicide or require temporary institutional care" (White, 1960), the question of the dangers of rigid control must be seriously considered.

4. Long-Term Prognosis

The concept of good control as defined by relative aglycosuria, normoglycemia, and absence of ketosis plus a rigidly weighed and controlled diet has been advocated for many years. It was thought that tight control would result in a diminution in the prevalence of complications, but recent evidence would suggest a contrary conclusion. Siperstein *et al.* (1968; Siperstein, 1970) noted that changes in muscle capillary basement membranes can occur even in pre-diabetics. Changes in nail fold capillaries have already occurred in newly diagnosed diabetic children (Kontras *et al.*, 1968). In addition, strong evidence against "rigid control" comes from the retrospective examination of patients after 20–30 years of diabetes (Larsson *et al.*, 1962; Knowles *et al.*, 1965; Deckert and Paulsen, 1968; Ahuja *et al.*, 1969). None of these investigators was impressed that the level of diabetic control was related to the prevalence or progression of any of the vascular or degenerative complications.

It is only possible to predict on a statistical basis what types of complications can occur. Paulsen and Colle (1969) have summarized those findings related to small blood vessels as well as to the other complications of this chronic disease. Retinopathy, according to Knowles *et al.* (1965), has a mean time of appearance of approximately 10 years after the time of the growth spurt. The risk data would suggest that retinopathy occurs in about 70% of all diabetics by the twentieth year of diabetes. The rate of progression and the ultimate severity are completely unpredictable. Lenticular opacities may appear around puberty. The mean time of appearance is in the tenth year of diabetes or later and the incidence varies from 2 to 47% depending upon the series reported. Glomerular sclerosis as evidenced by intermittent proteinuria, hypertension, an increase in the blood urea nitrogen, and edema can have a highly variable course as far as progression. Rarely seen during childhood or adolescence, the overall incidence is approximately 30% and the mean time of appearance is in the thirteenth to twentieth year of diabetes. The cumulative risk rate is about 33% or much lower than that for retinopathy. Hypertension in the absence of renal disease is unusual but can occur in the early twenties after 10 or 11 years of diabetes with an incidence of about 30% being reported. Vessel calcification may occur in the fifteenth to twentieth year of diabetes usually in the vessel of the leg, pelvis, and aorta.

Recent data would suggest that neuropathy is not unusual in juvenile growth onset diabetics. Electromyographic studies and measurements of nerve conduction velocity show abnormal changes early although clinical evidence of neuropathy is unusual before the eighth to tenth year of diabetes. Eeg-Olofsson and Petersen (1966) found a 5% incidence of pathological electromyographic patterns in the lower extremities of juvenile diabetics as well as a decreased conduction velocity of the ulnar and peroneal nerves in young children with diabetes. The prevalence of clinical neuropathy occurring at sometime during the life of a juvenile diabetic is between 5 and 50% depending upon the sensitivity of the screening device and the testing apparatus. In the presence of frequent hypoglycemic comas, abnormal electroencephalographic records were noted in a significant number of juvenile diabetics by the same authors. Larsson *et al.* (1962) reported a total death risk for juvenile diabetics in the fifteenth through forty-first year of diabetes of 22.7%.

C. Maturity Onset or Chemical Diabetes Mellitus

1. Clinical Manifestations

For a number of years, it has been known (Fajans *et al.*, 1969) that young children and adolescents may have glycosuria and intermittent symptoms of polyuria, polydipsia, and polyphasia and yet not require insulin nor become ketotic. During periods of acute stress such as infection, be it appendicitis or measles, a few of these children would lose their carbohydrate tolerance, become ketotic and acidotic, and require insulin for short periods of time. Ultimately, approximately 11% of these children may progress to insulin dependency after 1–17 years of observation (Rosenbloom *et al.*, 1972). The remainder, however, maintains sufficient insulin output to prevent ketosis and to maintain carbohydrate and fat homeostasis. The exact frequency of this problem in childhood is unknown. At one conference (Rosenbloom *et al.*, 1972) 198 children were reported from centers all over the United States. Clinical manifestations vary but are usually limited to a chance finding of glycosuria at the time of a routine urinalysis. These patients may have a variety of abnormalities of carbohydrate tolerance.

2. Pathogenesis

With the availability of normals for glucose tolerances in children by a variety of techniques reported by different investigators (Parker *et al.*, 1968; Pickens *et al.*, 1967; Cole and Bilder, 1968; Drash *et al.*, 1970;

Rosenbloom, 1973) (Table I), it is essential to establish this diagnosis by measuring both glucose and plasma insulin during a 4–5-hr glucose tolerance test done after 3 days of preparation on a high carbohydrate diet and an overnight fast. The current criterion for an abnormal oral glucose tolerance test has been defined by Rosenbloom *et al.* (1972) (Table II). It is critical that plasma insulin be assayed since the blood glucose may reflect a grossly abnormal tolerance test whereas the plasma insulin values may be as high or higher than those in the normal. This differentiates these children from those with insulin-deficient or growth onset juvenile diabetes mellitus.

It would appear that there are at least two distinct groups in that one secretes an excess of insulin (Rosenbloom, 1970; Drash *et al.*, 1970; Jackson *et al.*, 1969; Balsam *et al.*, 1973; Pildes, 1973) and the other a deficient or less than normal amount (Cerasi and Luft, 1970; Camerini-Davalos, 1965; Fajans *et al.*, 1969). Whether or not the excess of insulin represents a misspecified or proinsulin is yet to be determined.

Little is understood about the pathogenesis or etiology of this type of diabetes in children. A positive family history for diabetes has been present in over 75% of the patients reported, suggesting an underlying genetic basis. Whether or not the insulin assayed is a selectively inactive insulin, proinsulin, or an abnormal form of insulin remains unknown. The other possibility would suggest that peripheral resistance to insulin action may be present. However, without specific therapy, many of these children have gone through puberty and adolescence without requiring insulin for periods up to 17 years after onset.

3. Management

Data on treatment of these patients have been unfortunately limited to anecdotal reports or summaries of small groups of patients treated for variable periods with a variety of therapeutic agents including sulfonylurea drugs, insulin, or other oral hypoglycemic agents (Rosenbloom *et al.*, 1972). To date, there is no known therapy of proven value for the latent maturity onset child diabetic who is secreting insulin. Although tolbutamide has been used previously, the current prospectively controlled double blind study of the Combined University Study Group (Meinert *et al.*, 1970) would contraindicate its use in children as of now. There is no evidence that these children benefit from any of the oral hypoglycemic agents as evidenced by increased carbohydrate tolerance or a normalization of their insulin secretion. Current evidence is inadequate to determine whether or not the course of this latent maturity type of diabetes mellitus can be altered by any form of therapy.

TABLE I

Oral Glucose Tolerance Tests in Children (glucose in mg/100 ml)

Investigator:	Jackson et al. (1969)		Cole and Bilder (1968)		Drash et al. (1970)		Rosenbloom et al. (1972)		Parker et al. (1968)			
Age group:	1–14		0–12		4–16		1½–12		1–13			
Number:	200		159		55		54		15			
Fluid:	Capillary whole blood		Capillary whole blood		Capillary whole blood		Venous plasma		Venous whole blood		Venous plasma	
Method:	Somogyi-Nelson		Auto analyzer variable		Glucose oxidase		Auto analyzer		Glucose oxidase		Insulin (IRI)	
Glucose load:	1.75 gm/kg				1.75 gm/kg		1.75 gm/kg		1.75 gm/kg			
Time:	$\bar{x}$	+ 2SD	$\bar{x}$	+ 2SD	$\bar{x}$	+ 2SD	$\bar{x}$	+ 2SD	$\bar{x}$	+ 2SD	$\bar{x}$	+ 2SD
0	82	110	72	88	78	112	86	104	74	38	10	16
30	135	193	119	160	138	208	143	204	119	85	30	38
60	112	170	108	149	121	189	113	157	117	46	47	62
90									106	38	39	8
120	101	141	94	130	101	150	102	134	98	54	25	46
180	84	124	76	105	75	118	83	119	76	38	14	46
240					77	110	80	107	70	12	8	16
300									79	54	9	16

TABLE II

Criteria for Interpretation of Abnormality in Oral
Glucose Tolerance[a]

Investigator:	Jackson *et al.* (1969)	Rosenbloom (1973)
Population:	Children	Children
Fluid:	Capillary blood	Venous plasma
Time: 0	119	110
60	170	160
90	—	—
120	140	140
180	125	130
240	—	115

[a] Modified from Rosenbloom *et al.* (1972).

4. Long-Term Prognosis

In the largest series reported, about 10% of these patients will become insulin-dependent, ketotic juvenile type growth onset diabetics. What the ultimate prognosis is concerning vascular complications or life expectancy in these patients awaits a collaborative, long-term follow-up study.

D. Research Needs

In general, the specific etiology of diabetes mellitus, whether it is an autoimmune disease, a multifocal genetic disease, or the result of the effects of diet and exposure to refined sugars or other stresses of civilization remains to be elucidated. Certainly the understanding of the genetics of this condition requires precise definition. Recent evidence in studies of Fanconi's anemia and a concomitant increased incidence of diabetes might provide important markers for those interested in the genetic studies of autosomal recessive conditions and diabetes (Swift and Sholman, 1972).

1. Specific Areas of Research Needs

a. Transient Diabetes in the Neonate

More data are required to understand the relation of growth *in utero* and the insulin deficiency observed in transient diabetes in the neonate. The observation that these low birth weight infants may also have hypoglycemia as well as high levels of β-hydroxybutyrate and free fatty acids requires further investigation into the interrelationships between insulin action, growth, and metabolic homeostasis in the neonate. Long-term follow-up data are necessary to determine whether or not all neonates with transient diabetes recover or decompensate later in life into the various types of childhood diabetes. If the latter occurs, what are the precipitating factors? A recommendation that these babies be tested every 2–3 years with stress glucose tolerance tests may provide some answers to these questions.

b. Juvenile Growth Onset Insulin-Dependent Diabetes

There is great need to understand the pathophysiology of the highly unstable, emotional, ketotic diabetic. Specific recommendations for therapy by Baker *et al.* (1969) using beta-blockers require expansion and further documentation. Long-term prospective controlled studies of management in the evaluation of low fat diets, unsaturated fats, oral agents, etc., are critical.

Finally, the etiology of small vessel disease and the factors that influence the complications and the prognosis in diabetes require new insight and new approaches. Whether small vessel disease is the result of a closely associated genetic defect, is on an immunological basis, or on dietary or other indiscretions of management remains to be elucidated. The suggestion by Gabbay (see Chapter 28) that the etiology of the diabetes itself and the multiple complications may result from increased aldose reductase activity and accumulation of intracellular polyols requires further study and verification. The possible role of inhibitors of aldose reductase in the prevention and therapy in juvenile diabetics may be of major importance.

c. Chemical or Maturity Type of Onset

A precise definition, considering age, sex, race, and body build of the normal insulin and glucose response to glucose, tolbutamide, arginine, and glucagon is necessary in order to arrive at a precise definition of this type of carbohydrate intolerance in childhood. Studies to determine the incidence or frequency of this type of diabetes in the general population would be valuable. Data on all such patients should be centralized in one data bank in order that prospective controlled studies and prognosis can be determined over a number of years. A collaborative, prospective controlled study of oral hypoglycemic agents, insulin, or diet and outcome should be instituted. The search for an animal model for this type of diabetes appears important.

Although the fiftieth anniversary of the discovery and isolation of insulin is here, many areas of investigation remain for every discipline of science and of the humanities in the basic understanding and management of the pediatric patient with diabetes.

References

Adnitt, P. I., and Taylor, E. (1970). Progression of diabetic retinopathy: Relationship to blood sugar. *Lancet* **1,** 652.

Ahuja, M. M. S., Kumar, V., and Gossain, V. V. (1969). Interrelationship of vascular disease and blood lipids in young Indian diabetics. *Diabetes* **18,** 670.

Arey, S. L. (1953). Transient diabetes in infancy. *Pediatrics* **11,** 140.

Baker, L., Barcai, A., Kaye, R., and Haque, N. (1969). Beta adrenergic blockade and juvenile diabetes: Acute studies and long-term therapeutic trial. *J. Pediat.* **75,** 19.

Balsam, M. J., Kaye, R., and Baker, L. (1973). Chemical diabetes in children: Glucose tolerance tests. *Metab., Clin. Exp.* **22,** 283.

Berson, S. A., and Yalow, R. S. (1962). Immunoassay of plasma insulin. *Ciba Found. Colloq. Endocrinol.* [*Proc.*] **14,** 182.

Boden, G., Soeldner, J. S., Gleason, R. E., and Marble, A. (1968). Elevated serum human growth hormone and decreased serum insulin in prediabetic males after intravenous tolbutamide and glucose. *J. Clin. Invest.* **47,** 729.

Camerini-Dávalos, R. A. (1965). Biochemical and histological aspects of prediabetes. *In* "On the Nature and Treatment of Diabetes" (B. S. Leibel and G. A. Wrenshall, eds.), pp. 657–668. Excerpta Med. Found., Amsterdam.

Cerasi, E., and Luft, R., eds. (1970). "Pathogenesis of Diabetes Mellitus," 13th Nobel Symp. Wiley, New York.

Cole, H. S., and Bilder, J. H. (1968). Capillary blood standard values during oral glucose tolerance tests in infants and children. *Diabetes* **17,** 321.

Cornblath, M., and Schwartz, R. (1966). Transient Diabetes mellitus in early infancy. *In* "Disorders of Carbohydrate Metabolism in Infancy, Major Problems in Clinical Pediatrics," Vol. III, p. 105. Saunders, Philadelphia, Pennsylvania.

Danowski, T. S. (1957). "Diabetes Mellitus, with Emphasis on Children and Young Adults." Williams and Wilkins, Baltimore, Maryland.

Danowski, T. S., Lombardo, Y. B., Mendelsohn, L. V., Corredor, D. G., Morgan, C. R., and Sabeh, G. (1969). Insulin patterns prior to and after onset of diabetes. *Metab. Clin. Exp.* **18,** 731.

Deckert, T., and Paulsen, D. (1968). Prognosis for juvenile diabetics with late diabetic manifestation. *Acta. Med. Scand.* **183,** 351.

Drash, A., Field, J. B., Garces, L. Y., Kenny, F. M., Mintz, D., and Vazquez, A. M. (1968). Endogenous insulin and growth hormone response in children with newly diagnosed diabetes mellitus. *Pediat. Res.* **2,** 94.

Drash, A., Chiumello, G., and Hengstenberg, F. (1970). Studies in diabetes mellitus: Approaches to earlier diagnosis and therapy. *In* "Adolescent Endocrinology" (F. Heald and W. Hung, eds). pp. 51–77. Appleton, New York.

Eeg-Olofsson, O., and Petersen, I. (1966). Childhood diabetic neuropathy: A clinical and neurophysiological study. *Acta Paediat. Scand.* **55,** 163.

Fajans, S. S., Floyd, J. D., Jr., and Pek, S. (1969). The course of asymptomatic diabetes in young people as determined by levels of blood glucose and plasma insulin. *Trans. Ass. Amer. Physicians* **82**, 211–224.

Fajans, S. S., Floyd, J. D., Jr., Conn, J. W., and Pek, S. (1970). The course of asymptomatic diabetes of children, adolescents, and young adults. *In* "Early Diabetes" (R. A. Camerini-Dávalos and H. S. Cole, eds.), p. 377. Academic Press, New York.

Ferguson, A. W., and Milner, R. D. G. (1970). Transient neonatal diabetes mellitus in siblings. *Arch. Dis. Childhood* **45**, 80.

Gentz, J. C. H., and Cornblath, M. (1969). Transient diabetes of the newborn. *Advan. Pediat.* **16**, 345–363.

Hansen, A. P. (1970). Abnormal serum growth hormone response to exercise in juvenile diabetes. *J. Clin. Invest.* **49**, 1967.

Hutchinson, J. H., Keay, A. J., and Kerr, M. M. (1962). Congenital temporary diabetes mellitus. *Brit. Med. J.* **2**, 436.

Jackson, R. L., Guthrie, R. A., and Murthy, D. Y. N. (1969). Diabetes and prediabetes in children. *In* "Diabetes (J. Ostman, ed.), pp. 79–102. Exerpta Med. Found., Amsterdam.

Kaye, R., and Baker, L. (1965). Control of diabetes in childhood. *Postgrad. Med.* **38**, 515.

Klein, R. (1971). Diabetes mellitus. *In* "Current Pediatric Therapy" (S. S. Gellis and B. M. Kagan, eds.), Vol. 5, pp. 334–343. Saunders, Philadelphia, Pennsylvania.

Knowles, H. C., Jr., Guest, G. M., Lampe, J., Kessler, M., and Skillman, T. G. (1965). The course of juvenile diabetes treated with unmeasured diet. *Diabetes* **14**, 239.

Kohrman, A. F., and Weil, W. B. (1971). Juvenile diabetes mellitus. *Advan. Pediat.* **18**, 123–149.

Kontras, S. B., Bodenbender, J. G., Rettemnier, S. C., and Shaffer, T. E. (1968). Capillary microscopy in juvenile diabetes mellitus. *Amer. J. Dis. Child.* **116**, 135.

Larsson, Y., Sterky, G., and Christiansson, G. (1962). Long-term prognosis in juvenile diabetes mellitus. *Acta Paediat.* **51**, Suppl. 130.

Lawrence, R. D., and McCance, R. A. (1931). Gangrene in an infant associated with temporary diabetes. *Arch. Dis. Childhood* **6**, 343.

Meinert, C. L., Knatterud, G. L., Prout, T. E., and Klimt, C. R. (1970). The University Group Diabetes Program; a study of the effects of hypoglycemia agents on vascular complications in patients with adult-onset diabetes. *Diabetes* **19**, Suppl. 2.

Melani, F., Rubinstein, A. H., and Steiner, D. F. (1970). Human serum proinsulin. *J. Clin. Invest.* **49**, 497.

O'Brien, D. (1970). Evidence for an abnormal insulin in diabetes mellitus. *In* "Early Diabetes" (Camerini-Davalos, ed.). Academic Press, New York.

Pagliara, A. S., Karl, I. E., and Kipnis, D. M. (1973). Transient neonatal diabetes: Delayed maturation of the pancreatic beta cell. *J. Pediat.* **82**, 97.

Parker, M. L., Pildes, R. S., Chao, K. L., Cornblath, M., and Kipnis, D. M. (1968). Juvenile diabetes mellitus. A deficiency in insulin. *Diabetes* **17**, 27–32.

Paulsen, E. P., and Colle, E. (1969). Diabetes mellitus. *In* "Endocrine and Genetic Diseases of Childhood" (L. I. Gardner, ed.), pp. 808–823. Saunders, Philadelphia, Pennsylvania.

Pickens, J. N., Burkeholder, J. N., and Womack, W. N. (1967). The oral glucose tolerance test in normal children. *Diabetes* **16**, 11–14.

Pildes, R. S. (1973). Adult-onset diabetes mellitus in childhood. *Metab. Clin. Exp.* **22**, 307.

Posner, J. B., and Plum, F. (1967). Spinal fluid pH and neurologic symptoms in systemic acidosis. *N. Eng. J. Med.* **277**, 605–613.

Ramsey, W. R. (1926). Glucosuria of the newborn treated with insulin. *Trans. Amer. Pediat. Soc.* **38**, 100.

Renold, A. E., Stauffacher, W., and Cahill, G. F., Jr. 1972). Diabetes mellitus. *In* "The Metabolic Basis of Inherited Diseases" (J. B. Stanbury, J. B. Wyngaarden, and D. S. Frederickson, eds.), 3rd ed., pp. 83–118. McGraw-Hill, New York.

Rosenbloom, A. L. (1970). Insulin responses of children with chemical diabetes mellitus. *N. Engl. J. Med.* **282**, 1228–1231.

Rosenbloom, A. L. (1973). Criteria for interpretation of the oral glucose tolerance test and insulin responses with normal and abnormal tolerance. *Metab., Clin. Exp.* **22**, 301.

Rosenbloom, A. L., Drash, A., and Guthrie, R. (1972). Chemical diabetes mellitus in childhood: Report of a conference. *Diabetes* **21**, 45–49.

Rubinstein, A. H., and Steiner, D. F. (1970). Human proinsulin: Some considerations in the development of a specific immunoassay. *In* "Early Diabetes" (R. A. Camerini-Dávalos and H. S. Cole, eds.), p. 159. Academic Press, New York.

Siperstein, M. D. (1970). The relationship of carbohydrate derangements to the microangiopathy of diabetes. *In* (Pathogenesis of Diabetes Mellitus" (E. Cerasi and R. Luft, eds.), 13th Nobel Symp., Wiley, New York.

Siperstein, M. D., Unger, R. H., and Madison, L. L. (1968). Studies of muscle capillary basement membranes in normal subjects, diabetic, and prediatbetic patients. *J. Clin. Invest.* **47**, 1973.

Swift, M., and Sholman, L. (1972). Diabetes mellitus and the gene for Fanconi's anemia. *Science* **178**, 308–309.

Taubin, H., and Matz, R. (1968). Cerebral edema, diabetes insipidus and sudden death during treatment of diabetic ketoacidosis. *Diabetes* **17**, 108–109.

Unger, R. H. (1971). Glucagon physiology and pathophysiology. *N. Engl. J. Med.* **285**, 443–449.

White, P. (1960). Childhood diabetes. *Diabetes* **9**, 345.

White, P. (1965). The child with diabetes. *Med. Clin. N. Amer.* **49**, 1069.

Yde, H. (1969). Abnormal growth hormone response to ingestion of glucose in juvenile diabetes. *Acta Med. Scand.* **186**, 499.

Young, E., and Bradley, R. F. (1967). Cerebral edema with irreversible coma in severe diabetic ketoacidosis. *N. Engl. J. Med.* **276**, 665–669.

CHAPTER 28

The Sorbitol Pathway in the Nervous System

KENNETH H. GABBAY

Since the introduction of insulin therapy 50 years ago, ketoacidosis, infections and survival therefrom are no longer major problems in the care of diabetic patients. Today the diabetic patient faces the development of cataracts, retinopathy, nephropathy, neuropathy, and accelerated generalized atherosclerosis. The morbidity and incapacity associated with these complications is staggering: Diabetes is a leading cause of blindness in the United States and 60% of the death certificates mentioning renal or cardiovascular disease also carry the associated diagnosis of diabetes (Knowles *et al.*, 1965; Diabetes Source Book, 1964). Although the literature is divided on the role of good control of diabetes as a factor in reducing the incidence of complications, it is quite clear that the clinical expression and development of the various complications is a function of the duration of diabetes (Knowles *et al.*, 1965; Pirart, 1965).

The tissues bearing the brunt of diabetic manifestations (lens, retina, nerve, kidney, blood vessels, and islet cells) are freely permeable to glucose and do not require insulin for glucose penetration as do muscle and adipose tissue, and hence are exposed to the ambient blood glucose levels. Recent investigations of the process of cataract formation in galactosemia and diabetes (Kinoshita, 1965; Kinoshita *et al.*, 1968; Chylack and Kinoshita, 1969; Gabbay and Kinoshita, 1972) indicate a role for the sorbitol pathway in the metabolism of the excess hexose in lens tissue and suggest an involvement in some of the other diabetic tissue complications as well.

A. History and Enzymes of the Sorbitol Pathway

Historically, free fructose is an unusual sugar to be found in tissues or body fluids. Orr (1924) reported the presence of fructose as well as glucose in human fetal blood, then Hubbard and Russell (1937) reported the finding of free fructose in human cerebrospinal fluid. Mann (1946) characterized the main sugar of seminal fluid as fructose and demonstrated that the fructose was derived from blood glucose and was responsive to elevations of the latter in diabetes (Mann and Parsons, 1950). Mann and Parsons (1950) further showed that seminal fluid fructose content, was directly influenced by hormonal factors, virtually disappearing in castrated and hypophysectomized animals. Hers (1956) subsequently demonstrated that seminal fructose is formed in the seminal vesicles from blood glucose primarily through the activity of the "sorbitol pathway." Hers showed that this pathway consisted of two enzymes that were able to convert glucose to fructose via the intermediate step of sorbitol (Fig. 1). Subsequent work by Hers (1959) showed that the pathway is responsible for the formation of sorbitol in placenta, and that sorbitol is then converted to fructose by the fetal liver. Fructose is the main fetal blood sugar in sheep, disappearing after the separation of the fetus from the placenta (Huggett *et al.*, 1951).

Aldose reductase catalyzes the conversion of free glucose to its sugar alcohol sorbitol. It possesses broad substrate specificity for many aldoses and is characterized by low affinity for glucose and galactose (Hayman and Kinoshita, 1965; Moonsammy and Stewart, 1967; Gabbay and Tze, 1972; Gabbay, 1972b). Because of the low affinity for hexoses, the avail-

$$\text{D-GLUCOSE + TPNH} \xrightarrow{\text{Aldose Reductase}} \text{SORBITOL + TPN}^+$$

$$\text{SORBITOL + DPN}^+ \xrightarrow{\text{Sorbitol Dehydrogenase}} \text{FRUCTOSE + DPNH}$$

Fig. 1. The sorbitol pathway.

ability of large pools of free aldoses in diabetes and galactosemia causes increased formation of the sugar alcohols sorbitol and galactitol, respectively. Aldose reductase requires NADPH for its activity, and since NADPH is provided in the cell primarily by the action of the hexose-monophosphate shunt (HMP shunt) it is not surprising that a relationship between the HMP shunt and the sorbitol pathway has been described in a number of tissues (Kinoshita *et al.*, 1963; Gabbay, 1969).

An enzyme of the glucuronic acid-xylulose shunt, NADP-L-hexonate dehydrogenase, which has many similarities to aldose reductase, is also present in many tissues. Hexonate dehydrogenase (Mano *et al.*, 1961; Moonsammy and Stewart, 1967; Gabbay, 1972b) has a poor ability to convert hexoses to their respective sugar alcohols (K_m 0.5–2 M), while being more active in the reduction of uronic acids. Since the latter enzyme is present in many tissues, care must be taken to differentiate between this enzyme and true aldose reductase.

The further conversion of sorbitol to fructose is catalyzed by the sorbitol dehydrogenase enzyme. This ubiquitous enzyme has broad substrate specificities for many sugar alcohols, converting them to the respective keto sugars, i.e., sorbitol and mannitol to fructose, xylitol to D-xylulose, and ribitol to D-ribulose. Galactitol and D-arabitol are poor substrates for this enzyme. This latter property is of significance in galactosemia where the inability to further metabolize galactitol leads to enhanced accumulations of the latter in galactosemic tissues. Sorbitol dehydrogenase has been isloated and purified from a number of tissues (Blakely, 1951; McCorkindale and Edson, 1954; Smith, 1962) and has been shown to be hormonally sensitive during the development of the accessory sexual organs.

An aspect crucial to the understanding of the effect of sugar alcohol accumulation in tissues is that these sugar alcohols penetrate cell membranes poorly (Le Fevre and Davies, 1951; Wick and Drury, 1951). Once formed, these sugar alcohols are trapped intracellularly with the only disposition being either conversion to the respective keto sugar or slow leakage from the cell. Since fructose is poorly metabolized in some of these tissues and leaks out slowly, the net effect is an accumulation of solute inside the cell with resultant hypertonicity and osmotic consequences that result in tissue damage (see review of cataract formation by Kinoshita, Chapter 23). Although the sorbitol pathway has been demonstrated in many tissues, only three—lens (Kinoshita, 1965), sciatic nerve (Gabbay *et al.*, 1966; Stewart *et al.*, 1966), and renal papilla (Gabbay and O'Sullivan, 1968b)—have thus far been shown to contain sufficiently high concentrations of sugar alcohol in diabetes and galactosemia to be of possible osmotic significance (10–80 μmoles/gm of tissue). It should be noted that the accumulation of 1 μmole of solute per 1 gm of tissue is equivalent to 1 mosmole/liter in osmotic pressure.

The presence of sorbitol and fructose in a tissue is prima facie evidence for the presence of the pathway. However, as noted above, the enzymatic method using sorbitol dehydrogenase (Smith, 1962) for measuring sorbitol is not specific, and gas liquid chromatographic identification of sorbitol is necessary. At extremely high glucose concentrations it is also possible to form sorbitol via the action of NADP-L-hexonate dehydrogenase (Moonsammy and Stewart, 1967; Travis *et al.*, 1971); therefore, isolation and kinetic characterization of aldose reductase from the tissue is a more rigorous criterion for the presence of the pathway. DEAE-cellulose column chromatography clearly separates aldose reductase from NADP-L-hexonate dehydrogenase, and since NADP-L-hexonate dehydrogenase and aldose reductase are immunologically distinct (Gabbay and Cathcart, 1969; Gabbay, 1972b), it is now possible to obtain immunological confirmation that the isolated enzyme is indeed aldose reductase. If these criteria are satisfied, the existence of the sorbitol pathway in a tissue can be stated with confidence.

Aldose reductase appears to be highly localized to certain cell types within tissues; for instance, the highest concentration of aldose reductase is present in the lens epithelium (Kinoshita, 1965), while it is entirely localized to the Schwann cell in peripheral nerve (Gabbay and O'Sullivan, 1968a) to the kidney papilla (Gabbay and O'Sullivan, 1968b) and to the islets of Langerhans in pancreas (Gabbay and Tze, 1972). These localizations are important for correlating the site of sugar alcohol formation with the site of pathology and additionally imply much higher local concentrations of sugar alcohol than would be apparent from determining whole tissue levels.

B. The Sorbitol Pathway in the Nervous System

1. Peripheral Nervous System

Following the clarification of the mechanism of cataract formation in galactosemia and diabetes, the possible involvement of the sorbitol pathway in diabetic neuropathy was considered. From a clinical point of view, diabetic neuropathy is known to respond to good control of the blood sugar (Pirart, 1965). Pathologically, diabetic neuropathy is characterized by segmental demyelinization associated with Schwann cell abnormalities (Thomas and Lascelles, 1965). Physiologically, it is characterized by decreased nerve conduction velocity in both the sensory and motor nerves.

The clinical manifestations of diabetic neuropathy are protean (Ellenberg and Rifkin, 1970) with severe involvement of many organ systems.

Stewart and Passonneau (1964) showed the presence of free fructose in normal rabbit tibial nerve. The sorbitol pathway was subsequently demonstrated to be present in normal nerves (Gabbay *et al.*, 1966; Stewart *et al.*, 1966), and its activity was shown to be markedly elevated in diabetes. Table IA presents data obtained by gas liquid chromatography of the carbohydrate levels in streptozotocin diabetic rat sciatic nerves. It should be noted that the diabetic nerve contains approximately 17 μmoles of glucose/gm wet weight, indicating the availability of a large free glucose compartment in diabetic nerve. There is also a marked elevation of sorbitol and fructose, and a slight decrease in the inositol which is not statistically significant. A linear correlation between the blood glucose levels and the nerve fructose content was previously described (Gabbay *et al.*, 1966).

Further examination of Tables IB and IC indicates significant differences in the levels of sorbitol, fructose, and inositol between the peripheral nervous system and the central nervous system represented by spinal cord and brain. This difference is exaggerated in the diabetic state where the total sorbitol plus fructose level in nerve can reach 12 μmoles/gm wet weight in comparison with less than 0.4 μmole/gm wet weight in cord and brain tissues.

Studies of the distribution of the enzymes of the sorbitol pathway in nervous tissue demonstrated that (1) spinal and peripheral nerves contain aldose reductase only, (2) spinal cord contains mainly NADP-L-hexonate dehydrogenase, and (3) brain contains NADP-L-hexonate dehydrogenase and aldose reductase with the former predominating (Gabbay and O'Sullivan, 1968a; Moonsammy and Stewart, 1967). Sorbitol dehydrogenase is present in all three tissues.

Although all three tissues contain myelinated fibers, the myelinization process is carried on by different cells in the central and peripheral nervous systems. Central nervous system myelin (cord and brain) is formed by the oligodendroglial cell, while the Schwann cell forms the myelin in the spinal nerves and their extra-spinal extensions, the peripheral nerves. These differences, plus the aldose reductase distribution in the two systems, suggested a possible association of aldose reductase with Schwann cells. Indeed, in Wallerian degeneration experiments where the axons are resorbed and disappear in a few days while the Schwann cells begin to multiply, it was found that aldose reductase activity is actually increased, indicating the probable localization of aldose reductase to the Schwann cells. Conversely, approximately 90% of the sorbitol dehydrogenase activity disappears from Wallerian degenerated nerves, indicating

TABLE I

Sugar Content of Nervous Tissue (μmole/gm Wet Weight $\pm$ SD, $N = 12$)

	IA, Sciatic nerve		IB, Spinal cord		IC, Brain	
	Normal	Diabetic	Normal	Diabetic	Normal	Diabetic
Glucose	2.60 $\pm$ 1.30	17.03 $\pm$ 6.20	1.80 $\pm$ 0.1	12.60 $\pm$ 1.32	1.1 $\pm$ 0.36	6.92 $\pm$ 2.4
Sorbitol	0.34 $\pm$ 0.02	5.12 $\pm$ 1.92	<0.1	<0.1	<0.1	<0.1
Fructose	1.18 $\pm$ 0.24	6.73 $\pm$ 1.22	0.10 $\pm$ 0.04	0.56 $\pm$ 0.13	<0.1	0.27 $\pm$ 0.06
Inositol	3.21 $\pm$ 0.72	2.58 $\pm$ 0.42	5.80 $\pm$ 1.12	5.42 $\pm$ 0.82	5.31 $\pm$ 0.51	5.01 $\pm$ 0.62

probable association of this enzyme with the axonal elements (Gabbay and O'Sullivan, 1968a). Stewart *et al.*, (1965) have demonstrated that the fructose content rises sharply 4 days following the transection of a nerve, coincident with the time of Schwann cell proliferation.

The probable localization of aldose reductase (and hence sorbitol formation) to the Schwann cell cytoplasm can thus be construed to indicate much higher local sorbitol levels than indicated by *whole* nerve measurements. Furthermore, such localization allows a correlation with physiological and pathological changes in diabetic and galactosemic nerves.

a. Experimental Galactosemic Neuropathy

The galactose-fed rat model proved an excellent model for the study of the process of cataract formation. The model is also useful since it provides an opportunity to study the effects of sugar alcohol accumulation without the other attendant biochemical defects present in diabetes. As might be expected, the peripheral nerves of rats fed a 40% galactose diet contain large accumulations of galactitol. Stewart *et al.* (1967) found 17.4 μmoles/gm wet weight in nerves of rats fed the galactose diet for 5 weeks. The much higher levels of galactitol in galactosemic nerves than sorbitol in diabetic nerves are explained by the inability of sorbitol dehydrogenase to further metabolize galactitol. In fact, neither Stewart nor the present author were able to demonstrate the presence of tagatose in galactosemic nerves. Tagatose would be the expected product of sorbitol dehydrogenase action on galactitol.

The experimental galactosemic rat nerves appear swollen on gross examination, and the neural elements tend to be extruded from the cut surface of the collagenous sheath composing the outer cover of a peripheral nerve, suggesting that the swelling is occurring within the nerve fiber compartment. The water content of these nerves is increased by 25%.

The ability to accumulate galactitol and induce presumably osmotic swelling in the peripheral nerves of galactosemic animals (see review of cataract formation by Kinoshita, Chapter 23) enabled an examination of the possible relationship of these abnormalities to the physiological function of the nerve. Serial *in vivo* motor nerve conduction velocity (MNCV) studies indeed showed the rapid development of a MNCV decrease within a few days of feeding galactose to rats and the exacerbation of this defect when additional galactose is gavaged into the animals daily (Gabbay and Snider, 1972). Figure 2 demonstrates the correlation of the galactitol accumulation with the nerve swelling and the MNCV defect development. Withdrawal of galactose from the diet resulted in a substantial decrease of galactitol levels and water content, accompanied by

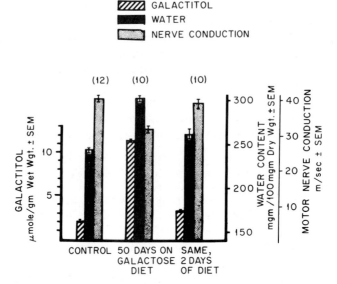

Fig. 2 Relationship of polyol and water accumulation to motor nerve conduction velocity in galactose-fed rats. Reproduced from Gabbay and Snider (1972) with permission of the publisher.

restoration of MNCV to normal. The causal relationship between galactitol accumulation and the MNCV defect development was also demonstrated by the use of an effective *in vivo* aldose reductase inhibitor in galactose-fed rats (Gabbay, 1972a). The development of the MNCV defect in rats pair-fed a 40% galactose diet was significantly delayed in the inhibitor treated group (Fig. 3). Cataract appearance in this group was also markedly delayed or prevented (Gabbay and Kinoshita, 1972). Thus, these experiments seem to confirm a causal relationship between the galactitol accumulations and neuropathy formation in the galactosemic model.

b. Diabetic Neuropathy

A causal relationship of sorbitol pathway activity to diabetic neuropathy formation is far from clear at the present time. Diabetic neuropathy appears to consist of two processes: an acute reversible metabolic damage (akin to that described above), and a more permanent nonreversible damage possibly reflecting Schwann cell loss and segmental demyelinization. Examination of teased nerve preparations of long-term diabetic patients shows evidence of patchy demyelinization and Schwann cell loss and evidence for attempts at regeneration of the myelin sheath.

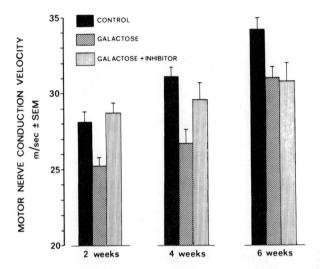

Fig. 3 The effect of an aldose reductase inhibitor (AY-22,284) on delaying the appearance of the motor nerve conduction velocity defect in rats pair-fed galactose diets (40%).

Clinically, it is possible to demonstrate improvement in MNCV in newly diagnosed diabetic patients brought under blood sugar control, and deterioration when the blood sugar is allowed to rise (Gregersen, 1968). However, long-term diabetic patients have a baseline irreversible motor and sensory conduction velocity defect, which is not completely amenable to blood glucose regulation. It should be emphasized that at any one point in time a continuum or a superimposition of the two processes is usually being observed in diabetic patients.

There are biochemical abnormalities that also appear to be of a permanent nature. A lipid synthesis defect is present in diabetic nerve (50% of normal) which Eliasson and Hughes (1960) were unable to reverse with prior treatment of the animals with insulin. Many of the enzymes that appear to be associated with Schwann cells are decreased. Adams and Field (1964) described a 40% decrease in acetic thiokinase activity in diabetic nerve, also not corrected with insulin. Aldose reductase activity is decreased by 30% in diabetic nerve. These various abnormalities may well reflect permanent alterations in the nerve, particularly the Schwann cells.

Figure 4 shows the appearance of the MNCV defect in streptozotocin-diabetic rats and the effect of blood sugar control on the nerve conduction. As expected, nerve glucose, sorbitol, and fructose content were reduced below normal levels in the "hypoglycemic controlled" rats and were intermediate in the mildly hyperglycemic animals (Gabbay, 1972a). Al-

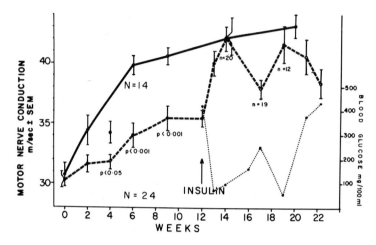

Fig. 4. The effect of streptozotocin-induced diabetes on motor nerve conduction velocity in rats. The solid line represents normal, the dashed line diabetic, and the dotted line blood glucose. Reproduced from Gabbay (1972a) with permission of the editor.

though a relationship is indicated between the sorbitol levels and the MNCV defect, it is not clear whether the elevated sorbitol levels are an index or a cause of the neuropathy. Acute treatment of diabetic rats with an aldose reductase inhibitor reduces the sorbitol and fructose levels by 52 and 64% respectively, and improves MNCV (Gabbay, unpublished observation), but a casual role for sorbitol accumulation awaits the long-term prevention of permanent nerve damage by aldose reductase inhibitors despite persistent hyperglycemia.

Although hyperglycemia is a major drive for sorbitol formation in diabetic nerve, it has become apparent that other factors may be involved as well. In *in vitro* incubations of normal and diabetic nerves at various glucose levels, it has been shown (Gabbay, 1969) that the capacity or ability of a diabetic nerve to form sorbitol is much greater than that of a normal nerve. Figure 5 demonstrates this phenomenon and additionally shows that 3,3-tetramethyleneglutaric acid (TMG), an aldose reductase inhibitor, is able to reduce sorbitol formation to baseline levels in both the normal and diabetic nerves. Lipid synthesis measured by the incorporation of [6-^{14}C] glucose into total nerve lipids was decreased by 50% in diabetic nerves and was not ameliorated by the aldose reductase inhibition. These data indicate a defect in the dissimulation of glucose into the normal metabolic pathways and a shunting into the accessory pathway of sorbitol metabolism. Such an effect could be produced by decreased phosphorylation of glucose in diabetic nerve, leading to accumu-

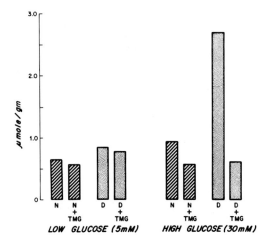

Fig. 5 Sorbitol content of normal (N) and diabetic (D) nerves incubated at indicated glucose levels for 2 hr. 3,3-Tetramethylene glutaric acid (TMG) concentration was 10 mM. Sorbitol was assayed enzymatically.

lation of free glucose and metabolism via the aldose reductase enzyme. Indeed, Eliasson and Hughes (1960) have shown that the addition of exogenous glucose-6-phosphate to incubated diabetic nerves completely restores lipid synthesis to normal. However, hexokinase levels (Types I and II) are elevated in whole diabetic nerves (Gabbay, unpublished observation), thus suggesting a possibly compartmentalized hexokinase deficiency (? Schwann cells).

2. Central Nervous System

The distribution of aldose reductase and NADP-L-hexonate dehydrogenase in brain and cord was described above; however, little is known regarding the localization of either enzyme. The low levels of sorbitol and fructose in these two tissues are explainable by (1) low levels of aldose reductase activity and (2) the demonstrated superior ability of these two tissues to utilize fructose (Stewart *et al.*, 1967) as compared to peripheral nerve.

Clements *et al.* (1968, 1971) and Prockop (1971) have proposed that the osmotic effects of sorbitol and fructose accumulations are responsible for the brain edema seen following the rapid lowering of blood glucose during the treatment of nonketotic hyperglycemic coma in patients and during the rapid lowering of blood glucose in dogs given prolonged glucose infusion. Since the levels of sorbitol and fructose in brain are negligible (<0.4 μmoles/gm or <0.4 mosmole/liter), it is difficult to explain the observed brain edema on this basis. A more likely possibility is a lag in CNS tissue glucose disappearance contrasted with the rapid blood glucose decrease. Such an effect was indeed observed in the study by Prockop.

C. Research Needs

The above discussion has been limited to defining the impact of long-term diabetes on patients today, and the possible role of an accessory pathway of glucose metabolism, the sorbitol pathway, in the development of some of the diabetic complications. The sorbitol pathway is present in many other tissues as well, and it may have an important but, as yet, undefined function in normal cellular metabolism. Recent investigations (Gabbay and Tze, 1972) indicate a role for this pathway in glucose-induced release of insulin from the beta cell. The role of this pathway in seminal fluid fructose formation and in placental metabolism (Hers, 1956) further suggests a possibly important function. For the moment, the sorbitol pathway has opened up an entirely new approach to the understanding of some of the complications of diabetes, and with the use of aldose reductase inhibitors offers a potential new means for the prevention and treatment of diabetes and its complications. Further studies in this area are necessary and the following problems appear to be current obstacles:

1. Cellular and ultrastructural localization of the enzymes of the pathway in various tissues.

2. Development of *sensitive* and *specific* assays for sorbitol and fructose. Currently, this is a major problem especially in tissues containing nanomolar amounts of these carbohydrates.

3. Development of direct and indirect techniques for studying sorbitol pathway activity in patients in a manner consistent with ethical medical research standards.

4. Development and testing of effective aldose reductase inhibitors suitable for human use.

References

Adams, L. C., and Field, R. A. (1964). Acetic thiokinase activity in extracts of sciatic nerves from normal and alloxan diabetic rabbits. *Abstr., Congr. Int. Diabetes. Fed. 5th, 1964* Int. Congr. Sed. No. 74, p. 50.

Blakley, R. L. (1951). The metabolism and antiketogenic effects of sorbitol. Sorbitol dehydrogenase. *Biochem. J.* **49,** 257–271.

Chylack, L. T., and Kinoshita, J. H. (1969). A biochemical evaluation of an experimental hyperglycemic cataract. *Invest. Ophthalmol.* **8,** 401–412.

Clements, R. S., Jr., Prockop, L. D., and Winegrad, A. I. (1968). Acute cerebral edema during treatment of hyperglycemia. *Lancet* **2,** 384–386.

Clements, R. S., Jr., Blumenthal, S. A., Morrison, A. D., and Winegrad, A. I. (1971). Increased cerebrospinal-fluid pressure during treatment of diabetic ketosis. *Lancet* **2,** 671–675.

Diabetes Source Book. (1964). Pub. Health Serv. Publ. No. 1168. U.S. Department of Health, Education, and Welfare, Washington, D.C.

Eliasson, S. G., and Hughes, A. H. (1960). Cholesterol and fatty acid synthesis in diabetic nerve and spinal cord. *Neurology* **10,** 143–147.

Ellenberg, M., and Rifkin, H., eds. (1970). "Diabetes Mellitus; Theory and Practice." McGraw-Hill, New York.

Gabbay, K. H. (1969). Factors affecting the sorbitol pathway in diabetic nerve. *Diabetes* **18,** 336.

Gabbay, K. H. (1972a). Role of sorbitol pathway in neuropathy. *In* "Vascular and Neurological Changes in Early Diabetes" (R. A. Camerini-Davalos and H. S. Cole, eds.), pp. 417–424. Academic Press, New York.

Gabbay, K. H. (1972b). Purification and immunological identification of bovine retinal aldose reductase. *Isr. J. Med. Sci.* **8,** 1626–1629.

Gabbay, K. H., and Cathcart, E. S. (1969). Purification of kidney aldose reductase. *Clin. Res.* **17,** 609.

Gabbay, K. H., and Kinoshita, J. H. (1972). Mechanism of development and possible prevention of sugar cataracts. *Isr. J. Med. Sci.* **8,** 1557–1561.

Gabbay. K. H., and O'Sullivan, J. B. (1968a). The sorbitol pathway: Enzyme localization and content in normal and diabetic nerve and cord. *Diabetes* **17,** 239–243.

Gabbay, K. H., and O'Sullivan, J. B. (1968b). The sorbitol pathway in diabetes and galactosemia: Enzyme localization and changes in kidney. *Diabetes* **17,** 300.

Gabbay, K. H., and Snider, J. (1972). Nerve conduction defect in galactose-fed rats. *Diabetes* **21,** 295–300.

Gabbay, K. H., and Tze, J. (1972). Inhibition of glucose induced insulin release by aldose reductase inhibitors. *Proc. Nat. Acad. Sci. U.S.* **69,** 1435–1439.

Gabbay, K. H., Merola, L. O., and Field, R. A. (1966). Sorbitol pathway: Presence in nerve and cord with substrate accumulation in diabetes. *Science* **151,** 209–210.

Gregersen, G. (1968). Variations in motor conduction velocity produced by acute changes of the metabolic state in diabetic patients. *Diabetologia* **4,** 273–277.

Hayman, S., and Kinoshita, J. H. (1965). Purification of lens aldose reductase. *J. Biol. Chem.* **240,** 877–882.

Hers, H. G. (1956). Le mécanisme da la transformation de glucose en fructose par les vesicules seminales. *Biochim. Biophys. Acta* **22,** 202–203.

Hers, H. G. (1959). Le mécanisme de la formation du fructose séminale et du fructose foetal. *Biochim. Biophys. Acta* **37,** 127–138.

Hubbard, R. S., and Russell, N. M. (1937). The fructose content of spinal fluid. *J. Biol. Chem.* **119,** 547–661.

Huggett, A., St. G., Warren, F. L., and Warren, N. V. (1951). The origin of the blood fructose of the foetal sheep. *J. Physiol. (London)* **113,** 258–275.

Kinoshita, J. H. (1965). Cataracts in galactosemia. *Inves. Ophthalmol.* **4,** 786–799.

Kinoshita, J. H., Futterman, S., Satoh, K., and Merola, L. O. (1963). Factors

482   *Kenneth H. Gabbay*

affecting the formation of sugar alcohols in ocular lens. *Biochem. Biophys. Acta* **74,** 340–350.

Kinoshita, J. H., Dvornik, D., Kraml, M., and Gabbay, K. H. (1968). The effect of an aldose reductase inhibitor on the galactose cataract. *Biochim. Biophys. Acta* **158,** 472–475.

Knowles, H. C., Guest, G. M., Lampe, J., Kissler, M., and Skillman, T. G. (1965). The course of juvenile diabetes treated with unmeasured diet. *Diabetes* **14,** 239–273.

LeFevre, P.G., and Davies, R. I. (1951). Active transport into human erythrocyte: Evidence from comparative kinetics and competition among monosaccharides. *J. Gen. Physiol.* **34,** 515–524.

McCorkindale, J., and Edson, N. L. (1954). Polyol dehydrogenases 1. The specificity of rat-liver polyol dehydrogenase. *Biochem. J.* **57,** 518–523.

Mann, T. (1946). Studies on the metabolism of semen. 3. Fructose as a normal constituent of seminal plasma. *Biochem. J.* **40,** 481–491.

Mann, T., and Parsons, U. (1950). Studies on the metabolism of semen. 6. Role of hormones, effect of castration, hypophysectomy and diabetes. Relation between blood glucose and seminal fructose. *Biochem. J.* **46,** 440–450.

Mano, Y., Suzuki, K., Yawata, K., and Schimazono, N. (1961). Enzymic studies on TPN l-hexonate dehydrogenase from rat liver. *J. Biochem. (Tokyo)* **49,** 618–634.

Moonsammy, G. I., and Stewart M. A. (1967). Purification and properties of brain aldose reductase and l-hexonate dehydrogenase. *J. Neurochem* **14,** 1187–1193.

Orr, A. P. (1924). Laevulose in blood of the human foetus. *Biochem. J.* **18,** 171–172.

Pirart, J. (1965). Diabetic neuropathy: A metabolic or a vascular disease? *Diabetes* **14,** 1–9.

Prockop, L. D. (1971). Hyperglycemia, polyol accumulation, and increased intracranial pressure. *Arch. Neurol.* **25,** 126–140.

Smith, M. G. (1962). Polyol dehydrogenase 4. Crystallization of the l-iditol dehydrogenase of sheep liver. *Biochem. J.* **83,** 135–144.

Stewart, M. A., and Passonneau, J. V. (1964). Identification of fructose in mammalian nerve. *Biochem. Biophys. Res. Commun.* **17,** 536–541.

Stewart, M. A., Passonneau, J. V., and Lowry, O. H. (1965). Substrate changes in peripheral nerve during ischaemia and Wallerian degeneration. *J. Neurochem.* **12,** 719–727.

Stewart, M. A., Sherman, W. R., and Anthony, S. (1966). Free sugars in alloxan diabetic nerve. *Biochem. Biophys. Res. Commun.* **22,** 488–491.

Stewart, M. A., Sherman, W. R., Kurien, M. M., Moonsammy, G. I., and Wisgerhof, M. (1967). Polyol accumulations in nervous tissues of rats with experimental diabetes and galactosemia. *J. Neurochem.* **14,** 1057–1066.

Thomas, P. K., and Lascelles, R. G. (1965). Schwann cell abnormalities in diabetic neuropathy. *Lancet* **1,** 1355–1357.

Travis, S. F., Morrison, A. D., Clements, R. S., Winegrad, A. I., and Oski, F. A. (1971). Metabolic alterations in the human erthrocyte produced by increases in glucose concentration: The role of the polyol pathway. *J. Clin. Invest.* **50,** 2104–2112.

Wick, A. N., and Drury, D. R. (1951). Action of insulin on the permeability of cells to sorbitol. *Amer. J. Physiol.* **166,** 421–423.

CHAPTER 29

Experimental Models of Diabetes

A. M. COHEN, A. TEITELBAUM, S. BRILLER,
L. YANKO, E. ROSENMANN, AND E. SHAFRIR

Despite the increasing body of knowledge and research, the syndrome of maturity-onset diabetes mellitus is still ill-defined. The syndrome has several outstanding characteristics. These prominent characteristics are (1) genetic predisposition, (2) interaction of genetic and environmental factors (obesity, diet, stress, etc.), (3) age dependence, (4) alterations in enzymatic activity, (5) insulin responses to glucose loading, and (6) vascular complications—mainly renal and retinal angiopathy. The problem of whether the control of hyperglycemia prevents the development of diabetic angiopathy has not yet been resolved.

The tendency toward metabolic decompensation, which is accompanied by ketoacidosis, although a characteristic of juvenile diabetes, is not an outstanding feature of the adult-onset diabetes, unless complicated, and will not be discussed here.

To elucidate some of these problems, the investigators sought animals with hyperglycemia as models for testing of hypotheses possibly applicable to man. It is of advantage that the laboratory animals may be studied in strictly controlled conditions, and in a short time relative to the lifetime of the investigator, which is not possible in the case of human diabetes.

The first experimental models of diabetes were obtained by pancreatectomy and by the time-honored method of chemically destroying the pancreas by alloxan (Dunn and McLetchie, 1943), and later by streptozotocin (Rakieten *et al.*, 1963). A prompt and transient diabeticlike syndrome may be produced by injections of insulin antiserum (Moloney and Coval, 1955; Armin *et al.*, 1960) or D-mannoheptulose (Simon *et al.*, 1961), whereas persistent diabetes is obtained after prolonged administration of growth hormone (Young and Corner, 1960). In later years, small rodents with various hyperglycemic syndromes were used, on which comprehensive reviews have been published during the last few years (Renold, 1968; Bray and York, 1971; Stauffacher *et al.*, 1971; Renold *et al.*, 1972).

As a model to investigate the effect of carbohydrate diets on the pathogenesis of diabetes, the common albino rat was used. In the following, we shall deal with some of the characteristics of nutritionally induced diabetic syndrome in our model and compare it with other experimental models of diabetes.

A. Metabolic Changes on Sucrose and Starch Diets

Yemenite and Kurdish Jewish immigrants to Israel were found to have an extremely low prevalence of diabetes. After a few years, the prevalence of diabetes has risen (Cohen, 1961) to a level comparable to the Western countries. In a survey, it was found that the diet of the Yemenite immigrants had contained almost no sucrose in their country of origin, while the quantity they consumed in Israel equaled that consumed by the immigrants from the Western countries; we have suggested that this dietary difference might be one of the reasons for the greater prevalence of diabetes in long-established Yemenite settlers and in the population in general (Cohen *et al.*, 1961). A similar effect of high sucrose intake on the incidence of diabetes has been found in South African Zulus (Campbell, 1963).

Feeding rats synthetic diets (Table I) resulted in impairment of the glucose tolerance and reduction of serum insulinlike activity when the carbohydrate given was sucrose in contrast to starch (Cohen and Teitelbaum, 1964). This was accompanied by an increased liver fat content and lipogenesis from acetate and other precursors (Portman *et al.*, 1956; MacDonald, 1962). Rises in the activity of some hepatic enzymes associated with glycolysis and lipogenesis have also been noted, which were greater on sucrose than on starch (Yudkin and Krauss, 1967; Bartley *et al.*, 1967; Bailey *et al.*, 1968). This was especially evident in long-term experiments (Cohen *et al.*, 1972a).

TABLE I

Percent Composition of Synthetic Diets by
Weight

Ingredient \ Diet	Starch	Sucrose
Casein	18	18
Cornstarch	72	0
Sucrose	0	72
Butter	5	5
Salts[a] and vitamins[b]	5	5

[a] USP salt mixture II.

[b] Thiamine hydrochloride, 0.6 mg; pyridoxo-hydrochloride, 0.5 mg; riboflavin, 0.6 mg; calcium pantothenate, 1.6 mg; biotin, 0.06 mg; vitamin B_{12}, 0.004 mg; folic acid, 0.6 mg; nicotinamide, 10 mg; vitamin A palmitate, 0.6 mg; vitamin D_2, 0.004 mg; inositol, 100 mg; choline chloride, 100 mg; per 100 gm of food.

The reason for the more marked response of liver enzymes of glycolysis and lipogenesis should be traced to the greater capacity of the liver to metabolize the fructose than the glucose component of sucrose (Pereira and Yangaard, 1971) since the activity of liver fructokinase exceeds severalfold the combined activities of hexokinase and glucokinase (Zakim et al., 1967). In addition, the carbohydrate load imposed on the liver in the case of the sucrose is higher since the extrahepatic metabolism of its fructose component is lower than that of glucose. Thus, the sucrose diet constitutes a greater stimulus for the induction of liver enzymes involved in the dissimulation of hexoses and their conversion into fat.

Another observation, possibly related to the development of intolerance to glucose, concerns the rise in the activity of hepatic glucose-6-phosphatase activity in rats on sucrose diets (Cohen *et al.*, 1972a). The enhanced activity of this enzyme indicates that coincident with glycolysis there is a substantial flow of fructose into gluconeogenesis. Increased activity of glucose-6-phosphatase and of glucokinase would imply an increased rate of recycling at the stage of glucose phosphorylation. This, in turn, may lead to a delay in the uptake from the circulation of the glucose moiety of sucrose and may set in motion, in susceptible animals, a chain of compensatory events such as hyperglycemia, hyperinsulinemia, and insulin resistance. Similar observations were also made

in spiny mice trapped in the desert and transferred to synthetic carbohydrate diets (Shafrir *et al.*, 1972a).

Human volunteers fed sucrose or starch for alternating periods of 5
weeks have also shown a reduced glucose tolerance and an increase in
serum cholesterol (Cohen *et al.*, 1966). At the time, an individual variation in men and experimental animals in the extent of reduction of the
glucose tolerance and rise in serum cholesterol, as a response to sucrose,
was noted (Cohen, 1967).

B. Interaction of Genetic Predisposition and Diet
in the Production of Hyperglycemia

To investigate more thoroughly the factors leading to glucose intolerance on sucrose diet, a group of 28 male and 19 female rats randomly
taken from the general stock (parent generation) were kept for 2 months
on the sucrose diet and then tested orally for glucose tolerance. The males
and females with the highest rises in blood glucose at 60 min following
the glucose load were mated, and their progeny was referred to as the
"upward" selection. The males and females with the lowest blood glucose
values were also mated, and their progeny was referred to as the "downward" selection (Cohen *et al.*, 1972c). The offspring of these two lines
were separated at the age of 21 days and kept apart, each litter being
divided into two groups, one group fed the sucrose diet, and their siblings
the starch diet. After 2 months, a glucose tolerance test was performed,
and the selection procedure outlined above was applied to the subsequent
generations S_2, S_3, S_4, and S_5.

Figure 1 shows the blood glucose values 60 min after an intragastric
glucose load in the upward and downward selected lines. In each line,
part of the siblings were placed on sucrose diet and another part on starch
diet. It is to be noted that in the succeeding generations in the upward
selected line, which were fed sucrose, the level of blood glucose rose gradually, and a considerable number of animals developed a diabeteslike
syndrome, whereas their siblings, which were fed starch, did not show a
rise in blood glucose. On the other hand, in the offspring of the downward
selected line, blood glucose did not rise significantly, neither on the sucrose nor the starch diet. This finding points to the need for interaction
between the genetic factor(s) and the dietary factor for expression of
the diabeteslike syndrome.

The appearance of diabetic features was studied with the aid of the
following parameters: serum lipids; obesity; activities of liver enzymes
of glycolysis, gluconeogenesis, and lipogenesis; serum insulin on fasting

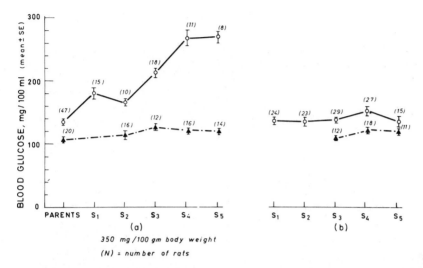

Fig. 1 Blood glucose values (mg %) at 60 min after a gastric glucose load (350 mg/100 gm body weight) in the parent and siblings of the (a) upward and (b) downward selected generations fed sucrose or starch: (○) sucrose-fed rats and (▲) starch-fed rats.

and after a glucose stimulus; insulin resistance; and retinal and renal angiopathy.

1. Obesity

Figure 2 presents the main body weight in the upward and downward selected lines. There was no significant difference in body weight between the hyperglycemic sucrose-fed rats of the upward selected line and the sucrose-fed normoglycemic rats of the downward selected line. In fact, the siblings of these two lines, which were fed starch and remained normoglycemic, did weigh more. Thus, there was no evidence of obesity accompanying the development of hyperglycemia. It is to be noted that in the succeeding generations, there was a tendency of weight increase both in the lines fed starch or sucrose. Again, this tendency was similar both in the hyperglycemic and in the normoglycemic rats and could not be related to hyperglycemia per se. Thus, obesity that usually accompanies genetically transmitted diabetes in rodents, except the Chinese hamster, is not present in our model.

2. Liver Enzyme Pattern

The activities of hepatic enzymes of the pathways of glycolysis, gluconeogenesis, and lipogenesis are shown in Table II. The activity of these

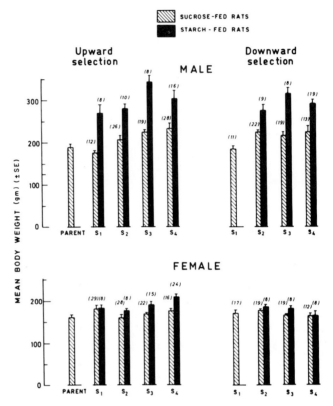

Fig. 2 Mean body weight of the parent and siblings of the upward and downward selected generations after 60 days of sucrose or starch feeding.

enzymes is expressed as a ratio of values in rats fed sucrose versus those fed starch, both in the parent generation and in the siblings of the selected generations, after 1, 5, and 10 months on the respective diets. Since the enzymatic activity changes with age, a separate comparison was made for each group of the sucrose-fed animals with respect to their starch-fed siblings, which were taken as 100%.

As mentioned before, in the parent generation, the activity of enzymes of glycolysis and lipogenesis increased in the sucrose-fed animals to a larger extent than that of the starch-fed controls (Cohen *et al.*, 1972a). There was no significant change in the activity ratio of the enzymes involved in gluconeogenesis. In the upward selected generations, 1 month of sucrose feeding already produced a relatively larger increase in the activity of enzymes involved in lipogenesis and glycolysis than a comparable period of starch feeding. However, the sucrose:starch activity ratio was lower than in the parent generation; with the prolongation of

TABLE II

Liver Enzyme Activity Ratio Sucrose: Starch[a] of Parents and Selected Generations

Group[b]	Age (month)	Pyruvate kinase	G6PDH	Malate enzyme	Acetyl-CoA carboxylase	Phosphoenol-pyruvate carboxykinase	Alanine amino-transferase (GPT)
Parents							
(9)	10	150	255	170	180	95	0
Selected generations							
(9)	1	120	160	185	115	40	108
(9)	5	460	315	320	130	70	115
(6)	10	98	150	115	80	150	155

[a] Activities of starch-fed animals taken as 100%.
[b] Numbers in parentheses indicate number of animals.

the sucrose feeding period to 6 months, the sucrose:starch activity ratio became even higher in the selected generations. On the other hand, the activity of enzymes involved in gluconeogenesis became depressed in the sucrose-fed animals. With continuation of sucrose feeding for 9 months, the activity ratio markedly decreased in the case of enzymes of glycolysis and lipogenesis; in contrast, the activity ratio in the case of enzymes associated with gluconeogenesis increased, indicating a shift of enzymatic activities into a diabeteslike pattern.

In the genetically transmitted hyperglycemia in rodents, the activities of liver enzymes of glycolysis and lipogenesis are generally elevated. The activity of glycolytic enzymes is markedly increased in the *obob* (Seidman *et al.*, 1967, 1970), *dbdb* (Chang and Schneider, 1970), and *KK* mice (Nakashima, 1969). The activity of lipogenic enzymes is also increased in the *obob* (Chang *et al.*, 1967; Lochaya *et al.*, 1963), the *dbdb* (Coleman and Hummel, 1967), and the *KK* mice (Kato, 1969).

This behavior is in agreement with the hyperinsulinemia concomitant with obesity in these strains. However, at the same time, there is an inconsistent increase in the activity of the phosphoenolypyruvate carboxykinase, fructose-1,6-diphosphatase, and glucose-6-phosphatase in *dbdb* mice (Chang and Schneider, 1970), *obob* mice (Seidman *et al.*, 1970), and *KK* mice (Nakashima, 1969).

Enhancement of enzymes of gluconeogenesis with a suppression of those of glycolysis and lipogenesis is characteristic of insulin deficiency, which ensues after alloxan treatment or pancreatectomy (Shrago *et al.*, 1963; Migliorini, 1971). Enhancement of enzymes of gluconeogenesis together with those of glycolysis and lipogenesis in mice with hereditary hyperglycemia, obesity, and hyperinsulinemia may result from the failure of insulin to suppress liver gluconeogenesis and may represent an expression of a selective insulin resistance.

In our model, the first few months of carbohydrate feeding are associated with a rise in the activity of glycolytic and lipogenic enzymes and a decrease in the gluconeogenic enzymes of the dicarboxylic acid shuttle (Cohen *et al.*, 1972a). This is not an adaptive response to obesity or overfeeding since the hyperglycemic sucrose-fed rats of the upward selected lines do not gain weight when compared to their normoglycemic starch-fed siblings. Thus, the pattern of response in enzymatic activities seems dependent not only on the amount of carbohydrate intake but also on the molecular properties of the carbohydrate consumed. Furthermore, in our model, after longer periods of feeding, and particularly in succeeding generations, the pattern of hepatic enzymatic activity changes toward a typically diabetic one, most probably in connection with the developing general insulin resistance of the tissues.

3. Plasma Insulin

The fasting plasma insulin levels (Fig. 3) of the downward selected line (normoglycemic), sucrose- or starch-fed, were not statistically different from those of the starch-fed animals of the upward selected line. On the other hand, the fasting insulin levels in the succeeding generations of the sucrose-fed animals in the upward selected line (hyperglycemic) were significantly greater.

The dynamics of insulin release at 30 and 60 min following a glucose load is shown in Fig. 4. On glucose stimulation, the sucrose- and starch-fed siblings of the downward selected line exhibit no difference in the insulin response, whereas there is a greater response in both sucrose- and starch-fed animals of the upward selected line.

It is to be noted that there is no direct relationship between the plasma glucose level and the insulin response and that different animals with high glucose values may have high, normal, or low insulin responses.

4. Insulin Resistance

Figure 5 demonstrates the insulin resistance, expressed as the extent of blood glucose decrease from the fasting level at 15 and 30 min after an intravenous injection of 30 milliunits of insulin/100 gm body weight. The figure shows that the upward and downward selected lines differ in

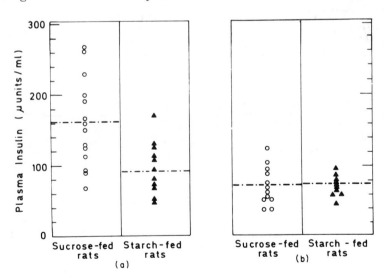

Fig. 3 Fasting plasma insulin levels of the siblings of the (a) upward and (b) downward selected lines after 30 days of sucrose and starch feeding.

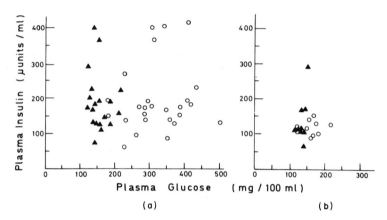

Fig. 4 The correlation between the plasma insulin and the plasma glucose at 60 min after a glucose load (350 mg/100 gm body weight) in the siblings of the (a) upward and (b) downward selected generations after 60 days of (○) sucrose or (▲) starch feeding.

their sensitivity to insulin. In the downward selected line, both the sucrose- and starch-fed animals show a decreased insulin resistance in relation to that of the parent generation fed the standard laboratory chow. In the upward selected line, the starch-fed animals show an insulin resis-

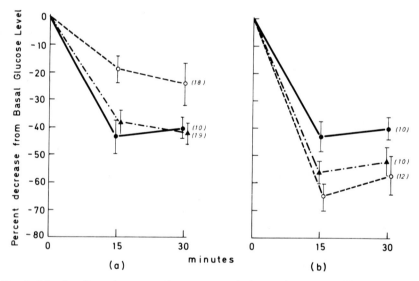

Fig. 5 The insulin resistance to intravenous administration of 30 milliunits/100 gm body weight of the siblings of the (a) upward and (b) downward selected generations after 60 days of sucrose or starch feeding: (○) sucrose-fed rats, (▲) starch-fed rats, and (●) parent generation.

tance comparable to that of the parent generation on the chow, whereas sucrose feeding markedly increased the insulin resistance in their siblings.

Differences in insulin resistance may explain the normoglycemia and the low plasma insulin levels observed in the downward selected line, even when fed sucrose. It appears that the low circulating insulin level in the presence of low insulin resistance is sufficient to maintain normal blood glucose levels, even when the rats are challenged by the sucrose diet.

The starch-fed upward selected line has a greater insulin resistance than the downward selected line. The resistance is similar in magnitude to that of the parent generation and is accompanied by a greater insulin response, which is sufficient to maintain a normal blood glucose level. Sucrose feeding causes the insulin resistance to increase, forcing insulin production to its peak; apparently, the pancreas is unable to produce sufficient insulin to meet the continued challenge of the high sucrose diet and hyperglycemia ensues.

The upward selected line may have a common genetic background expressing itself by the fact that on the challenge of a high sucrose diet, it (a) is more liable to develop an insulin resistance and (b) produces a greater amount of insulin. In these two respects, it differs from the downward selected line, which has a decreased insulin resistance, accompanied by normal amounts of plasma insulin, neither of which is affected by a high sucrose diet. Whether the nature or action of insulin in the upward selected line is normal or additional factors are present modifying the responsiveness of the tissues to insulin remains to be determined. However, the question of whether a similar genetic difference is present in humans, making some individuals prone to diabetes upon the long-lasting challenge of a high carbohydrate diet, especially one which is rich in sucrose, has likewise to be studied.

C. Vascular Complications

The renal lesions, intercapillary glomerulosclerosis, the occurrence of microaneurysms in the posterior region of the retina, and a progressive thickening of the capillary basement membrane in different organs are considered specific for diabetes mellitus.

1. Renal Lesions

Specimens from the kidney, taken for light and electron microscopy (Cohen and Rosenmann, 1971; Rosenmann *et al.*, 1971), showed that the renal alterations (Figs. 6 and 7) in the sucrose-fed rats consisted of

494   *Cohen, Teitelbaum, Briller, Yanko, Rosenmann, and Shafrir*

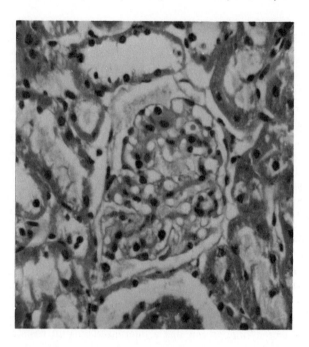

Fig. 6 Renal corpuscle of a control starch-fed rat of the upward selected line. Note normal glomerular tuft. Hematoxylin and eosin. ×420.

diffuse intercapillary glomerulosclerosis. In several instances, exudative or lipohyaline glomerular lesions were conspicuous. Arterial and arteriolar sclerosis were prominent in a few cases. The electron microscopic examination corroborated and extended the findings observed by the light microscope. The glomerular mesangial matrix was markedly increased (Fig. 8). The peripheral basement membrane was thickened, as evaluated by numerous measurements. In a few mesangial areas, collagen fibrils were observed. The exudative lesions consisted of extracellular fibrinoid deposited between the basement membrane and the endothelial cells, the cytoplasm of which contained many lipid droplets. Such changes were not observed in any of the starch-fed siblings. These renal changes described herein are similar to those found in alloxan diabetic rats (Orskov *et al.*, 1965; Mann *et al.*, 1951), golden hamsters (Beaser *et al.*, 1964), dogs (Bloodworth, 1965), and monkeys (Gibbs *et al.*, 1966); pancreatectomized rats (Foglia *et al.*, 1950) and dogs (Ricketts *et al.*, 1959); after administration of pituitary extract in the dog (Ricketts *et al.*, 1959); after immunization with insulin in guinea pigs (Andreev, 1970); hereditary diabetes mellitus in the Chinese hamster (Lawe, 1962; Shirai *et al.*, 1967), *obob* mice (Hellman, 1965); the Geneve colony of spiny mice

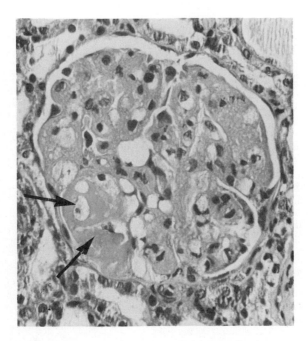

Fig. 7 Representative renal corpuscle of a sucrose-fed rat. There is diffuse glomerulosclerosis and typical exudative (lipohyalin) lesions (arrows). Hematoxylin and
eosin. ×420.

(Gonet *et al.*, 1965b); *KK* mice (Camerini-Dávalos *et al.*, 1970); dogs
(Bloodworth, 1965); and man (Dachs *et al.*, 1964).

2. Retinal Changes

The retinal changes consisted of loss of mural and endothelial cells
of the capillaries, strand formation (Figs 9 and 10), and microaneurysms
(Fig. 11). The vascular changes of the retinas of these animals were
studied in a digested pepsin-trypsin flat preparation (Cohen *et al.*, 1972b).
These changes were observed only very rarely in other experimental models. In surgically pancreatectomized rats, retinal changes have not been
observed (Levine *et al.*, 1963). In neither the streptozotocin diabetic
Chinese hamster (Sibay *et al.*, 1971) nor in the growth hormone with hydrocortisone diabetic Chinese hamster (Pometta *et al.*, 1966), was retinal
changes noted. Mural cell loss and acellularity of the capillaries were
observed in spontaneously diabetic dogs (Patz *et al.*, 1965). They were
also observed in 1 alloxan diabetic dog after 53 months of diabetes (Engerman and Bloodworth, 1965) and in alloxan diabetic rats (Lundboek,

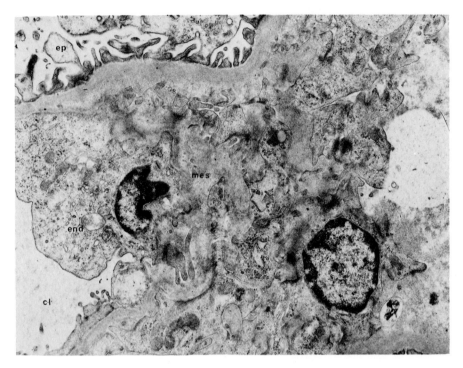

Fig. 8 Electron micrograph of a glomerular mesangial region of a sucrose-fed rat. There is a marked increase of the mesangial matrix (mes). Abbreviations: end, endothelial cell; ep, epithelial cell; and cl, capillary lumen. ×20235.

et al., 1967). In somatotropin-induced diabetes, it was observed in 2 dogs 67 and 69 months after the induction of diabetes (Engerman and Bloodworth, 1965), and in a 10-year-old dog with diabetes which lasted 8 years (Hausler *et al.*, 1964). They were also observed in the albino rat with diabetes induced by growth hormone and hydrocortisone (Agarwal and Agarwal, 1965).

Retinal microaneurysms were observed in 2 out of 12 spontaneously diabetic dogs (Patz *et al.*, 1965; Sibay and Hausler, 1967). Microaneurysms are exceptional in the experimentally induced diabetic animal. While in alloxan diabetic rabbits treated with ACTH, retinal microaneurysms were described (Becker, 1952), this observation has not been confirmed (Engerman *et al.*, 1964). Similarly, in a few metasomatotropin-induced diabetic Chinese hamsters, aneurysms of the capillaries were observed (Hausler *et al.*, 1963); however, this observation could not be reproduced in a similar experiment (Pometta *et al.*, 1966). In 5 alloxan diabetic rats injected with growth hormone and hydrocortisone, micro-

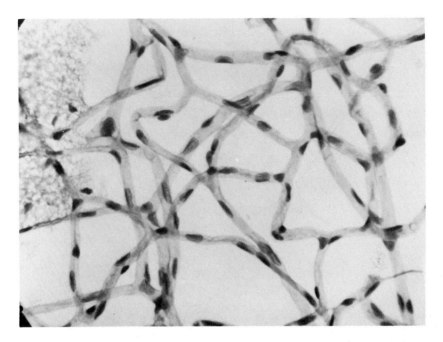

Fig. 9 Digested retina of a starch-fed (control) rat showing normal cellularity of the capillaries. Periodic acid-Schiff and hematoxylin. ×250.

aneurysms were observed (Agarwal and Agarwal, 1965). In 1 alloxan diabetic dog, microaneurysms were also described (Engerman *et al.*, 1964). Similarly, in the metasomatotropin- and corticotropin-induced diabetic dogs, whose diabetes lasted for more than 4 years, capillary aneurysms have been observed in 3 dogs (Engerman *et al.*, 1964; Hausler *et al.*, 1967).

3. Relation of Microangiopathy to Metabolic or Genetic Factor(s)

From the standpoint of the development of microvascular complications, especially retinopathy and nephropathy, and its relationship to effective control of the metabolic disturbances in diabetes, there are two main viewpoints. According to Siperstein (Siperstein *et al.*, 1968), muscular capillary basement membrane thickening results from the genetic defect responsible for diabetes mellitus and is found independently of the metabolic changes. However, this was not confirmed in a similar study (Williamson *et al.*, 1971). No reference will be made here as to

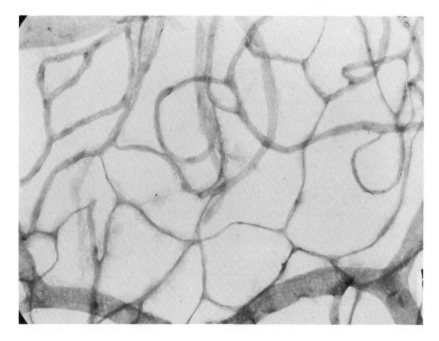

Fig. 10 Digested flat retina of a sucrose-fed rat showing gross diminution of mural and endothelial cells and strand formation. Periodic acid-Schiff and hematoxylin. ×250.

the clinical reports of the effect of treatment and prevention of the vascular complications in diabetes since this is not within the scope of this paper.

In the data obtained from the Chinese hamster, KK mice, sand rats, *obob* and *dbdb* mice (Siperstein, 1970; Siperstein *et al.*, 1968), and in the spiny mice of the Geneve colony (Creutzfeldt *et al.*, 1970), no increase in the thickening of the basement membrane was reported as a hereditary state or in the diabetic state.

In order to answer the question whether angiopathy is a result of the metabolic changes accompanying the state of diabetes or is independent and solely related to a genetic factor, sucrose- and starch-fed siblings of the upward and downward selected lines were studied.

Figure 12 shows the incidence of diffuse glomerulosclerosis in sucrose-fed animals and their starch-fed siblings. In the sucrose-fed animals of the upward selected line, the incidence of nephropathy varied from 8 to 14% in all age groups, whereas no nephropathy was found in the sucrose-fed animals of the downward line. Neither the starch-fed animals of the upward nor the downward selected line showed renal pathology. Further-

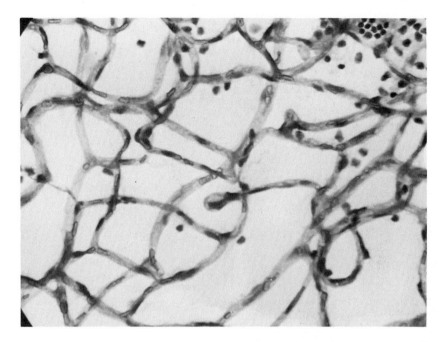

Fig. 11 Digested flat retina of a sucrose-fed rat showing microaneurysms. Periodic acid-Schiff and hematoxylin. ×420.

more, in the upward selected line, the nephropathy appeared as early as the age of 3–5 months.

Figure 13 shows that the incidence of the vascular retinopathy in the upward selected line of the sucrose-fed rats was as high as 32% while no vascular changes were observed in their starch-fed siblings. Vascular retinopathy in the sucrose-fed rats was also observed as early as the age of 4 months. These findings clearly demonstrate that the sucrose-fed animals of the upward selected line which developed hyperglycemia also developed retinal and renal angiopathy. The starch-fed siblings with the same genetic pattern remained normoglycemic and did not develop angiopathy. This indicated that the genetic factor(s) alone is not sufficient to induce the pathological angiopathy unless the animal develops "diabetes," i. e., hyperglycemia with its accompaning metabolic changes. This experiment supports the view that the development of diabetic angiopathy is not independent of the metabolic changes and that it is not related solely to the genetic factor(s).

By preventing the metabolic changes in an individual with the genetic tendency to develop vascular angiopathy, one may be able to prevent or delay the appearance of vascular changes.

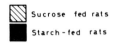

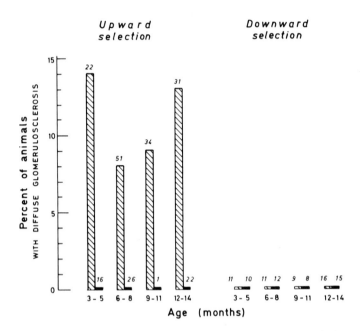

Fig. 12 Incidence of glomerulosclerosis at different ages in the siblings of the upward and downward selected lines on sucrose or starch feeding.

D. Comparison of the Sucrose-Induced Diabetes in Selected Rats with Other Experimental Models

To compare our model with other experimental models, the outstanding characteristics of the experimentally diabetic models have been listed. These models have been categorized into three groups: (a) genetically hyperglycemic and/or obese, (b) environmentally induced hyperglycemia, and (c) hyperglycemia induced by surgical or chemical pancreatectomy or by hormone treatment.

Table III lists the characteristics of the genetically transmitted hyperglycemia in rodents, which, as was said before, excluding the Chinese hamster, are associated with obesity. A change in environment has no effect on the development of the syndrome since it is genetically transmitted to them. In obese mice *obob* (Coleman and Hummel, 1968) and in New Zealand obese mice (NZO) (Sneyd, 1964), *increased* levels of insulin-

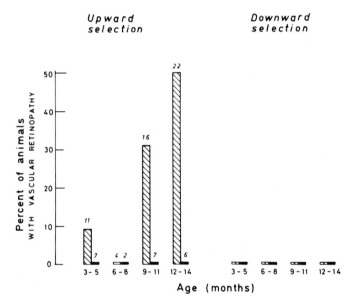

Fig. 13 Incidence of retinal vascular changes at different ages in the siblings of the upward and downward selected lines on sucrose or starch feeding.

like activity (ILA) in serum has been exhibited. Insulin resistance was observed in the *obob* mouse (Batt and Mialhe, 1966), in the Yellow A^y mouse (Weitze 1940), and in the Chinese hamster (Carpenter *et al.*, 1967). In these experimental models, renal vascular lesions have been observed only in the Chinese hamster (Lawe, 1962; Shirai *et al.*, 1967), in *obob* mice (Hellman, 1965), and in *KK* mice (Camerini-Dávalos *et al.*, 1970), but no retinal changes have been noted although in a few metasomatotropin-induced diabetic Chinese hamsters (Hausler *et al.*, 1963), diabetic retinal changes, including microaneurysms, have been noted, but this observation could not be reproduced (Pometta *et al.*, 1966). In several of these rodents, hyperinsulinemia (Coleman and Hummel, 1967; Hellerstrom and Hellman, 1963) and insulin resistance (Batt and Mialhe, 1966) were observed, even in the absence of hyperglycemia, which usually precedes the advent of frank hyperglycemia (Coleman and Hummel, 1967, 1968; Genuth, 1969; Stauffacher *et al.*, 1967) and is related to hyperphagia (Chlouverakis and White, 1969; Malaisse *et al.*, 1968). The activity of the glycolytic, gluconeogenic, and

TABLE III

Genetic Hyperglycemia and/or Obesity Syndrome

Animals	Inheritance	Hyper-insulinemia	Insulin resistance	Obesity	Enzymatic activity			Vascular	
					Glycolysis	Gluconeo-genesis	Lipo-genesis	Renal	Retinal
Single gene mutants									
Mice									
Yellow Aᵛ	Autosomal dominant	+	+	+	?	?	?		−
Obese (*obob*)	Autosomal recessive	+	+	++	←	←	←	+	−
Diabetes (*dbdb*)	Autosomal recessive	+	+	++	←	←	←	+	−
Adipose (*adad*)	Autosomal recessive	+	+	++					−
Inbred strains and hybrids									
Mice									
New Zealand obese (NZO)	Inbred polygenic	+	−	+					−
KK	Inbred dominantᵃ	+	+	+	←	←	←	+	−
C₃HfXIF₁ (Wellesley)	Hybrid polygenic	+		++	←	←-	←	+	(
Chinese hamster, *Cricetus griseus*	Inbred recessive polygenic	+	+	−	?	?	?	+	−

ᵃ With revived penetrance.

lipogenic pathways of the liver is generally elevated in them. The increased enzymatic activity, which is not typically diabetic, reverses to normal on pair feeding (Bray *et al.*, 1970; Cahill *et al.*, 1967; Chang and Schneider, 1970), which could be interpreted as a result of overfeeding and obesity.

In the Chinese hamster, where diabetes is inherited (Yergenian, 1964), the disease appears to be quite similar to the so-called juvenile diabetes in man, in that no islet hypertrophy and hypersecretion of insulin have been demonstrated (Carpenter *et al.*, 1967) ; it is not inducible by changes in diet and is not accompanied or preceded by obesity (Yergenian, 1964). In these animals, mild insulinemia and insulin resistance have been described (Dulin and Gerritsen, 1967). Although increased basement membranelike material has been observed in the mesangial portion of the renal glomeruli in the diabetic animals, it also seems to occur as well in nondiabetic animals with increasing age (Shirai *et al.*, 1967). No retinal vascular lesions have been described.

Table IV lists the characteristics of hyperglycemic animals affected by the environmental conditions. The sand rat *Psammomys obesus*, has aroused particular interest as an experimental animal because of an apparent similarity to certain human diabetic prone populations. Studies of the various human population groups are, in general, consistent with the fact (Campbell, 1963; Cohen *et al.*, 1961) that diabetes is related to food intake. In a similar fashion, the sand rat, in its natural habitat, has a food supply that is relatively low in caloric content. When fed the high caloric laboratory diet, it develops obesity and overt diabetes mellitus, including cataract, glycosuria, hyperglycemia, hyperinsulinemia (Hackel *et al.*, 1966; Haines *et al.*, 1965; Schmidt-Neilsen *et al.*, 1964), and insulin resistance (De Fronzo *et al.*, 1967). Until now, renal and retinal lesions were not reported in the hyperglycemic sand rat which develops cataracts. On the other hand, when the sand rat is maintained on a restricted caloric intake, it loses weight, maintains normal glucose tolerance, and maintains a low plasma insulin (Schmidt-Neilsen *et al.*, 1964).

Spiny mice, *Acomys cahirinus*, bred in Geneve, tend to be obese and to develop hyperglycemia, but do not exhibit elevated levels of circulating insulin or exaggerated glucose-induced insulin responses (Renold *et al.*, 1972). However, congenital hyperplasia of the islets of Langerhans (Gonet *et al.*, 1965a) and high pancreatic insulin content (Stauffacher *et al.*, 1970) are present. The early response to several stimuli, including glucose, arginine, glucagon, isoprotorenol, aminophylline, and dibutyryl cyclic AMP is decreased in all instances, suggesting the existence of a disturbance in the insulin release mechanisms (Renold *et al.*, 1972). Thickening of the glomerular capillary basement membrane with advancing age has

TABLE IV

Environmentally Influenced Hereditary Hyperglycemia

Animals	Inheritance	Hyper-insulinemia	Insulin resistance	Obesity	Enzymatic activity			Vascular	
					Glycolysis	Gluconeo-genesis	Lipo-genesis	Renal	Retinal
sand rat *Psammomys obesus*	Polygenic (?)	++	+	++	−	−(?)	?		
spiny mouse *Acomys cahirinus* (Geneve strain)	Polygenic (?)	−?	?	+				+	

been observed in them (Gonet *et al.*, 1965b). On the other hand, spiny mice directly collected from their native habitat in Israel do not develop diabetes, obesity, or pancreatic changes, although they are markedly susceptible to high carbohydrate diets with respect to hyperlipidemia and glucose intolerance (Shafrir *et al.*, 1972a,b).

Table V shows the characteristics of experimental diabetes induced by pancreatectomy, chemicals, and hormones. The blood insulin level is reduced (Morgan and Lazarow, 1965). Renal lesions have been described in this group of animals as mentioned above. Although retinal lesions have been demonstrated in this group, as stated above, they are rare and exceptional.

Our model represents a selected out-population segment with genetic predisposition, which, on interaction with high sucrose intake, leads to the appearance of overt diabetes. There is no obesity; insulin resistance is not carried genetically; high, low, or normal levels of insulin have been noted at the appearance of hyperglycemia. When first put on the sucrose diet, our model demonstrated increased lipogenesis, glycolysis, and decreased gluconeogenesis (except an increase in glucose-6-phosphatase activity which is specific to fructose metabolism). With time, the enzymatic pattern turns into a diabetic one, i.e., increased gluconeogenesis and decreased lipogenesis and glycolysis. In this respect, our model differs from the rodents with the hereditary or environmentally induced hyperglycemic syndomes. In rodents with genetic hyperglycemic obesity, the Chinese hamster and the environmentally induced diabetic animals such as *Acomys cahirinus*, renal, but not retinal, lesions have been demonstrated. In animals with diabetes resulting from pancreatectomy, chemicals, or growth hormone, retinal lesions have been only rarely observed. In our model, both kinds of vascular changes have been shown frequently, pointing out a difference from the other experimental animals.

The need is stressed for the interaction of both genetic factor(s) and metabolic change in diabetes to induce the onset of diabetic nephropathy and retinopathy. By preventing the metabolic changes in an individual with the genetic tendency to develop vascular angiopathy, one may prevent the appearance of vascular changes.

It appears that our model, with hyperglycemia, closely resembles human adult-onset diabetes. It is compatible with the observations on the increased prevalence of diabetes in populations undergoing transition from a limited caloric and sucrose intake to the high caloric density food rich in sucrose in the so-called developed countries.

The genetic difference in insulin sensitivity and insulin response to a high sucrose diet between the upward and downward selected lines with the resultant hyper- and normoglycemia may serve as a model for the

TABLE V

Hyperglycemia Induced by Surgical Pancreatectomy, Chemicals, and Hormones

Animals	Hyper-insulinemia	Insulin resistance	Obesity	Enzymatic activity			Vascular		
				Glycolysis	Gluconeo-genesis	Lipo-genesis	Renal	Retinal S & A[a]	MA[b]
Surgical pancreatectomy									
Rat	↓		−	↓	↑	↓	+	+	−
Dog							+	−	−
Chemicals									
Alloxan									
Golden hamster									
Rat	↓		−	↓	↑	↑	+++	++	+
Dog (+somatotropin)							+++	++	+
Monkey							++		
Streptozotocin									
Chinese hamster								−	−
Growth hormone and hydrocortisone									
Rat	↑	↑					+	+	−
Chinese hamster			±	±	↑	±	−	−	±
Insulin antibodies									
Guinea pig	↓	?	−				+	?	?

[a] Strand formation and acellularity.
[b] Microaneurysms.

varying prevalence of diabetes in a particular or in different populations. Individuals or groups with different genetic sensitivity to a high sucrose intake may develop, accordingly, high or normal blood glucose values, thus suggesting why, in the community consuming the same amount of sucrose, a certain percentage of the population will develop diabetes, while others will not. Furthermore, a genetically prone individual may be prevented from developing diabetes by avoiding the challenge of a high sucrose diet. The necessity of long periods of feeding with the noxious diet for the animals to develop the impaired tolerance to carbohydrates and the accompanying metabolic and vascular changes, explains, in part, the age-dependent rise in the incidence of diabetes.

References

Agarwal, L. P., and Agarwal, P. K. (1965). The retinal vascular pattern: General consideration. *ENT Mon.* **44,** 94–101.

Andreev, D. (1970). Diabetes-like vascular lesions in the kidneys of guinea pigs immunized with an insulin-adjuvant mixture. *Acta Diabetol. Lat.* **7,** 243–259.

Armin, J., Grant, R. T., and Wright, P. H. (1960). Acute insulin deficiency provoked by single injections of anti-insulin serum. *J. Physiol. (London)* **153,** 131–138.

Bailey, E., Taylor, C. B., and Bartley, W. (1968). Effect of dietary carbohydrate on hepatic lipogenesis in the rat. *Nature (London)* **217,** 471–472.

Bartley, W., Dean, D., Taylor, C. B., and Bailey, E. (1967). The effect on some enzymes of rat tissue of diets low in fat content. *Biochem. J.* **103,** 550–555.

Batt, R., and Mialhe, P. (1966). Insulin resistance of the inherited obese mouse— obob. *Nature (London)* **212,** 289–290.

Beaser, S. B., Matthew, F. S., Donaldson, G. W., McLaughlin, R. J., and Sommers, S. C. (1964). Alloxan diabetes in the golden hamster *Mesocricetus auratus. Diabetes* **13,** 49–53.

Becker, B. (1952). Diabetic retinopathy. *Ann. Intern. Med.* **37,** 273–289.

Bloodworth, J. M. B., Jr. (1965). Experimental diabetic glomerulosclerosis. II. The dog. *Arch. Pathol.* **79,** 113–125.

Bray, G. A., and York, D. A. (1971). Genetically transmitted obesity in rodents. *Physiol. Rev.* **51,** 598–646.

Bray, G. A., Barry, W. S., and Mothon, S. (1970). Lipogenesis in adipose tissue from genetically obese rats. *Metab., Clin. Exp.* **19,** 839–848.

Cahill, G. F., Jr., Jones, E. E., Lauris, V., Steinke, J., and Soeldner, J. S. (1967). Studies on experimental diabetes in the Wellesley hybrid mouse. II. Serum insulin levels and response of peripheral tissues. *Diabetologia* **3,** 171–174.

Camerini-Dávalos, R. A., Opperman, W., Mittle, R., and Ehreneich, T. (1970). Studies of vascular and other lesions in *KK* mice. *Diabetologia* **6,** 324–329.

Campbell, G. D. (1963). Diabetes in Asians and Africans in and around Durban. *S. Afr. Med. J.* **37,** 1195–1208.

Carpenter, A. M., Gerritsen, G. C., Dulin, W. E., and Lazarow, A. (1967). Islet and beta cell volumes in diabetic Chinese hamsters and their nondiabetic siblings. *Diabetologia* **3,** 92–96.

Chang, A. Y., and Schneider, D. I. (1970). Abnormalities in hepatic enzyme activities during development of diabetes in *db* mice. *Diabetologia* **6,** 274–278.

Chang, H. C., Seidman, I., Teebor, G. W., and Lane, M. D. (1967). Liver acetyl CoA carboxylase and fatty acid synthetase: Relative activities in the normal state and in hereditary obesity. *Biochem. Biophys. Res. Commun.* **28**, 682–686.

Chlouverakis, C., and White, P. A. (1969). Obesity and insulin resistance in the obese hyperglycemic mouse (obob). *Metab., Clin. Exp.* **18**, 998–1006.

Cohen, A. M. (1961). Prevalence of diabetes among different ethnic Jewish groups in Israel. *Metab., Clin. Exp.* **10**, 50–58.

Cohen, A. M. (1967). Effect of dietary carbohydrate on the glucose tolerance curve in the normal and the carbohydrate-induced hyperlipemic subject. *Amer. J. Clin. Nutr.* **20**, 126–130.

Cohen, A. M., and Rosenmann, E. (1971). Diffuse glomerulosclerosis in sucrose-fed rats. *Diabetologia* **7**, 25–28.

Cohen, A. M., and Teitelbaum, A. (1964). Effect of dietary sucrose and starch on oral glucose tolerance and insulin-like activity. *Amer. J. Physiol.* **206**, 105–108.

Cohen, A. M., Bavly, S., and Poznanski, R. (1961). Change of diet of Yemenite Jews in relation to diabetes and ischemic heart disease. *Lancet* **1**, 399–401.

Cohen, A. M., Teitelbaum, A., Balogh, M., and Groen, J. J. (1966). Effect of interchanging bread and sucrose as main source of carbohydrate in low fat diet on the glucose tolerance curve of healthy volunteer subjects. *Amer. J. Clin. Nutr.* **19**, 59–62.

Cohen, A. M., Briller, S., and Shafrir, E. (1972a). Effect of long-term sucrose feeding on the activity of some enzymes regulating glycolysis, lipogenesis, and gluconeogenesis in the rat liver and adipose tissue. *Biochim. Biophys. Acta* **279**, 129–138.

Cohen, A. M., Michaelson, I. C., and Yanko, L. (1972b). Retinopathy in rats with disturbed carbohydrate metabolism following a high sucrose diet. I. Vascular changes. *Amer. J. Ophthalmol.* **73**, 863–869.

Cohen, A. M., Teitelbaum, A., and Saliternik, R. (1972c). Genetics and diet as factors in development of diabetes mellitus. *Metab., Clin. Exp.* **21**, 235–240.

Coleman, D. L., and Hummel, K. P. (1967). Studies with the mutation diabetes, in the mouse. *Diabetologia* **3**, 238–248.

Coleman, D. L., and Hummel, K. P. (1968). *In* "Journées annuelles de diabétologie Hôtel-Dieu" (M. Dérot, ed.), pp. 19–30. Editions Médicales, Flammarion, Paris.

Creutzfeldt, W., Mende, D., Williams, B., and Söling, H. D. (1970). Vascular basement membrane thickness on muscle of spiny mice and activities of glycolysis and gluconeogenesis in the liver of animals with spontaneous and experimental diabetes and of untreated human diabetics. *Dibetologia* **6**, 356–360.

Dachs, S., Churg, J., Mautner, W., and Grishman, E. (1964). Diabetic nephropathy. *Amer. J. Pathol.* **44**, 155–168.

De Fronzo, R., Miki, E., and Steinke, J. (1967). Diabetic syndrome in sand rats. III. Observation on adipose tissue and liver in the nondiabetic stage. *Diabetologia* **3**, 140–142.

Dulin, W. E., and Gerritsen, G. C. (1967). *Proc. Cong. Int. Diabetes Fed., 6th., 1900* Int. Congr. Ser. No. 140, p. 107.

Dunn, J. S., and McLetchie, N. G. B. (1943). Experimental alloxan diabetes in the rat. *Lancet* **2**, 384–387.

Engerman, R. L., and Bloodworth, J. M. B., Jr. (1965). Experimental diabetic retinopathy in dogs. *Arch. Opthalmol.* **73**, 205–210.

Engerman, R. L., Meyer, R. K., and Buesseler, J. A. (1964). Effects of alloxan diabetes and steroid hypertension on retinal vasculature. *Amer. J. Ophthalmol.* **58**, 965–978.

Foglia, V. G., Mancini, R. E., and Cardeza, A. F. (1950). Glomerular lesions in the diabetic rat. *AMA Arch. Pathol.* **50**, 75–83.

Genuth, S. M. (1969). Hyperinsulinism in mice with genetically determined obesity. *Endocrinology* **84**, 386–391.

Gibbs, G. E., Wilson, R. B., and Gifford, H., Jr. (1966). Glomerulosclerosis in the long term alloxan diabetic monkey. *Diabetes* **15**, 258–261.

Gonet, A. E., Mougin, J., and Renold, A. E. (1965a). Hyperplasia and hypertrophy of the islets of Langerhans, obesity and diabetes mellitus in the mouse *Acomys dimidiatus. Acta Endocrinol. Suppl.* **100**, 135.

Gonet, A. E., Stauffacher, W., Pictet, R., and Renold, A. E. (1965b). Obesity and diabetes mellitus with striking congenital hyperplasia of the islets of Langerhans in spiny mice (*Acomys cahirinus*). I. Histological findings and preliminary metabolic observations. *Diabetologia* **1**, 162–171.

Hackel, D. B., Frohman, L., Mikat, E., Lebovitz, H. E., Schmidt-Nielsen, K., and Kinney, T. D. (1966). Effect of diet on the glucose tolerance and plasma insulin levels of the sand rat (*Psammomys obesus*). *Diabetes* **15**, 105–114.

Haines, H. B., Hackel, D. B., and Schmidt-Neilsen, K. (1965). Experimental diabetes mellitus induced by diet in the sand rat. *Amer. J. Physiol.* **208**, 297–300.

Hausler, H. R., Sibay, T. M., and Stachowska, B. (1963). Observations of retinal microaneurysms in a metahypophyseal diabetic Chinese hamster. *Amer. J. Ophthalmol.* **56**, 242–244.

Hausler, H. R., Sibay, T. M., and Campbell, J. (1964). Retinopathy in a dog following diabetes induced by growth hormone. *Diabetes* **13**, 122–126.

Hausler, H. R. *et al.* (1967). Retinopathy in a dog following diabetes induced by growth hormone. *Diabetes* **13**, 122–126.

Hellerstrom, C., and Hellman, B. (1963). The islet of Langerhans in yellow obese mice. *Metab., Clin. Exp.* **12**, 527–536.

Hellman, B. (1965). Studies in obese-hyperglycemic mice. *Ann. N.Y. Acad. Sci.* **131**, 541–558.

Kato, K. (1969). Studies on lipogenesis in hereditary obese hyperglycemic mice (*KK* strain). *Nagoya J. Med. Sci.* **32**, 129–141.

Lawe, J. E. (1962). Renal changes in hamsters with hereditary diabetes mellitus. *Arch. Pathol.* **73**, 88–96.

Levine, R., Lazzarini-Robertson, A., Jr., Foglia, V. G., and Singer, J. (1963). The retina in experimental diabetic rats. *Arch. Ophthalmol.* **70**, 252–255.

Lochaya, S., Hamilton, J. C., and Mayer, J. (1963). Lipase and glycerokinase activities in the adipose tissue of obese hyperglycemic mice. *Nature (London)* **197**, 182–183.

Lundboek, K., Steen Olsen, T., Ørskov, H., and Østerby-Hansen, R. (1967). Long-term experimental insulin deficiency diabetes—a model of diabetic angiopathy? *Acta Med. Scand. Suppl.* **476**, 159–173.

MacDonald, I. (1962). Some influences of dietary carbohydrate on liver and depot lipids. *J. Physiol. (London)* **162**, 334–344.

Malaisse, W. J., Malaisse-Lagae, F., and Coleman, D. L. (1968). Insulin secretion in experimental obesity. *Metab., Clin. Exp.* **17**, 802–807.

Mann, G. U., Goddard, J. W., and Adams, L. (1951). The renal lesions associated with experimental diabetes in the rat. *Amer. J. Pathol.* **27**, 857–864.

Migliorini, R. H. (1971). Early changes in the levels of liver glycolytic enzymes after total pancreatectomy in the rat. *Biochim. Biophys. Acta* **244**, 125–128.

Moloney, P. J., and Coval, M. (1955). Antigenicity of insulin: Diabetes induced by specific antibodies. *Biochem. J.* **59**, 179–185.

Morgan, C. R., and Lazarow, A. (1965). Immunoassay of pancreatic and plasma insulin following alloxan injection of rats. *Diabetes* **14**, 669–671.

510   *Cohen, Teitelbaum, Briller, Yanko, Rosenmann, and Shafrir*

Nakashima, K. (1969). Glycolytic and gluconeogenic metabolites and enzymes in the liver of obese hyperglycemic mice (*KK*) and alloxan diabetic mice. *Nagoya J. Med. Sci.* **32**, 143–158.

Ørskov, H., Steen Olsen, T., Nielsen, K., Rafaelson, O., and Lundbaek, K. (1965). Kidney lesions in the rats with severe long-term alloxan diabetes: Influence of age, alloxan damage and insulin administration. *Diabetologia* **1**, 172–179.

Patz, A., Berkow, J. W., Maumenee, A. E., and Cox, J. (1965). Studies on diabetic retinopathy and nephropathy in spontaneous canine diabetes. *Diabetes* **14**, 700–708.

Pereira, J. N., and Yangaard, N. O. (1971). Different rates of glucose and fructose metabolism in rat liver tissue. *In vitro. Metab., Clin. Exp.* **20**, 392–400.

Pometta, D., Taton, J., Rees, S. B., and Kuwabara, T. (1966). Retinal vascular change in the Chinese hamster. *Diabetologia* **2**, 215.

Portman, O., Lowey, E. Y., and Bruno, D. (1956). The effect of dietary carbohydrates on experimentally induced hypercholesterolemia and hyperbetalipoproteinemia in rats. *Proc. Soc. Exp. Biol. Med.* **91**, 321–323.

Rakieten, N., Rakieten, M. L., and Nadkarni, M. V. (1963). Studies on the diabetic action of streptozotocin (NSC-37917). *Cancer Chemother. Rep.* **29**, 91–98.

Renold, A. E. (1968). Spontaneous diabetes and/or obesity in laboratory rodents. *Advan. Metab. Disord.* **3**, 49–84.

Renold, A. E., Cameron, D. P., Amherdt, M., Stauffacher, W., Marliss, E., Orci, L., and Rouiller, C. (1972). Endocrine-metabolic anomalies in rodents with hyperglycemic syndromes of hereditary/or environmental origin. *Is. J. Med. Sci.* **8**, 189–206.

Ricketts, H. T., Test, C. E., Petersen, E. S., Lints, H., Tupikova, N., and Steiner, P. E. (1959). Degenerative lesions in dogs with experimental diabetes. *Diabetes* **8**, 298–306.

Rosenmann, E., Teitelbaum, A., and Cohen, A. M. (1971). Nephropathy in sucrose-fed rats. Electron and light microscopic studies. *Diabetes* **20**, 803–810.

Schmidt-Nielsen, K., Haines, H. B., and Hackel, D. B. (1964). Diabetes mellitus in the sand rat induced by standard laboratory diet. *Science* **143**, 689–691.

Seidman, I., Horland, A. A., and Teebor, G. W. (1967). Hepatic glycolytic and gluconeogenic enzymes of the obese-hyperglycemic mouse. *Biochim. Biophys. Acta* **146**, 600–603.

Seidman, I., Horland, A., and Teebor, G. W. (1970). Glycolic and gluconeogenic enzyme activities in the hereditary obese-hyperglycemic syndrome and in acquired obesity. *Diabetologia* **6**, 313–316.

Shafrir, E., Gutman, A., and Cohen, A. M. (1974). Metabolic adaptation of spiny mouse (*Acomys cahirinus*) transferred from desert to laboratory diets. *Horm. Metab. Res.* **6**, 103–111.

Shafrir, E., Teitelbaum, A., and Cohen, A. M. (1972). Hyperlipidemia and impaired glucose tolerance in *Acomys cahirinus* maintained on synthetic carbohydrate diets. *Is. J. Med. Sci.* **8**, 990–992.

Shargo, E., Landy, H. A., Nordlie, R. C., and Foster, D. O. (1963). Metabolic and hormonal control of phosphoenolpyruvate carboxykinase and malic enzyme in rat liver. *J. Biol. Chem.* **238**, 3188–3192.

Shirai, T., Welch, G. W., 3rd, and Sims, A. E. (1967). Diabetes mellitus in the Chinese hamster. II. The evolution of renal glomerulopathy. *Diabetologia* **3**, 266–286.

Sibay, T. M., and Hausler, H. R. (1967). Eye findings in two spontaneously diabetic related dogs. *Amer. J. Ophthalmol.* **63**, 289–294.

Sibay, T. M., Hausler, H. R., and Hayes, J. A. (1971). The study and effect of streptozotocin (NSC-379 17) rendered diabetic Chinese hamsters. *Ann. Ophthalmol.* **3**, 596–601.

Simon, E., Scow, R. O., and Chernick, S. S. (1961). Effects of D-mannoheptulose and D-sedoheptulose on blood glucose and ketone bodies in the rat. *Amer. J. Physiol.* **201**, 1073–1077.

Siperstein, M. D. (1970). The relationship of carbohydrates derangements to the microangiopathy of diabetes. *In* "Pathogenesis of Diabetes Mellitus" (E. Cerasi and R. Luft, eds.), 13th Nobel Symp., pp. 81–96. Wiley, New York.

Siperstein, M. D., Unger, R. N., and Madison, L. L. (1968). Studies of muscle capillary basement membranes in normal subject, diabetic, and prediabetic patients. *J. Clin. Invest.* **47**, 1973–1999.

Sneyd, J. G. T. (1964). Pancreatic and serum insulin in the New Zealand strain of obese mice. *J. Endocrinol.* **28**, 163–172.

Stauffacher, W., Lambert, A. E., Vecchio, D., and Renold, A. E. (1967). Measurement of insulin activity in pancreas and serum of mice with spontaneous ("obese" and "New Zealand obese") and induced (gold thioglucose) obesity and hyperglycemia with consideration on the pathogenesis of the spontaneous syndrome. *Diabetologia* **3**, 230–237.

Stauffacher, W., Orci, L., Amherdt, M., Burr, I. M., Balant, L., Froesch, E. R., and Renold, A. E. (1970). Metabolic state, pancreatic insulin content and blood cell morphology of normoglycemic spiny mice (*Acomys cahirinus*): Indications for an impairment of insulin secretion. *Diabetologia* **6**, 330–342.

Stauffacher, W., Orci, L., Cameron, D. P., Burr, I. M., and Renold, A. E. (1971). Spontaneous hyperglycemia and/or obesity in laboratory rodents: An example of the possible usefulness of animal disease models with both genetic and environmental components. *Recent Progr. Horm. Res.* **27**, 41–91.

Weitze, M. (1963). Hereditary adiposity in mice and the cause of this anomaly (1940). Store Nordiske Videnskaboghandel Copenhagen. (Hellerstorm, C. and Hellman, B., eds.) *Metab. Clin. Exp.* **12**, 527.

Williamson, J. R., Vogler, N. J., and Kilo, C. (1971). Microvascular disease in diabetes. *Med. Clin. N. Amer.* **55**, 847–860.

Yergenian, G. (1964). Spontaneous diabetes mellitus in the Chinese hamster. (*Cricetulus griseus*). IV. Genetic aspects. *Ciba Found. Colloq. Endocrin.* [*Proc.*] **15**, 25–48.

Young, F. G., and Corner, A. (1960). Growth hormones. *In* "Diabetes" (R. H. William, ed.), p. 216. Harper, (Hoeber) New York.

Yudkin, J., and Krauss, R. (1967). Dietary starch, dietary sucrose, and hepatic pyruvate kinase in rat. *Nature* (*London*) **215**, 75.

Zakim, D., Pardini, R. S., Herman, R. H., and Sauberlich, H. E. (1967). Mechanism for the differential effects of high carbohydrate diet on lipogenesis in the rat liver. *Biochim. Biophys. Acta* **144**, 242–251.

CHAPTER 30

Adult Diabetes

OSCAR B. CROFFORD

Rather than attempt a general review of the field of diabetes as it occurs in adults, two approaches to the management of diabetic outpatients will be described that are not now in widespread use but that are currently being tested in the Diabetic Clinics of the Vanderbilt Medical Center. Both of these approaches pertain to nutrition in the management of diabetics because good nutrition is the key to the successful control of the metabolic components of this disease. The first concerns the application of computer techniques to nutrition in diabetics, while the second concerns the use of breath acetone measurements in assessing the clinical status of diabetic outpatients.

A. Application of Computer Techniques

For the physician to be successful in achieving adequate metabolic control in patients with diabetes, the first step is that he convince himself and then convince the patient that good nutritional habits are absolutely essential. Unless the patient has regular and consistent eating habits, it is impossible to achieve adequate metabolic control with the use of either insulin or any of the oral agents used in the treatment of diabetes.

The importance of consistent eating habits can be attributed to the failure in diabetics of the control mechanisms that enable nondiabetics to maintain blood glucose concentrations within rather narrow limits de-

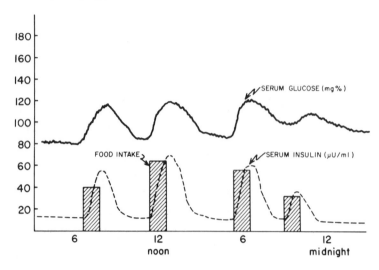

Fig. 1 Temporal relationships between food intake, serum glucose, and serum insulin concentrations in nondiabetics.

spite highly variable patterns of eating. Figure 1 is a diagram of the temporal relationships between food intake, serum glucose, and serum insulin concentrations in nondiabetics during a 24-hr day. The height of the bars represents the quantity of food eaten. The broken line represents the insulin concentrations that are achieved in the plasma subsequent to the release of insulin by the islets of Langerhans of the pancreas. The line at the top of the figure shows the serum glucose concentrations. The figure is intended to illustrate that the intake of food is followed promptly by a rise in the serum insulin concentration and that the serum glucose concentration remains relatively constant. Figure 2 illustrates that if the nondiabetic elects to omit lunch, nothing serious happens. Since there is no food intake at 12 noon, there is no rise in the serum insulin concentration, and the calories that have been stored from previous meals suffice to maintain the blood sugar at the same level as if the patient had eaten his usual noontime meal. Figure 3 shows how different the situation is in patients with diabetes. The event which initiates the day is not the intake of food but the injection of insulin before breakfast. Once that injection has been made the patient is committed to a predetermined serum insulin level throughout the day. For such a patient to achieve serum glucose concentrations that even approach those seen in nondiabetics, it is essential that this pattern of intake be appropriate for the quantity and type of insulin preparation that he has taken. Figure 4 illustrates that the diabetic cannot indulge in dietary indiscretions like omission of a meal.

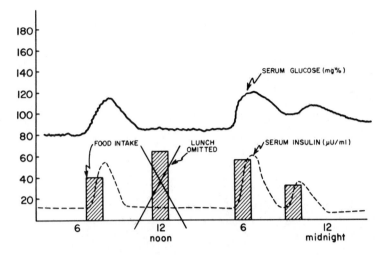

Fig. 2 Temporal relationships between food intake, serum glucose, and serum insulin concentrations in nondiabetics.

The insulin level in the serum is high and the liver and peripheral tissues continue to conserve foodstuffs rather than to release them. In the mid-afternoon the blood sugar falls to hypoglycemia level and the patient has an insulin reaction which he aborts by the intake of candy or orange juice. The intake of concentrated sweets then leads to extreme hyperglycemia, and the patient exhibits the wide fluctuations of blood sugar

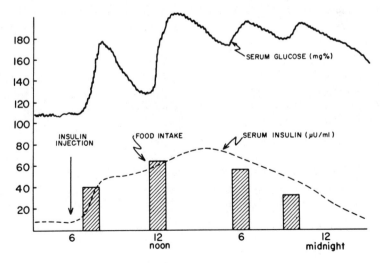

Fig. 3 Temporal relationships between food intake, serum glucose, and serum insulin concentrations in diabetics.

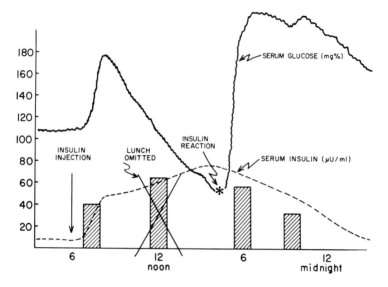

Fig. 4 Temporal relationships between food intake, serum glucose, and serum insulin concentrations in diabetics.

which are so typical of the insulin-requiring diabetic that fails to eat in a consistent way.

Having discussed the importance of good nutrition in managing diabetics, let us now consider the method that the physicians and nutritionists use to instruct the patient as to the proper diet that he should attempt to follow. Most clinics, including our own, use the so-called exchange system, meaning that foodstuffs are collected in nutritionally equivalent groups so that one member of the group can be substituted for any other member of that same group (Meal Planning with Exchange Lists, 1956). Thus, the patient would not be instructed to eat an orange every morning for breakfast the rest of his life; rather he would be instructed to eat *one fruit exchange*, which means that he can select one portion from any of the more than 30 items listed in the fruit section of the exchange book. If his diet calls for *one bread exchange*, he can select one portion of any of the items listed on the bread exchange list. This allows the patient a certain degree of variability in his diet without changing the quantity or quality of foods that he eats. This is a very effective system both for the patient and for the dietitian and although there are numerous modifications of this basic plan, no one has proposed that it be eliminated. For the physician, however, the exchange system provides more dietary details than are necessary for the management of the patient, and usually it is impossible to teach this system to medical students and to young

```
1) Diet order:

        1800 cal. diabetic diet

   1800 x 0.40 = 720 cal ÷ 4 cal/g = 180g CHO (40%)

   1800 x 0.40 = 720 cal ÷ 9 cal/g =  80g Fat (40%)

   1800 x 0.20 = 360 cal ÷ 4 cal/g =  90g PRO (20%)

   Divide into portions of 1/3, 1/3 and 1/3. Take 1 milk
   and 1 bread exchange from breakfast and give at 3:00 PM.
   Take 1 milk, 2 bread and 1 meat exchange from supper
   and give as a bedtime snack.
```

Fig. 5 A traditional way for writing the diet order for a patient with diabetes mellitus.

physicians in training. Figure 5 shows the impracticality of the system as illustrated in the kind of diet order that must be written by the physician in order for the dietitian to translate the dietary prescription into the portions of foodstuff that the patient must eat. Although most physicians think in terms of calories, the exchange system is based on grams, and the physician must calculate the number of grams in the various foodstuffs that he intends the patient to eat. The total food to be used must then be subdivided into portions and the correct number of exchanges shifted from the main meals to the between meal feeding periods in order to achieve the desired food intake pattern during the 24-hr period. One can see that the detail required is more time-consuming than most physicians are willing to devote to the writing of the diet prescription, and the result is that the diet prescription is either not written or is written in a careless and often totally unsatisfactory way. The main point is that although the exchange system is useful in enabling the dietitian to translate the physician's prescription into portions of foodstuffs and although the exchange system is useful in allowing the patient to have variability in his diet, it is unnecessarily complicated and therefore not learned by a majority of practicing physicians. Figure 6 illustrates that the problem does not stop for the physician having to write the dietary prescription, for if another physician examines the hospital chart of the patient and tries to determine the diet that the patient is eating, he encounters the formidable looking table that appears in this figure in which calories are totally lacking and which can be understood by the physician only if he is familiar with the exchange system. Most physicians looking at this translated version of the diet prescription would still be totally unable to have any feel for what the patient is actually supposed to eat.

The initial approach to this difficulty was to try and identify the minimal amount of information that the physician must know about the diet

DIET AS GIVEN TO PATIENT:

	EXCHANGES				GRAMS			
	Bkft	Noon	Night	Total	Pro	Fat	Cho	Cals
MILK		/1	/1	2	16	20	24	
PROTEIN	2	3	2/1	8	56	40		
VEGETABLE		A	1B	1	2		7	
FRUIT	1	1	1	3			30	
CHO	2	2/1	1/2	8	16		120	
FAT	1	1	2	4		20		
TOTAL					90	80	181	

COMMENTS:

Fig. 6 A traditional way for translating the diet order into the exchange format.

in order to manage the patient intelligently. The results are shown in Fig. 7. Since we can assume that the diet for all diabetics should avoid the use of concentrated sweets and that the dietitians will strive to have each meal balanced with respect to carbohydrates, proteins, and fats, the essential dietary information can be reduced to the total number of calories and to the distribution of those calories throughout the six feeding periods of the day. In this example, the patient·is on an 1800 cal diet with 25% of the calories consumed at breakfast, no midmorning snack, 35% of the calories for lunch, 5% of the calories as a midafternoon snack, 25% for supper, and 10% of the total caloric content of the diet consumed as a bedtime snack. The way that this format is actually used is illustrated in Fig. 8. Thus the physician prescribes the diet using only the total calories and the set of six numbers which indicate the percentage

ASSUMPTIONS: 1) Avoidance of concentrated sweets

2) Balanced with respect to Carbohydrate, Proteins and Fats

(40% CHO, 20% P, 40% F)

1800 Calorie Diabetic Diet

25 - 0 - 35 - 5 - 25 - 10

Fig. 7 Necessary assumptions and proposed style for writing diet orders for patients with diabetes mellitus.

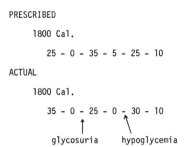

PRESCRIBED

   1800 Cal.

      25 - 0 - 35 - 5 - 25 - 10

ACTUAL

   1800 Cal.

      35 - 0 - 25 - 0 - 30 - 10

        ↑   \
   glycosuria   hypoglycemia

Fig. 8 Functional aspects of the diet order as written in the proposed format (see text).

distribution of the diet throughout the day. This figure illustrates that on the basis of a nutritional history obtained at the time of a return visit, the dietitian determined that the actual diet differed from that prescribed in that 35% of the calories were being consumed for breakfast, only 25% for lunch, and that the midafternoon snack was entirely eliminated. Based on this information and on the knowledge that the patient was taking a morning injection of one of the intermediate acting insulin preparations, the physician could surmise that the glycosuria occurring before lunch was probably because the number of calories consumed at breakfast were excessive and that the hypoglycemic episodes occurring in the middle of the afternoon were because insufficient calories were being consumed during the time of the peak action of the insulin preparation. Proper treatment, of course, would be re-emphasis and reinstruction on the diet as prescribed.

Since this abbreviated form of dealing with diabetic diets seemed to be so convenient for the physicians, the dietitians were asked to convert all of the diets on all of the patients to this format. Although this could be done, it was a rather time-consuming manual procedure that could be simplified greatly by using the most elementary of computer techniques that are available in most laboratories. Figure 9 shows an example of the computer print-out on the Olivetti desk top calculator showing how the exchanges are entered into the computer directly from the exchange lists that have already been worked out for the patients and the fraction of the total calories consumed at each feeding period is given in the print-out.

Being completely naive about computers, the present author could see no reason why the program could not be inserted into the machine backwards and have the computer print out the exchange list when the physician entered the total number of calories and the fraction of calories to be eaten at each of the six feeding periods. Although it is not quite

Fig. 9 Computer print-out of diet prescription.

that simple, it is a trivial task for a full size computer and Fig. 10 illustrates the computer print-out as it now exists. Thus, this patient's dietary prescription calls for 1800 cal with 25% for breakfast, no midmorning snack, 35% for lunch, a 5% midafternoon snack, 25% for supper, and a 10% bedtime snack. The computer then calculates the number of exchanges required to fulfill the diet which is then printed out in the standard exchange list format. Even though this looks somewhat formidable, it is no more complicated than the manual system which has traditionally been used.

There are at least two benefits that have resulted from the application of these very simple computer techniques to nutrition for diabetics. First, it relieves the dietitian from a number of very time-consuming manual

```
GARY STEWERT
   1800 CALØRIES
  BKFST    25%
  MM SN     0%
  LUNCH    35%
  MA SN     5%
  SUPPER   25%
  BT SN    10%
  M/V 2 2 0 0 1
    2 1
          3
```

```
  PLAN    1
  MEAL    INDX   WM  SM  P  V  R  C   F      CAL     %
  BKFST    363    0  0+  2  0  0  2+  2      446  25.33
  LUNCH    468    0   2  1  0  0  3   4      617  35.04
  MA SN     10    0   1  0  0  0  0   0       80   4.54
  SUPPER   351   1+   0  1  1  0  1   0      432  24.53
  BT SN     54    0   0  1  0  0  1   1      186  10.56

  %P= 20.90   %F= 38.33   %C= 40.77   CAL= 1761   %CAL=   97.83
```

```
  PLAN    2
  MEAL    INDX   WM  SM  P  V  R  C   F      CAL    %
  BKFST    376    0   1  2  0  2  1   2      464  25.62
  LUNCH    478    0   0  4  0  1  3   2      626  34.57
  MA SN     11   0+   0  0  0  0  0   0       85   4.69
  SUPPER   366    0   0  2  1  1  2   2      448  24.74
  BT SN     60    0   1  0  0  1  1   0      188  10.38

  %P= 20.32   %F= 37.27   %C= 42.41   CAL= 1811   %CAL= 100.61
```

Fig. 10 Computer print-out of two meal plans for a hypothetical patient with diabetes mellitus.

calculations allowing time for what this person has been trained to do; namely, teach the patient how to use the exchange system in order to achieve consistent eating habits. Second, in first using the computer it was learned that for a given diet, there was not just one way that the exchange list could be filled out but that there were always dozens and sometimes hundreds of ways that one could combine the various food groups and still fulfill the prescribed diet. It is now the practice to print out not simply one meal plan but 20 meal plans for each patient. When one realizes that every time the meal plan calls for one exchange there are at least 30 specific food items that are listed in the exchange booklet for that particular food group, one begins to understand that the diet has the potential of a great deal more variety than most patients realize. Thus, the notion that a diabetic diet is any way restrictive is simply not correct. Armed with 20 meal plans handily calculated by the computer, the dietitian can then sit down with the patient and select the ones that are most appropriate for his eating habits and instruct him in the importance in the use of the exchange system.

At the present time this service is available to the health care professionals within the Tennessee Mid-South region and, in due course, the success of the program will be assessed in contributing to the clinical-management of patients. Thus far, the results have been encouraging.

B. Use of Breath Acetone Analyses

The second topic concerns the use of breath acetone analyses in the management of diabetic outpatients. Elevated levels of acetoacetate in blood and urine and, in a semiquantitative way, the odor of acetone in breath have long been hallmarks of uncontrolled diabetes (Henderson *et al.*, 1952). The tablets, powders, and dip sticks used for these determinations are not sensitive enough to detect the presence of acetoacetate or acetone in the blood, urine, or breath of the diabetic who is under even moderately good control. With gas chromatography (Jansson and Larsson, 1969), however, acetone can be measured quantitatively in the breath of well-nourished nondiabetics and becomes elevated grossly in diabetics who are inadequately treated with insulin (Stewart and Boettner, 1964; Sulway *et al.*, 1971). Therapeutic decisions can be made on a more rational basis if the metabolic state of the patient is assessed with both blood glucose and breath acetone measurements.

Breath is an ideal sampling medium. An end-tidal breath sample is collected by having the patient exhale into a 50-cm³ glass syringe. The syringe serves for collection, transport, and storage of the gas which can be injected directly into the sample inlet port of the gas chromatography instrument. The sample collection method is ideally suited for use in children who are told to "blow up the syringe like a balloon." The analysis is rapid ($2\frac{1}{2}$ min), specific for acetone, sensitive (down to one nanomole per liter of alveolar gas) and, potentially, of very low cost. In our diabetic clinics the results of the breath acetone measurement are known before the fasting blood sugar.

Our results in nondiabetics as well as in diabetics in ketoacidosis agree with those reported previously by others (Henderson *et al.*, 1952; Stewart and Boettner, 1964; Sulway *et al.*, 1971). The normal range is from 10 to 50 nmolar with the sample collected in the morning after a standard overnight fast. Although it is beyond the scope of this discussion, our studies have included factors other than insulin insufficiency that can produce breath acetone elevations. The most important, of course, is an inadequate dietary carbohydrate intake. This and other factors (exercise, stress, etc.) have not made interpretation of the results prohibitively complicated.

For clinical purposes we classify diabetics with hyperglycemia as being in one of two categories depending entirely upon the breath acetone concentration. The scheme is shown in Fig. 11. It is undoubtedly oversimplified but nevertheless useful. Patients with hyperglycemia and a normal breath acetone (<50 nmolar) are considered to have hyperglycemia of poor nutrition—usually overeating—and are treated with more vigorous dietary measures. Although it had been our practice in the past to treat hyperglycemic patients with "more insulin," the results were usually disappointing if overeating was the major problem. More insulin led to more overeating and to more obesity. More obesity led to more insulin resistance and in most instances the net result was that the hyperglycemia was not improved. Our emphasis today is: "You don't treat overeating with insulin but with better dietary management." Our general policy in well-nourished adult diabetics is not to increase the insulin dose unless the breath acetone is elevated.

If the patient has hyperglycemia and an elevated breath acetone, he is considered to be inadequately treated with insulin and the dose is adjusted appropriately.

Still another situation exists if the patient has an elevated breath acetone and a normal blood sugar. This results from an intake of dietary carbohydrate that is insufficient to meet the metabolic needs of the patient. It the patient is trying to lose weight, this is a very useful index of success and can be used to encourage the patient. In some instances, however, especially in children, the diet is malapportioned with the per-

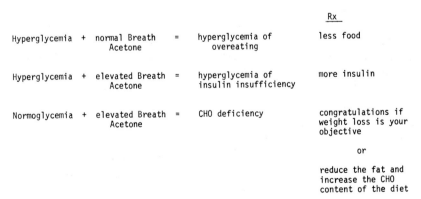

Fig. 11 Scheme for subclassifying patient with hyperglycemia according to the breath acetone levels (see text).

centage of carbohydrate in the diet being too low. This can easily result if the child or his parents are unduly fearful that "starchy foods make his diabetes worse."

Although it is still too early to assess the long-term results of the program, we have found this systematic approach to the management of diabetic outpatients extremely useful in developing the skills of our entire health care team.

Acknowledgment

This work was supported by USPHS Grant HL 08195 and a grant from the Justin and Valere Potter Foundation.

References

Henderson, J., Karger, B. A., and Wrenshall, G. A. (1952). Acetone in the breath. *Diabetes* **1**, 188–193.

Jansson, B. O., and Larsson, B. T. (1969). Laboratory methods. *J. Lab. Clin. Med.* **74**, 961–965.

Meal Planning with Exchange Lists. (1956). Booklet prepared by Committees of American Diabetes Association, Inc., New York, N.Y. and the American Dietetic Association, Chicago, Ill., in cooperation with Chronic Disease Program— Public Health Service Department of Health, Education and Welfare.

Stewart, R. D., and Boettner, E. A. (1964). Expired-air acetone in diabetes mellitus. *N. Engl. J. Med.* **270**, 1035–1038.

Sulway, M. J., Trotter, E., Trotter, M. D., and Malins, J. M. (1971). Acetone in uncontrolled diabetes. *Postgrad. Med. J., Suppl.* pp. 382–387.

Therapeutic Use
of Sugars

CHAPTER 31

Sugars in Parenteral Nutrition

H. C. MENG

The field of parenteral nutrition has advanced rapidly during the last few years. The International Society of Parenteral Nutrition has been formed, and numerous symposia on the subject have been held in the United States as well as abroad. In addition, parenteral nutrition units have been established in many medical centers throughout the United States. With careful management and frequent monitoring of the biochemical and physiologic parameters and clinical changes, the procedure as an adjunct of patient care and medical management has been considered one of the most important advances in medicine.

Parenteral nutrition concerns the administration of nutrients by routes other than the gastrointestinal tract. Since the volume of solutions and the amount of nutrients needed are large, it is not convenient to use subcutaneous, intramuscular, or intraperitoneal route. In addition, the absorption of the three major foodstuffs—carbohydrate, protein, and fat—is slow, time-consuming, and not without discomfort. Thus, intravenous administration is the only practical means for the introduction of all nutrients.

From the standpoint of cellular nutrition it makes little difference whether the nutrients are administered enterally or parenterally; with the latter, consideration must be given concerning the sterility and nonpyrogenicity of the preparations for intravenous administration. In addition, the rate and total amount of nutrient given must be carefully regulated and the form of certain nutrients must be modified, e.g., fat in the form of an emulsion, protein as a partially hydrolyzed hydrolysate or a mixture of crystalline amino acids, and carbohydrate as monosaccharides.

The indications for parenteral nutrition are many (Elman, 1948; Geyer, 1960; Meng and Law, 1970, Meng, 1971). It may be briefly stated that parenteral nutrition is indicated in patients whose clinical condition dictates that they cannot or should not eat by mouth. Examples of such conditions are severe burns, major surgery, dysfunction of the gastrointestinal tract, cachexia resulting from cancer, renal failure, coma, and anorexia nervosa. The energy expenditure and nitrogen loss in these patients are greater than those in normal individuals. Conventional intravenous therapy with 5–10% glucose, saline, and occasionally hydrolysate is grossly inadequate to meet such nutritional needs. It has been seen too often that a patient dies of starvation or complications of malnutrition rather than his primary disease. The only way to prevent starvation and catabolism of the body cell mass is to institute parenteral nutrition at the earliest date with all nutrients in adequate amounts.

In the present communication, results of the use of various sugars as a calorie source in parenteral nutrition are presented in three parts: (1) glucose alone or glucose–fructose mixture, (2) xylitol, and (3) sorbitol. Glucose or glucose–fructose was used in patients and sorbitol or xylitol was given to dogs.

A. The Use of Glucose or Glucose–Fructose in Patients

1. Composition of Solutions for Parenteral Nutrition

The composition of solutions used at the Vanderbilt University School of Medicine and elsewhere in the United States usually consists of a carbohydrate as hypertonic glucose (20–40%) and a protein in the form of either partially hydrolyzed protein or crystalline amino acid mixture along with electrolytes and vitamins. This is exemplified by the solutions shown in Tables I and II, used at Vanderbilt University Medical Center, for adult and pediatric patients, respectively. One liter of the solution shown in Table I furnishes 900 kcal and 4.17 gm of nitrogen or 26 gm of protein equivalent; 3 liters are usually given to an adult patient daily with a total caloric intake of 2700 kcal and 12.5 gm of nitrogen or 78 gm of protein equivalent. In general, the volume of the solution (Table II) given to infants and children is 125 ml/kg/day which furnishes 0.52 gm

TABLE I

Composition of Solution for Parenteral Nutrition in Adults[a]

Nutrient	Amount	Vitamin[b]	Amount
Glucose	250 gm	Ascorbic acid	155 mg
Nitrogen[c]	4.2 gm	Vitamin A	1111 U.S.P. units
Sodium	23.3 mEq	Vitamin D	111 U.S.P. units
Potassium	40.0 mEq	Thiamine HCl	5.5 mg
Magnesium	5.0 mEq	Riboflavin	1.1 mg
Chloride	35.0 mEq	Pyridoxine HCl	1.7 mg
Acetate	52.0 mEq	Niacinamide	11.1 mg
Zinc sulfate	5.0 mg	Vitamin E	0.55 IU
Kilocalories[d]	900	Folate	100 μg
Distilled water	to 1000 ml	B$_{12}$	10 μg
		Panthenol	2.8 mg

[a] Calcium as calcium gluconate and phosphorus as potassium (dibasic and acid) phosphate were added as needed, and iron was injected separately when required. The currently used solution contains 5 mEq of calcium as calcium gluseptate and 260 mg of phosphorus as potassium (dibasic) phosphate per liter.

[b] Vitamins were supplied as aqueous multivitamin infusion (MVI), a product of U.S. Vitamin and Pharmaceutical Corporation, folate and B$_{12}$ were added.

[c] Nitrogen was given as a mixture of crystalline amino acids.

[d] Nonprotein calories.

TABLE II

Composition of Solution for Parenteral Nutrition in Infants

Nutrient	Amount/liter	Vitamin	Amount/liter
Carbohydrate[a]	200 gm	A	4000 U.S.P. units
Nitrogen[b]	4.1 gm	D	400 U.S.P. units
Kilocalories	800	Ascorbic acid	200 mg
Sodium	25 mEq	Thiamine HCl (B$_1$)	20 mg
Potassium	30 mEq	Riboflavin (B$_2$)	4 mg
Magnesium	10 mEq	Pyridoxine HCl (B$_6$)	6 mg
Calcium	10 mEq	Niacinamide	40 mg
Acetate	20 mEq	Panthenol	10 mg
Phosphorus	6 mEq	E	2 IU
ZnSO$_4$	10 mg	B$_{12}$	100 μg
		Folic acid	1 mg

[a] Carbohydrate is supplied either as invert sugar or as glucose.

[b] Nitrogen is supplied as crystalline amino acids.

of nitrogen or 3.2 gm of protein equivalent, 25 gm of carbohydrate and
110 kcal.

2. Route and Rate of Administration

Because of the hyperosmolality of the solutions, the technique of Wilmore and Dudrick (1969) using central venous delivery was adapted. Our routine procedure was to place a catheter into an external jugular or subclavian vein percutaneously or through a cutdown (Van Way *et al.*, 1973). The catheter was then directed into the superior vena cava with the location of the tip verified by fluoroscopy or by X-ray. Scrupulous attention must be paid to the subsequent care of the catheter and its entry site to avoid infection.

The rate of infusion is one of the factors affecting utilization of the intravenously administered nutrients and the incidence of adverse effects. Infusion of glucose at rates of 0.4–0.5 gm/kg body weight/hr continuously during the 24 hr period seems to be a reasonable rate in most of the adult patients. However, our usual practice is to start the infusion at a slower rate (50–75 ml/kg/hr) or to administer a diluted solution, e.g., one-half strength. The volume and/or the concentration of the solution are increased slowly in a stepwise manner during the first 4–5 day period for the development of metabolic adaptation. A solution of fully prescribed volume (3 liters or so) and concentration was then given daily with constant monitoring of biochemical and clinical parameters such as serum electrolytes, urea, proteins, osmolality, blood glucose, and ammonia. Liver function tests were also followed at appropriate intervals. Continuous 24-hr urine collections were made for total nitrogen, sodium, and potassium determination for obtaining their intake–output balances. A test for urinary sugar was carried out every 4 hr. Body weight was recorded daily or at frequent intervals.

Three cases are presented to illustrate the effectiveness of parenteral nutrition in correcting negative nitrogen balance and in promoting wound healing, weight gain, and even growth using glucose or glucose–fructose mixture as a calorie source along with crystalline amino acids, minerals, and vitamins.

Case 1. K. C., a 74-year-old woman, was admitted to Vanderbilt University Hospital with a 6-month history of vague epigastric pain. She was thin and debilitated and was found to have a large gastric mass. Total parenteral nutrition was begun 2 days before the day of, and continued for 15 days after, operation. At exploration she was found to have a large gastric lymphosarcoma. She underwent a total gastrectomy, splenectomy, distal pancreatectomy, partial colectomy, and gastric replace-

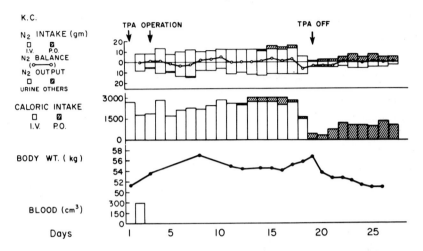

Fig. 1 Caloric intake, body weight, and nitrogen balance in K.C. From Van Way *et al.* (1973), with kind permission of the publisher.

ment with a Hunt-Lawrence pouch (Scott *et al.*, 1968). Her daily caloric intake, body weight, and nitrogen balance are shown in Fig. 1. She was in positive nitrogen balance throughout except on the second, third, and fourth postoperative days when a negative nitrogen balance of 5 gm or less was observed. With an average daily nitrogen and caloric intake of 11.6 gm and 2439 kcal, respectively, she gained a total of 6 kg during the period on total parenteral nutrition. She lost most of the weight gained after being taken off total parenteral nutrition. Figure 2 shows the fluid, sodium, potassium, and magnesium balance. It can also be seen that nitrogen and potassium balances correlated with each other. Magnesium was always in positive balance. The sodium balance was variable depending for the most part on intake. Fluid balance was always positive. She left the hospital weighing more than she did on admission. At follow-up examination 7 months later, she was found to have maintained her weight and was clinically well.

This patient benefited by the use of total parenteral nutrition. She withstood an extensive surgical procedure and gained weight while on parenteral nutrition therapy. In addition, only a slight negative nitrogen balance was observed on the second, third, and fourth postoperative days. Despite the nitrogen loss, a net average daily nitrogen retention of 0.52 gm was achieved during the 15 days following operation. The short postoperative period of negative nitrogen balance resulted, in part, from difficulties encountered in regulating her insulin dosage. The need of exogenous insulin in this patient was undoubtedly because of her age and partial

532   H. C. Meng

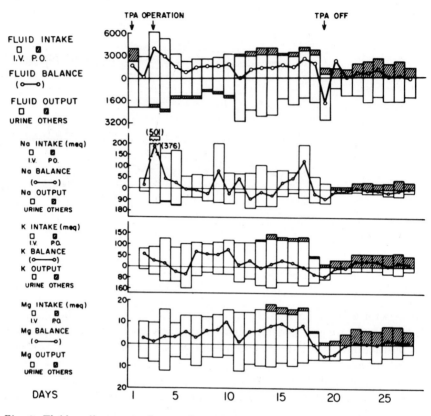

Fig. 2 Fluid, sodium, potassium, and magnesium balance in K.C. From Van Way *et al.* (1973), with kind permission of the publisher.

pancreatectomy rendering the endogenous secretion of this hormone inadequate. The subsequent weight loss of about 4 kg observed after parenteral nutrition was discontinued resulted, in part, from inadequate oral food intake and perhaps some water loss. While this may represent excess body water, frank edema or symptoms of circulatory overload did not occur in this patient.

Case 2. D. B. was a 28-year-old man with a 5-year history of regional enteritis involving duodenum, jejunum, ileum, and colon. He also had a draining fistula from the perirectal area to the back of the right thigh. A gastrojejunostomy was performed for the posterior duodenal ulcer about 3 years prior to the present admission. Oral intake of food was intolerable because of frequent diarrhea. The patient was anorexic and mentally depressed. He had lost more than 40 kg of body weight since the beginning of his illness. Total parenteral nutrition was given for 38 days during

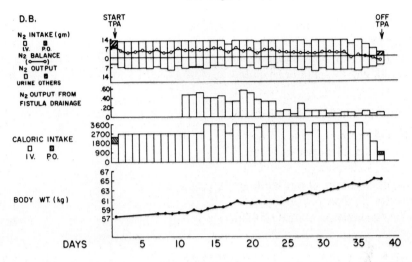

Fig. 3 Caloric intake, body weight, nitrogen balance, and nitrogen loss from fistula drainage of D.B. during the period of total parenteral alimentation. From Meng and Sandstead (1972), with kind permission of the publisher.

which time he received nothing by mouth except water. The total intake of amino acids and glucose by the intravenous route was 3.06 and 29.04 kg, respectively. The average daily nitrogen intake was 11.85 gm and average daily caloric intake was 3057 kcal. Figure 3 shows the caloric intake, body weight gain, nitrogen balance, and the progressive decrease in nitrogen loss from the fistula drainage. It can be noted that he gained about 1.5 kg of body weight after 14 days of parenteral nutrition furnishing 2700 kcal/day from non-nitrogen calories (47 kcal/kg/day). Increasing non-nitrogen calories to 3600 (60 kcal/kg/day) seemed to improve the curve of body weight gain. The total weight gained was 7.0 kg with an average daily gain of 0.19 kg. The nitrogen balance was maintained positive throughout the period of total parenteral nutrition except during the last 3 days when caloric and nitrogen intakes were reduced. The daily average of nitrogen retention was 4.14 gm. The nitrogen loss from the fistula drainage was progressively decreased through the period of total parenteral nutrition. The volume of fistula drainage was also greatly decreased. Figure 4 is a photograph of the fistula taken immediately before and after total parenteral nutrition, showing the improved appearance; in fact, it was closing.

It is seen from Fig. 5 that, in general, there was considerable water retention. The average daily water retained was 1652 ml without the consideration of insensible water loss. Sodium, potassium, and magnesium balances were maintained positive except for the last 3 days when intakes

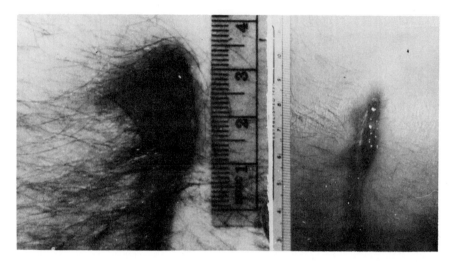

Fig. 4 Size and appearance of fistula from the perirectal area to the back of the right thigh: (A) prior to parenteral nutrition and (B) at the end of parenteral nutrition.

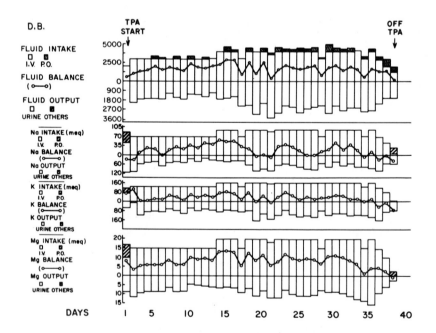

Fig. 5 Fluid, sodium, potassium, and magnesium balances of D.B. during the total parenteral alimentation period. From Meng and Sandstead (1972), with kind permission of the publisher.

of these electrolytes were reduced. The daily average of sodium, potassium, and magnesium retention was 17.8, 36.2, and 8.1 mEq, respectively.

Serum sodium, potassium, chloride, and bicarbonate were maintained within the normal ranges throughout the period of parenteral nutrition. Serum urea remained at the levels of 10–12 mg %. A slight increase in total plasma proteins and plasma albumin was observed during the later part of parenteral nutrition period. The hemoglobin and hematocrit increased from 9.9 gm % and 32.4% to 11.2 gm % and 37.0%, respectively, during the entire period of parenteral nutrition.

Table III shows the progressive increase in total body potassium during parenteral nutrition. The serum potassium determined on the corresponding dates of total body potassium measurements remained fairly constant throughout. The increase in total body potassium is indicative as the increase in lean body mass.

Case 3. Baby Girl M. (B. G. M.) was the product of a 28-week gestation to a 23-year-old gravida 3, para 2, abortus 0 mother. Pregnancy was uncomplicated until approximately 3 weeks before delivery when the mother developed preeclampsia with 4+ proteinuria, generalized edema, and hypertension (BP 150/100). The mother was admitted to the hospital 4 days before delivery and treated with bed rest, high-protein low-sodium diet, hydrochlorothiazide, phenobarbital, and bendroflumethiazide without control of the hypertension. Membranes ruptured 4 hr before delivery and labor was then augmented with intravenous oxytocin (Pitocin). The infant was a breech delivery with a one minute Apgar score of 2. The child was intubated and given assisted ventilation for approximately

TABLE III

Total Body and Serum Potassium[a]

Days on TPA[b]	Potassium	
	Total body (mEq)	Serum (mEq/liter)
3	2,367	4.6
14	2,758	3.8
31	2,859	4.1
38	3,191	3.8

[a] Total body potassium was measured by the method of Brill *et al.* (1972). From Meng and Sandstead (1972), with kind permission of the publisher.

[b] Here, TPA stands for total parenteral alimentation.

30 min. After extubation the child breathed spontaneously without difficulty.

At birth the child weighed 850 gm and had a head circumference of 26 cm, length 36.5 cm, and chest circumference 20.5 cm. When the infant's condition stabilized, a No. $3\frac{1}{2}$ F, radiopaque open-ended catheter was inserted into the umbilical vein and passed through the ductus venosus into the vena cava above the diaphragm. The position was confirmed either by X-ray and/or pressure tracings. Another catheter was inserted through an umbilical artery into the aorta to a level below the renal arteries. The latter catheter was used for sampling, while the former was used only for infusion of the nutrients. As a final precaution to prevent bacterial contamination and infusion of particulate matter, an in-line Millipore filter (No. SXGS-025-05-0.22 μm) was placed in the infusion line just before entry into the patient.

To facilitate urine collection, the child was placed on a metabolic bed. After establishing that the child was stable clinically, preliminary blood chemistry including liver and kidney function tests, serum electrolytes, blood culture, pH, hematocrit, total serum proteins with electrophoresis and serum osmolality was measured as base line studies. The solution shown in Table II was infused at a constant rate over a 24 hr period using the Harvard peristaltic pump at a beginning rate of 125 ml/kg/24 hr. The first 24-hr infusion was a half-strength formula. This was slowly increased to full strength, as tolerated, over the next 48–72 hr.

The infant's status was monitored by daily measurements of electrolytes, pH, hematocrit, and serum osmolality until the stability was established, and then twice a week. Weight was recorded daily on the same scale, under the same conditions, and at the same time of day. Intravenous solutions were changed and cultured every 24 hr and blood cultures were taken weekly. No antibiotics were used during the infusion unless indicated for specific reasons. Each voided urine specimen was tested and recorded for glucose, pH, protein, and ketones.

Figures 6 and 7 show the balance studies. Throughout the 26-day period of study, there was a positive balance of nitrogen (7.608 gm), potassium (34.76 mEq), sodium (19.94 mEq), and a steady weight gain. She was discharged at 2 months of age weighing 2.24 kg. On follow-ups, she gained weight, developed, was eating well and appeared to be a thriving infant.

In plotting the steady weight gain of this infant, two significant features present themselves:

1. This 850 gm infant at 21 days weighed 1169 gm, a gain of 319 gm over its birth weight, and has already crossed the predicted curve of an infant with a birth weight of 1000 gm. This weight gain is much

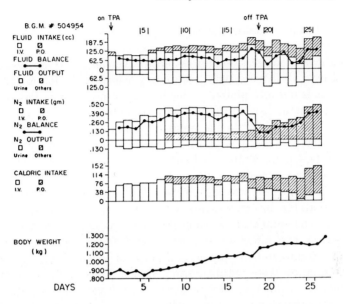

Fig. 6 Fluid and nitrogen balance, caloric intake, and body weight of B.G.M. From Dolanski *et al.* (1973), with kind permission of the Southern Medical Association.

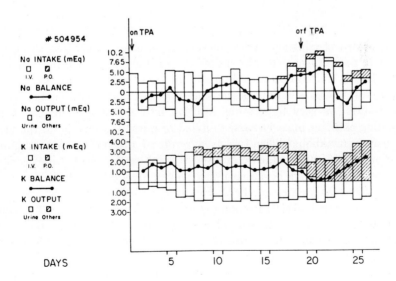

Fig. 7 Sodium and potassium balances of B.G.M. From Dolanski *et al.* (1973), with kind permission of the Southern Medical Association.

greater than the predicted weight gain of premature infants fed by conventional feedings, either by gavage or orally.

2. When this infant's weight is plotted versus weeks of gestation as reported by Lubchenco (1963), the infant consistently follows the curve on which it is initially plotted at birth, suggesting that she has continued to gain weight at her intrauterine growth rate as opposed to that predicted in extrauterine life for premature infants conventionally fed.

During the period of parenteral nutrition the infant experienced no adverse effects. On the third and fourth day there was only mild transient acidosis which was easily overcome by changing the concentration of the infusate. This acidosis was also expressed by a decreased pH in the urine showing that the infant was able to excrete excess hydrogen ions into the urine. The infant at no time had urinary sugars greater than 1+ by Tes-Tape and blood sugars were over 200 mg/100 ml on only rare occasions. Serum sodium, potassium, proteins, calcium, BUN, and alkaline phosphatase were within normal limits. Thus, parenteral nutrition was effective in quickly achieving positive nitrogen balance and normal growth while the infant was receiving an average of 0.4–0.5 gm/kg/day of nitrogen and a total caloric intake of 100–118 kcal/kg/day.

B. The Use of Xylitol in Parenteral Nutrition in Dogs

As shown in clinical studies, the use of parenteral nutrition as an adjunct in patients' care has been applied to almost all specialties of medicine. In the United States, practically all investigators administer a concentrated glucose solution as the sole source of nonprotein energy. Since the solution is very hypertonic, infusion into a central vein via an indwelling catheter must be used to avoid or minimize damage to the vein. However, metabolic difficulties as a result of the intravenous administration of hypertonic glucose and sepsis resulting from the use of central venous catheter have been experienced (Van Way *et al.*, 1973; Meng and Law, 1970; Dudrick *et al.*, 1972; Rea *et al.*, 1970; Heird *et al.*, 1972a,b; Filler and Eraklis, 1970; Brennan *et al.*, 1972; Dillon *et al.*, 1973). Thus, studies to use other sugars as glucose substitutes have been conducted, and limited attempts have also been made to infuse solution which may be tolerated by the peripheral veins.

Xylitol and sorbitol have been suggested as the possible substitutes for glucose in parenteral nutrition. Of these sugars, xylitol has received much attention recently. Although much work has been done concerning its metabolism and utilization (Horecher *et al.*, 1969; Meng and Law,

1970; Lang and Fekl, 1971) and its clinical use in parenteral nutrition (Lang and Fekl, 1971; Bessert and Rittmeyer, 1971; Mehnert, 1966), studies of the effects of long-term continuous intravenous infusion of xylitol on metabolism, liver function, nitrogen balance, etc., have not been carried out, even in animals. Thus, the long-term use of xylitol with the "true meaning of parenteral nutrition" remains to be studied. Furthermore, the toxicity observed following the intravenous administration of xylitol which has been reported (Donahoe and Powers, 1970; Thomas *et al.*, 1970, 1972a,b; Schumer, 1971) should be verified. For these reasons, this work has been undertaken to study the effects of long-term intravenous infusion of xylitol in parenteral nutrition in dogs. Comparison was also made between xylitol and glucose under the same experimental conditions.

1. Animal, Diet and Experimental Protocol

Five healthy young adult beagle dogs were used. The animals were confined in individual metabolic cages which permitted a quantitative collection of urine. In Period 1 they were fed a complete synthetic diet with 2.0 gm of protein as casein, 10 gm of carbohydrate as sucrose, and 3.0 gm of fat as Wesson Oil/kg/day (Table IV). This diet furnished 75 kcal/kg/day of which 53% came from carbohydrate, 11% from protein, and 36% from fat. Adequate amounts of minerals and known vitamins were also added to the diet according to that described in a previous publication (Meng and Early, 1949).

In Period 2, Dogs 15, 24, and 27 were given complete parenteral nutri-

TABLE IV

Composition of Synthetic Diet

Nutrient	gm/kg/day	kcal/kg/day	% Total cal
Carbohydrate (sucrose)	10.0	40	53
Fat (lard)	3.0	27	36
Protein (casein)	2.0	8	11
Alpha cell	0.90	—	—
Salt mixture	0.63	—	—
Vitamin mixture	0.04	—	—
Vitamin A	100 U.S.P. units	—	—
Vitamin D	10 U.S.P. units	—	—
Vitamin E (as α-tocopherol)	0.5 IU	—	—
Total	16.75 gm	75	100

tion with the same amounts of nutrients as those given in Period 1, but infused intravenously as glucose, protein hydrolysate, and soybean oil emulsion along with electrolytes and vitamins. Dog 13 was also given complete parenteral nutrition with 6.6 gm of glucose, 1.82 gm of sorbitol, and 1.82 gm of xylitol, crystalline amino acids, and soybean oil emulsion. Another dog (Dog 28) was given a carbohydrate-free diet by mouth, while carbohydrate as xylitol at a dosage of 10 gm/kg/day was infused by vein; the same regimen was maintained in Periods 3, 4, and 5. In Period 3, Dogs 15, 24, and 27 were given complete parenteral nutrition as in Period 2 except that fat was omitted and additional glucose was infused to maintain isocaloric intake. This was also true for Dog 13. In Period 4, Dogs 15, 24, and 27 were given complete parenteral nutrition as in Period 2 except xylitol instead of glucose was infused. Dog 13 was also given xylitol in the place of glucose. In Period 5, an increased amount of xylitol was given with the omission of fat in Dogs 15, 24, and 27, and 13. In Periods 6 and 7, Dogs 15, 24, 27, and 13 were given the same regimens for parenteral nutrition as in Periods 5 and 4, respectively, except that the protein was supplied as amino acids in Dogs 15, 24, and 27 and as protein hydrolysate in Dog 13. Thus, all dogs were given various amounts of xylitol during the last four periods. Isocaloric intake was maintained and the same amounts of minerals and vitamins were given during all periods. The regimens given are shown in Table V.

Infusions were given into the superior vena cava via an external jugular vein by way of an indwelling catheter continuously during a 24-hr period. The infusion rate of xylitol was 0.35–0.7 gm/kg/hr depending on the daily total dosage. The xylitol concentration ranged from 12 to 15%. During infusion, the animals remained in the individual metabolic cages. The swivel of Jacobs (1931) with some modification as reported previously was again used (Meng and Early, 1949); the arrangement permitted the animal complete freedom of movement at all times. The dogs were killed at the end of the experiment for histochemical studies.†

During the entire experimental period, the dogs were weighed and blood samples were withdrawn about 3 hr after the completion of the previous, and just before the beginning of the following day's, dietary regimen for measurements of packed cell volume, plasma glucose, xylitol, proteins, and triglycerides and serum bilirubin. The liver function tests including sulfbromophthalein retention, serum alkaline phosphatase, and serum GPT were also done at the same intervals. In some dogs plasma lactate, pyruvate and insulin, and blood pH and pCO_2 were also measured. Daily

---

† The microscopic histopathological studies were done by Major L. J. Ackerman and Dr. L. D. Jones, Pathology Division, U.S. Army Medical Research and Nutrition Laboratory, Denver, Colorado.

TABLE V

Regimens for Oral and Parenteral Alimentation[a]

Period	Dogs 15, 24, and 27			Dog 13	Dog 28		
	Route of administration	Dietary regimen	Nutrient (gm/kg/day)	Nutrient (gm/kg/day)	Route of administration	Dietary regimen	Nutrient (gm/kg/day)
1	Oral	Complete synthetic diet	Carbohydrate: sucrose 10 Fat: Wesson Oil 3 Protein: casein 2	Carbohydrate: sucrose 10 Fat: Wesson Oil 3 Protein-casein 2	Oral	Complete synthetic diet	Carbohydrate: sucrose 10 Fat: Wesson Oil 3 Protein: casein 2
2	Intravenous	Total parenteral alimentation	Carbohydrate: glucose 10 Fat: soybean oil[b] 3 Protein: hydrolysate[c] 2	Carbohydrate: glucose 6.6 xylitol 1.82 sorbitol 1.82 Fat: soybean oil[b] 3.0 Protein: amino acids[d] 2.0	Oral Intravenous	Carbohydrate-free diet Carbohydrate	Fat: Wesson Oil 3 Protein: casein 2 xylitol 10
3	Intravenous	Total parenteral alimentation	Carbohydrate: glucose 16.75 Protein: hydrolysate[c] 2	Carbohydrate: glucose 13.4 xylitol 1.82 sorbitol 1.82 Protein: amino acids[d] 2		As Period 2	
4	Intravenous	Total parenteral alimentation	Carbohydrate: xylitol 10 Fat: soybean oil[b] 3 Protein: hydrolysate[c] 2	Carbohydrate: xylitol 8.18 sorbitol 1.82 Fat: soybean oil[b] 3		As Period 2	
5	Intravenous	Total parenteral alimentation	Carbohydrate: xylitol 16.75 sorbitol 1.82 Protein: amino acids[d] 2	Carbohydrate: xylitol 14.93 sorbitol 1.82 Protein: amino acids[d] 2		As Period 2	
6	Intravenous	Total parenteral alimentation	Carbohydrate: xylitol 14.93 sorbitol 1.82 Protein: amino acids[d] 2	Carbohydrate: xylitol 16.75 Protein: hydrolysate[c] 2			
7	Intravenous	Total parenteral alimentation	Carbohydrate: xylitol 8.18 sorbitol 1.82 Fat: soybean oil[b] 3 Protein: amino acids[d] 2	Carbohydrate: xylitol 10 Fat: soybean oil[b] 3 Protein: hydrolysate[c] 2			

[a] All dogs received isocaloric intake and the same amount of vitamins and electrolytes. From Meng and Anderson (1973).
[b] Soybean oil was given as a 10% emulsion, Intralipid, Vitrum
[c] Hyprotigen, McGaw Laboratories.
[d] Crystalline amino acid solution, Aminofusin, J. Pfrimmer Co.

fluid balance, nitrogen balance, and urinary loss of xylitol were measured in all periods; the results are presented as the daily means of each individual period.

Plasma and urinary xylitol was determined by the method of Baily (1959) adapted for xylitol. Methods described in Sigma Technical Bulletin, 726/826 UV-10-68 were used (Marbach and Weil, 1967) for plasma pyruvate and lactate determination. Plasma insulin[+] was measured by the method of Hales and Randle (1963) with slight modification. Standard methods were used for other determinations.

C. Results

1. General Appearance and Body Weight

All dogs which received parenteral nutrition remained healthy, lively, and in good spirits except when large dosage of xylitol was infused; during these periods, the animals were less active and became somewhat apathetic. However, they recovered gradually when the dosage of xylitol was reduced. The body weight of these dogs is shown in Fig. 8. In general, there was a trend of weight loss during the last two periods. However, Dog 27 maintained its weight throughout.

2. Plasma Glucose, Xylitol, Lactate, Pyruvate, and Insulin

In general, plasma glucose levels were maintained within the normal limits except on two occasions: (a) The plasma glucose was reduced to 64 mg % (average of Dogs 15, 24, and 27) at the end of Period 6 during which 14.93 gm of xylitol/kg/day were given; (b) at the end of Period 5 when the xylitol dosage was 14.93 gm/kg/day in Dog 13. The plasma xylitol levels ranged from 8.5 to 11.8 mg % when xylitol dosage was 8.2–10 gm/kg/day, while they ranged from 12.5–15.7 mg % when the animals were given large dosages of xylitol (14.93–16.75 gm/kg/day) (Table VI).

A trend of a slight increase in plasma lactate was observed in dogs receiving xylitol. However, there was no significant change in plasma pyruvate. An increase in plasma insulin level was observed only in Dog 27 given a large dosage (16.75 gm/kg/day) of glucose (Period 3). Xylitol did not affect the plasma insulin levels (Table VII).

[+] The author wishes to express his thanks to Dr. Ian Burr, Associate Professor of Pediatrics, Vanderbilt University School of Medicine for the measurements of plasma insulin.

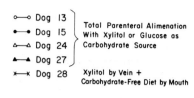

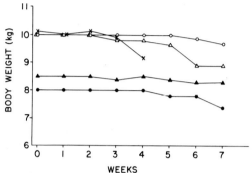

Fig. 8 Body weight of Dogs 15, 24, 27, and 13 given intravenous infusion of glucose (Periods 2 and 3) or of xylitol (Periods 4–7) in combination with amino acids, minerals, and vitamins. Dog 28 received intravenous xylitol in Periods 2–5 and a carbohydrate-free diet by mouth. All dogs were given a complete diet by mouth in Period 1. See Table V for detailed nutritional regimens. From Meng and Anderson (1973).

TABLE VI

Plasma Glucose and Xylitol[a]

	Mean glucose (mg %)			Mean xylitol (mg %)		
Period	Dogs 15, 24, 27	Dog 13	Dog 28	Dogs 15, 24, 27	Dog 13	Dog 28
0	97	82	119	1.2	0.9	—
1	90	63	137	1.4	0.9	1.4
2	93	80	132	2.5	1.1	1.4
3	98	79	113	4.8	1.1	2.2
4	79	83	110	11.8	10.5	11.3
5	83	50	—	15.2	14.8	11.2
6	82	80	—	12.5	15.7	—
7	64	—	—	10.2	8.5	—

[a] From Meng and Anderson (1973).

TABLE VII

Plasma Lactate, Pyruvate, and Insulin[a]

Period	Lactate (μM/ml)		Pyruvate (μM/liter)		Insulin (mμg/ml)	
	Dog 27	Dog 28	Dog 27	Dog 28	Dog 27	Dog 28
1	0.9	0.9	80	55	0.6	1.21
2	0.7	0.9	82	23	0.2	0.21
3	0.9	1.2	38	40	1.8	0.40
4	1.0	1.7	44	60	0.3	0.20
5	1.1	1.4	39	21	0.4	0.20
6	1.8	—	74	—	0.2	—
7	1.1	—	22	—	0.2	—

[a] From Meng and Anderson (1973).

3. Blood pCO_2 and pH

As shown in Table VIII, there was a slight decrease in venous blood pCO_2 in dogs given xylitol. Significant changes in blood pH were not observed.

4. Changes in Liver Function Tests

The sulfobromophthalein retention was not significantly changed in dogs receiving xylitol or glucose. The SGPT was significantly increased in dogs given xylitol; the increase was less when xylitol dosage was re-

TABLE VIII

pCO_2 and pH[a]

Period	pCO_2		pH	
	Dog 27	Dog 28	Dog 27	Dog 28
1	33	38	7.33	7.41
2	31	35	7.35	7.41
3	30	35	7.40	7.45
4	30	27	7.38	7.43
5	27	30	7.41	7.41
6	25	—	7.43	—
7	29	—	7.37	—

[a] From Meng and Anderson (1973).

duced. A slight increase in SGPT was also observed in Dogs 15, 24, 27, and 13 (Period 3) given large dosage of glucose (13.4–16.75 gm/kg/day). The serum alkaline phosphatase was elevated in dogs receiving xylitol. Again, this elevation of serum alkaline phosphatase was apparently dosage-related (Table IX).

5. Fluid Intake, Output, and Balance

It can be seen in Fig. 9 that glucose or xylitol infused intravenously in combination with amino acids or protein hydrolysate actually decreased the water intake by mouth (Dogs 15, 24, 27, and 13). Oral intake of water in the volume similar to that during the period of complete oral feeding was observed in Dog 28 which received xylitol solution by vein plus a carbohydrate-free diet by mouth. The urinary output was not significantly increased in dogs given xylitol. The urinary excretion in dogs receiving glucose and xylitol showed no appreciable difference. The fluid balance expressed as the difference between intake (intravenous plus oral) and urinary output was always positive in all dogs during the periods on parenteral nutrition. However, in general the fluid balance was more positive during the periods on parenteral nu-

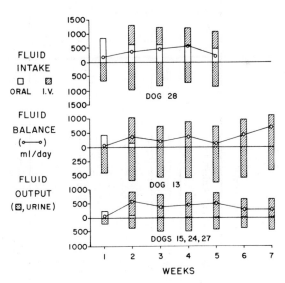

Fig. 9 The average daily fluid intake, output as urine, and fluid balance. Bars above the 0 line represent intake: □, oral and ▨, intravenous. Bars below the 0 line represent the urinary output, ▣. (○—○) fluid balance which was positive in all dogs. The nutritional regimens given to all dogs are as described in Fig. 8. From Meng and Anderson (1973).

TABLE IX

Changes in Liver Function Tests[a]

Period	Sulfobromophthalein (%) Dogs			SGPT (Wroblewski units) Dogs			Serum alkaline phosphatase (B.U.) Dogs		
	15, 24, 27	13	28	15, 24, 27	13	28	15, 24, 27	13	28
0	1.8	2.0	1.7	29.6	40.1	20.1	2.5	1.3	—
1	2.2	2.5	1.9	31.5	52.6	22.5	2.7	1.3	3.4
2	1.7	1.5	1.7	56.7	38.2	24.6	4.1	3.3	2.7
3	2.5	1.1	1.6	67.3	103.4	31.4	5.7	5.3	16.9
4	2.8	2.6	2.1	102.6	125.0	30.0	9.7	10.1	17.3
5	3.0	3.0	—	106.4	139.0	—	23.7	25.1	19.0
6	4.3	2.5	—	92.9	87.6	—	15.8	58.2	—
7	4.2	3.4	—	72.0	42.4	—	10.5	41.0	—

[a] From Meng and Anderson (1973).

trition than that of the period on complete oral feeding. There was no apparent difference in the degree of fluid retention in dogs given intravenous infusion of glucose and xylitol.

6. Xylitol Intake, Urinary Loss, and Balance

During administration of xylitol at a dosage of about 10 gm/kg/day (Dog 28) or during the first two periods of parenteral nutrition including xylitol (Dog 15, 24, 27, and 13), the urinary loss of this substance was about 10% or less of the administered dose. The urinary loss of xylitol for a prolonged period (Dogs 15, 24, 27, and 13 during Period 6). Further increase in urinary loss was observed in the same dogs during Period 7 (Fig. 10).

7. Urinary Nitrogen Loss and Nitrogen Balance

In general, the nitrogen intake was fairly constant throughout except during the last two periods when it was slightly decreased. The urinary nitrogen loss during complete oral feeding (Period 1), parenteral nutrition with glucose (Periods 2 and 3), and parenteral nutrition with xylitol

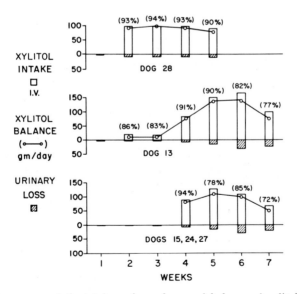

Fig. 10 The average daily intake, urinary loss, and balance of xylitol. Bars above the 0 line represent the amount of xylitol given intravenously in grams, □, and bars below the 0 line represent the urinary loss of xylitol, ▨. (O—O) xylitol retained. Figures in parentheses are the percent of the administered dose retained. The nutritional regimens are as described in Fig. 8. From Meng and Anderson (1973).

(Periods 4 and 5) were similar with nitrogen maintained in balance. The urinary nitrogen loss during Periods 6 and 7 was increased in most of the animals, and the nitrogen balance was negative although it became balanced in Dogs 13 and 28 during Period 7 (Fig. 11).

8. Other Findings

Significant changes in hematocrit, total plasma proteins, plasma trigly-cerides, or serum bilirubin levels were not observed. There were no significant changes observed at autopsy. Microscopic examinations revealed no abnormality attributable to xylitol infusion. Intravenous fat pigments as previously described were found in the liver and spleen (Thompson, 1970). Oxalate crystals were not found in the kidney.

D. The Use of Sorbitol in Parenteral Nutrition in Dogs

Sorbitol is oxidized to fructose and then converted to glucose; it is metabolized as such thereafter (Embden and Griesbach, 1914; Wich *et*

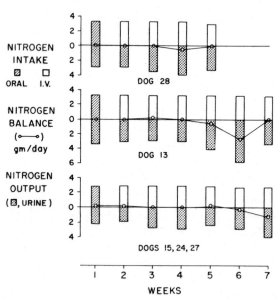

Fig. 11 The nitrogen intake, output in the urine, and balance in Dogs 15, 24, 27, and 13 given intravenous infusion of glucose (Periods 2 and 3) and xylitol (Periods 4–7) in combination with amino acids, minerals, and vitamins in parenteral nutrition. Dog 28 was given xylitol intravenously in Periods 2–5 plus a carbohydrate-free diet by mouth. Bars above the 0 line represent the nitrogen intake: ▨, by mouth, and ▢, by vein. Bars below the 0 line represent the nitrogen loss in the urine, ▨. From Meng and Anderson (1973).

al., 1951; Stetten and Setten, 1951; Blakley, 1957; Seeburg *et al.*, 1955; Adcock and Gray, 1957; Murisasco *et al.*, 1966). It has been suggested that sorbitol is also converted to glucose directly (Hers, 1955; Muntz and Vanko, 1962; Bye, 1969). Sorbitol has also been used in parenteral nutrition as a carbohydrate source in recent years (Seeburg *et al.*, 1955; Murisasco *et al.*, 1966; Bye, 1969; Huguenard and Blaise, 1956; Stuhlfauth *et al.*, 1960; Griem and Lang, 1960; Lee *et al.*, 1972). The reasons for the use of sorbitol are the suggested evidence of less insulin dependence, less initiating effect to the vein, better utilization in patients with glucose intolerance, more antiketogenic than glucose and none reactive in producing browning or Maillard reaction when combined with amino acids. However, although sorbitol has been used in some European countries, it is not given in parenteral nutrition as an energy source in the United States. This is in part because some of the claimed advantages over glucose remain to be substantiated. In addition, some workers are concerned with its diuretic effect since it is said to be not reabsorbed by the kidney (Burget *et al.*, 1937; Leimdorfer, 1954; Smith *et al.*, 1940). Our work was to study the tolerance of the intravenously administered sorbitol at high daily dosages and to compare sorbitol with glucose as an energy source in long-term parenteral nutrition in dogs.

1. Animal, Diet and Experimental Protocol

Five healthy young adult beagle dogs were also used in this study. The period of environmental adaptation and confinement of the dogs in the individual metabolic cages were as described in the study of xylitol.

Table X shows the regimens given for oral and parenteral nutrition. It can be seen that in Period 1 all dogs were fed a synthetic diet (Table IV) by mouth; this diet furnishes 75 kcal/kg/day of which 53% came from carbohydrate (sucrose), 11% from protein (casein), and 36% from fat (Wesson oil). Vitamins and minerals were also added to the diet as described (Meng and Early, 1949). In Period 2, Dogs 16, 19, and 21 were given a carbohydrate-free diet by mouth plus a sorbitol solution by vein. The sorbitol dosage was 10 gm/kg/day. Dogs 17 and 20 received total parenteral nutrition with crystalline amino acids, glucose, and Intralipid (a 10% soybean oil emulsion). In Period 3, Dogs 16, 19, and 21 received the same nutrients through the same routes as in Period 2. Dogs 17 and 20 received all nutrients by vein as in Period 2 except that sorbitol was given instead of glucose. In Period 4, Dogs 16, 19, and 21 received a carbohydrate and fat-free diet by mouth. Dogs 17 and 20 received total parenteral nutrition except that there was no fat. All 5 dogs were given 16.75 gm/kg/day of sorbitol by vein. In Period 5, the nutrients given to Dogs 16, 19, and 21 and routes of administration were the same as

TABLE X

Regimens for Oral and Parenteral Alimentation[a]

Period[b]	Dogs 16, 19, and 21			Dogs 17 and 20		
	Dietary regimen	Nutrient (gm/kg/day)	Route of administration	Dietary regimen	Nutrient (gm/kg/day)	Route of administration
1	Complete synthetic diet	Sucrose 10, Wesson Oil 3, Casein 2	Oral	Complete synthetic diet	Sucrose 10, Wesson Oil 3, Casein 2	Oral
2	Carbohydrate-free diet; Carbohydrate	Wesson Oil 3; Sorbitol 10	Oral; Intravenous	Total parenteral alimentation	Soybean oil 3, Amino acids 2, Glucose 10	Intravenous
3	As Period 2	As Period 2	As Period 2	As Period 2 except sorbitol (10 gm) given instead of glucose		
4	Carbohydrate and fat-free diet; Carbohydrate	Sorbitol 16.75	Oral; Intravenous	Total parenteral alimentation	Amino acids 2, Sorbitol 16.75, Soybean oil 0	Intravenous
5	As Period 4	As Period 4	As Period 4	As Period 4 except glycose (16.75 gm) given instead of sorbitol		

[a] All dogs received isocaloric intake and the same amounts of vitamins and electrolytes.
[b] Eight days per period.

in Period 4. Dogs 17 and 20 were given total parenteral nutrition as in Period 4 except that sorbitol (Period 4) was replaced by glucose at the same dosage (16.75 gm/kg/day). Thus, all dogs were given an isocaloric intake with the same amounts of nutrients throughout the experimental periods. Each period (Period 1–5) lasted for 8 days. The rate of sorbitol infusion was 0.42–0.7 gm/kg/hr. The daily infusions were given into the superior vena cava via an indwelling catheter continuously during the 24-hr period as described in the study of xylitol. The final concentration of sorbitol solution administered was from 12 to 15%.

The biochemical and physiologic studies and fluid and nitrogen balance measurements were as described in the study of xylitol.

E. Results

1. Body Weight and General Appearance

It can be seen in Fig. 12 that all dogs maintained their weight throughout except Dog 19 which lost 1.7 kg during the 32 days of study of sorbitol infusion at dosages of 10 gm (Periods 2 and 3) and 16.75 gm/kg/day (Periods 4 and 5). It is not known why this dog lost weight. Perhaps

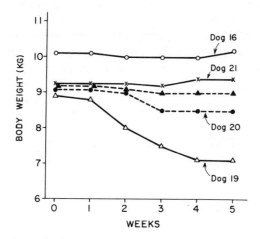

Fig. 12 Body weight of Dogs 16, 19, 21, 17, and 20. Dogs 16, 19, and 21 were given intravenous infusion of sorbitol (10 gm/kg/day) plus a carbohydrate-free diet by mouth (Periods 2 and 3), and sorbitol at 16.75 gm/kg/day by vein plus a carbohydrate-fat-free diet by mouth (Periods 4 and 5). Dogs 17 and 20 received intravenous infusion of glucose during Periods 2 and 3, and sorbitol during Periods 4 and 5 in combination with amino acids, minerals, and vitamins. See Table X for detailed nutritional regimens.

it had some liver abnormality initially since its serum alkaline phosphate and SGPT were elevated prior to sorbitol administration. It may also result from the loss of sorbitol in the urine which reached an average of 41.5 gm/day or 31% of the administered sorbitol during Period 5. Only Dog 19 appeared somewhat apathetic during Period 5.

2. Plasma Sorbitol and Glucose

The method used for sorbitol measurement was the same one used for xylitol (Baily, 1959). The highest serum sorbitol levels were 13.9 mg % (Dog 16) and 26.1 mg % (Dog 20) during which period 10 and 16.75 gm/kg/day were given, respectively. The serum sorbitol was 21 mg % in Dog 19 during Period 5 when its urinary loss was greater than that in any other dogs. Most of the measurements of blood glucose were within the normal range except in Dogs 16 and 17 in which the lowest value of 47.5 mg % was observed at the end of Period 4. The average plasma sorbitol and glucose levels are shown in Table XI.

3. Changes in Liver Functions

Table XII shows the average values of liver function tests including sulfobromophthalein retention, SGPT, and serum alkaline phosphatase. There was no significant change in sulfobromophthalein retention in any of the 5 dogs. SGPT was slightly increased in some animals (Dogs 19 and 20). However, this change was reversible after discontinuation of sorbitol infusion; in some cases, it declined in spite of continued infusion. The serum alkaline phosphatase was increased in Dog 16 at the end of Periods 4 and 5 and in Dog 17 at the end of Periods 3 and

TABLE XI

Plasma Glucose and Sorbitol

Period	Mean glucose (mg %)		Mean sorbitol (mg %)	
	Dogs 16, 19, 21	Dogs 17, 20	Dogs 16, 19, 21	Dogs 17, 20
0	75.2	77.8	1.84	2.59
1	93.2	81.5	2.14	2.22
2	85.7	79.3	7.67	2.28
3	72.2	72.3	11.32	9.53
4	61.5	69.0	17.79	21.51
5	72.5	68.2	18.44	7.09

TABLE XII

Changes in Liver Function Tests

Period	Sulfobromophthalein retention (%)		SGPT (Wroblewski units)		Serum alkaline phosphatase (B.U.)	
	Dogs 16, 19, 21	Dogs 17, 20	Dogs 16, 19, 21	Dogs 17, 20	Dogs 16, 19, 21	Dogs 17, 20
0	2.7	2.0	35.6	28.4	5.2	3.6
1	3.0	1.6	40.1	31.4	6.4	4.0
2	3.1	2.5	53.2	24.5	6.3	4.9
3	2.8	2.8	34.2	43.0	6.7	11.7
4	3.2	2.0	32.3	51.0	7.2	11.7
5	3.3	1.6	32.9	32.4	8.5	8.6

4 during which time sorbitol was given. The finding in Dog 19 was high even during oral feeding (Period 1) but showed no further increase during the periods receiving sorbitol. The serum alkaline phosphatase remained within the normal range throughout in Dog 21.

4. Fluid Intake, Output, and Balance

Figure 13 shows the average daily fluid intake, urinary output, and fluid balance. It can be seen that in dogs (Filler and Eraklis, 1970; Horecker *et al.*, 1969; Lang and Fekl, 1971) receiving intravenous administration of sorbitol plus oral feeding of a carbohydrate-free diet (Periods 2 and 3) or of a carbohydrate-fat-free diet (Periods 4 and 5), little or no water was taken by mouth; total fluid intake was greater than that of oral feeding in Period 1 because of sorbitol infusion. The average daily urinary output was slightly increased in these dogs during the same periods. The average daily fluid balance or retention in these 3 dogs was also slightly increased; however, Dog 19 had a slight negative fluid balance

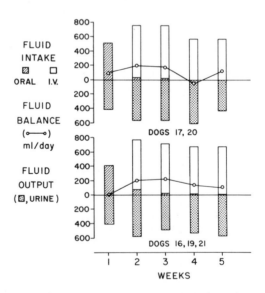

Fig. 13 The average daily fluid intake, output (urine), and fluid balance. Bars above the 0 line represent intake: ▨, oral and ▢, intravenous. Bars below the 0 line represents the urinary output, ▦. (○—○) fluid balance: Above the 0 line equals positive balance or water retention, below the 0 line represents negative balance or water loss. The nutritional regimens given to all dogs are as described in Fig. 12.

during Period 5. Similar findings in fluid balance were observed in Dogs 17 and 20 which received total parenteral nutrition with sorbitol (Periods 3 and 4) or with glucose (Periods 2 and 5) as the carbohydrate source. It can be noted that the average daily fluid balance was slightly negative during Period 4; this was because of the negative balance observed in Dog 20 when it received 16.75 gm of sorbitol/kg/day. The urinary volume was somewhat decreased and the fluid balance became positive when glucose (16.75 gm/kg/day) instead of sorbitol was given (Period 5).

5. Sorbitol Intake, Urinary Loss, and Balance

In dogs 16, 19, and 21 which were given 10 gm sorbitol/kg/day by vein with an infusion rate of 0.42 gm/kg/hr (Periods 2 and 3), the urinary loss of sorbitol was from 7–10% of the total amount administered; thus, the amount retained was 90–93% of the dosage given. When the total dosage was increased to 16.75 gm/kg/day and administered at a rate of 0.7 gm/kg/hr, there was an increase in the urinary output of sorbitol. However, the amount retained was still about 90% of the dosage given (Period 4). The urinary loss of sorbitol was further increased and the percent of the total amount administered was decreased to 81 in Period 5. The low average daily retention partly resulted from the high urinary loss in Dog 19. In the other 2 dogs (16 and 21) the daily urinary loss of sorbitol was not as high. In dogs (17 and 20) receiving total parenteral nutrition, the sorbitol retained was about 85% of the total amount administered whether the dosage was 10 or 16.75 gm/kg/day (Periods 3 and 4) (Fig. 14).

6. Nitrogen Intake, Urinary Nitrogen Loss, and Nitrogen Balance

The nitrogen intake was fairly constant in all dogs throughout except that in Dog 19 which was slightly decreased because of its weight loss. It can be seen in Fig. 15 that the average daily urinary nitrogen loss was variable. In Dogs 16, 19, and 21, which received sorbitol by vein and other nutrients by mouth, moderate to marked increase in the average daily urinary nitrogen loss was observed; the excessive loss or urinary nitrogen occurred in all three dogs during Period 3. The daily average urinary nitrogen loss in Dogs 17 and 20 which received total parenteral nutrition was similar to that in Dogs 16, 19, and 21. The nitrogen balance was negative in all 5 dogs whether sorbitol was given by vein with oral

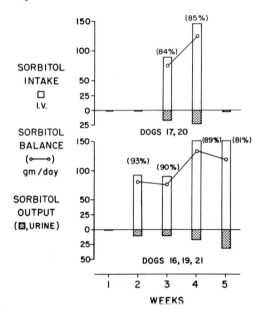

Fig. 14 The average daily intake, urinary loss, and balance of sorbitol. Bars above the 0 line represent the amount of xylitol given intravenously in grams, □, and bars below the 0 line represent the urinary loss, ▣. (O—O) represents xylitol retained. Figures in parentheses are percent of the administered dose retained. The nutritional regimens are as described in Fig. 12.

feeding of other nutrients or sorbitol was infused intravenously in total parenteral nutrition. The average daily nitrogen balance was more negative during Period 3 in Dogs 16, 19, and 21. It is surprising that the negative nitrogen was also observed in Dogs 17 and 20 which were given parenteral nutrition with glucose as a carbohydrate source (Periods 2 and 4).

7. Other Findings

Packed cell volume and serum proteins, triglycerides, and bilirubin showed no significant changes in all dogs throughout except the serum triglyceride in Dog 19 was increased (130 mg %) at the end of Period 2. The histopathological examinations of various tissue (e.g., liver, spleen, lung, kidney, and heart) showed no significant changes attributable to the long-term infusion of sorbitol. However, a mild degree of intravenous fat pigments was again observed in the liver and spleen in Dogs 17 and 20 which also received soybean oil emulsion.

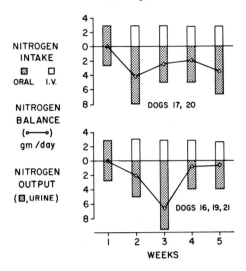

Fig. 15 The average daily nitrogen intake, output, and balance. Bars above the 0 line represent the nitrogen intake: ▨, oral, and ☐, intravenous. Bars below the 0 line represent the nitrogen output in the urine, ▦. (O—O) nitrogen balance; below the line represents negative balance. Dogs 16, 19, and 21 were given sorbitol intravenously at 10 gm/kg/day in Periods 2 and 3, and 16.75 gm/kg/day in Periods 4 and 5, respectively; other nutrients were given by mouth as a carbohydrate-free (Periods 2 and 3) or a carbohydrate-fat-free (Periods 4 and 5) diet. Dogs 17 and 20 received total parenteral nutrition. In Periods 2 and 5, glucose was the carbohydrate source. Sorbitol was given in Periods 3 and 4.

F. Discussion

Glucose is considered as a physiologic sugar. With adequate nitrogen intake as amino acids or protein hydrolysate along with minerals and vitamins, glucose can serve as the sole energy source in total parenteral nutrition. It is well established that total parenteral nutrition with adequate calories, nitrogen, and other nutrients is capable of correcting negative nitrogen balance, promoting wound healing, and achieving body weight gain, growth, and development (Meng and Law, 1970; Filler and Eraklis, 1970; Coats, 1967; Dudrick *et al.*, 1968, 1969, 1970; Filler *et al.*, 1969; Wilmore, *et al.*, 1969; Heller, 1969; Borrensen *et al.*, 1970; Jacobson, 1970; Johnston and Spivey, 1970). However, the use of a hypertonic glucose solution (20–40%) may induce hyperglycemia. In our studies, we have found it helpful and sometimes necessary to start the solutions for parenteral nutrition (Tables I and II) at a slow rate of infusion and small total daily dosage or one-half to three-quarter strength concentration of these preparations; the amount or concentration of the solution is gradually increased to the fully prescribed daily dosage and/or concentration. In general, most of the adult patients tolerate infusion of 3 liters of the solution (Table I) furnishing 2700 kcal/day as glucose without raising the blood glucose over 200 mg/100 ml. However, debilitated, elderly, diabetic, partially pancreatized, postoperative, or trauma patients often develop hyperglycemia. This can lead to glycosuria and osmotic diuresis, which, if allowed to proceed unchecked, produce the clinical picture of the non-ketotic hyperosmolar syndrome. As water is lost in excess of electrolytes, the patient becomes dehydrated and serum osmolality rises over 350 mOsm/liter. Thus, frequent monitoring of blood and urinary sugar, especially during the early days of infusion, is necessary. If hyperglycemia develops, insulin should be administered. Insulin dosage should be based on the blood glucose levels. Usually, we give 10 units of regular insulin every 4–6 hr at the beginning; it may be increased in steps of 5–10 units per dose. Recently, we have been adding 10–20 units of insulin per liter of solution and find this rather satisfactory in controlling blood sugar levels. Occasionally, it has been found very difficult to control the progressive elevation of blood sugar and marked increase in urinary loss. In this case, the practice is to temporarily decrease the total daily intake and the rate of infusion. If discontinuation is necessary, a 5–10% glucose solution should be infused to prevent hypoglycemia. Thus, it may be readily recognized that an effective safe glucose substitute as a calorie source would be of paramount value in parenteral nutrition. It may be stated, however, that glucose should not be replaced completely by any other single sugar in total.

Some degree of glucose intolerance in neonates and infants after surgery has been reported (Elphick and Wilkinson, 1968). Many reports have been published to indicate that fructose alone or as invert sugar is better utilized than glucose, but these findings have been disputed. The readers are referred to some review articles (Geyer, 1960; Thoren, 1963). In the solution for parenteral nutrition used in infants and children (Table II), 20% carbohydrate as a mixture of glucose (10%) and fructose (10%) was added. The reason for the use of invert sugar is that most of the premature infants and neonates could not tolerate 20% glucose; elevation of blood sugar levels and marked spillage of sugar in the urine were almost invariably observed. This is especially true in postoperative or septic cases. The use of glucose (10%)–fructose (10%) mixture seems capable of correcting these difficulties without the administration of insulin. In fact, insulin was never given to infants receiving parenteral nutrition. Either the glucose concentration or the solution volume or both should be reduced when hyperglycemia is encountered. However, except for the subjective clinical observations, evidence is lacking to demonstrate the advantages of glucose–fructose over glucose alone. It should be added that lactic acidosis has been reported in children (Anderson, *et al.*, 1969) and in adults (Bergstrom *et al.*, 1968). However, acidosis was not observed in infants receiving parenteral nutrition solution containing fructose. Further study on the use of glucose–fructose or invert sugar in parenteral nutrition is warranted.

The results of this study indicate that the intravenously administered xylitol was apparently utilized for energy whether given singly or in combination with a nitrogen source along with minerals and vitamins in parenteral nutrition. Even with a caloric intake of 75 kcal and protein intake of 2.0 gm/kg of body weight/day which are considered low for beagle dogs, nitrogen balance and body weight were maintained for 14–30 days. From the point of view of nitrogen balance, body weight, urinary volume, and fluid balance, it is difficult to detect any difference between the use of glucose and of xylitol as a carbohydrate source in parenteral nutrition when one considers the daily intravenous infusion of xylitol is limited to 14–20 days at a daily dosage of not more than 10 gm/kg/day.

The rate of infusion of xylitol is of paramount importance to achieve maximal utilization and to avoid adverse effects. This is true for most nutrients administered intravenously. In the case of xylitol, a total daily dosage of 10 gm/kg/day with an infusion rate of 0.42 gm/kg/hr seems to be the maximal amount and rate that can be tolerated by dogs. It should be pointed out also that at the infusion rate of 0.42 gm/kg/hr, the urinary loss of xylitol was about 6 or 7% of the total administered dose; it was increased to 10% loss at the end of the fourth period. Thus, it seems important to consider the length of the period during which xylitol is being infused.

It is also interesting to note that at a xylitol concentration of 12–15% given in this study, diuresis and excessive oral water intake were not observed. However, van Eys and Wang (1972) and H. C. Meng and G. E. Anderson (unpublished data) have found that a too concentrated xylitol solution (20–50%) was not well tolerated in rabbits and dogs, respectively.

Intravenous infusion of xylitol has been found to increase serum lactate resulting in lactic acidosis (Schumer, 1971; Thomas *et al.*, 1972b; Forster, 1972) and to decrease pCO_2 (Schumer, 1971; Thomas *et al.*, 1972b). In the present study in dogs given large dosages of xylitol, only a slight rise in serum lactate and a slight decrease in pCO_2 were observed. In addition, there was no significant change in blood pH. Thus, it is reasonable to state that lactic acidosis probably did not occur. It may be postulated that perhaps the toxicities following xylitol infusion observed by other investigators primarily resulted from the high rate of infusion, large total dosage, and highly concentrated solution. Forster (1972) reported that under the same experimental conditions other sugars including glucose also increased the serum lactate levels.

Elevation of plasma insulin following intravenous administration of xylitol has been observed in patients (Spitz *et al.*, 1970) and in dogs (Kuzuya *et al.*, 1966). However, in these studies a short-term infusion or injection was given. In the present study, xylitol was infused continuously during the 24-hr period. This may, in part, be the reason why elevation of plasma insulin was not observed.

Intravenous infusion of xylitol did elevate SGPT and serum alkaline phosphatase in some dogs; again, these changes seem to be dose dependent and occur after administration for a prolonged period. Three points may be commented on further:

1. The changes are reversible.
2. Total parenteral nutrition using xylitol in combination with amino acids, vitamins, and minerals may aggravate the effect of xylitol; thus, it may be possible that amino acids may contribute to the changes in liver function tests.
3. Glucose given to dogs receiving total parenteral nutrition also produced a slight elevation af SGPT and serum alkaline phosphatase.

The fact that sulfobromophthalein retention and serum bilirubin showed no significant changes seems to suggest that the elevation of SGPT and serum alkaline phosphatase resulted from some metabolic alterations.

Intravenous administered sorbitol given at a dosage of 10–16.75 gm/kg/day and at an infusion rate of 0.42–0.7 gm/kg/hr singly or in combination with amino acids and other nutrients in total parenteral nutrition seems to be utilized for energy. In dogs given sorbitol by vein plus a carbohydrate-free or carbohydrate-fat-free diet by mouth, the percent of the total amount of sorbitol retained was decreased when the experimental period was prolonged (Fig. 14); this decrease in sorbitol retention results, in part, from increased urinary loss. It is not known if the increased sorbitol loss in the urine results from the prolonged period of intravenous infusion or increased sorbitol dosage or both. Although it has been reported (Burget *et al.*, 1937; Leimdorfer, 1954; Smith *et al.*, 1940) that sorbitol is not reabsorbed by the kidney, diuretic effect of sorbitol was not encountered as judged by the urinary volume of Dogs 16, 19, and 21 in Periods 2–5. The slight increase in urinary volume during sorbitol infusion periods over that of the oral feeding period probably resulted from the increased administration of fluid without increasing water intake by mouth. It is also interesting to note that the percent of sorbitol retained is lower in dogs receiving total parenteral nutrition than in those given sorbitol alone by vein. It seems that the urinary sorbitol loss may be further increased by the intravenous administration of other nutrients, particularly amino acids.

It is surprising to observe the moderate to marked negative nitrogen balance in dogs receiving sorbitol whether given alone by vein· plus a carbohydrate-free or carbohydrate-fat-free diet by mouth or in combination with other nutrients in total parenteral nutrition. It has been reported that intravenously administered sorbitol can serve as an energy source and achieve positive nitrogen balance (Griem and Lang, 1960; Lee *et al.*, 1972). The reason for the finding of negative nitrogen balance is not known. It is possible that this results, in part, from the loss of calories as sorbitol. However, the number of calories lost as sorbitol was not large, except in Dog 19 in Period 5. Furthermore, the sorbitol loss does not seem to correlate with the degree of negative nitrogen balance. In addition, the negative nitrogen balance in Dogs 17 and 20 during total parenteral nutrition with glucose as a calorie source in Periods 2 and 5 was similar to that in Periods 3 and 4 when sorbitol was used. However, the negative nitrogen balance was only slight in Dogs 16, 19, and 21 during Periods 4 and 5 in comparison with that during Period 1 when a synthetic diet was given by mouth.

The blood glucose level was decreased in 2 dogs to as low as 48 mg % when 16.75 gm of sorbitol were given in Period 4. However, none of the animals showed any increase in blood glucose. Unfortunately, plasma insulin and fructose were not measured, and it cannot be ascertained that the decrease in blood glucose results from increased insulin secretion or decreased conversion from sorbitol → fructose → glucose. The observed low levels of plasma sorbitol and glucose might have been the result of the time of weekly blood withdrawal which was 2–3 hr after the discontinuation of the previous day's infusion.

As in the dogs given xylitol, sorbitol also elevated SGPT and serum alkaline phosphatase in some dogs. Sulfobromophthalein retention and serum bilirubin remained within normal limits. Again the elevation of SGPT and serum alkaline phosphatase were reversible, and these changes seemed somewhat more marked in dogs receiving total parenteral nutrition with sorbitol as a carbohydrate source than in those given sorbitol by vein plus a carbohydrate-free or carbohydrate-fat-free diet by mouth. Thus, the intravenous administration of other nutrients, especially amino acids, may contribute to the alteration of these liver function tests.

Based on the results of these studies obtained in our laboratory and those reported by other investigators, it may be suggested that a mixture of two or even three sugars should be used in total parenteral nutrition. Glucose in various amounts should be a constant component in the solution. Other sugars such as fructose, sorbitol, xylitol, or fructose–xylitol may be given in combination with glucose. However, further studies are necessary to determine which combination and what ratio of sugars may be used in patients with specific pathological conditions.

G. Conclusion

Long-term total parenteral nutrition with adequate nitrogen, calories, minerals, and vitamins administered continuously into a central vein through an indwelling catheter is capable of achieving weight gain, wound healing, and closure of enterocutaneous fistula, reversing negative nitrogen balance and promoting growth and development. In most of the adult patients with a variety of diseases including surgery except severe burns, 12–13 gm of nitrogen or about 78 gm of protein equivalent as crystalline amino acids, 2700 kcal and a total fluid volume of 3 liters/day are found adequate and well tolerated. In infants and young children, 0.52 gm of nitrogen or 3.2 gm of protein equivalent and 113 kcal with a fluid volume of 125 ml/kg/day are given. Various sugars have been used as a calorie source. In all adult patients a hypertonic solution (25%) of glucose is used. Nonketotic hyperosmolar syndrome has been encountered, especially in elderly patients and in those with diabetes mellitus, partial pancreatectomy, trauma, and sepsis. Insulin is necessary in some these cases; the regulation of insulin dosage and rate of glucose infusion have been found difficult. A 20% sugar solution containing glucose (10%) and fructose (10%) along with other nutrients is given to infants and children as the source of calorie; this is well tolerated and insulin has not been necessary.

Long-term intravenous administration of xylitol or sorbitol either singly or in combination with crystalline amino acids, vitamins, and minerals in total parenteral nutrition at dosages of 8–16.75 gm/kg/day and with an infusion rate of 0.35–0.7 gm/kg/hr is apparently tolerated, and seems to be utilized for energy in dogs. However, SGPT and serum alkaline phosphatase are elevated, especially when xylitol or sorbitol is infused for a prolonged period or in combination with other nutrients in total parenteral nutrition; these changes are found to be reversible. Diuretic effect has not been encountered although urinary loss of these sugars, particularly sorbitol, is high in some dogs. Histopathological changes in tissues have not been observed which may be attributable to the infusion of xylitol or sorbitol. It is suggested that the use of various sugars in combination in total parenteral nutrition should be studied, especially in patients who have glucose intolerance.

Acknowledgment

This investigation was supported by a research grant (GM-16432) from the National Institutes of Health, Department of Health, Education and Welfare.

References

Adcock, L. D., and Gray, C. H. (1957). The metabolism of sorbitol in the human subject. *Biochem. J.* **65,** 554.

Anderson, G., Brohult, J., and Sterner, G. (1969). Increasing metabolic acidosis following fructose infusion in two children. *Acta Paediat. Scand.* **58,** 301.

Bailey, J. M. (1959). A microcolorimetric method for the determination of sorbitol, mannitol and glycerol in biological fluids. *J. Lab. Clin. Med.* **54,** 158.

Bergström, J., Hultman, E., and Roch-Norlund, A. E. (1968). Lactic acid accumulation in connection with fructose infusion. *Acta Med. Scand.* **184,** 359.

Bessert, B., and Rittmeyer, P. (1971). Klinische Erfahrungen mit hochdosierter infusion. *Z. Ernaehrungswiss., Suppl.* **11,** 29.

Blakley, R. L. (1957). The metabolism and antigenic effects of sorbitol. Sorbitol dehydrogenase. *Biochem. J.* **49,** 257.

Borrensen, H. C., Coran, A. G., and Knutrud, O. (1970). Metabolic results of parenteral feedings in neonatal surgery. *Ann. Surg.* **172,** 291.

Brennan, M. F., Goldman, M. H., O'Connell, R. C., Kundsin, R. B., and Moore, F. D. (1972). Prolonged parenteral alimentation: Candida growth and the prevention of candidemia by Amphotericin instillation. *Ann. Surg.* **176,** 265.

Brill, A. D., Sandstead, H. H., Price, R., Johnston, R. E., Law, D. H., and Scott, H. W. (1972). Changes in body composition after jejunoileal bypass in morbidly obese patients *Amer. J. Surg.* **123,** 49.

Burget, G. E., Todd, W. R., and West, E. S. (1937). Sorbitol as a diuretic. *Amer. J. Physiol.* **119,** 283.

Bye, P. A. (1969). The utilization and metabolism of intravenous sorbitol. *Brit. J. Surg.* **56,** 653.

Coats, D. (1967). Long-term parenteral nutrition with amino acid solution. *In* "Symposium of the International Society of Parenteral Nutrition" (N. Henning and G. Berg, eds.), p. 55. Pallas, Munich.

Dillon, J. D., Jr., Schaffner, W., Van Way, C. W., III, and Meng, H. C. (1973). Septicemia and total parenteral nutrition—distinguishing catheter-related from other septic episodes. *J. Amer. Med. Ass.* **223,** 134.

Dolanski, E. A., Stahlman, M. T., and Meng, H. C. (1973). Parenteral alimentation of premature infants under 1,200 grams. *S. Med. J.* **66,** 41.

Donahoe, J. F., and Powers, R. J. (1970). Biochemical abnormalities with xylitol. *N. Engl. J. Med.* **282,** 690.

Dudrick, S. J., MacFadyen, B. V., Van Buren, C. T., Rubert, R. L., and Maynard, A. T. (1972). Parenteral hyperalimentation—metabolic problems and solutions. *Ann. Surg.* **176,** 259.

Dudrick, S. J., Wilmore, D. W., Vars, H. M., and Rhoads, J. E. (1968). Long-term total parenteral nutrition with growth, development and positive nitrogen balance. *Surgery* **64,** 134.

Dudrick, S. J., Wilmore, D. W., Vars, H. M., and Rhoads, J. E. (1969). Can in-

travenous feeding as the sole means of nutrition support growth in the child
and restore weight loss in an adult? *Ann. Surg.* **169**, 974.

Dudrick, S. J., Wilmore, D. W., Steiger, E., Mackie, J. A., and Fitts, W. T., Jr.
(1970). Spontaneous closure of traumatic pancreatoduodenal fistula with total
intravenous nutrition. *J. Trauma* **10**, 542.

Elman, R. (1948). "Parenteral Alimentation in Surgery with Special Reference
to Protein and Amino Acids." Harper, New York.

Elphick, M. C., and Wilkinson, A. W. (1968). Glucose intolerance in newborn infants
undergoing surgery for alimentary tract anomalies. *Lancet* **2**, 539.

Embden, G., and Greisbach, W. (1914). Uber Milchsaure-und Zuckerfilding in der
isoberter Leber. *Hoppe-Seyler's Z. Physiol. Chem.* **91**, 251.

Filler, R. M., and Eraklis, A. J. (1970). Care of the critically ill child: Intravenous
alimentation. *Pediatrics* **46**, 456.

Filler, R. M., Eraklis, A. J., Rubin, V. G., and Das, J. B. (1969). Long-term
total parenteral nutrition in infants. *N. Engl. J. Med.* **281**, 589.

Forster, H. (1972). Safety in xylitol. *N. Engl. J. Med.* **286**, 790.

Geyer, R. P. (1960). Parenteral nutrition. *Physiol. Rev.* **40**, 150.

Griem, W., and Lang, K. (1960). Versuche zur parenteralen Ernahrung mit Amino-
sauren-Sorbit-Losungen. *Klin. Wochenschr.* **38**, 336.

Hales, C. N., and Randle, P. J. (1963). Immunoassay of insulin with insulin-antibody
precipitate. *Biochem. J.* **88**, 137.

Heird, W. C., Driscoll, J. M., Jr., Schullinger, J. N., Grebin, B., and Winters,
R. W. (1972a). Intravenous alimentation in pediatric patients. *J. Pediat.* **80**,
351.

Heird, W. C., Nicholson, J. F., Driscoll, J. M., Jr., Schullinger, J. N., and Winters,
R. W. (1972b). Hyperammonia resulting from intravenous alimentation using
a mixture of synthetic L-amino acids. *J. Pediat.* **81**, 162.

Heller, L. (1969). Clinical and experimental studies on complete parenteral nutrition.
Scand. J. Gastroenterol. **4**, Suppl. 3, 7.

Hers, H. G. (1955). The conversion of fructose-1-C^{14} and sorbitol-1-C^{14} to liver and
muscle glycogen in the rat. *J. Biol. Chem.* **214**, 373.

Horecker, B. L., Lang, K., and Takagi, Y., eds. (1969). "Metabolism, Physiology
and Clinical Uses of Pentoses and Pentitols." Spinger-Verlag, Berlin and New
York.

Huguenard, P., and Blaise, J. (1956). L'apport énergétique réalisé par les solutés
de sorbitol. *Presse Med.* **64**, 2114.

Jacobs, H. R. D. (1931). An apparatus for constant intravenous injection into unrestrained animals. *J. Lab. Clin. Med.* **16**, 901.

Jacobson, S. (1970). Complete parenteral nutrition in man for seven months. *In*
"Advances in Parenteral Nutrition" (G. Berg, ed.), p. 6. Thieme, Stuttgart.

Johnston, I. D. A., and Spivey, J. (1970). The use of long-term parenteral
nutrients in alimentary failure. *In* "Advances in Parenteral Nutrition" (G.
Berg, ed.), p. 483. Thieme, Stuttgart.

Kuzuya, T., Kanazawa, Y., and Kosaka, K. (1966). Plasma insulin response to
intravenously administered xylitol in dogs. *Metab., Clin. Exp.* **15**, 1149.

Lang, K., and Fekl, W., eds. (1971). "Xylit in der Infusionstherapie," *Z. Ernaehrungswiss.*, Suppl. 11. Steinkopff Verlag, Darmstadt.

Lee, H. A., Morgan, A. G., Waldram, R., and Bennett, J. (1972). Sorbitol: Some
aspects of its metabolism and role as an intravenous nutrient. *In* "Parenteral
Nutrition" (A. W. Wilkinson, ed.), p. 121. Churchill, London.

Leimdorfer, A. (1954). The diuretic effect of sorbitol. *Arch. Int. Pharmacodyn. Ther.* **100**, 161.

Lubchenco, L. O. (1963). Intrauterine growth as estimated from liveborn birth weight data at 24 to 42 weeks of gestation. *Pediatrics* **32**, 793.

Marbach, E. P., and Weil, M. H. (1967). Rapid enzymatic measurement of blood lactate and pyruvate. *Clin. Chem.* **13**, 314.

Mehnert, H. (1966). Die Verwertung von Xylit bei parentralar Ernahrung. *In* "Parenterale Ernahrung, Anaestesiologic und Wiederbelebung" (K. Lang, R. Frey, and M. Halmagyi, eds.), Vol. 6, p. 28.

Meng, H. C. (1971). Principles of parenteral nutrition. *Hosp. Med.* p. 102.

Meng, H. C., and Anderson, G. E. (1973). The use of xylitol in long-term parenteral nutrition in dogs. *Supple. 15, Z. Ernährungswissenschaft,* **10**, 54.

Meng, H. C., and Early, F. (1949). Studies on complete parenteral alimentation on dogs. *J. Lab. Clin. Med.* **34**, 1.

Meng, H. C., and Law, D. H., eds. (1970). "Parenteral Nutrition." Thomas, Springfield, Illinois.

Meng, H. C., and Sandstead, H. H. (1972). Long-term total parenteral nutrition in patients with chronic inflammatory diseases of the intestines. *In* "Parenteral Nutrition" (A. W. Wilkinson, ed.), p. 213. Churchill, London.

Meng, H. C., Law, D. H., and Sandstead, H. H. (1970). Some clinical experience. *In* "Advances in Parenteral Nutrition." (G. Berg, ed.), p. 64. Thieme, Stuttgart.

Muntz, J. A., and Vanko, O. M. (1962). The metabolism of sorbitol using C^{14}-labeled sorbitol. *J. Biol. Chem.* **273**, 3583.

Murisasco, A., Unal, D., Jauffret, P., and de Belsunce, M. (1966). The clinical use of sorbitol 30% solution for I.V. infusion. *Aggressologie* **7**, 253.

Rea, W. J., Wyrick, W. J., Jr., McClelland, R. E., and Webb, W. R. (1970). Intravenous hyperosmolar alimentation. *Arch. Surg.* **100**, 393.

Schumer, W. (1971). Adverse effect of xylitol in parenteral nutrition. *Metab., Clin. Exp.* **20**, 345.

Scott, H. W., Jr., Gobble, W. G., Jr., and Law, D. H., IV. (1965). Clinical experience with jejunal pouch (Hunt-Lawrence) as substitute stomach after total gastrectomy. *Surg., Gynecol. Obstet.* **121**, 1231.

Scott, H. W., Jr., Law, D. H., IV., Gobble, W. G., Jr., and Sawyers, J. L. (1968). Clinical and metabolic studies after total gastrectomy with a Hunt-Lawrence jejunal food pouch. *Amer. J. Surg.* **115**, 148.

Seeburg, V. P., McQuarrie, E. B., and Secor, C. C. (1955). Metabolism of intravenously infused sorbitol. *Proc. Soc. Exp. Biol. Med.* **80**, 303.

Smith, W. W., Finkelstein, N., and Smith, H. W. (1940). Renal excretion of hexitols and their derivatives and of endogeneous creatinine-like chromogen in dogs and man. *J. Biol. Chem.* **135**, 231.

Spitz, A. H., Rubenstein, A H., Bersohn, I., and Bassler, K. H. (1970). Metabolism of xylitol in healthy subjects and patients with renal disease. *Metab., Clin. Exp.* **10**, 24.

Stetten, M. R., and Setten, D., Jr. (1951). Metabolism of sorbitol and glucose compared in normal and alloxan-diabetic rats. *J. Biol. Chem.* **193**, 157.

Stuhlfauth, K., Mehnert, H., and Pette, C. (1960). Das Verhalten des Glukose-Fruktose-und-Sorbit-stoffwechsels bei leberkranken und lebergesunden Patienten vor, wahrend und nach i.v. Infusion von Sorbitit. *Medizinische Welt,* 1367.

Thomas, D. W., Edwards, J. B., and Edwards, R. G. (1970). Examination of xylitol. *N. Engl. J. Med.* **283**, 437.

Thomas, D. W., Edwards, J. B., Gilligan, J. E., Lawrence, J. R., and Edwards, R. G. (1972a). Complications following intravenous administration of solutions containing xylitol. *Med. J. Aust.* **2,** 1238.

Thomas, D. W., Gilligan, J. E., Edwards, J. B., and Edwards, R. G. (1972b). Lactic acidosis and osmotic diuresis produced by xylitol infusion. *Med. J. Aust.* **2,** 1247.

Thompson, S. W. (1970). Histologic and ultrastructural changes following intravenous administration of fat emulsions. *In* "Parenteral Nutrition" (H. C. Meng and D. H. Law, eds.), p. 408. Thomas, Springfield, Illinois.

Thoren, L. (1963). The use of carbohydrate and alcohol in parenteral nutrition. *In* "Colloquium on Intravenous feeding" (A. Wretlind, ed.), Separatum "Nutritio et Dieta," Vol. 5, p. 304. Karger, Basel.

van Eys, J., and Wang, Y. M. (1972). Xylitol toxicity. *N. Engl. J. Med.* **286,** 1163.

Van Way, C. W., III, Meng, H. C., and Sandstead, H. H. (1973). An assessment of the role of parenteral alimentation in the management of surgical patients. *Ann. Surg.* **177,** 103.

Wick, A. N., Almen, M. C., and Joseph, L. (1951). The metabolism of sorbitol. *J. Amer. Pharm. Ass.* **40,** 542.

Wilmore, D. W., and Dudrick, S. J. (1969). Safe long-term venous catheterization. *Arch. Surg.* **98,** 256.

Wilmore, D. W., Groff, D. B., Bishop, H. C., and Dudrick, S. J. (1969). Total parenteral nutrition in infants with catastrophic gastrointestinal anomalies. *J. Pediat. Surg.* **4,** 181.

CHAPTER 32

Toxicity of Parenteral Xylitol

D. W. THOMAS, J. B. EDWARDS,
and R. G. EDWARDS

A. Clinical Uses of Xylitol

1. Intravenous Nutrient and Other Uses

Xylitol was first identified as a product of the metabolism of L-xylulose which accumulates in the genetic disorder essential pentosuria (Touster *et al.*, 1956). It was later realized to be a normal constituent of the glucuronate–xylulose cycle (Touster, 1960). Since then it has entered clinical practice as an oral and intravenous nutrient. In addition, several other properties of this pentitol have been claimed to be clinically useful. Antiketotic effects of small intravenous doses have been described (Yamagata *et al.*, 1965). More recently, a possible use in the treatment of hemolytic anemia resulting from glucose-6-phosphate dehydrogenase deficiency has been proposed (Wang *et al.*, 1971). However, the most enthusiastic use of xylitol has been as an intravenous nutrient. Its claimed effectiveness is based on its rapid entry into cells, said not to require insulin, and conversion to useful metabolites (Keller and Froesch, 1971).

A number of reports have been published on the effective use of xylitol for nutritional purposes in a variety of clinical settings. It is effective as a glucose substitute for the provision of energy because it is metabolized by an insulin-independent pathway. It can be converted to liver glycogen and has an antiketotic action. This perhaps gives xylitol an

567

advantage over glucose in that it can be a useful source of calories in conditions characterized by carbohydrate intolerance and insulin resistance. Several groups have used this pentitol for such purposes with encouraging effects (Bässler *et al.*, 1962; Spitz *et al.*, 1970a). Furthermore, xylitol has been infused intravenously for specific investigational purposes in man (Spitz *et al.*, 1970b) and in animals (Wilson and Martin, 1970) with no obvious adverse reactions being described.

Xylitol was introduced into clinical practice in Australia for its use as an intravenous nutrient. It became available during the first half of 1969 as 10, 20, and 50% solutions in water (w/v). Doses recommended ranged from 200 to 500 gm/day, but no instructions relating to rates of administration were issued. It was recommended for use in a wide range of clinical situations which included metabolic acidosis. No contraindications to its use were given except states of insulin-induced hypoglycemia. Slight diuresis and mild laxative action were the only side effects listed. As a source of calories it was stated to be superior to glucose, presumably on the basis of its rapid incorporation into major metabolic pathways independent of insulin. Thus, it was particularly recommended for insulin-deficient states such as diabetes mellitus (including diabetic hyperglycemic ketoacidotic coma) and in the nutritional management of post-traumatic and postsurgical patients.

2. Reported Toxicity of Parenteral Xylitol

Shortly after its introduction in South Australia it became apparent to those responsible for producing laboratory data used to assess the metabolic status of patients in hospital that infusions of solutions containing xylitol were associated with certain adverse reactions (Thomas *et al.*, 1970, 1972a). These were also reported to the Australian Drug Evaluation Committee who promptly recommended that xylitol be withdrawn from clinical use throughout Australia and prohibited as an import into the country. Subsequently, other national health authorities were informed of these adverse reactions through the facility of the World Health Organization (WHO Drug Bulletins Nos. 73 and 77, 1970). Table I shows the occurrence of adverse reactions observed in 10 patients studied retrospectively in South Australia of a group of 22 patients known to have received intravenous solutions containing xylitol. These adverse reactions have been grouped, in the light of studies to be described, into two—effects thought to be produced by the intravenous administration of hypertonic sugar solutions, and effects thought to be peculiar to the infusion of solutions containing xylitol. The adverse reactions within the second group are possibly associated with tissue damage resulting from wide-

TABLE I

Adverse Reactions Associated with Intravenous Administration of Solutions Containing Xylitol[a]

Case	Effects of hypertonic carbohydrate solution				Effects possibly peculiar to xylitol			
	Diuresis	Acidosis	Hyper-uricemia	Cerebral disturbances	Hepatic disturbances	Oliguria	Azotemia	Crystal deposition
1	+	+	ID	ID	–	ID	–	–
2	–	–	ID	+	++	–	+	ID
3	–	–	–	–	++	–	–	–
4	–	–	ID	+	–	–	–	–
5	ID	–	–	–	–	–	+	–
6	–	–	ID	ID	–	ID	++	ID
7	ID	–	ID	–	+++	–	+	ID
8	–	–	ID	ID	++	ID	–	ID
9	–	–	ID	–	+	ID	++	ID
10	–	+	ID	–	ID	ID	+	ID

[a] (–) Adverse reaction; (+), no adverse reaction; (ID), inadequate or insufficient data available.

spread deposition of crystals. The crystals observed in tissues from 5 patients given intravenous solutions containing xylitol have been positively identified as crystals of calcium oxalate (Evans *et al.*, 1972).

These reports were a surprise to many who had been using xylitol in apparently similar circumstances for some time. However, some others have also described adverse reactions associated with intravenous infusions of xylitol. Increased serum concentration of bilirubin and uric acid with increased urinary excretion of uric acid were seen in volunteers infused with 5 and 10% solutions in water (w/v) at rates of 2.34–2.90 gm/kg of body weight (Donahoe and Powers, 1970). Increases in serum lactate, uric acid, and bilirubin concentrations and alkaline phosphatase activities in patients given xylitol intravenously in doses of 1.5 gm/kg* of body weight have also been described (Schumer, 1971). Xylitol loading tests in 2 normal adult male volunteers in doses of 4.5 gm/kg* of body weight produced pain in the right upper quadrant of the abdomen, vertigo, headache, nausea, and vomiting. Serum and urine uric acid, serum bilirubin, lactate and phosphate concentrations, and serum glutamic-oxalacetic transaminase and alkaline phosphatase activities were increased as well. Toxicity experiments in rabbits demonstrated hyperosmolality effects of infusing hypertonic solutions of xylitol and suggested that some adverse reactions observed in human patients may be explained on this basis (Wang *et al.*, 1972; van Eys and Wang, 1972).

In an attempt to define the etiology of these adverse reactions observed in South Australia two studies were attempted. At first, closer examination was made of effects following controlled infusions of xylitol into a human patient who previously exhibited some adverse reactions during therapeutic infusions of a solution containing xylitol. Further observations on other human patients were prevented by the withdrawal of xylitol from clinical usage. Animal experiments were then undertaken, and results of these studies will be described in some detail. Some comparisons have been made with effects produced by other sugars used for nutritional purposes and infused intravenously as hypertonic solutions.

B. Studies of Xylitol Toxicity

1. Studies on a Human Patient

A patient who had previously exhibited adverse reactions to intravenous solutions containing xylitol was infused with a solution containing 50% xylitol in water (w/v) at a rate of 1.0 gm/kg of body weight for 7 hr (Thomas *et al.*, 1972b). This study demonstrated the production

* Dosages calculated to be 0.24 and 0.36 gm/kg/hr, respectively.

of extracellular hyperosmolality accompanied by an initial fall in the serum sodium concentration. Increased urine output occurred with the production of a fluid deficit exceeding 2 liters, and an increase in the urinary excretion of solute presumed to be xylitol. A rise in blood lactate concentration with increased blood lactate:pyruvate ratio occurred together with a decrease in serum bicarbonate concentration. These disturbances were accompanied by little change in arterial carbon dioxide partial pressure.

Two distinct effects were thus produced. An accumulation of xylitol in the extracellular compartment produced a state of extracellular hyperosmolality. Xylitol passed with the glomerular filtrate into the renal tubules and, overwhelming what mechanisms existed for its reabsorption, it remained in the tubule fluid in such high concentration to promote an osmotic diuresis. The second obvious effect was the production of a metabolic acidosis that appeared to result from an accumulation of excess lactate.

These effects appeared suitable for further investigations using animals infused with hypertonic solutions of xylitol at similar rates and for similar periods of time.

2. Animal Studies

a. Description of Experimental Procedures

Native Australian dogs (*Canis familiaris dingo*) weighing from 8 to 18 kg were sedated with 20 mg of xylazine intramuscularly, anesthesia induced with intravenous pentothal in amounts of 50–200 mg and maintained by intravenous chloralose [α —D(+)glucochloralose] given as a single loading dose of 100 mg/kg of body weight. Constant mechanical ventilation was provided via a cuffed endotracheal tube. Solutions were infused intravenously at a set rate controlled by a constant speed peristaltic pump. During a control period of $\frac{1}{2}$–1 hr a solution of 0.9% sodium chloride in water (w/v) was infused. Thereafter 50% solutions (w/v) of xylitol (from various sources and prepared freshly from xylitol powder dissolved in pyrogen-free sterile water), glucose, fructose, or sorbitol were infused at rates to give a dose of 1.0 gm/kg of body weight per hour for 5–5$\frac{1}{2}$ hr. Additional amounts of sodium chloride solution were infused to replace the blood removed for various analyses as described. Central body temperature and arterial blood pressure were constantly monitored. Urine was collected in $\frac{1}{2}$ hr aliquots and analyses performed as shown in Table II. Samples of arterial blood were removed every $\frac{1}{4}$ hr and analyses performed as shown in Table III.

TABLE II

Animal Studies—
$\frac{1}{2}$ Hour Urinalyses

Volume
Osmolality
Electrolytes: Sodium
                Potassium
                Phosphate
Urea
Uric acid
Creatinine
Reducing substances
Glucose
Fructose

A variety of effects were observed. Some of these were similar to those adverse reactions described in human patients, others were not. Certain effects seen in human patients were not observed in dogs. Results are expressed as the means of observations made in a number of dogs. Eight dogs were infused with xylitol solutions, three with glucose, three with fructose, and a further three with sorbitol.

TABLE III

Animal Studies—Blood and Plasma Analyses

$\frac{1}{4}$ Hour samples
   pH
   PaCO₂
   PaO₂

$\frac{1}{2}$ Hour samples		
Sodium	Lactate	Total protein
Potassium	Pyruvate	Albumin
Chloride	Glucose	Bilirubin
Total CO₂	Insulin	Cholesterol
Urea	Phosphate	Alkaline phosphatase
Uric acid	Triglyceride	Lactate dehydrogenase
Creatinine		Glutamic oxalacetic
Osmolality		transaminase

2 Hour samples
   Hemoglobin
   Packed cell volume
   Red blood cell count
   White blood cell count

---

b. Effects of Hypertonic Solutions

As observed in most of the original patients given solutions containing xylitol all dogs displayed an osmotic diuresis whether infused with xylitol, glucose, sorbitol, or fructose. A typical example is shown in Fig. 1. After

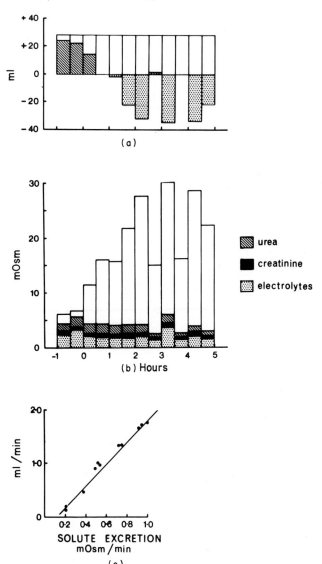

Fig. 1 (a) Fluid balance, (b) solute excretion, and (c) urine flow of a dog infused with xylitol at a rate of 1.0 gm/kg of body weight per hour.

the commencement of infusing intravenously hypertonic sugar solutions all animals not only exhibited an osmotic diuresis but also developed a fluid deficit with a negative fluid balance. This was most severe with infusions of glucose. Infusions of xylitol, fructose, and sorbitol produced negative fluid balances of decreasing magnitude. The relationship between urine flow rate and solute excretion rate shows that there is not an increase in urinary free water clearance but an obligatory excretion of water resulting from the increased excretion of solute. Analyses of the solutes excreted in urine show a relative constant excretion of sodium and potassium (and accompanying anions), creatinine, and urea, but an increasing excretion of other solutes during the infusion of hypertonic sugar solutions as shown in Fig. 1. During the infusions of glucose and fructose these solutes were identified as glucose and fructose, respectively. The solutes excreted in large amounts during infusions of xylitol and sorbitol were neither glucose nor other reducing sugars and were assumed to be xylitol and sorbitol, respectively.

Measured plasma osmolality showed an increase during the infusion of these hypertonic sugar solutions. These changes are shown in Fig. 2. No significant difference was observed between the hyperosmolality effects of xylitol, glucose, or sorbitol, but fructose had a significantly smaller effect. This is probably accounted for by the more rapid intracellular uptake of fructose particularly by the liver (Kjerulf-Jensen, 1942). It was also observed, as shown in Fig. 2, that during the infusion

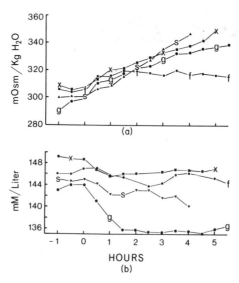

Fig. 2 (a) Mean plasma osmolality and (b) sodium concentration in dogs infused with xylitol (x), glucose (g), fructose (f), or sorbitol (S).

of these hypertonic sugar solutions the plasma concentration of sodium was reduced. In each case this could not be adequately accounted for by the urinary excretion of sodium or by the dilution of the extracellular volume by the infused water, as part of the hypertonic sugar solution. It would appear that a significant cause of this effect is the movement of water from the intracellular space into the extracellular space promoted by the hyperosmolality of the extracellular fluids. Just such an effect has been produced in dogs by intravenous infusions of hypertonic solutions of sodium chloride and mannitol (Winters *et al.*, 1964). This could be prevented by a slower infusion of sugar to match its cellular uptake and metabolism and urinary excretion. Then a large osmolar gradient would not develop between the intracellular and extracellular compartments. Although no significant changes in plasma potassium concentrations and certain plasma enzymatic activities (in particular lactate dehydrogenase and glutamic oxalacetic transaminase) were observed in dogs after intravenous infusions of hypertonic sugar solutions such changes reported in rabbits (Wang *et al.*, 1972) may have a similar explanation. Intracellular constituents may be released into the extracellular fluids as water moves out of the cell in response to the osmotic gradient produced by the infusion of hypertonic solutions.

c. *Dilutional and Metabolic Acidosis*

In the patients described developing adverse reactions associated with intravenous infusions of xylitol-containing solutions an acidosis was defined as a reduction in the serum bicarbonate concentration. This occurred in all dogs infused with hypertonic sugar solutions. As shown in Fig. 3 this was most noticeable with infusions of sorbitol and xylitol. Such changes have been reported with intravenous infusions of hypertonic sodium chloride and mannitol solutions (Winters *et al.*, 1964) and may result from several effects. It must be remembered that as plasma sodium concentrations are reduced during infusions of hypertonic sugar solutions other plasma constituents such as bicarbonate concentrations may fall for similar reasons. In addition to this dilutional effect other reasons may exist for these falls in plasma bicarbonate concentration. Certain intravenous solutions used for parenteral nutrition—such as 50% dextrose and 5% protein hydrolysate solution in water—have produced similar effects (Chan *et al.*, 1972b) as a result of the infusion of significant quantities of hydrogen ion contained in these solutions (Chan *et al.*, 1972a) and of their body buffer depleting effect (Fortner, 1972). Measurements of blood concentrations of hydrogen ion (hydrion) during the infusions of hypertonic sugar solutions into dogs are shown in Fig. 3. Significant in-

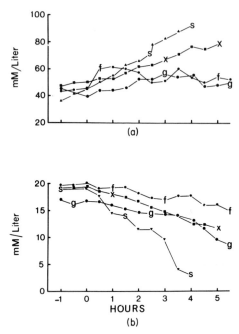

Fig. 3 (a) Mean blood hydrogen ion (hydrion) and (b) plasma bicarbonate concentration in dogs infused with xylitol (x), glucose (g), fructose (f), or sorbitol (S).

creases occurred during the infusions of sorbitol and xylitol. These changes occurred in animals that were constantly ventilated and had a constant arterial partial pressure of carbon dioxide ($PaCO_2$), reflecting a constant alveolar ventilation, and can only be explained by the occurrence of a metabolic acidosis. Figure 4 shows that indeed all dogs infused with hypertonic sugar solutions developed increases in blood lactate concentration. This was most marked with sorbitol and xylitol. Changes in blood pyruvate concentration were also observed during infusions of these four sugars suggesting these substrates were being utilized and converted in part to pyruvate. In particular, there were marked increases in pyruvate concentration following intravenous administration of fructose and glucose with smaller increases occurring with sorbitol and xylitol. Toward the end of infusions with sorbitol, despite an increase in blood lactate concentration, there was a fall in blood pyruvate concentration. Calculations of the blood lactate:pyruvate ratios shown in Fig. 4 demonstrate that significant increase in this parameter occurred only during the infusions of xylitol and sorbitol, being more obvious with sorbitol. Similar changes have been described following infusions of glucose in rabbits

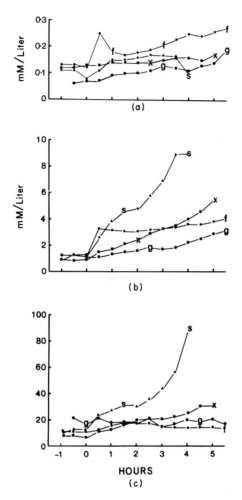

Fig. 4 (a) Mean blood pyruvate, (b) lactate concentrations, and (c) blood lactate:pyruvate ratios of dogs infused with xylitol, glucose, fructose, or sorbitol (symbols as in Fig. 3).

(Ainsworth and Allison, 1971). In the present dog experiments during infusions of glucose and fructose, blood lactate:pyruvate ratios did not significantly change. It has been suggested that the ratio of blood lactate:pyruvate concentrations reflects the cytoplasmic NADH:NAD ratio. These changes are considered to be the result of depletion of NAD and accumulation of NADH in the cytoplasm following the initial NAD-linked dehydrogenase reactions involving xylitol and sorbitol as illustrated in Fig. 5. Xylitol has been described as a particularly strong re-

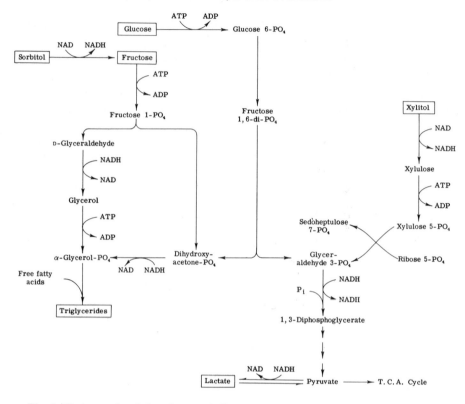

Fig. 5 Pathways involving the metabolism of xylitol, glucose, fructose, and sorbitol.

ductant of cytoplasmic nucleotides (Williamson *et al.*, 1971). This state is reflected by an increased conversion of pyruvate to lactate, facilitated by the increased cytoplasmic concentration of NADH (see Fig. 5). Furthermore, such a change in the redox potential of cells utilizing xylitol (in particular the liver and kidney) would also impair their ability to utilize lactate. A similar course of events has been described during the metabolism of ethanol (Kreisberg, 1972), with the development of a lactic acidosis. Similar observations have been made in dogs and rabbits infused with xylitol in doses of 1.0 gm/kg of body weight or greater (Wade and Bostrom, 1971). Impaired lactate tolerance after infused loads of lactate was also demonstrated in these animals. Less is known of the ability of xylitol when infused in man to produce an accumulation of lactate. However, experiences with other sugars, in particular fructose, are well known. Lactate accumulation has been described with infusions of fructose (Elliot *et al.*, 1967; Bergström *et al.*, 1968) and severe metabolic acidosis has also been described (Andersson *et al.*, 1969).

d. Increased Uric Acid Excretion

The association between intravenous infusions of fructose and hyper-uricemia were first described by Perheentupa and Raivio in 1967. Several mechanisms have been invoked including the increased degradation of purine ribonucleotides. Fructose is rapidly phosphorylated to fructose 1-phosphate (see Fig. 5) with an associated rapid depletion of intracellular adenine nucleotides, in particular ATP, and inorganic phosphate (Woods *et al.*, 1970). Since 5′-nucleotidase is normally inhibited by ATP, and AMP deaminase is inhibited by inorganic phosphate, the reduction in intracellular concentrations of these compounds would be expected to stimulate the catabolism of AMP to inosine with rapid degradation to uric acid in man and to allantoin in dogs.

In only two of the patients exhibiting adverse reactions to solutions containing xylitol were serum uric acid determinations made. In both patients considerable increases were observed. Although the end product of purine catabolism in the dog is allantoin, small amounts of uric acid can be detected in their plasma. However, during the infusing of solutions containing xylitol, glucose, fructose, or sorbitol no significant changes in plasma uric acid were observed (see Fig. 6). Nevertheless, measuring the ratio of uric acid clearance in the urine to urinary creatinine clearance showed a dramatic increase, especially during the infusions of sorbitol, fructose, and xylitol. These results suggest that during the infusion of these compounds there is a significant increase in the production of uric acid. Such a situation as described above may be seen after the phosphorylation of any of these compounds as they are metabolized as shown in Fig. 5. Decreased intracellular concentrations of ATP and inorganic phosphate have been documented in liver tissue of dogs and rabbits infused with xylitol at rates of 1.0 gm/kg of body weight or greater (D. N. Wade, personal communication). The rates of these phosphorylation reactions are probably quite rapid, except for glucose, whose entry into most cells is controlled by insulin. This may account for the observed much smaller effect glucose has on the clearance of uric acid in the dog. This evidence of increased uric acid excretion may underestimate the extent of purine breakdown in dogs since allantoin is the end product rather than uric acid.

Similar studies have been made in man during infusions of fructose (Fox and Kelley, 1972). Increased uric acid production with hyperuricemia was found to be the result of increased breakdown of purine ribonucleotides. It may be added that in situations where increased serum uric acid concentration and hyperlactatemia coexist, the presence of lactate or other organic acids may compete with urate for urinary excretion.

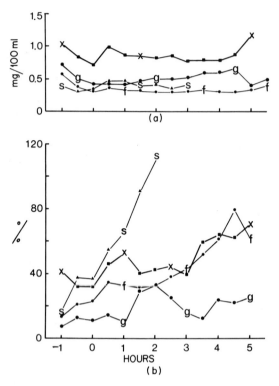

Fig. 6 (a) Mean plasma uric acid concentration and (b) mean urinary uric acid clearance:creatinine clearance ratios of dogs infused with xylitol, glucose, fructose, or sorbitol (symbols as in Fig. 3).

This competition may further increase serum uric acid concentration (Goldfinger *et al.*, 1965).

e. Other Effects Not Seen in Human Patients

(1) HYPOGLYCEMIA. The intravenous infusions of glucose, sorbitol, and fructose were associated with increasing concentrations of glucose in the plasma as shown in Fig. 7. Xylitol in contrast was associated with a decrease in plasma glucose concentrations, and in some dogs a marked hypoglycemia occurred. Similar findings have been observed in dogs elsewhere (Wade and Bostrom, 1971).

(2) INSULIN RELEASE. Infusions of xylitol and sorbitol were accompanied by considerable increases in plasma insulin concentration. This may well explain the reciprocal changes in plasma glucose observed in

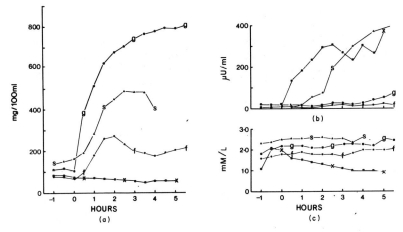

Fig. 7 (a) Mean plasma glucose, (b) insulin, and (c) phosphate concentrations in dogs infused with xylitol, glucose, fructose, or sorbitol (symbols as in Fig. 3).

dogs infused with xylitol. However, in dogs infused with sorbitol high plasma glucose concentrations persisted despite high concentrations of insulin in the plasma. It appears that a state of insulin resistance existed. The poor response of insulin to intravenous glucose in the dog may be another example of the importance of the oral route of administration of this sugar for the appropriate release of insulin. The ability of xylitol when infused intravenously into dogs to release large amounts of insulin is well known. It has a far more potent effect on insulin release than glucose has (Kuzuya, 1969; Kosaka, 1969; Wilson and Martin, 1970).

An adequate explanation of this effect has not yet been provided although there is some suggestion that glucose and xylitol mediate insulin release through a common metabolite in the pentose phosphate pathway (Wilson and Martin, 1970). In man xylitol has a much less potent action on insulin release (Hirata *et al.*, 1969). Its effect has been shown to be less than that of glucose, both being administered by an intravenous route (Spitz *et al.*, 1970a). The ability of sorbitol to release large amounts of insulin with coexisting high concentrations of plasma glucose remains unexplained.

(3) HYPOPHOSPHATEMIA. During infusions of xylitol a slow but significant reduction of plasma phosphate was observed, as shown in Fig. 7. Such a response is considered to reflect the transport of glucose into cells, especially muscle and other sites of utilization, mediated by insulin and accompanied by phosphate (Pollack *et al.*, 1934). Similar changes

have been observed after xylitol infusions in man (Spitz *et al.*, 1970a). The lack of similar responses after sorbitol would again support the failure of action of insulin in mediation of glucose uptake by cells and suggest a state of insulin resistance. Lack of responses during infusions of glucose and fructose are consistent with the observed low concentrations of insulin in the plasma.

The fall in serum concentrations of phosphate has been described after prolonged infusion of glucose in man (Travis *et al.*, 1971) and the implication of these changes, especially in relation to oxygen transport by red blood cells, is discussed by those authors.

(4) HYPERTRIGLYCERIDEMIA. Figure 8 illustrates the observed increases in plasma triglyceride concentration observed only after infusions of xylitol. This response is considered to reflect several abnormalities in the metabolism of xylitol which need to be confirmed by appropriate studies. During the metabolism of xylitol glyceraldehyde 3-phosphate is formed (see Fig. 5). But instead of entering the glycolytic pathway and forming 1,3-diphosphoglycerate, it is suggested that conversion to dihydroxyacetone phosphate predominates. With the presumed reduction in cytoplasmic NAD, as a result of the NAD-linked dehydrogenation of xylitol to D-xylulose, and the eventual lack of inorganic phosphate in cells metabolizing xylitol, the conversion of glyceraldehyde 3-phosphate to 1,3-diphosphoglycerate may be impaired favoring the formation of dihydroxyacetone phosphate. The relative abundance of NADH would favor the further formation of α-glycerophosphate. Such a situation would thus favor triglyceride synthesis from free fatty acids and offer an explanation of the antiketotic effect of xylitol. A similar situation may be expected during infusions of sorbitol. However, the conversion of glycerol to α-glycerophosphate may be impaired by the falling intracellular concen-

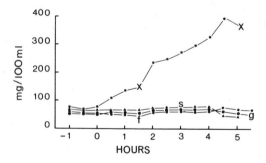

Fig. 8   Mean plasma triglyceride concentrations in dogs infused with xylitol, glucose, fructose, or sorbitol (symbols as in Fig. 3).

trations of ATP, resulting from the phosphorylation of fructose to form fructose 1-phosphate (see Fig. 5). Although the mechanisms to explain these observations are theoretical, several important alterations in the metabolism of xylitol are suggested and should be confirmed or denied by appropriate studies.

3. Changes Observed in Human Patients but Not Observed in Dogs

a. Cerebral Disturbances

Obviously these effects could not be observed in such short-term non-survival experiments in anesthetized dogs. Many possible explanations can be offered as discussed previously (Thomas *et al.*, 1972a). Not lightly to be dismissed is the possibility that some of these effects may again result from the hyperosmolality state induced by infusions of hypertonic solutions producing a state of cerebral dehydration. Together with a superadded metabolic acidosis, this gives a metabolic situation not unlike that seen in diabetic hyperglycemic ketoacidosis with altered state of consciousness. Cerebral dehydration with an increase in the concentration of intracellular constituents occurring secondarily to extracellular hyperosmolality induced in dogs by infusions of hypertonic sodium chloride and mannitol has been associated with increases in cerebrospinal fluid bicarbonate concentration (Winters *et al.*, 1964). A similar mechanism operating in human patients infused with hypertonic xylitol may contribute to the cerebral disturbances observed. More serious intracerebral metabolic defects may be responsible, and again this is an effect that needs further careful evaluation.

b. Hepatic Damage

In dogs infused with xylitol, glucose, fructose, and sorbitol there was no evidence of hepatic damage based on determinations of concentrations of plasma proteins, bilirubin, and cholesterol and plasma activities of lactate dehydrogenase and glutamic oxalacetic transaminase. Liver biopsies taken at the conclusion of every experiment involving infusions of xylitol showed completely normal morphology and no evidence of hepatic cell damage or cellular infiltration. Nevertheless, evidence of hepatic damage was not observed in human patients infused with xylitol until the second or third day of intermittent or continuous infusions. In these short-term animal experiments the lack of evidence of hepatic damage after only 5–5½ hr of infusion with xylitol may not necessarily be significant.

c. Renal Disturbances

Significant reductions in urine flow or creatinine clearance were not observed in any of the dogs infused with xylitol, glucose, fructose, or sorbitol. Furthermore, no changes occurred in the plasma concentrations of urea or creatinine. Thus, in these short-term infusion studies no evidence of disturbances in renal function, in particular oliguria or azotemia, was observed. Renal biopsies taken at the conclusion of every experiment involving infusions of xylitol showed completely normal morphology, and no evidence of renal parenchymal damage or cellular infiltration was seen.

d. Oxalate Crystal Deposition

Liver, renal, and cerebral tissues taken from dogs infused with xylitol were carefully examined for crystal deposition. Care was taken to scan all sections using Nicol prisms crossed to near extinction. This procedure makes the identification of calcium oxalate crystals easier since they are strongly birefringent. Ordinary tissue stains such as hematoxilin and eosin do not stain these crystals which may be easily missed if examined with ordinary light microscopy.

4. Summary of Studies

The animal studies and limited human studies described support the concept that osmotic diuresis, metabolic acidosis, and hyperuricemia are a result of the rapid infusion of hypertonic solutions of xylitol. The ability of hypertonic solutions of glucose, fructose, or sorbitol to produce similar changes supports this concept. Furthermore, the administration of xylitol was associated with insulin release, and decreases in the plasma concentrations of glucose and phosphate, suggesting the rapid movement of these substances into certain tissue cells. The administration of xylitol was also associated with increased triglyceride concentrations in the plasma. In the dogs studied no evidence of hepatic, renal, or cerebral damage was observed and in particular no evidence of calcium oxalate crystal deposition seen. Although one cannot directly extrapolate these findings to explain events observed in human patients after infusions of solutions containing xylitol, certain conclusions can be proposed, if only to form the basis for a careful reevaluation of xylitol metabolism in human patients.

The osmotic diuresis, acidosis, and increased uric acid concentration in the blood appear to be effects that can be produced by infusing hypertonic sugar solutions intravenously. Xylitol and sorbitol were different from glucose and fructose in that the acidosis occurring during infusion was more marked. This can be explained by two effects. First, the altered NADH:NAD ratios, resulting from the dehydrogenation reactions that these sugar alcohols undergo, shift the equilibrium between pyruvate and lactate to form an excess of lactate. The pyruvate produced as a result of the further metabolism of these sugars will tend to be converted to lactate with a further excess accumulating. Second, the altered redox potential of tissues responsible for clearing lactate, especially the liver, reduces the ability of these tissues to clear lactate—again favoring its accumulation.

It would appear that calcium oxalate crystal deposition may, in part at least, explain the occurrence of renal damage and disturbed function. A satisfactory explanation of the origin of oxalate has yet to be made. Certainly large amounts were produced as the extent of oxalate crystal deposition seen in the kidneys of these patients was extensive. Although the normal metabolism of xylitol is unlikely to yield oxalate, it is possible that for some reason the metabolism of xylitol did not follow expected pathways, with the production of unusual metabolites leading in some way to the formation of oxalate. The transketolase reaction by way of transferring "active glycolaldehyde" from xylulose 5-phosphate to ribose 5-phosphate to form sedoheptulose 7-phosphate and glyceraldehyde 3-phosphate, as shown in Fig. 5, may be a reaction to be considered in this respect. It has been suggested that glycolaldehyde may be released with subsequent conversion to glycolic, glyoxalate, and, finally, oxalic acid (Thomas *et al.*, 1972a). Experimental evidence of such an event has yet to be provided.

The hepatic abnormalities observed in human patients infused with solutions containing xylitol remain unexplained. Certainly, other users elsewhere have not experienced these adverse reactions except for reports of increased bilirubin concentrations in volunteers infused with xylitol solutions (Donahoe and Powers, 1970; Schumer, 1971). At the present time this adverse reaction has been included among those possibly peculiar to xylitol.

The cerebral disturbances also remain to be explained. Certainly, in one case calcium oxalate crystal deposition was observed in blood vessels of the midbrain. Nevertheless, the significance of such a finding remains obscure. Vascular crystal inclusions have been observed in other circumstances and this may be a coincidental finding (Glynn, 1940). The possibility that changes in cerebral function are contributed to by hyperosmolality of extracellular fluids cannot be ignored, and for the time being this effect has been placed with others thought to be the result of carbohydrate-induced hyperosmolality.

The reason why solutions containing xylitol used in South Australia during the first half of 1969 should have produced these unusual collections of adverse reactions is not known. Several cases with similar adverse reactions were reported in other parts of Australia, but to our knowledge no other users elsewhere have described similar occurrences.* A theory that the solutions were in some way different, while appearing attractive, has no convincing evidence to sustain it. Although there is some evidence that some of these solutions contained traces of unidentified material (WHO Drug Bulletin No. 77), it has not yet been demonstrated either *in vitro* or *in vivo* that such material interferes with the metabolism of xylitol or is otherwise responsible for these unusual adverse reactions.

C. Recommendations for Usage

1. Uses of Hypertonic Sugar Solutions for Parenteral Feeding

Substances able to provide therapeutic benefits in certain diseased states are of considerable value, and simple nutrients fulfilling this use will be highly valued. But certain words of caution should be sounded as one progresses from the laboratory to the feeding of man, even to the treating of his diseases with newly identified compounds. Touster himself sounded words of caution when he said, ". . . the use of xylitol has to rest ultimately on very practical clinical considerations" (Touster, 1970). When new substances available for nutritional purposes are in-

---

* Since the 1972 International Conference on Sugars in Nutrition it has come to our attention that similar adverse effects, including osmotic diuresis, altered state of consciousness and oxalate deposition, have been observed in patients receiving intravenous infusions of solutions containing xylitol in the Federal Republic of Germany. In six patients studied massive deposits of oxalate crystals were found in the kidneys, and in five of the six patients crystal deposits were found adjacent to blood vessels in a variety of locations within the brain substance and leptomeninges. In four of these patients oxalate crystals were also found in the lungs (W. Büngener, personal communication).

A further six cases are described in *Acta Neuropathologica*, **27**, No. 2, February, 1974.

fused intravenously into human patients certain factors must be carefully considered beforehand. This applies especially to substances that are actively metabolized and enter vital metabolic pathways in many tissues. The rate of administration should be such that it reasonably matches the rate of utilization and hence prevents the accumulation of the infused substance in the extracellular fluids and establishment of an osmotic gradient between the extracellular and intracellular compartments. The rate of cellular uptake and mode and extent of metabolism of substances administered should be thoroughly understood. The nature and properties of the enzymes necessary for the metabolism of infused substances should be characterized in healthy and diseased individuals.

The user should have a thorough understanding of the effects preexisting disease may have on the metabolism of infused substrates. Before infusing substances in large doses one should undertake an assessment of cardiovascular efficiency and the adequacy of tissue perfusion, and of hepatic and renal function. The extent of trauma (including surgery) with consequent increase in circulating adrenal cortical steroids and catecholamines, and the presence of insulin deficiency or resistance—as seen in diabetes mellitus, potassium depletion, and states of stress—should also be assessed. The effects of the administration of drugs such as diuretics, steroids, sympathetic blockers or sympathomimetics, phenformin or parenterally administered amino acids, for example, may place certain requirements on the nature of substances infused and the rate of infusion. Not to be forgotten is the nutritional state of the individual who is to receive the infused substance. The mass of muscle tissue, and extent of adipose tissue depots, and the adequate availability of vitamins, minerals, essential fatty acids, and amino acids must also be assessed. All these factors should be taken into consideration when choosing not only the type of sugar to be infused into a patient but also in selecting the infusion rate—weighing on one hand the benefit to the patient of such an infusion and on the other hand the ability of the patient to handle effectively this additional metabolic load.

Perhaps if the use of xylitol in Australia had followed such considerations certain of these adverse reactions may not have occurred. These unfortunate events, however, have highlighted these important principles and promoted a more serious examination of the processes involved, and to that end some benefit has been derived from these experiences.

2. Future Uses of Xylitol

The effects of long-term infusions in conscious animals of hypertonic xylitol solutions need to be assessed taking into consideration the effects observed with short-term infusions and those observed in human patients. Certain of the metabolic changes observed in animals need further elucidation, in particular the mechanisms of insulin release, hypertriglyceridemia, the effects of prolonged intracellular depletion of ATP and inorganic phosphate, as well as the effect of reduced concentrations of extracellular inorganic phosphate.

Needing urgent attention, however, is the occurrence of extensive calcium oxalate crystal deposition and the problem of its origin and metabolic formation. The causes of hepatic and cerebral disturbances need further elucidation as well. Finally, the nature and *in vivo* as well as *in vitro* effects of the material present in trace amounts in certain xylitol solutions must be established.

By the time these problems have been solved considerable information will be available for the appropriate use of xylitol. Then careful reassessment of the effects of infusing xylitol intravenously into healthy as well as sick human patients should be carried out. Such assessment will be assisted by the knowledge gained from effects observed in animals.

Acknowledgments

The studies described were assisted by grants from Eisai Co. Ltd., Tokyo, Japan, and the National Health and Medical Research Council of Australia.

References

Ainsworth, S. K., and Allison, F. (1971). Effects of hypertonic solutions infused intravenously in rabbits. I. Metabolic acidosis and lactic acidemia. *J. Lab. Clin. Med.* **78**, 619–632.

Andersson, G., Brohult, J., and Sterner, G. (1969). Increasing metabolic acidosis following fructose infusion in two children. *Acta Paediat. Scand.* **58**, 301–304.

Bässler, K. H., Prellwitz, W., Unbehaun, V., and Lang, K. (1962). Xylitol metabolism in man: On the value of xylitol as sugar substitute for diabetic patients. *Klin. Wochenchr.* **40**, 791–793.

Bergström, J., Hultman, E., and Roch-Norlund, A. E. (1968). Lactic acid accumulation in connection with fructose infusion. *Acta Med. Scand.* **184**, 359–364.

Chan, J. C. M., Malekzadeh, M., and Hurley, J. (1972a). pH and titratable acidity of amino acid mixtures used in hyperalimentation. *J. Amer. Med. Ass.* **220**, 1119–1120.

Chan, J. C. M., Asch, M. J., Lin, S., and Hays, D. M. (1972b). Hyperalimentation with amino acid and casein hydrolysate solutions. Mechanism of acidosis. *J. Amer. Med. Ass.* **220**, 1700–1705.

Donahoe, J. F., and Powers, R. J. (1970). Biochemical abnormalities with xylitol. *N. Engl. J. Med.* **282**, 690.

Elliot, W. C., Cohen, L. S., Klein, M. D., Lane, F. J., and Gorlin, R. (1967). Effects of rapid fructose infusion in man. *J. Appl. Physiol.* **23**, 865–869.

Evans, G. W., Phillips, G., Mukherjee, T. M., Snow, M. R., Lawrence, J. R., and

Thomas, D. W. (1972). Identification of crystals deposited in brain and kidney after xylitol administration by biochemical, histochemical and electron diffraction methods. *J. Clin. Pathol.* **26**, 32–36.

Fortner, C. L. (1972). Reduction in serum CO_2 during hyperalimentation. *J. Amer. Med. Ass.* **221**, 716.

Fox, I. H., and Kelley, W. N. (1972). Studies on the mechanism of fructose-induced hyperuricaemia in man. *Metab. Clin. Exp.* **21**, 713–721.

Glynn, L. E. (1940). Crystalline bodies in the tunica media of a middle cerebral artery. *J. Path. Bact.* **51**, 445–446.

Goldfinger, S., Klinenberg, J. R., and Seegmiller, J. E. (1965). Renal retention of uric acid induced by infusion of β-hydroxy butyrate and acetoacetate. *N. Engl. J. Med.* **272**, 351–355.

Hirata, Y., Fujisawa, M., and Ogushi, T. (1969). Effect of intravenous injection of xylitol on plasma insulin. *In* "Metabolism, Physiology, and Clinical Uses of Pentoses and Pentitols" (B. L. Horecker, K. Lang, and Y. Takagi, eds.), pp. 226–229. Springer-Verlag, Berlin and New York.

Keller, U., and Froesch, E. R. (1971). Metabolism and oxidation of U-^{14}C-glucose, xylitol, fructose and sorbitol in the fasted and in the streptozotocin-diabetic rat. *Diabetologia* **7**, 349–356.

Kjerulf-Jensen, K. (1942). The phosphate esters formed in the tissue of rats and rabbits during assimilation of hexoses and glycerol. *Acta Physiol. Scand.* **4**, 249–258.

Kosaka, K. (1969). Stimulation of insulin secretion by xylitol administration. *In* "Metabolism, Physiology, and Clinical Uses of Pentoses and Petitols" (B. L. Horecker, K. Lang, and Y. Takagi, eds.), pp. 212–225. Springer-Verlag, Berlin and New York.

Kreisberg, R. A. (1972). Glucose-lactate inter-relations in man. *N. Engl. J. Med.* **287**, 132–137.

Kuzuya, T. (1969). Some recent observations on xylitol-induced insulin secretion. *In* "Metabolism, Physiology, and Clinical Uses of Pentoses and Pentitols" (B. L. Horecker, K. Lang, and Y. Takagi, eds.), pp. 230–233. Springer-Verlag, Berlin and New York.

Perheentupa, J., and Raivio, K. (1967). Fructose-induced hyperuricaemia. *Lancet* **2**, 528–531.

Pollack, H., Millet, R. F., Essex, H. E., Mann, F. C., and Bollman, J. L. (1934). Serum phosphate changes induced by injections of glucose into dogs under various conditions. *Amer. J. Physiol.* **110**, 117–122.

Schumer, W. (1971). Adverse effects of xylitol in parenteral alimentation. *Metab. Clin. Exp.* **20**, 345–347.

Spitz, I. M., Rubenstein, A. H., Bersohn, I., and Bässler, K. H. (1970a). Metabolism of xylitol in healthy subjects and patients with renal disease. *Metab., Clin. Exp.* **19**, 24–34.

Spitz, I. M., Bersohn, I., Rubenstein, A. H., and Van As, M. (1970b). The response of growth hormone to xylitol administration in man. *Amer. J. Med. Sci.* **260**, 224–229.

Thomas, D. W., Edwards, J. B., and Edwards, R. G. (1970). Examination of xylitol. *N. Engl. J. Med.* **283**, 437.

Thomas, D. W., Edwards, J. B., Gilligan, J. E., Lawrence, J. R., and Edwards, R. G. (1972a). Complications following intravenous administration of solutions containing xylitol. *Med. J. Aust.* **1**, 1238–1246.

Thomas, D. W., Gilligan, J. E., Edwards, J. B., and Edwards, R. G. (1972b). Lactic acidosis and osmotic diuresis produced by xylitol infusion. *Med. J. Aust.* **1**, 1246–1248.

Touster, O. (1960). Essential pentosuria and the glucuronate-xylulose pathway. *Fed. Proc., Fed. Amer. Soc. Exp. Biol.* **19**, 977.

Touster, O. (1970). *In* "Parenteral Nutrition" (H. C. Meng and D. H. Law, eds.), pp. 131–138. Thomas, Springfield, Illinois.

Touster, O., Reynolds, V. H., and Hutcheson, R. M. (1956). The reduction of L-xylose by guinea pig liver mitochondria. *J. Biol. Chem.* **221**, 697–709.

Travis, S. F., Sugerman, H. J., Ruberg, R. L., Dudrick, S. J., Delivoria-Papadopoulos, M., Miller, L. D., and Oski, F. A. (1971). Alterations of red-cell glycolytic intermediates and oxygen transport as a consequence of hypophosphataemia in patients receiving intravenous hyperalimentation. *N. Engl. J. Med.* **285**, 763–768.

van Eys, J., and Wang, Y. M. (1972). Xylitol toxicity. *N. Engl. J. Med.* **286**, 1162–1163.

Wade, D. N., and Bostrom, J. (1971). The toxicity of intravenous xylitol infusions. *Aust. N. Z. J. Med.* **1**, 278–279.

Wang, Y. M., Patterson, J. H., and van Eys, J. (1971). The potential use of xylitol in glucose-6-phosphate dehydrogenase deficiency anaemia. *J. Clin. Invest.* **50**, 1421–1428.

Wang, Y. M., Patterson, J. H., and Van Eys, J. (1972). Xylitol toxicity in the rabbit. *Fed. Proc., Fed. Amer. Soc. Exp. Biol.* **31**, 726.

Williamson, J. R., Jakob, A., and Scholz, R. (1971). Energy cost of gluconeogenesis in rat liver. *Metab. Clin. Exp.* **20**, 13–26.

Wilson, R. B., and Martin, J. M. (1970). Plasma insulin concentrations in dogs and monkeys after xylitol, glucose or tolbutamide infusion. *Diabetes* **19**, 17–22.

Winters, R. W., Scaglione, P. R., Nahas, G. G., and Verosky, M. (1964). The mechanism of acidosis produced by hyperosmotic infusions. *J. Clin. Invest.* **43**, 647–658.

Woods, H. F., Eggleston, L. V., and Krebs, H. A. (1970). The cause of hepatic accumulation of fructose 1-phosphate on fructose loading. *Biochem. J.* **119**, 501–510.

Yamagata, S., Goto, Y., Ohneda, A., Anzai, M., Kawshima, S., Chiba, M., Maruhama, Y., and Yamauchi. Y. (1965). Clinical effects of xylitol on carbohydrate and lipid metabolism in diabetes. *Lancet* **2**, 918–921.

CHAPTER 33

The Safety of Oral Xylitol

M. BRIN AND O. N. MILLER

During the period in which xylitol was being considered as a special dietary additive with unusual qualities for various abnormal states in man, adverse findings were reported of the Australian (Thomas *et al.*, 1970, 1972) and the Chicago (Donohoe and Powers, 1970) parenteral studies. While these findings were highly significant, per se, it was felt that the implications were not necessarily applicable to orally administered xylitol. First, it was suggested that the adverse findings may have resulted from a contaminant in the parenteral solutions. Second, the dose levels used appeared to be in excess of those found to be safe, parenterally, in other countries (Horecker *et al.*, 1969). Third, it was believed that tissue oxalate at autopsy is not a direct specific metabolic product from xylitol as inferred by the Australian group; rather, a review of the literature revealed that oxalate is a relatively common histopathological finding in subjects succumbing to various conditions of hemodynamic shock such as a ruptured aneurism and various types of uremia. Therefore, oral use of xylitol would obviate the possibility of both contamination as well as dosage since the rate of absorption would be an intrinsic limiting factor.

A. Deposition of Oxalate in Human Disease

As shown in Table I, oxalate is found in a wide variety of human tissues, i.e., liver, kidney, and brain, from persons succumbing to ruptured

591

TABLE I

Pathology of Oxalate Crystals (Stones)

I. Oxalate crystals found in tissues of nonxylitol-treated subjects as follows:

A. Tissues	Ref.	B. Terminal illness	Ref.
Vascular walls[a]	Glynn (1940); Bednar et al. (1971)	Rupture of aneurism[a]	Glynn (1940)
Heart	Bennett and Rosenblum (1961)	Uremia[a]	Bennett and Rosenblum (1961), Bennington et al. (1964), Bednar et al. (1971)
Kidney[a]	Bennett and Rosenblum (1961), Bennington et al. (1964)	Hepatic disease[a]	Bennington et al. (1964)
Liver[a]	Bednar et al. (1971)	Neoplasia	Bennington et al. (1964)
Thyroid	Bednar et al. (1971)	Transfusion	Bennington et al. (1964)
Testes	Bednar et al. (1971)	Ileal (Crohn's) disease	Anonymous
Brain[a]	Bednar et al. (1971)		

II. In Australian subjects treated with xylitol, oxalate crystals were found as follows:

A. Tissues	Ref.	B. Australian subjects admitted for	Ref.
Kidney[a]	Thomas et al. (1972)	Ruptured aneurism[a]	Thomas et al. (1972)
Midbrain[a]	Thomas et al. (1972)	Carotid artery thrombosis[a]	Thomas et al. (1972)
		Obstructing carcinoma[a]	Thomas et al. (1972)
		Pancreatitis (renal transplant)	Thomas et al. (1972)
		Multiple fractures	Thomas et al. (1972)
		Lung abscess	Thomas et al. (1972)
		(Uremic hepatic death)[a]	Thomas et al. (1972)

[a] Similar findings in xylitol and nonxylitol treatment.

aneurism, uremia, hepatic disease, etc., but without the administration of xylitol solutions. Certain reports suggest its occurrence in at least 30 out of 50 cases of uremia (Bennett and Rosenblum, 1961). The precise precursor of the oxalate has not been identified, although the number of known precursors is limited, and at least *in the past in these cases has not included xylitol.* In fact, there is no known basis for the conversion of xylitol to oxalate other than its being a precursor for normal intermediates such as pyruvate and glucose just as any other sugar may be such a precursor. Note that in the lower section of Table I, the same

Joint FAO/WHO Food Standards Programme

Codex Committee on Foods for Special Dietary Uses

Fourth Session, Cologne, 3-7 November 1969

Proposed Draft Standard for Foods, for Use

in a Diet for Diabetics

2.   DESCRIPTION

The composition of foods for use in a diet for diabetics takes care of the fact that the disturbed metabolism of diabetics reacts differently to (harmful carbohydrates +) carbohydrates and polyalcohols such as sorbitol, mannitol and xylitol, which are chemically related to carbohydrates. Fat content as well as total calory-content are reduced as against those of comparable normal foods as far as possible.

(Harmful carbohydrates +) include:

d-glucose, invert sugar, disaccharides, starch, starch degradation products.

4.   ADDITIONS and ADDITIVES:

4.2   Sweeteners

4.2.1 (Sugar replacing substances +)
        fructose,
        sorbitol,
        mannitol,
        xylit. . . .

4.2.2 (Non-nutritive sweeteners ++)

        saccharine, sodium and potassium salt,
        cyclamates,
        (cyclohexysulfamine acid and sodium,
        potassium and calcium salts)

Fig. 1 Joint FAO/WHO food standards program.

tissues and organs were affected whether or not a xylitol solution was given.

B. WHO/FAO Proposed Approval of Xylitol in Special Dietary Foods

Xylitol is proposed for use as a dietary additive for diabetics by the WHO/FAO Joint Foods Program (1969), as shown in Fig. 1. In this

<u>Summary of a Study on the</u>

<u>"Effects of the Long-Term Oral</u>

<u>Administration of Xylitol"</u>

by
Professor M. Mosinger

In this study the author indicated the following:

1. Xylitol has been administered orally during 11 months to 40 rats and during 24 months to 30 wistar rats at a level of 100mg/kg/day.

    The weight curve and the general clinical condition were studied and at the end of the study an autopsy with a detailed macroscopic and microscopic study had been made. The experiments were carried out with two generations and the fertility was also checked.

2. The results are compared with those obtained from 450 control animals kept under similar experimental conditions.

3. Results:

    (a)   Not one single malignant tumor could be discovered in the animals treated with xylitol.

    (b)   Xylitol does not stimulate the growth of tumors which are known to occur spontaneously in control animals. Therefore, xylitol is not co-carcinogenic.

    (c)   Xylitol does not have a harmful influence on reproduction in these animals.

    (d)   Xylitol does not produce any pathological changes in the animals under investigation.

In addition, we have been advised that this study has been extended to three generations and that the conclusions described above are still valid.

It has also been indicated that Professor Mosinger is planning a continuation of the xylitol study with higher dose levels.

Fig. 2   Summary of rat study done by Mosinger (1971).

it is classified as a sugar-replacing substance. It is sweeter than mannitol and sorbitol, and although fructose is a carbohydrate, the metabolism of which requires less insulin, its conversion to lipid reduces its desirability as discussed elsewhere in this book.

C. Animal Tolerance of Xylitol as a Food Additive

Table II gives a summary of a number of studies on the oral tolerance of xylitol in animals. The only generally limiting factor in xylitol use is that dosage without adaptation may result in soft stools or osmotic diarrhea depending upon dosage and/or lack of adaptation. Most investigators reported that tolerance was markedly increased by administering

A Thirteen Week Oral Feeding Study

in Rats with Xylitol

(Swarm and Banziger, unpublished)

SUMMARY

Three groups of 16 rats (8 of each sex) were fed xylitol as a dietary admix at levels of 5, 10 and 20 grams kg/day. A fourth group, the control group, received only the basal diet of Purina laboratory chow.

Each rat was observed daily to determine general behavior, food and water consumption. Body weights and food consumption were determined and recorded weekly. The concentration of compound in the diet was adjusted accordingly each week. Ophthalmic and neurologic examinations were performed on all rats prior to treatment and at 4, 8 and 12 weeks. Hematological studies, blood glucose determinations and urinalyses at 4, 8 and 12 weeks and serum alkaline phosphatase, SGOT, BUN, bilirubin and uric acid were determined at 13 weeks.

All animals fed compound at the level of 10 and 20 grams/kg/day survived the entire period of the study. Some reduced weight gains were dose dependent. Except for transient diarrhea in a few, no clinical abnormalities that were related to treatment were noted. Clinical laboratory studies did not distinguish treated animals from controls.

Autopsy failed to demonstrate any lesions related to the level of compound given in the diet.

In conclusion, rats fed xylitol as a dietary admix at levels of 5, 10 and 20 grams/kg/day tolerated the feeding well except for transient diarrhea and slightly reduced weight gains at higher dose levels.

Fig. 3   Rat study done by Swarm and Banzinger (1970).

TABLE II

Oral Dosage of Xylitol to Animals

Ref.	Species	Dosage	Findings
Foglia *et al.* (1963)	(a) Rat diabetic	5 mg/kg body weight/day	No effect on fasting blood sugar, glucose tolerance or pancreatic histology.
	(b) Rat	9 gm/kg body weight/day 30 days	No adverse effects.
Kiekebusch *et al.* (1961)	(a) Rat	10% diet, 12 weeks 20% diet, 5 weeks	No effect on weight gain, fertility, electrolytes or histology of liver, kidney, and heart.
	(b) Rat	50% of diet (37.5 gm/kg body weight/day	All animals succumbed in 1–2 weeks.
Mosinger (1971)	Rat	100 mg/kg body weight/day 2 years	No malignant tumors, no effect on fertility or reproduction, and no pathology in any tissues studied.
Pool (1971)	Dog	10 gm/kg body weight/day	No change in pH or pCO_2 in blood.
Swarm and Banziger (1970)	Rat	5, 10, 20 gm/kg body weight/day 16 per group 13 weeks	All survived at high levels, some transient diarrhea. Clinical chemistries normal. No pathology.
Banziger (1970)	Monkey	1, 3, 5 gm/kg body weight/day 4 per group 13 weeks	All survived with weight gains. Intermittent soft stools. No compound related effects on behavior, appetite, weight, ophthalmic, or neurologic exam, clinical chemistries or tissue pathology.
Hosoya and Iitoyo (1969)	Rat	20% of diet 4 months	Growth normal in adapted rats. Reproduction and lactation normal and pup growth normal if adapted.

divided doses at 2–3 hr intervals. This holds both for experimental animals and man and also for sorbitol as well (Dubach *et al.*, 1969).

The rat study of longest duration, by Mosinger (1971), although done at the low dose level of 100 mg/kg of body weight, was continued for a period of 2 years, as shown in Fig. 2. It also included a study of fertility and reproduction, and carcinogenicity for which there were no adverse findings, as summarized in Fig. 2.

Also, two studies were done by Swarm and Banziger (1970) in rats and monkeys. The summary for rats is shown in Fig. 3 and dosage levels were 5, 10, and 20 gm/kg of body weight per day for 13 weeks. There was some reduction in growth at the higher level subsequent to some diarrhea, but no adverse effects were seen in clinical chemistries, tissue pathology, or in ophthalmic or neurological examination (Banziger, 1970).

A summary of the monkey study is shown in Fig. 4 (Banziger, 1970). Xylitol was intubated gastrically at dose levels of 1, 3, and 5 gm/kg of

13 Week Oral Toxicity Study

Monkeys Lot No. 008

Project No. 131–118

Sponsor: Hoffmann–La Roche Inc.,   September 25, 1970

SUMMARY

Xylitol was administered orally by gastric intubation, twice daily six days a week for 13 weeks, to each of three groups of two male and two female rhesus monkeys at dosage levels of 5.0, 3.0 and 1.0 g/kg/day. A fourth group served as a control and received daily doses of distilled water.

Soft feces and/or diarrhea were generally present during week 1 in all test groups and continued intermittently with prevalence almost solely among the high level group (5.0 g/kg/day). The control group exhibited an occasional soft feces and a one-day diarrhea.

No compound-related effects were observed with respect to behavior, appetite, body weight, ophthalmoscopic and neurologic examinations, clinical laboratory studies, organ weights and ratios, and gross and microscopic pathology.

Fig. 4   Monkey study done by Banziger (1970).

body weight per day. Soft stools were present during the first week at all levels but remained almost exclusively at the high level. Otherwise, there were no subjective or objective adverse findings in extensive studies including ophthalmic and neurological examination, clinical chemistries, and histopathology.

A more extensive rat study was done by Hosoyo and Iitoyo (1969) (Fig. 5). Levels of xylitol up to 20% of the diet were fed for up to 4 months with no significant adverse findings. Growth rate was normal. Adaptation to high dosage was shown to be important. Xylitol had no effects on the number or size of pups or dams fed xylitol, and the pups adapted readily to the supplement at 5 and 10% in diet, after weaning,

<u>Effects of Xylitol Administration on The</u>

<u>Development (Growth) of Rats</u>

Hosoya and Iitoyo, 1969
(translated)

<u>CONCLUSIONS</u>

Effects of xylitol administration in rats were studied in terms of the increment of body weight, the reproduction behavior and the hepatic XDH activity.

1.   When more than 10% of the diet was xylitol, rats had some diarrhea. But, if the xylitol content was increased by 5% every week till it reached 20%, and then maintained at 20% for 4 months, there were no significant differences between the increment of body weight and that of the control group in rats.

2.   When rats adapted to xylitol their hepatic XDH (NAD) activity was also induced.

3.   Administration of xylitol to female rats during conception and pregnancy had no effects on birth performance and on the growth of offspring.

    When newborn offspring were fed xylitol containing diets from the day they started to eat, there was observed the same need for the adaptation of offspring to xylitol as that observed with parent rats.

Fig. 5   Rat study done by Hosoya and Iitoyo (1969).

as instituted following adverse effects with the 20% xylitol level, initially.

D. Human Tolerance of Orally Administered Xylitol

Table III summarizes some relevant oral tolerance studies in man for xylitol. Amador and Eisenstein (1970) administered either xylitol or glucose at a dose level of 20 or 40 gm/m² body surface, and studied blood chemistries and insulin levels. Biochemical findings are shown in Table IV. Values for LDH, SGOT, and alkaline phosphatase, the measures of liver function remained normal throughout. Uric acid and bilirubin were slightly elevated but returned to normal in 8 hr and remained so 24 hr after ingestion. It is well documented that uric acid increases wherever blood lactic acid increases (Jeandet and Lestradet, 1961).

Also, findings with xylitol should not be compared to "normal blood value ranges" because normal values are not "normal" in the sense that they are values obtained on postabsorptive fasting blood samples. Therefore, as shown in Table V, Amador and Eisenstein observed that, as measured by areas under the plotted curves, far less insulin was released following xylitol dosing than after glucose at the two dosage levels.

Another tolerance study is outlined in Fig. 6. Amador and Eisenstein (1971) adapted 5 persons with increments of 30 gm of xylitol per day in three individual doses at 3 day intervals up to 120 gm/day. Liver func-

1. Method: 5 adults took up to 120gm/day
   starting at 30gm/day in 3 divided doses
   and increased 30gm/day at 3 day intervals.

2. Findings:

   (a) Subjective: 3 of 5 had a small effect on
   stools (1 with diarrhea at 90g/d, 1 with loose
   stool at 30g/d and other on final day).

   (b) Objective: transient increase in plasma
   lactate in 2, and urate in 4. Five liver function
   tests normal throughout.

3. Conclusions:

   (a) Xylitol is well tolerated and produces
   virtually no gastrointestinal distress at less
   than 90g/d.

   (b) Elevation of blood urate is probably due
   to rise in lactate, which also rises after
   fructose.

Fig. 6 Oral tolerance of xylitol in man (Armador and Eisenstein, 1971).

TABLE III

The Oral Dosage of Xylitol To Man

Ref.	Species	Dosage	Findings
Amador and Eisenstein (1970)	Man	40 g xylitol/m² body surface	No effect on SGOT, LDH, alkaline phosphatase. Slight elevation of plasma glucose and insulin, uric acid and bilirubin (latter 2 normal in 8 hr).
Amador and Eisenstein (1971)	Man	10 gm × 3 times/day increasing to 40 gm × 3 times/day or 120 gm/day	Slight (but rapidly reversible) elevation of plasma uric acid and lactate at high dose levels. One subject had diarrhea and flatus at high dose levels.
Manenti and Della Casa (1965)	Man	20 gm/day to diabetics	No reported adverse effects.
Dubach et al. (1969)	Man	75 gm/day for 21 days also up to 220 gm/day, 19 students	No diarrhea at less than 130 gm/day. At 220 gm/day there was aversion to sweets. Weight and fasting blood sugar normal. Xylitol caused less meteorism and flatus than sorbitol and was preferred by the students. Adaptation is maintained on intermittent dosage.
Asano et al. (1972)	Man	5–30 gm/dose	Absorption was 90% at low dose to 66% at high dose. Bilirubin, SGOT, alkaline phosphatase, uric acid, and blood sugar remained normal, and there was no diarrhea. No obvious adverse effects.

TABLE IV

Biochemical Findings in Normal Subjects Following Oral Administration of 40 gm of Xylitol/m² of Body Surface[a]

Item	Fasting	After 3 hr	After 8 hr	After 24 hr
Uric acid (mg/100 ml)	5.0 ± 0.99[b]	6.8 ± 0.99	5.8 ± 1.1	5.5 ± 0.31
Total bilirubin (mg/100 ml)	0.7 ± 0.30	1.1 ± 0.18	0.7 ± 0.5	0.6 ± 0.22
Lactic dehydrogenase (IU/liter)	132 ± 14.0	119 ± 7.8	111 ± 9.3	143 ± 29.0
SGOT (IU/liter)	39.0 ± 5.7	42 ± 8.6	38 ± 2.0	34 ± 2.1
Alkaline phosphatase (IU/liter)	37.0 ± 7.7	35 ± 8.4	28 ± 3.6	34 ± 6.9

[a] From Amador and Eisenstein (1970).

[b] Mean value ± SEm.

tion tests were normal throughout, while there was transient increase in plasma lactate and urate. There was no diarrhea below 90 gm/day.

An oral tolerance study (Asano *et al.*, 1972) shown in Fig. 7 revealed that xylitol absorption was variable over the dosage range 5–30 gm, ranging between 99 and 66%, respectively. There were no objective or subjective adverse effects, as shown.

Dubach *et al.* (1969) observed large tolerances for xylitol in young adults (Fig. 8). Up to 220 gm/day were tolerated for up to 3 weeks, with no major adverse effects. Tolerance for oral xylitol or sorbitiol was

TABLE V

Serum Glucose and Insulin Responses Following Oral Administration of Glucose and Xylitol[a]

Test substance and dose	Serum glucose response[b]	Serum insulin response[b]
a. 40 gm/m² of body surface		
Glucose	11,024	19,580
Xylitol	3,276	7,390
p =	<0.05	<0.05
b. 20 gm/m² of body surface		
Glucose	6,448	19,116
Xylitol	2,787	5,728
p =	<0.05	<0.05

[a] From Amador and Eisenstein (1970).

[b] Area under curve in square millimeters.

Xylitol Absorption in Healthy Man
[Asano   et   al.(1972)]

1.   Dose: 5-30 g xylitol

2.   Absorption: 90% at 5g.
                  76% at 15g.
                  66% at 30g.

3.   Chemistries: Bilirubin, SGOT,
     Alkaline phosphatase, uric acid,
     blood sugar - all normal.

4.   Clinical Findings: No diarrhea or
     other symptoms.

5.   Conclusion ". . . . up to 30gm xylitol
     was found to be well absorbed by human
     subjects and to have no adverse effect,
     judging by laboratory tests and symptoms.

Fig. 7   Oral tolerance study (Asano *et al.*, 1972).

also compared to levels up to **75 gm/day** for up to **2 weeks**. The conclusions were that there were no significant adverse effects with xylitol except for loose stools which could be controlled by appropriate dosing schedule. Also, because of less flatulence and meteorism the students preferred xylitol over sorbitol.

E. Xylitol in Dental Caries

Another aspect of orally administered xylitol of interest is the effect on dental caries. Previous work by many suggested that xylitol reduced

TABLE VI

The Effect of Xylitol on the Growth of NIH Enterococci[a]

Culture	Basal no CH_2O	Basal + 0.2% xylitol	Basal + 0.1% sucrose 0.1% glucose	Basal + all 3
		MEDIA[b]		
HS4	0	0	4+	4+
HS6	0	0	4+	4+
E49	0	0	4+	4+
SLI 175	0	0	4+	4+

[a] From Cort (1966).
[b] 0, no growth; 4+, excellent growth. Adaptation studies were unsuccessful.

Oral Tolerance for Xylitol in Volunteers

with Normal Metabolism

[Dubach et al.(1969)]

1. Tolerance for 5 to 75gm per day.

   (a) Procedure: 19 adults received 75gm/day (5gm increments) to 28 days.

   (b) Findings: Stable body weight, and blood sugar, slightly positive urine reducing sugar. Few soft stools and meteorism.

2. Tolerance for 40–220gm per day.

   (a) Procedure: 19 adults received 40gm/day (10gm increments) to 100gm/day; 18 to 150gm/day in second week; 6 to 220gm/day in third week.

   (b) Findings: Occasional meteorism above 130gm/day where doses not divided properly, aversion to sweets. Weight and blood sugar constant, some positive reducing in sugar urine.

3. Comparison between Xylitol and Sorbitol.

   (a) Procedure: 26 adults received up to 75gm/day beginning with 5gm in increments 5gm/day to 75gm in 2 weeks, as xylitol or sorbitol.

   (b) Findings: More meteorism and flatulence on sorbitol than xylitol, and xylitol preferred by 21 of 26. Urinary sugar comparable, with low urine levels of dose.

4. Conclusions:

   (a) No significant adverse findings with xylitol.
   (b) Meteorism controlled by spacing doses 3-4 hrs.
   (c) Aversion to sweets at high dose levels.
   (d) Xylitol and sorbitol differ by more meteorism and flatulence with sorbitol.

Fig. 8 Oral tolerance study (Dubach *et al.*, 1969).

dental caries, and this will be elaborated upon later by Russell (Chapter 36). However, it is felt pertinent to present two studies done in Roche Laboratories. Cort (1966) observed that various NIH strains of cariogenic enterococci did not grow on xylitol alone, although they grew when sugar was present (Table VI). Furthermore, she was unsuccessful in attempting to get the strains to adapt to xylitol alone. Also, Grunberg *et al.* (1972) showed that although sorbitol and mannitol reduced dental caries one-third to one-half that observed with a cariogenic diet, the caries was virtually eliminated with xylitol (Table VII). In addition, the caries observed with xylitol was less severe (Table VIII).

TABLE VII

Incidence of Caries in Rats Fed Normal or Cariogenic Diets
Supplemented with Sugars or Sugar Alcohols

Group	Type of diet	Percentage of rats showing caries left lower molars		
		First	Second	Third
1	Normal	0	0	0
2	Cariogenic (C)	46	92	69
3	C + 10% glucose	73	82	64
4	C + 10% sucrose	50	63	38
5	C + 10% mannitol	23	54	15
6	C + 10% sorbitol	38	63	38
7	C + 10% xylitol	0	14	0

TABLE VIII

Lesion Scores for Molars of Rats Fed Normal or Cariogenic Diets
Supplemented with Sugars or Sugar Alcohols

Group	Type of diet	Average lesion score[a] left lower molars		
		First	Second	Third
1	Normal	0	0	0
2	Cariogenic (C)	0.8	2.6	1.2
3	C + 10% glucose	1.5	2.5	1.4
4	C + 10% sucrose	1.5	2.5	0.9
5	C + 10% mannitol	0.6	1.2	0.4
6	C + 10% sorbitol	0.8	1.1	0.5
7	C + 10% xylitol	0	0.3	0

[a] The average lesion score was obtained by adding the value of 0 to 4 scored for each rat's tooth in each group of rats and then dividing by the total number of rats in that group.

F. Summary

Evidence was presented that xylitol, when appropriately administered orally with adaptation, is well tolerated and safe to levels of at least 90 gm/day, with no subjective or objective adverse findings. Less insulin was released into blood from xylitol than from glucose, and xylitol was more protective against dental caries than sorbitol or mannitol. The slight metabolic abberations seen after proper oral xylitol dosage were transient and were not different from those seen after dosage with "acceptable" materials including certain common sugars. Therefore, it is felt strongly that xylitol has an important place in the human diet as a special dietary additive, in certain metabolic situations, and to reduce dental caries.

Acknowledgment

The information presented is the product of a group effort involving a number of co-investigators from various divisions of Hoffmann-La Roche.

References

Amador, F., and Eisenstein, A. (1970). The effect of xylitol on insulin secretion in man. (unpublished observations).

Amador, F., and Eisenstein, A. (1971). The effects of oral xylitol administration in human subjects. (unpublished observations).

Anonymous. (1972). Hyperoxaluria and ileal disease. *Nutr. Rev.* **30,** 73–74.

Asano, T., Levitt, M. D., and Goetz, F. C. (1972). Xylitol absorption in healthy men. *Diabetes* **21,** 350–351.

Banziger, R. (1970). A thirteen week oral toxicity study of xylitol in monkeys. Hoffmann-La Roche, Inc. (unpublished observations).

Bednar, B., Jirasek, A., Sbejskal, J., and Chybil, M. (1971). Secondary uremic acidosis. *Zentralbl. Allg. Pathol. Pathol. Anat.* **102,** 289–297.

Bennett, B., and Rosenblum, C. (1961). Identification of calcium oxalate crystals in the myocardium in patients with uremia. *J. Lab. Invest.* **10,** 947–955.

Bennington, J. L., Haber, S. L., Smith, J. V., and Warner, N. E. (1964). Crystals of calcium oxalate in human kidneys. *Amer. J. Clin. Pathol.* **41,** 8–14.

Cort, W. M. (1966). Xylitol and dental caries. Hoffmann-La Roche, Inc. (unpublished observations).

Donahoe, J. F., and Powers, R. J. (1970). Biochemical abnormalities with xylitol. *N. Engl. J. Med.* **282,** 690.

Dubach, V. C., Feiner, E., and Forgo, I. (1969). Oral tolerance for xylitol in volunteers with normal metabolism. *Schweiz. Med. Wochenschr.* **99,** 190–194.

Foglia, V. G., Yabo, R., Bernaldez, J. P., and Aguire, L. M. (1963). Xylitol and pancreatic diabetes of the rat. *Folia Endocrinol.* **16,** 240–245, *Chem. Abstr.* **60,** 2170g (1964).

Glynn, L. E. (1940). Crystalline bodies in the tunica media of a middle cerebral artery. *J. Pathol. Bacteriol.* **51,** 445–446.

Grunberg, E., Beskid, G., and Brin, M. (1972). Efficacy of xylitol in reducing dental caries in rats. *Abstr., Int. Cong. Nutr. 9th, 1972.*

Horecker, B. L., Lang, K., and Takagi, Y., eds. (1969). "Metabolism, Physiology and Clinical Uses of Pentoses and Pentitols." Springer-Verlag, Berlin and New York.

Hosoya, N., and Iitoyo, N. (1969). Effects of xylitol on administration on the development (growth) of rats. *J. Jap. Soc. Food Nutr.* **22,** 17–20 (transl.).

Jeandet, J., and Lestradet, H. (1961). L'hyperlactacidémie, cause probable de l'hyperuricémie dans la glycogenose hépatique. *Rev. Fr. Edud. Clin. Biol.* **6,** 71–72.

Kiekebusch, W., Griem, W., and Lang, K. (1962). The applicability of xylitol as a dietary carbohydrate and its applicability. *Klin. Wochenschr.* **39,** 447–448, *Chem. Abstr.* **55,** 17781g (1961).

Manenti, F., and Della Casa, L. (1965). Effects of xylitol on the carbohydrate balance in diabetics. *Boll. Soc. Med.-Chir. Modena* **65,** 1309–1316.

Mosinger, M. (1971). Effects of long term administration of xylit. Center of Investigation and Medical Research, Marseille (unpublished observations) (transl.).

Pool, W. (1971). Hoffmann-La Roche Inc. (unpublished observations).

Swarm, R. L., and Banziger, R. (1970). Thirteen week oral feeding study in rats and monkeys with xylitol. Hoffmann-La Roche Inc. (unpublished observations).

Thomas, D. W., Edwards, J. B., and Edwards, R. G. (1970). Examination of xylitol. *N. Engl. J. Med.* **283,** 437.

Thomas, D. W., Edwards, J. B., Gilligan, J. E., Lawrence, J. R., and Edwards, R. G. (1972). Complications following intravenous administration of solutions containing xylitol. *Med. J. Aust.* **1,** 1238–1248.

WHO/FAO Joint Foods Program. (1969). "Codex Committee on Foods for Special Dietary Uses," Agenda item No. 7 (B). World Health Organ., Geneva.

CHAPTER 34

The Renal Metabolism and Uses of Mannitol

H. EARL GINN

A. Absorption and Metabolism

Mannitol is a six carbon alcohol with a molecular weight of 182 daltons. It is the reduced form of the sugar mannose. Many studies of intestinal and renal physiology in man and animals have assumed that mannitol is a nonabsorbable molecule which remains intact in the intestine if administered orally (Kameda et al., 1968; Fordtean, et al., 1965) and that it is not metabolized (Stuart et al., 1970). Yet there have been contradictory reports concerning the absorption and metabolism of mannitol. Jaffe, in 1883, demonstrated absorption by feeding dogs mannitol and recovering large quantities in the urine. Several investigators subsequently have demonstrated absorption by the gut (Carr and Krantz, 1938; Hindle and Code, 1962) and metabolism (Wick et al., 1954; Voegtlin et al., 1925; Lafon, 1937) of the alcohol in animals.

Perhaps the variability in earlier experiments was related to the difficulties of applying the Corcoran and Page (1947a) mannitol analytical method, particularly to stool samples. In this reaction, formaldehyde is generated from mannitol and subsequently is used to develop a specific color. Nonmannitol formaldehyde generation is significant in urine, blood, and especially stool. The formaldehydegenic material in stool samples is so variable as to render correction by the blank method unreliable.

Recently, Nasrallah and Iber (1969) demonstrated that mannitol is absorbed and metabolized in man. [U-¹⁴C] Mannitol was administered orally to one group of subjects and intravenously to another. Two independent methods were employed for the detection of mannitol, one chemical and the other radioactive. A mean of 17.5% of the orally administered mannitol was recovered in the urine. As much as 18% of the orally administered radioactivity was estimated to be in the form of expired $^{14}CO_2$ in 12 hr. On the other hand, after [¹⁴C]mannitol was administered intravenously, almost all was recovered in the urine and stool. Only about 1% of the administered dose was recovered as expired $^{14}CO_2$. These findings suggest that the amount of mannitol oxidized by the body organs (possibly the liver) may be proportional to both the concentration reaching that organ and duration of exposure. Intravenously administered doses usually persist for a shorter period of time because it is excreted at the full glomerular filtration rate by the kidney. In contrast, the oral dose rises and falls more slowly. Wick *et al.*, in a study of mannitol oxidation in rats, found 2–3% of the mannitol as expired CO_2 after intravenous administration, 50% as expired CO_2 after oral administration, and 70% after intrasplenic administration, i.e., directly into the portal venous system (Wick *et al.*, 1954). These investigators concluded that mannitol is oxidized by the liver following exposure of that organ to high concentrations.

When given intravenously mannitol equilibrates rapidly in the extracellular fluid space with virtually no entry into the body cell mass unless very high concentrations are maintained for prolonged periods of time (Stuart *et al.*, 1970; Buckell, 1964). It is relatively nontoxic and filters through the renal glomerulus without being reabsorbed from the nephron. Within the tubular lumen the particles of mannitol prevent obligatory water reabsorption and carry electrolytes with the unreabsorbed water (Kaplan *et al.*, 1952).

B. Use as a Diuretic Agent

In the past mannitol was used, as well as inulin, in renal function studies. Smith (1951) noted that some of the patients with post-traumatic oliguria who were undergoing renal clearance studies developed a diuresis when given mannitol. From this fortuitous observation, he speculated that mannitol could have prevented the expected incidence of anuria, and he referred to this relationship as the "osmotic accident." In 1947, Corcoran and Page (1947b) recommended the use of mannitol in treating acute renal failure. Subsequently, it has been extensively used to treat acute oliguric renal failure (Barry and Malloy, 1962; Barry,

1963; Barry and Crosby, 1963; Bourne and Cerny, 1964), to accelerate the elimination of drugs in the urine (Cirksema *et al.*, 1964; Setter *et al.*, 1966) to reduce intracranial pressure, and to correct hyponatremic overhydration (Hammond *et al.*, 1967; Moore, 1963).

Most of the treatment regimes aim to achieve a urine flow of more than 5 ml/min. Since it is obvious that such high rates of urine flow can cause rapid depletion of body water and electrolytes, they also make recommendations concerning the rapid administration of intravenous electrolyte solutions (Barry and Malloy, 1962; Barry, 1963; Bourne and Cerny, 1964; Cirksema *et al.*, 1964; Setter *et al.*, 1966; Wise and Chater, 1962; Hammond *et al.*, 1957; Moore, 1963). Otherwise, severe or even lethal dehydration can result.

If such a high throughput of fluid and electrolyte is to be safely maintained in sick patients, great skill and also supervision are imperative because there are hazards which no rigid regime can be relied upon to avert. If there is initial over- or under-hydration this abnormality should be treated first. Otherwise the strict maintenance of fluid balance would be undesirable. If the patient has an impaired capacity to conserve or excrete certain electrolytes, experience may not be sufficient guide to the correct composition of the intravenous replacement fluid. If there are factors present which disturb the normal internal distribution of water or electrolytes, a correct external balance may not prevent large internal shifts. To avoid these hazards it is necessary to understand the mechanisms by which mannitol can produce a diuresis (Fig. 1).

The administration of osmotic agents such as mannitol, particularly

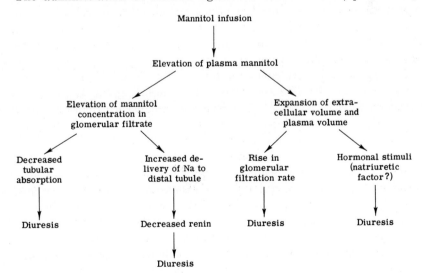

Fig. 1 Mechanism by which mannitol causes diuresis.

when hypertonic solutions are used, is likely to cause them to accumulate at least temporarily in extracellular fluid and attract water into this compartment. If external balance is maintained, the bulk of this water must come from cells. Extracellular expansion and cellular dehydration therefore result from the retention of osmotic diuretics. These changes may have little clinical significance if the rate of mannitol administration were low or the total dose small.

Recently, Morgan *et al.* (1968) studied the degree of mannitol retention that results from the intravenous administration of 10% mannitol. Their patients were being treated for barbiturate and salicylate overdosage. These drugs are known to cause a state of antidiuresis. There was marked retention of mannitol, ranging from 27 to 314 mg in their 23 patients. This arose because, although the urinary output never fell far behind the fluid input, the rate of mannitol excretion usually fell short of its rate of infusion by some 10–40 gm/hr. This short fall was predictable, since the urinary concentration of osmotic diuretics tends to approach isotonicity with plasma as the urine flow rises. The retained mannitol can be expected to be primarily confined to the extracellular space (Newman *et al.*, 1944) where it will cause a rise in osmolarity. The resulting osmotic gradient is dissipated by the movement of water into the extracellular fluid from the cells, causing cellular dehydration. When external water and electrolyte balance is maintained, the retention of 100 gm of mannitol by an average adult will draw theoretically nearly 2 liters of water from the cells into the extracellular fluid. Measurements of extracellular volume were more often in the neighborhood of 40% of body weight than the accepted average normal value of 20%. Fortunately, most patients tolerate well such large internal movements of fluid for short periods. Nevertheless, the potential hazards to people with heart disease or kidney failure are obvious.

The expansion in extracellular volume and plasma volume often causes a rise in glomerular filtration rate and an accompanying diuresis. Furthermore, there is increasing evidence that expansion of plasma volume can cause a diuresis which does not depend upon a rise in glomerular filtration rate (Rector *et al.*, 1964). This may result from release of a hormone called "natriuretic factor."

Since mannitol is not reabsorbed from the nephron, the elevation of its concentration in glomerular filtrate results in a significant concentration gradient for sodium between tubular lumen and interstitium. Although recent evidence suggests that active sodium transport from the proximal tubules remains constant during mannitol diuresis (Kiil *et al.*, 1971) there is increased passive influx of sodium along this concentration gradient (Kiil *et al.*, 1961; Kjekshus *et al.*, 1969). Net electrolyte and obligatory water reabsorption are thereby reduced resulting in diuresis. It is also probable that the increased sodium delivery to the distal convoluted tubule causes a reduction in release of renin from the juxtoglomerular apparatus.

Mannitol within the tubular lumen may increase the intratubular pressure. Direct micropunctute studies indicate that mannitol, through increased intratubular pressure, can result in compression of the peritubular veins. This venous compression can lead to diversion of medullary blood to the cortex, a shift that probably has a protective value for patients with insipient tubular necrosis.

Electron microscopy studies of renal tissue from patients who had received mannitol revealed dense osmiophilic bodies in the cytoplasm of proximal and distal epithelial cells. These bodies, which have been referred to as "osmotic nephrosis," appear to be cytolysosomes (Muehrcke, 1969). They are not associated with loss of renal function and disappear soon after cessation of mannitol therapy (Stuart *et al.*, 1970).

In conclusion, if the mechanisms of mannitol diuresis are thoroughly understood and the potential hazards diligently avoided mannitol may be, using the words of F. D. Moore, "a solution for several old problems."

References

Barry, K. G. (1963). Post-traumatic renal shutdown in humans: Its prevention and treatment by the intravenous infusion of mannitol. *Mil. Med.* **128,** 224.

Barry, K. G., and Crosby, W. G. (1963). Prevention and treatment to renal failure following transfusion reactions. *Transfusion* **3,** 35.

Barry, K. G., and Malloy, J. P. (1962). Oliguric renal failure. Evaluation and therapy by intravenous infusion of mannitol. *J. Amer. Med. Ass.* **179,** 510.

Bourne, C. W., and Cerny, J. C. (1964). The role of mannitol in acute trauma. *Univ. Mich. Med. Cent. J.* **30,** 109.

Buckell, M. (1964). Blood changes on intravenous administration of mannitol or urea for reduction of intracranial pressure in neurosurgical patients. *Clin. Sci.* **27,** 223.

Carr, C. J., and Krantz, J. C. Jr. (1938). Sugar alcohols. XII. The fate of polygalitol and mannitol in the animal body. *J. Biol. Chem.* **124,** 221.

Cirksema, W. J., Bastian, R. C., Malloy, J. P., and Barry, K. G. (1964). Use of mannitol in exogenous and endogenous intoxications. *N. Engl. J. Med.* **270,** 161.

Corcoran, A. C., and Page, I. H. (1947a). A method for the determination of mannitol in plasma and urine. *J. Biol. Chem.* **170,** 165.

Corcoran, A. C., and Page, I. H. (1947b). Crush-syndrome: Post traumatic anuria: Observations on genesis and treatment *J. Amer. Med. Ass.* **134,** 436.

Fordtran, J. S., Rector, F. C., Jr., Ewton, M. F., Soter, N., and Kinney, J. (1965). Permeability characteristics of human small intestine. *J. Clin. Invest.* **44,** 1935.

Hammond, W. G., Carter, R. C., Davis, J. M., and Moore, F. D. (1957). Osmotic diuresis as treatment in severe hyponatremia. *Surg. Forum* **7,** 52.

Hindle, W., and Code, C. F. (1962). Some differences between duodenal and ileal sorption. *Amer. J. Physiol.* **203,** 215.

Jaffe, M. (1883). Ueber das vorkommen von mannit im normalen hunderharm. *Z. Physiol. Chem.* **7,** 297.

Kameda, H., Abei, T., Nasrallah, S. M., and Iber, F. L. (1968). Functional and histological injury to intestinal mucosa produced by hypertonicity. *Amer. J. Physiol.* **214,** 1090.

Kaplan, S. A., Foman, S. J., and Rapoport, S. (1952). Effects of epinephrine and I-nor-epinephrine on renal excretion of solutes during mannitol diuresis in the hydropenic dog. *Amer. J. Physiol.* **169,** 588.

Kiil, F., Aukland, K., and Refsum, H. E., (1961). Renal sodium transport and oxygen consumption. *Amer. J. Physiol.* **201,** 511.

Kiil, F., Johannesen, J., and Aukland, K. (1971). Metabolic rate in renal cortex and medulla during mannitol and saline infusion. *Amer. J. Physiol.* **220,** 565.

Kjekshus, J., Aukland, K., and Kiil, F. (1969). Oxygen cost of sodium reabsorption in proximal and distal parts of the nephron. *Scand. J. Clin. Lab. Invest.* **23,** 307.

Lafon, M., (1937). Sur l'utilisation alimentaire des heritols par la souris. *C. R. Soc. Biol.* **126,** 1147.

Moore, F. D., (1963). Tris buffer, mannitol and low visions dextran: Three new solutions for old problems. *Surg. Clin. N. Amer.* **43,** 577.

Morgan, A. G., Bennett, J. M., and Polak, A. (1968). Mannitol retention during diuretic treatment of barbiturate and salicylate overdosage. *Quart. J. Med.* **37,** 589.

Muehrcke, R. C. (1969). "Acute Renal Failure: Diagnosis and Management." Mosby, St. Louis, Missouri.

Nasrallah, S. M., and Iber, F. L. (1969). Mannitol absorption and metabolism in man. *Amer. J. Med. Sci.* **25,** 80.

Newman, E. V., Bordley, J., and Winternitz, J. (1944). The interrelationships of glomerular filtration rate (mannitol clearance), extracellular fluid volume, surface area of the body, and plasma concentration of mannitol. *Bull. Johns Hopkins Hosp.* **75,** 253.

Rector, F. C., Jr., Van Giesen, G., Kiil, F., and Seldin, D. W. (1964). Influence of expansion of extracellular volume on tubular reabsorption of sodium independent of changes in glomerular filtration rate and aldosterone activity. *J. Clin. Invest.* **43,** 341.

Setter, J. G., Maher, J. F., and Schreiner, G. E., (1966). Barbiturate intoxication: Evaluation of therapy including dialysis in a large series selectively referred because of severity. *Arch. Intern. Med.* **117,** 224.

Smith, H. W. (1951). "The Kidney, Structure and Function in Health and Disease." Oxford Univ. Press, London and New York.

Stuart, F. P., Torres, E., Fletcher, R., Crocker, D., and Moore, F. D. (1970). Effects of a single, repeated and massive mannitol infusion in the dog: Structural and functional changes in kidney and brain. *Ann. Surg.* **172,** 190.

Voegtlin, C., Dunn, E. R., and Thompson, J. W., (1925). Antagonistic action of certain sugars, amino acids and alcohols on insulin intoxication. *Amer. J. Physiol.* **71,** 574.

Wick, A. N., Morita, T. N., and Joseph, L. (1954). The oxidation of mannitol. *Proc. Soc. Exp. Biol. Med.* **85,** 188.

Wise, B. L., and Chater, N. (1962). Use of hypertonic mannitol solution in decreasing brain mass and lowering cerebrospinal fluid pressure. *J. Neurosurg.* **19,** 1038.

CHAPTER 35

Xylitol as a Therapeutic Agent in Glucose-6-Phosphate Dehydrogenase Deficiency

J. VAN EYS, Y. M. WANG, SAM CHAN,
VORAVARN S. TANPHAICHITR, and
SYNTHIA M. KING

A. Erythrocyte Metabolism

The mammalian red cell is a unique cell since it has a highly specialized function in carrying oxygen and assisting in the transport of carbon dioxide. To this end, the cell has a limited metabolic capacity which is geared toward maintenance of functional integrity of the cell membranes and the hemoglobin molecule. The metabolic capacity for these functions is largely limited to glycolysis as the only energy yielding pathway and supplier of 2,3-diphosphoglycerate (2,3-DPG), which is necessary for modulating normal hemoglobin functions. The pentose phosphate shunt is present and has a specialized function in maintaining cellular integrity. The mature cell cannot synthesize proteins and thus has a finite life-span both *in vivo* and *in vitro*.

The pentose phosphate shunt generates reduced nicotinamide-adenine dinucleotide phosphate (NADPH) which in the red cell is to a large part used to keep the tripeptide γ-glutamylcysteinylglycine, glutathione

613

(GSH) in the reduced state. GSH is oxidized by peroxide generating compounds to oxidized glutathione (GSSG). If the rate of GSH oxidation is exceeded by peroxide present, cell membrane damage results. Thus, the GSH system is primarily responsible for the protection of the red cell membrane. When GSSG accumulates, the excess will leave the red cell, and a constant replenishing of the supply of GSH is required in a two-step biosynthetic pathway. The red cell is capable of effecting that synthesis. Fig. 1 illustrates the pertinent pathway and the associated enzymes. There are several recent reviews available of glutathione metabolism in the red cell (Prins and Loos, 1969; Srivastava, 1971).

When any one of the enzymes of glutathione reduction or synthesis is diminished in activity, the red cell is susceptible to oxidative damage and hemolysis will result. Such diminished activity may be acquired as is seen nutritionally through riboflavin deficiency, which results in a deficiency of the coenzyme concentration of the enzyme glutathione reductase (cf. Sauberlich *et al.*, 1792). Alternatively, and more commonly, the defects are genetically determined. There are genetic disorders described for every enzyme in the sequence. However, some of those disorders

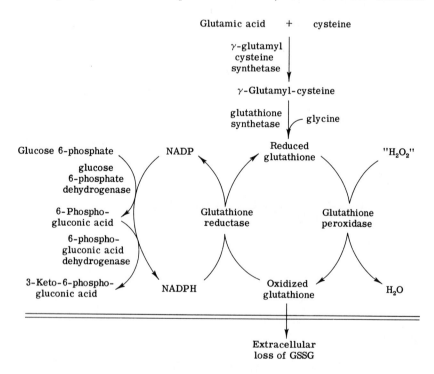

Fig. 1 Summary of the glutathione metabolism in red cells.

are rarities: for example, only two patients have so far been reported to suffer from the disorder for γ-glutamylcysteine synthetase deficiency (Konrad *et al.*, 1972). On the other hand, glucose-6-phosphate dehydrogenase (G6PD) affects an estimated 3% of the world population (Marks and Banks, 1965).

B. The Clinical Spectrum of Glucose-6-Phosphate Dehydrogenase Deficiency

While 3% of the world population has significantly diminished red cell G6PD activity, this does not represent in all cases an enzyme change through the same unique mutation. There are many genetic variants known thus, wide differences exist between the actual *in vivo* intracellular activity and stability of the enzyme among different patients with G6PD deficiency. Within this variation a number of clinical syndromes can be recognized (Keller, 1970).

1. Drug-Induced Hemolytic Anemias

In drug-induced hemolytic anemia ordinarily there is sufficient enzymatic activity to maintain a normal red cell mass. However, exposure to oxidant drugs or infections will result in significant hemolysis. This is a common form, and is seen in the United States in 8% of the black population. This type of enzyme is usually designated as the A— variant. The disorder is thought to be benign when no drug exposure occurs, but recent evidence suggests a shortened life expectancy of such individuals (Petrakis *et al.*, 1970).

2. Neonatal Jaundice

Any variant, even when producing little or no morbidity at a mature age, can be the etiology for neonatal jaundice. This is often spontaneous, but many times aggravated by drugs such as certain vitamin K preparations. On occasion exchange transfusions may be required.

3. Favism

Favism is expressed by the acute hemolysis which follows raw or partially cooked fava bean ingestion. On occasion, in particularly sensitive persons, hemolysis may follow pollen inhalation. This disorder is only seen in Caucasian patients with G6PD deficiency, but not all G6PD-deficient patients demonstrate this sensitivity. There may be an additional genetic function responsible for the susceptibility. This form of G6PD deficiency, which is especially prevalent in the Mediterranean area, can be quite severe and can result in significant mortality.

4. Congenital Nonspherocytic Hemolytic Anemia

Congenital nonspherocytic hemolytic anemia occurs when the deficiency of the enzyme is so severe that a chronic hemolysis results. Because of a chronic anemia, there can be significant morbidity from this form of expression of the disorder. These patients are in addition susceptible to episodes of excessive hemolysis from drugs and have all the complications of a chronic hemolytic anemia, such as aplastic crises, gallbladder disease, and splenomegaly.

5. Secondary Chronic Hemolysis

On many occasions patients with otherwise mild or only drug-induced anemia will show a chronic hemolytic anemia because there is a chronic interference with the red cell production–breakdown balance. Such conditions may arise from accelerated destruction because of chronic drug ingestion (e. g., aspirin) or from diminished production as a result of marrow depression secondary to infections or uremia.

C. Xylitol as Potential Therapeutic Agent

1. Theory of a Rational Therapy

Under several conditions it would be desirable to have a specific safe therapy, such as during acute hemolytic crises, for congenital nonspherocytic hemolytic anemia from G6PD deficiency and for neonatal jaundice. If the pathophysiological concept of a lack of NADPH generation as the cause of hemolysis is correct, then alternative ways of generating NADPH would be corrective.

Xylitol was early considered as possible therapeutic agent because it is known to potentially generate NADPH through oxidation to L-xylulose (Fig. 2).

To be certain that xylitol could be a therapeutic agent it had to be established that red cells are capable of metabolizing this pentitol.

Early reports by Asakura and his co-workers showed rapid uptake of red cells and suggested the presence of xylitol dehydrogenase activity

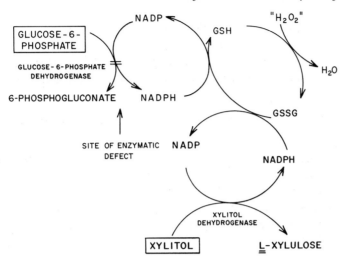

Fig. 2 The rationale for the use of xylitol in G6PD deficiency.

in erthrocytes (Asakura *et al.*, 1967, 1969, 1970; Yoshikawa, 1970). Further demonstration of NADP-linked xylitol dehydrogenase and utilization of xylitol by red cells was thought desirable.

2. Xylitol Metabolism in Red Cells

a. Xylitol Dehydrogenase in Red Cells

To investigate the polyol dehydrogenation in the human erythrocytes glutathione reduction was first evaluated in toluene-treated hemolysates as previously reported (Wang and van Eys, 1970). The method couples the polyol dehydrogenation with GSSG reduction via the reduced pyridine nucleotide to the readily measurable GSH. Through such a reaction sequence we were able to demonstrate that xylitol, sorbitol, ribitol, D-mannitol, galactitol, L-arabitol, and D-arabitol could be metabolized in the hemolysates with NAD as coenzymes but not glycerol, D-erythritol, or inositol. However, only xylitol could be utilized substantially when NADP was used to replace NAD. Sorbitol was found to be oxidized slowly. Table I shows the summary of the polyol oxidation by red cells and the products identified.

b. The Partial Resolution of the Polyol Dehydrogenase Activities

The preliminary observations combined with literature reports suggested multiple enzymes for the oxidation of polyols. In an attempt to

TABLE I

Polyol Product Formed by Pyridine Nucleotide-
Linked Dehydrogenases in the Erythrocyte

Substrate[a]	Coenzyme	Product[b]
Xylitol	NAD	D-Xylulose
Sorbitol	NAD	D-Fructose
Ribitol	NAD	D-Ribulose
Galactitol	NAD	(D)-Tagatose[c]
L-Arabitol	NAD	L-Ribulose
D-Mannitol	NAD	D-Fructose
D-Arabitol	NAD	D-Ribulose
Xylitol	NADP	L-Xylulose
Sorbitol	NADP	D-Glucose

[a] The reaction mixtures contained, in a final volume of 8.0 ml, 400 μmoles tris-hydroxymethylaminomethane, pH 8.0, 50 μmoles of nicotinamide, 25 μmoles of $MgCl_2$, 10 mg of pyridine nucleotide (NAD or NADP), 750 μmoles of sodium pyruvate, 90 IU of crystalline rabbit muscle lactic acid dehydrogenase (LDH), and for NAD-linked reactions a polyol concentration of 0.075 M. For the NADP–xylitol reaction a xylitol concentration of 0.30 M was used, while for the NADP–sorbitol reaction a polyol concentration of 0.60 M was used, and GSSG (10 μmoles) was used as acceptor system rather than the pyruvate-LDH couple.

[b] The products were isolated by deproteinization of the reaction mixture with 2 ml 30 % TCA, after which the solutions were deionized by passing through a 1×40 cm Amberlite MB-3 resin. The solutions were then lyophilized. The products were identified by (a) paper chromatography, (b) specific color reactions, (c) utilization of the product as substrates for specific enzymes, and (d) demonstration of the reverse reactions.

[c] The configuration is assumed.

separate these proteins, the toluene-treated hemolysate was brought to 30% saturation in $(NH_4)_2SO_4$. Then a fraction was obtained by raising the saturation to 40% followed by five further fractions obtained by raising the saturation by 5% for each one. There was almost no protein left in the supernatant after reaching the final 65% saturation. The precipitated proteins were redissolved in a tris buffer (0.001 M tris-HCl at pH 8.0 containing 2.5×10^{-6} M mercaptoethanol) and dialyzed against the same buffer before assaying the enzymatic activities. Figure 3 shows the specific activity of the enzyme fractions toward various polyols in a repre-

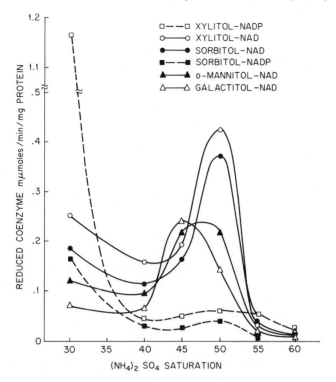

Fig. 3 The separation of human erythrocyte polyol dehydrogenases with ammonium sulfate. Substrate concentration for NAD-dependent reactions: Xylitol 0.019 *M;* sorbitol 0.15 *M;* galactitol, 0.15 *M,* and D-mannitol, 0.60 *M* NADP-dependent reactions: xylitol 0.30 *M* and sorbitol 0.60 *M.*

sentative experiment. Thirty-five percent of the NADP-linked xylitol dehydrogenase activity was found in the 0–30% fraction with a specific activity of 1.17 mμmoles/mg/min. However, the major NAD-linked activities were in the fractions between 40 and 45%, in which 38% of the total protein precipitated. The specific activity toward the galactitol and D-mannitol was high in this fraction, while the highest specific activity peak for xylitol and sorbitol occurred in the 45–50% fraction, where only 6% of the protein precipitated. The experiment was closely reproducible in multiple trials. This partial separation of the enzymatic activities indicated that at least an NADP-linked polyol dehydrogenase and an NAD-linked polyol dehydrogenase exist. The NAD-lined enzymatic activity could be tentatively divided into at least two enzymes: an NAD-linked xylitol dehydrogenase and an NAD-linked galactitol dehydrogenase. It is possible that an NAD-linked D-mannitol dehydrogenase is present or that D-mannitol could be utilized by both the NAD-linked xylitol and galactitol dehydrogenases.

c. The NADP-Linked Xylitol (L-Xylulose) Dehydrogenase

In the 45% ammonium sulfate precipitate xylitol was the only polyol which could be oxidized readily with NADP as the coenzyme, though a slow reaction with sorbitol was seen, which was additive to the rate with xylitol and resulted in glucose rather than a ketose. The reverse reaction was also demonstrated with commercial L-xylulose. The specific activity with xylitol as substrate was increased about 77-fold in this fraction over the hemolysate. 3-Acetylpyridine adenine dinucleotide phosphate, and thionicotinamide-adenine dinucleotide phosphate could not replace NADP in the enzymatic reaction. Magnesium as well as manganese were found to stimulate the enzymatic activity, but high levels of divalent ions, especially magnesium, were inhibitory. A broad pH optimum was found, and small variations in optimum pH were observed in different preparations. The K_m values for the substrates were estimated to be 3.2×10^{-2} M for xylitol, which was slightly different from that found in the hemolysates, and 6×10^{-5} for NADP as published previously (Wang and van Eys, 1970). The monovalent cations had no stimulatory effect. Moreover, no substrate inhibitory effect could be observed in the xylitol-NADP system. The properties of the enzyme are summarized in Table II. This is the enzyme that is kinetically altered in pentosuria, where a decreased affinity for NADP exists (Table III). It thus is likely that the enzyme is genetically the same as the liver enzyme.

d. Utilization of Xylitol by Intact Cells

The presence of red cell enzymes demonstrated the possibility of xylitol utilization by intact erythrocytes. The demonstration that it can occur physiologically was obtained from storage experiments. Venous blood was drawn from healthy nonsmoking Chinese and Caucasian students in acid citrate dextrose (ACD) as preservative. Blood was stored in sterile plastic tubes at 4°C for 2–3 weeks. Sampling was done on individual tubes each

TABLE II

The Properties of the NADP-Linked Specific
Xylitol Dehydrogenase

K_m, xylitol (M)	3.2×10^{-2}
K_m, NADP (M)	6×10^{-5}
Optimum pH	7.5–8.5
Mg^+, optimal concentration (M)	4×10^{-3}
Mn^+, optimal concentration (M)	1×10^{-2}

TABLE III

The Michaelis Constant of NADP for Xylitol Dehydrogenase in Normal and Pentosuric Subjects[a]

Subjects	$K_m(M \times 10^{-5})$
Normal (4)[b]	4.9 ± 1.6[c]
M.F., homozygous	50
G.W., homozygous	100
I.B., homozygous	67
I.B. Senior, heterozygous[d]	4.0;50
E.W., heterozygous[e]	4.7;80

[a] The assays were performed as reported previously (Wang and van Eys, 1970).
[b] Number of observations.
[c] Mean ± 1 SD.
[d] Father of I.B.
[e] Brother of G.W.

time every 4–7 days. At the end of 2–3 weeks, recovery experiments were performed on the stored blood in incubation at room termperature for 5 hr after addition of xylitol or inosine to the appropriate tubes.

Recovery experiments were also performed on recently outdated Red Cross blood stored under standard blood bank conditions. The addition of xylitol to blood stored in ACD had a marked effect on the maintenance of 2,3-DPG levels in red cells. This effect correlated well with the utilization of xylitol and was far greater than extra added glucose but inferior to inosine. Inosine and xylitol were not additive in their effects. Figure 4 demonstrates these findings.

Xylitol addition to ACD stored blood also results in ATP and other high-energy phosphate regeneration (Table IV). Xylitol and inosine are additive in that effect. A beneficial effect of xylitol on the osmotic fragility and ATP levels of stored blood was previously noted by Asakura's group (Asakura, *et al.*, 1969).

3. Model Systems for Xylitol in G6PD Deficiencies

a. Effect of Xylitol on Red Cell Integrity in Model Systems

There is no animal that clearly has a clinical syndrome resembling G6PD deficiency. However, the rabbit is very drug sensitive, which is at least in part the result of the decreased activity of glutathione reduc-

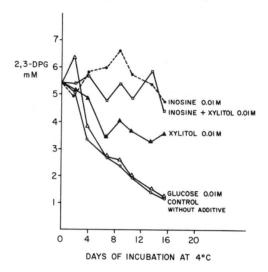

Fig. 4 The effect of xylitol and other additives on the 2,3-DPG content of stored human red cells. Freshly drawn blood was collected in ACD solutions under sterile conditions and stored at 4°C. All points are the average of triplicate determinations. 2,3-DPG concentration was determined by an automated method (Loos and Prins, 1970).

tion. Rabbits were challenged *in vivo* using aceytlphenylhydrazine (APH) as a hemolyzing drug. As shown in Fig. 5, injection of xylitol protected the animals against excessive hemolysis (Wang *et al.*, 1971).

TABLE IV

Regeneration of High Energy Phosphate in Maximally Stored Blood[a]

Additions	ATP (mmoles/liter whole blood)	~P
None	−0.06	−0.46
Glucose	+0.30	+0.46
Xylitol	+0.18	+0.94
Inosine	+0.24	+2.05
Inosine + xylitol	+0.30	+4.02

[a] ATP was assayed by the firefly luciferase method (Strehler and McElroy, 1957). High energy phosphate was assayed by the 7-min hydrolyzable phosphate (Fiske and Subbarow, 1925). Outdated Red Cross blood was used, and the additives were allowed to act for 4 hr. The additives were in a final concentration of 0.01 *M*.

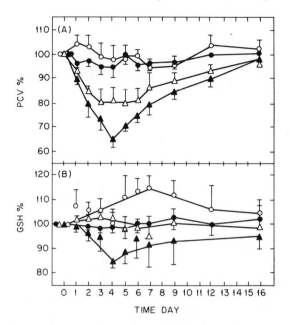

Fig. 5 Hematological changes in rabbits expressed as percent of initial values: (A) PCV % and (B) GSH %. Each point represents an average of 3 animals ±1 SD. The following symbols are used: (●) control, (○) xylitol treated, (▲) APH-treated, and (△) experimental group to which both APH and xylitol were given. APH at dose of 10 mg/kg of body weight was injected intraperitoneally at zero time. Xylitol, in a 20% solution, was injected through an ear vein, four times a day for 6 days, in a dose of 0.5 gm of xylitol per kilogram of body weight, over 4 min each time. Two to three milliliters of blood were taken daily. GSH was determined by the method of Beutler *et al.* (1963) [Reproduced by permission of the *Journal of Clinical Investigation* (Wang *et al.*, 1971)].

Such experiments were encouraging so that *in vitro* tests on human blood were performed.

b. The Effect of Xylitol on G6PD-Deficient Cells in Vitro

Red blood cells were incubated with methylphenyldiazene carboxylate to oxidize approximately 80–90% of the GSH to GSSG, after which GSH recovery was evaluated. Alternatively, the cells were challenged with APH, and GSH preservation was measured. The methodology has been described previously (Wang *et al.*, 1971).

All G6PD-deficient patients so far examined have normal NADP-dependent xylitol dehydrogenase activity. Table V shows that the cells from the

TABLE V

Recovery and Maintenance of Red Blood Cell GSH by Xylitol *in vitro* in Patients with Drug-Induced Hemolytic Anemia

Patient	G6PD activity (IU/100 ml RBC)	G6PD[a] variant	NADP-XDH[b] activity (μmoles/ hr/100 ml RBC)	GSH[c] recovery (%)	GSH[c] maintenance (%)
Black (4)[d]	17.4 ± 2.4[e]	A⁻	22.0 ± 2.5	141.1 ± 25.6 (3)	130.0 ± 13.0 (3)
Chinese (1)	2.0	Taipei (?)	17.9	182	358.4
Normal (5)	145.0 ± 42.5	B	18.8 ± 2.6	58.4 ± 11.1	56.8 ± 19.2

[a] Variants were determined through the criteria set forth by the WHO Committee on glucose-6-phosphate dehydrogenase (World Health Organization, 1967; Yoshida *et al.*, 1971).
[b] XDH stands for xylitol dehydrogenase.
[c] GSH recovery and maintenance are expressed as percent of the effect of glucose.
[d] Figures in parentheses indicate the number of patients.
[e] Mean ± 1 SD.

drug-sensitive variants, not clinically apparent between hemolytic episodes, are clearly better protected or repaired by xylitol than by glucose. The same holds true for patients who have congenital nonspherocytic hemolytic anemia (Table VI).

D. Xylitol Toxicity

1. Intravenous Xylitol

Recent reports suggested that there is a serious xylitol toxicity (Thomas *et al.*, 1972a,b), which could preclude its clinical use even when there are specific medical indications rather than just caloric supplementation. The dosages effective in the rabbit were a 20% solution injected four times a day at 0.5 gm/kg of body weight, each dose administered as an intravenous push. This dose gives a peak plasma level of a 150 mg % of xylitol with a half-life of approximately 20 mins. We attempted to evaluate whether such doses could be within the safe range.

a. Adult Rabbits

Xylitol is toxic in the rabbit (Wang *et al.*, 1972, 1973). It can result in a rise of the serum glutamic oxalacetic transaminase (SGOT) levels and results in other evidence of liver malfunction. However, hyperuricemia, which has been reported in humans (Donahoe and Powers, 1972; Schumer, 1971) is not seen, nor can oxalic acid deposition be found. But a lactic acidosis may occur. The toxicity is dose response related in the rabbit, similar to that observed in humans (Igarashi *et al.*, 1971). The degree depends on (a) the achieved osmolality in the serum and (b) the total dose. These parameters can be modified through rate of administration of solutions of varying concentration. Figure 6 illustrates the dependence of the toxicity on these parameters. This can be summarized by saying that a great deal of the toxicity seen in the rabbit is the result of hyperosmolarity (Wang *et al.*, 1972, 1973; van Eys and Wang, 1972) as was observed in human toxicity in Australia (Thomas *et al.*, 1972a). A safe dose for the rabbit is a blood level of 50 mg %, which can be achieved by a total dose of 1–2 gm/kg/day which in turn can be achieved readily as a 5–10% solution. These levels are effective for increasing *in vivo* red cell GSH levels.

TABLE VI

Recovery and Maintenance of Red Blood Cell GSH by Xylitol *in vitro* in Patients with Congenital Nonspherocytic Hemolytic Anemia[a]

Patient	G6PD activity (IU/100 ml RBC)	G6PD variant	NADP-XDH activity (μmoles/ hr/100 ml RBC)	GSH recovery (%)	GSH maintenance (%)
L.Y.	9.2	Chicago	18.9	133.3	115.5
S.Y.	0	Chicago	22.6	184.0	153.7
R.P.	0		17.0	173.4	135.0
A.B.	0		20.8		199.3
K.S.	2.4	Chicago(?)	23.9	110.0	173.2
Average[b]	2.3		20.6 ± 2.8	150.2 ± 34.6	155.3 ± 32.6
Normal[b]	145.0 ± 42.5	B	18.8 ± 2.6	58.4 ± 11.1	56.8 ± 19.2

[a] See footnotes in Table V for explanation of G6PD variants, GSSG recovery and maintenance, and meaning of abbreviations.
[b] Figures are average ± 1 SD.

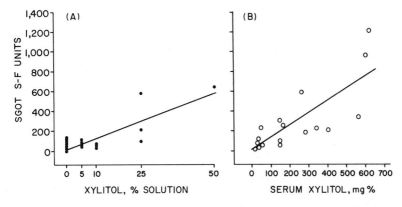

Fig. 6 The correlation between parenterally administered xylitol and observed SGOT levels: (A) Concentration of xylitol solution administered at a dose of 1 gm/kg of body weight and (B) serum xylitol concentration. Xylitol was determined chemically (Bailey, 1959), SGOT was determined by standard clinical chemistry methodology.

b. Newborn Rabbits

Preliminary experiments also suggest that the newborn rabbit has a great deal of tolerance for xylitol. Doses up to 8 gm/kg in a 50% solution injected intraperitoneally had no effect on mortality or growth. The serum xylitol concentration rose in some animals to as high as 365 mg %. No significant SGOT elevations were seen in these animals. Packed cell volume did rise, indicating a very modest hemoconcentration. Human infants have also been reported to tolerate parenteral xylitol well (Erdmann, 1969; Kumagai, 1969).

2. Oral Xylitol

a. Oral Xylitol in the Rabbit

Oral xylitol is tolerated by the rabbit and yields excellent blood levels. Figure 7 shows the result of single administration of xylitol to a group of rabbits at varying dose levels. Doses of 10 gm/kg do give SGOT elevations, but lower doses do not. Furthermore, gradually increasing the amount administered seems to protect against this toxicity. This is illustrated in Fig. 8, where a single animal tolerated up to 10 gm/kg of body weight with only a slight SGOT increase. While Fig. 7 shows peak values after acute administration of xylitol, the SGOT rise in these animals is clearly apparent after 1 hr far in excess over the levels found in the animal shown in Fig. 8. Adaption to xylitol has been observed previously (Bässler, 1969).

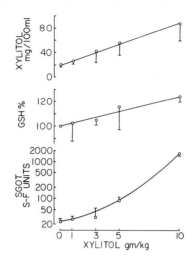

Fig. 7 Dose responses of plasma xylitol, SGOT, and whole blood GSH levels after administration of a 50% xylitol solution orally by gastric intubation. Each point represents an average of the peak values of the observations in 3 animals ± 1 SD. Sigma-Frankel unit is abbreviated as S-F unit.

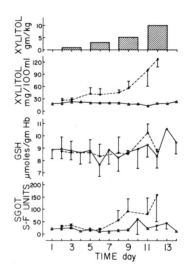

Fig. 8 Plasma xylitol, SGOT, and whole blood GSH levels after long-term oral xylitol administration. The upper panel indicates the oral dosage. Each point represents an average value of 3 animals ±1 SD: (○) levels before each administration, and (●) levels 1 hr after administration.

b. Oral Xylitol in Humans

The literature suggests that xylitol orally is well tolerated with less toxic effects than intravenous administration (Bässler, 1969; see also Brin and Miller, Chapter 33). This difference between the intravenous and oral route is very likely the result of the lack of an immediate osmolality increase.

E. Future Needs

The three syndromes in glucose-6-phosphate dehydrogenase deficiency that require therapy are: (a) favism and other extreme drug-induced hemolyses, (b) neonatal jaundice when exchange transfusion seems indicated, and (c) congenital nonspherocytic hemolytic anemia. It would be highly desirable for the latter group to develop oral therapy since the treatment requirement is clearly chronic. For neonatal jaundice therapy would only take a short time, but very likely would require intravenous therapy. The risk of an exchange transfusion is significant, and if xylitol toxicity were less dangerous than the finite mortality of the exchange this form of therapy could be indicated even if some human toxicity were to be present. Finally, favism is sporadic and intermittent but occurs in areas of the world where blood transfusions are not as readily organized as they are in some urbanized societies.

It is clear that the next needs are (a) human clinical trials under rigorously controlled conditions, (b) clear elucidation of the human tolerance for xylitol, and (c) explanation of the actual mechanism of toxicity seen in Australia.

If xylitol indeed holds hope for a rational therapy for this common human disease, it would be very unfortunate if geographically isolated and unexplained claims of toxicity were to withhold xylitol from the medical armamentarium.

Acknowledgment

The original work was supported by grants of the John A. Hartford Foundation, Inc. and the World Health Organization.

References

Asakura, T., Adachi, K., Minakami, S., and Yoshikawa, H. (1967). Non-glycolytic sugar metabolism in human erythrocytes. Part I. Xylitol metabolism. *J. Biochem. (Tokyo)* **62,** 184–193.

Asakura, T., Ninomiya, H., Minakami, S., and Yoshikawa, H. (1969). Utilization of xylitol in animals and man. *In* "Metabolism, Physiology, and Clinical Use of Pentoses and Pentitols" (B. L. Horecker, K. Lang, and Y. Takagi, eds.), pp. 158–173. Springer-Verlag, Berlin and New York.

Asakura, T., Adachi, K., and Yoshikawa, H. (1970). Reduction of oxidized glutathione by xylitol. *J. Biochem.* (*Tokyo*) **67**, 731–733.

Bailey, J. M. (1959). A microcolorimetric method for the determination of sorbitol, mannitol, glycerol in biological fluids. *J. Lab. Clin. Med.* **54**, 158–162.

Bässler, K. H. (1969). Adaptive process concerned with absorption and metabolism of xylitol. *In* "Metabolism, Physiology, and Clinical Use of Pentoses and Pentitols" (B. L. Horecker, K. Lang, and Y. Takagi, eds.), pp. 190–196. Springer-Verlag, Berlin and New York.

Beutler, E., Duron, O., and Kelley, B. M. (1963). Improved method for the determination of blood glutathione. *J. Lab. Clin. Med.* **61**, 882–888.

Donahoe, J. F., and Powers, R. J. (1972). Biochemical abnormalities with xylitol. *N. Engl. J. Med.* **282**, 690–691.

Erdmann, G. (1969). The use of polyols in pediatrics. *In* "Metabolism, Physiology, and Clinical Use of Pentoses and Pentitols" (B. L. Horecker, K. Lang, and Y. Takagi, eds.), pp. 341–347. Springer-Verlag, Berlin and New York.

Fiske, C. H., and Subbarrow, Y. (1925). The colorimetric determinations of phosphorous. *J. Biol. Chem.* **66**, 375–400.

Igarashi, T., Shoji, T., Yamamoto, H., Tanabe, Y., Miyakawa, T., Nakajima, Y., and Kobayashi, M. (1971). Infusionsgeschwindigkeit und Verträglichkeit der Xylitlösung. *Z. Ernährungswiss., Suppl.* **11**, 20–28.

Keller, D. F. (1970). Glucose-6-phosphate dehydrogenase deficiency. A pharamacogenetic prototype. *CRC Crit. Rev. Clin. Lab. Sci.* **1**, 247–302.

Konrad, P. N., Richards, F., Valentine, W. N., and Paglia, D. E. (1972). γ-Glytamylcysteine synthetase deficiency, a cause of hereditary hemolytic anemia. *N. Engl. J. Med.* **286**, 557–561.

Kumagai, M. (1969). The clinical use of xylitol in pediatrics. *In* "Metabolism, Physiology, and Clinical Use of Pentoses and Pentitols" (B. L. Horecker, K. Lang, and Y. Takagi, eds.), pp. 348–355. Springer-Verlag, Berlin and New York.

Loos, J. A., and Prins, H. K. (1970). A mechanized system for the determination of ATP + ADP, 2,3-diphosphoglycerate, glucose-1,6-diphosphate and lactate in small amounts of blood cells. *Biochim. Biophys. Acta* **201**, 185–195.

Marks, P. A., and Banks, J. (1965). Drug-induced hemolytic anemias associated with glucose-6-phosphate dehydrogenase deficiency: A genetically heterogenous trait. *Ann. N.Y. Acad. Sci.* **128**, 198–206.

Petrackis, N. L., Wiesenfeld, S. L., Sams, B. J., Collen, M. F., Cutler, J. L., and Siegelaub, A. B. (1970). Prevalence of sickle cell trait and glucose-6-phosphate dehydrogenase deficiency. *N. Engl. J. Med.* **282**, 767–770.

Prins, H. K., and Loos, J. A. (1969). Glutathione. *In* "Biochemical Methods in Red Cell Genetics" (J. J. Yunis, ed.), pp. 115–137. Academic Press, New York.

Sauberlich, H. E., Judd, J. H., Nichoalds, G. E., Broquist, H. P., and Darby, W. J. (1972). Application of the erythrocyte glutathione reductase assay in evaluating riboflavin nutritional status in a high school student population. *Amer. J. Clin. Nutr.* **25**, 756–762.

Schumer, W. (1971). Adverse effects of xylitol in parenteral alimentation. *Metab., Clin. Exp.* **20**, 345–347.

Srivastava, S. K. (1971). Metabolism of red cell glutathione. *Exp. Eye Res.* **11**, 294–305.

Strehler, B. L., and McElroy, W. D. (1957). Assay of adenosine triphosphate. *In* "Methods in Enzymology" (S. P. Colowick and N. O. Kaplan, eds.), Vol. 3, pp. 871–873. Academic Press, New York.

Thomas, D. W., Edwards, J. B., Gilligan, J. E., and Lawrence, J. R., and Edwards, R. G. (1972a). Complications following intravenous administration of solutions containing xylitol. *Med. J. Aust.* **1**, 1238–1246.

Thomas, D. W., Gilligan, J. E., Edwards, J. B., and Edwards, A. G. (1972b). Lactic acidosis and osmotic diuresis produced by xylitol infusion. *Med. J. Aust.* **1**, 1246–1248.

van Eys, J., and Wang, Y. M. (1972). Xylitol toxicity. *N. Engl. J. Med.* **286**, 1162–1163.

Wang, Y. M., and van Eys, J. (1970). The enzymatic defect in essential pentosuria. *N. Engl. J. Med.* **282**, 892–896.

Wang, Y. M., Patterson, J. H., and van Eys, J. (1971). The potential use of xylitol in glucose-6-phosphate dehydrogenase deficiency. *J. Clin. Invest.* **50**, 1421–1428.

Wang, Y. M., King, S. M., Patterson, J. H., and van Eys, J. (1972). Xylitol toxicity in the rabbit. *Fed. Proc., Fed. Amer. Soc. Exp. Biol.* Abstr. No. 2873.

Wang, Y. M., King, S. M., Patterson, J. H., and van Eys, J. (1973). The mechanism of xylitol toxicity in the rabbit. *Metab., Clin. Exp.* **22**, 885–894.

World Health Organization. (1967). Standardization of procedures for the study of glucose-6-phosphate dehydrogenase. *World Health Organ., Tech. Rep. Ser.* **366**.

Yoshida, A., Beutler, E., and Motulsky, A. G. (1971). Human glucose-6-phosphate dehydrogenase variants. *Bull. W. H. O.* **45**, 245–253.

Yoshikawa, H. (1970). Non-glycolytic metabolism of carbohydrates in red blood cells. In "Metabolism and Membrane Permeability of Erythrocytes and Thrombocytes" (E. Deutsch, E. Gerlach, and K. Moser, eds.), pp. 37–42. Thieme, Stuttgart.

Sugars in the Oral Cavity

CHAPTER 36

Carbohydrates as a Causative Factor in Dental Caries: Epidemiological Evidence

A. L. RUSSELL

A. Mechanisms of Dental Caries

The initial lesion of dental caries always occurs on the external surface of the tooth. The basic mechanism is dissolution of the inorganic portion of tooth substance by acids evolved by bacteria from a suitable substrate, usually a carbohydrate (Kesel, 1948). To produce a lesion of caries *in vivo*, some defense must be provided against the powerful buffering action of saliva, and the organism must be one which can colonize at the site of attack. The principal thrust of dental caries research over the past few years has been in the direction of organisms capable of attaching themselves, by means of a sticky, adherent matrix called plaque, to the external surfaces of the teeth, particularly to the "smooth" surfaces, and most particularly in human caries to those smooth surfaces between abutting teeth. Some human tooth surfaces are rarely attacked by caries. The most susceptible areas are, in order: pits and fissures; interproximal tooth surfaces; and, later in life, the root surface after it has been exposed by recession of the gingivae (Backer Dirks, 1965). Once the lesion

635

has penetrated the enamel or cementum into the underlying dentin, the mechanism becomes complex, including attack and dissolution of the acid-resistant organic material (Engel, 1950) which makes up about 20% (dry weight) of the substance of dentin (Zipkin, 1969).

B. Studies with Animals

Forty years ago there was hope that the essential etiologic agent in caries had been discovered in *Lactobacillus acidophilus,* of the family Lactobacteriaceae (Jay 1938). This hope slowly faded, to be rekindled on discoveries that another member of the lactic acid family, *Streptococcus mutans,* could produce smooth surface caries in monoinfected experimental animals with production of a plaque composed principally of dextrans, and that these dextrans were evolved from sucrose and from no other substrate (Fitzgerald and Jordan, 1968).

As studied *in vitro,* or in laboratory animals born germfree and monoinfected with a strain of *S. mutans,* the processes involved may be outlined as follows: (a) the sucrose molecule is split into its glucose and fructose moities, which are polymerized into dextrans and levans (Ebert and Schenk, 1968); (b) the glucans principally implicated in the plaque of smooth surface caries are high molecular weight dextrans and have the property of agglutinating *S. mutans* in suspension (Gibbons and Fitzgerald, 1969); (c) adherance of this mass to the surface of tooth enamel seems to be mediated by a high molecular weight mucinlike polymer in saliva (Gibbons, 1970); and (d) acids (principally lactic), formed under the plaque, attack the mineral portion of tooth substance (Carlsson, 1968). Although dextrans are produced by *S. mutans* only from sucrose, the organisms can utilize other sources of energy; many or most strains can hydrolyze high molecular weight levan, (da Costa and Gibbons, 1968), and many or most evolve intercellular amylopectins which can be utilized in the absence of extracellular substrate (Gibbons and Banghart, 1968b). The organisms were first cultivated in glucose media (Edwardsson, 1968), and can maintain themselves on other sugars, particularly monosaccharides, for long periods, regaining their dextran-producing capability promptly upon the addition of sucrose to the media. This characteristic has led Gibbons and Banghart (1968b) to warn that

. . . individuals ingesting small amounts of dietary sucrose may harbour potentially cariogenic bacteria in an avirulent, non-encapsulated phase. A dietary alteration resulting in increased sucrose ingestion could then favor the selection of extracellular polysaccharide producing cells, which would lead to gelatinous plaque formation and rampant caries . . .

It must be reemphasized that the foregoing outline is a summary of the caries mechanism as it attacks smooth enamel surfaces in experimental animals. Less attention has been directed toward the processes involving caries of the exposed root surface. The exposed tissue is cementum, bonelike in its physical character, which can be colonized more readily than enamel by means of a less adherent polymer, such as a levan, by organisms such as *Odontomyces viscosus* which are not dependent upon sucrose for polymer formation (Gibbons and Banghart, 1968a).

The presence of a protective plaque is necessary to prevent the removal of the bacterial colony from an exposed surface of a tooth by the scouring actions of chewing and movements of the muscles of the cheeks, lips, and tongue. Less protection is necessary for maintenance of the bacterial colony lodged in the deep pits and fissures of tooth enamel found principally on the chewing surfaces of the molars and premolars. It would seem that dental caries in pits and fissures can be produced by any acidogenic, aciduric organism from any substrate which can be degraded to a pH of 4.5 or below.

It is clear that the principal agent in the etiology of dental caries is the cariogenic microflora. The role of carbohydrates in the disease process is that of supplying energy at the site of attack to the active organisms. When refined dextrose solution was fed only one mouth of albino rats joined in parabiosis, blood sugar was equivalent in each of the paired animals, but dental caries was rampant only in the mouth receiving the sugar (Kamrin, 1954). Dental caries cannot be induced in germfree animals no matter how "cariogenic" the diet on which they subsist (Orland *et al.*, 1954).

C. Studies with Human Beings

The epidemiological evidence relevant to the role of sugars in human dental caries is given below, along with an evaluation of the role of microflora in caries and, so far as possible, needs of the organisms for specific types of substrate. Most of the evidence which can be adduced is presumptive.

1. Variations in Caries Prevalence

There are variations, at least sixty-fold in magnitude, in the prevalence of dental caries. These variations are related to remoteness in both time and space (Russell, 1971). Lesions of caries are rare in remains of ancient man; a typical finding is that of Leigh, who studied 230 Egyptian skulls surviving from about 4000 B.C. (Leigh, 1935). Reports that tooth enamel of ancient man was no less defective than the enamel of current humans are likewise typical (Bodecker, 1930; Falin, 1961). When present-day populations are discovered with low caries prevalence, they are usually isolated from men of the Western world (Russell, 1966). In the words of Rosebury and Waugh (1939):

> . . . dental caries is distinctly more prevalent among natives from settlements in which contact with white men is relatively unrestricted and least prevalent among more primitive groups with a minimum of such contact. . .

If man is the vector of the disease, it would be expected that caries prevalence would increase in these former isolates in frequent contact with men of the outside world. Caries prevalence seems to be increasing, in fact, in many such isolated communities (Russell, 1966, 1971; Russell *et al.*, 1961). Although patterns of prevalence are incomplete, they are compatible with the hypothesis that dental caries is spreading via human carriers throughout the world, much as El Tor cholera is spreading today, but at a much slower rate of speed (Russell, 1971).

2. Dental Caries and Sugar

The relationship between general nutrition and dental caries, worldwide, is inverse. In general, those groups with the fewer nutritional deficiencies have the greater problems with dental caries (Russell, 1966). Small isolated groups with minimal caries experience are usually poorly fed (Mehta and Shroff, 1965; Winter, 1968). No consistent relationship between nutritional states and dental caries could be developed in the series of international nutrition surveys conducted by the Interdepartmental Committee on Nutrition for National Development (ICNND) during the 1960's (Russell, 1966). On the other hand, there was a clear relationship between national usage of sugar in the diet and dental caries experience in the populations surveyed by ICNND teams. Table I is an illustration of the observed relationship between numbers of decayed, missing, and filled (DMF) permanent teeth in civilians aged 20–24 years (Russell, 1966), and table usage of sugar in the populations from which they were drawn (Hand *et al.*, 1962). The finding is consistent with the conclusions of a nearly endless list of studies and observations (Winter, 1968) dating back at least to the time of Aristotle, who is said to have asked:

> Why do figs, when they are soft and sweet, produce damage to the teeth? (Guerini, 1909).

TABLE I

Mean Number of Decayed, Missing, and Filled
Permanent Teeth in Civilians Aged 20–24 Years[a]

Country	Mean DMF per person aged 20–24
Far East—6–16 kg/person/year	
Burma	0.9
Vietnam: hill tribes	0.6
coastal peoples	1.8
Thailand	1.1
Malaya	5.0
Near East—13–19 kg/person/year	
Jordan	1.5
Lebanon: refugees	1.1
civilians	3.0
South America—23–44 kg/person/year	
Ecuador	8.4
Trinidad	9.2
Chile	11.6
Colombia	12.6

[a] Examined in a series of international nutrition surveys (Russell, 1966) with per capita consumption of sugar per person per year (Hand *et al.*, 1962).

But dental caries can and does occur in peoples who have never used sugar or any other processed foodstuff (Anonymous, 1968; Sinclair *et al.*, 1950). Prevalence in such groups tends to be relatively light with most lesions occurring in pits and fissures. The addition of sugar to the diet of native children of Naurua seems to have had little effect, over a generation, upon their relatively low caries experience (Cadell, 1959). High caries experience has been reported for native adults of Easter Island, despite the fact that "candy and soft drinks are normally unavailable . . . ." (Taylor, 1966). Such reports as these underscore the importance of sugars in dental caries, but are inconsistent with the concept that caries may result from a single specific etiologic agent dependent upon one single type of substrate.

3. Prevalence of S. mutans

Attempts to associate the presence of *S. mutans* in the mouths of human subjects with caries experience have yielded equivocal results. Krasse and his co-workers found a positive relationship between the relative percentages of "caries-inducing" streptococci and caries activity in only one of three groups studied (Krasse *et al.*, 1968). "Cariogenic" streptococci were recovered from 69 and 59%, respectively, of children aged 12–14 years in Don Matias and Heliconia, Colombia, with a three-fold difference in caries experience (12.4 and 4.1 mean DMF teeth, respectively) (Jordan *et al.*, 1969). *Streptococcus mutans*-induced caries in animals can be prohibited absolutely with penicillin, but lesions of caries were found in 69 of 196 first molars which had erupted into the mouths of 49 children after they had begun a prophylactic regime of 200,000 units of penicillin per day by mouth (Littleton and White, 1964).

Still other findings are more compatible with the hypothesis that the cariogenic flora in humans can be heterogenous. Dental plaque can be inhibited by a dextranase in laboratory animals monoinfected with a strain of *S. mutans* (Fitzgerald *et al.*, 1968), or in human subjects selected because they "harbored the pathogens in sufficiently high numbers" (Keyes *et al.*, 1971), but dextranase appeared to be ineffective against plaque in subjects who had not been screened for presence of *S. mutans* strains (Caldwell *et al.*, 1971; Lobene, 1971). Human plaque is not only composed of "extremely complex, heterogeneous, dynamic microenvironments," but "many of the organisms present in plaque are very different morphologically and metabolically from similar organisms grown in culture" (Critchley and Saxton, 1970). Some recent evidence seems to reaffirm the role of lactobacilli as an etiologic agent in caries (Steinle *et al.*, 1967). It may be true that numbers of lactobacilli in saliva are not necessarily related to the numbers active in the carious lesion, but it is sometimes forgotten that the organisms studied in the early lactobacillus programs came not from saliva but from "direct inoculation of scrapings from carious teeth" (Hadley *et al.*, 1930).

Sucrose is clearly important in the metabolism of *S. mutans* and may be essential to its implantation in the mutans-free mouth (Krasse *et al.*, 1967), but human dental plaque can be formed from such varied sugars as glucose, xylitol, and fructose (Scheinin and Mäkinen, 1971). The bacterial mass of plaque is essential to the caries mechanism, but not all plaque is cariogenic. Littleton and his associates found ample plaque but low caries experience in 12 children so retarded that they were fed by stomach tube. When plaque material from their mouths was added to sucrose, there was very little drop in pH of the mixture over 30 min (Littleton *et al.*, 1967).

The evidence from laboratory and *in vivo* studies clearly implicates oral microorganisms as the primary etiologic factor in dental caries. The general relationship between the usage of sucrose and dental caries in populations is unmistakable. But this relationship might be a function of the popularity of sucrose as a dietary constituent rather than its essentiality as a cariogenic substrate. If another sugar were used more frequently and in greater amount than sucrose, that other sugar might well relate more closely to human dental caries. The substrate most frequently used in laboratory caries studies has been glucose. It must be concluded that oral plaque can be evolved from many sugars other than sucrose. Caries does occur in populations subsisting principally on starches from cereal foods or tubers, although the prevalence is usually low in such populations and the lesions tend to be limited to pits and fissures. Trisaccharides are considered not to be important in the etiology of caries.

D. Research Needs

The fact that so little can be concluded on the basis of epidemiological study is in itself a clear indication of the need for expanded research with human populations. Very little is known about the typical oral flora in caries-free individuals, partly because few such individuals occur in Western countries. Only a score or so of caries-free recruits come into such installations as the Great Lakes Naval Training Station during a given year, and about half of them incur one or more lesions of caries during their first 12 months in service. It would be helpful to know whether the typical oral flora is essentially different in those populations in Alaska, Asia, and Africa known to be relatively free of dental caries. There is no evidence to this point. Dental research has been done almost exclusively in those geographic areas where dental caries poses the greater problem. Equipment for anaerobic culture developed over the past few years now makes it quite practicable to isolate the "cariogenic" streptococci under austere field conditions.

Some questions will probably need to be answered through longitudinal studies and studies with adults, both more expensive and difficult than single-survey observations with children. Further investigations of cariogenic flora and its substrate requirements are clearly required to spell out the role of other substrates and other plaque polymers under the complex conditions found in the human oral cavity.

There are no epidemiologic data whatever on the prevalence or attack patterns of caries of the root surface. Conventional caries indices were evolved through studies with children; cemental caries occurs during the middle and later years of life after recession of gingivae has exposed a root surface. New indices, or modifications of the old, should be developed and applied in studies of middle-aged adults who still retain their own natural teeth.

E. Summary

Dental caries is the result of the action of oral microflora. It is most prevalent in the world's best fed populations and with those with the greatest amount of sucrose in the diet. Sucrose is not the only cariogenic carbohydrate. Most conclusions about the importance of one carbohydrate or another in living people must be tentative, pending more complete information on the distribution and energy requirements of cariogenic organisms in broad human populations, particularly those in which caries is now relatively low in prevalence.

References

Anonymous. (1968). Discovery of prehistoric cemetary remains reveals Ohio Indians with arthritis and bad teeth. *J. Amer. Dent. Ass.* **77**, 792–793.

Backer Dirks, O. (1965). The distribution of caries resistance in relation to tooth surfaces. *Caries-Resist. Teeth, Ciba Found. Sympo., 1964* pp. 66–83.

Bodecker, C. F. (1930). Concerning defects in the enamel of teeth of ancient American Indians. *J. Dent. Res.* **10**, 313–322.

Cadell, P. B. (1959). Dental conditions amongst native Nauruans. *Aust. Dent. J.* **4**, 389–394.

Caldwell, R. C., Sandham, H. J., Mann, W. V., Jr., Finn, S. B., and Formicola, A. J. (1971). The effect of dextranase on human dental plaque. 1. The effect of a dextranase mouthwash on dental plaque in young adults and children. *J. Amer. Dent. Ass.* **82**, 124–131.

Carlsson, J. (1968). Plaque formation and streptococcal colonization on teeth. *Odont. Revy* **19**, Suppl. 14.

Critchley, P., and Saxton, C. A. (1970). The metabolism of gingival plaque. *Int. Dent. J.* **20**, 408–425.

daCosta, T., and Gibbons, R. J. (1968). Hydrolysis of levan by human plaque streptococci. *Arch. Oral Biol.* **13**, 609–617.

Ebert, K. H., and Schenk, G. (1968). Mechanisms of biopolymer growth: The formation of dextran and levan. *Advan. Enzymol.* **30**, 179–221.

Edwardsson, S. (1968). Characteristics of caries-inducing human streptococci resembling *Streptococcus mutans*. *Arch. Oral Biol.* **13**, 637–646.

Engel, M. B. (1950). The softening and solution of the dentin in caries. *Amer. Dent. Ass. J.* **40**, 284–294.

Falin, L. I. (1961). Histological and histochemical studies of human teeth of the bronze and stone ages. *Arch. Oral Biol.* **5**, 5–13.

Fitzgerald, R. J., and Jordan, H. V. (1968). Polysaccharide-producing bacteria and caries. *In* "The Art and Science of Dental Caries Research" (R. S. Harris ed.), p. 79. Academic Press, New York.

Fitzgerald, R. J., Keyes, P. H., Stoudt, T. H., and Spinell, D. M. (1968). The effects of a dextranase preparation on plaque and caries in hamsters, a preliminary report. *J. Amer. Dent. Ass.* **76**, 301–304.

Gibbons, R. J. (1970). Microbial ecological models and dental diseases. "Research Perspectives in Dentistry," No. 9, pp. 72–88. Dental Research Institute and School of Dentistry, University of Michigan, Ann Arbor.

Gibbons, R. J., and Banghart, S. (1968a). Induction of dental caries in gnotobiotic rats with a levan-forming streptococcus and a streptococcus isolated from subacute bacterial endocarditis. *Arch. Oral Biol.* **13,** 297–308.

Gibbons, R. J., and Banghart, S. (1968b). Variations in extracellular polysaccharide synthesis by cariogenic streptococci. *Arch. Oral Biol.* **13,** 697–701.

Gibbons, R. J., and Fitzgerald, R. J. (1969). Dextran induced agglutination of *Streptococcus mutans* and its potential role in the formation of dental plaque. *Int. Ass. Dent. Res.* **47,** 55 (abstr.).

Guerini, V. (1909). "A History of Dentistry from the Most Ancient Times Until the End of the Eighteenth Centruy." Lea & Febiger, Philadelphia, Pennsylvania.

Hadley, F. B., Bunting, R. W., and Delves, E. A. (1930). Recognition of *Bacillus acidophilus* associated with dental caries: A preliminary report. *J. Amer. Dent. Ass.* **27,** 2041–2058.

Hand, D. B., Schaefer, E. E., and Wilson, C. E. (1962). A comparative study of food consumption patterns in Latin America, Middle Eastern and Far Eastern countries. Presented before *Int. Congr. Food Sci. Technol. 1st 1960.*

Jay, P. (1938). *Lactobacillus acidophilus* and dental caries. *Amer. J. Pub. Health Nat. Health* **28,** 6,759–61.

Jordan, H. V., Englander, H. R., and Lim, S. (1969). Potentially cariogenic streptococci in selected populations in the western hemisphere. *J. Amer. Dent. Ass.* **78,** 1331–1335.

Kamrin, B. B. (1954). Local and systemic cariogenic effects of refined dextrose solution fed to one animal in parabiosis. *J. Dent. Res.* **33,** 824–829.

Kesel, R. (1948). Caries of the enamel: Evidence for the decalcification theory. *J. Amer. Dent. Ass.* **37,** 381–90.

Keyes, P. H., Hicks, M. A., Goldman, B. M., McCabe, P. M., and Fitzgerald, R. J. (1971). The effect of dextranase on human dental plaque. 3. Dispersion of dextranous bacterial plaques on human teeth with dextranase. *J. Amer. Dent. Ass.* **82,** 136–141.

Klatsky, M. (1948). Studies in the dietaries of contemporary primitive peoples. *J. Amer. Dent. Ass.* **36,** 385–391.

Krasse, B., Edwardsson, S., Svenssen, I., and Trell, L. (1967). Implantation of caries-inducing streptococci in the human oral cavity. *Arch. Oral Biol.* **12,** 231–236.

Krasse, B., Jordan, H. V., Edwardsson, S., Svensson, I., and Trell, L. (1968). The occurrence of certain "caries-inducing" streptococci in human dental plaque material with special reference to frequency and activity of caries. *Arch. Oral Biol.* **13,** 911–918.

Leigh, R. W. (1935). Notes on the somatology and pathology of ancient Egypt. *J. Amer. Dent. Ass.* **22,** 199–222.

Littleton, R. W., and White, C. L. (1964). Dental findings from a preliminary study of children receiving extended antibiotic therapy. *J. Amer. Dent. Ass.* **68,** 520–525.

Littleton, R. W., Carter, C. H., and Kelley, R. T. (1967). Studies of oral health in persons nourished by stomach tube. I. Changes in the pH of plaque material after the addition of sucrose. *J. Amer. Dent. Ass.* **74,** 119–123.

Lobene, R. R. (1971). The effect of dextranase on human dental plaque. 2. A clinical study of the effect of dextranase on human dental plaque. *J. Amer. Dent. Ass.* **82,** 132–135.

Mehta, F. S., and Shroff, B. C. (1965). Aspects of dental diseases in the Indian aborigines. *Int. Dent. J.* **15,** 182–189.

Orland, F. J., Blayney, J. R., Harrison, R. W., Reyniers, J. A., Trexler, P. C., Wagner, M., Gordon, H. A., and Luckey, T. D. (1954). Use of the germfree animal technic in the study of experimental dental caries. I. Basic observations on rats reared free of all microorganisms. *J. Dent. Res.* **33,** 147–174.

Rosebury, T., and Waugh, L. M. (1939). Dental caries among Eskimos of the Kuskokwin area of Alaska. I. Clinical and bacteriological findings. *Amer. J. Dis. Child.* **4,** 871–893.

Russell, A. L. (1966). World epidemiology and oral health. *In* "Environmental Variables in Oral Disease," Publ. No. 81, pp. 21–39. Amer. Ass. Advan. Sci., Washington, D.C.

Russell, A. L. (1971). World survey on the epidemiology of dental caries, with emphasis on isolated populations. Presented before the Canadian Medical Expedition to Easter Island Symposium on Microbiology, University of Montreal.

Russell, A. L., Consolazio, C. F., and White, C. L. (1961). Dental caries and nutrition in Eskimo scouts of the Alaska National Guard. *J. Dent. Res.* **40,** 594–603.

Scheinin, A., and Mäkinen, K. K. (1971). The effect of various sugars on the formation and chemical composition of dental plaque. *Int. Dent. J.* **21,** 302–321.

Sinclair, B., Cameron, D. A., and Goldsworthy, N. E. (1950). Some observations on dental conditions in Papua-New Guinea, 1947, with special reference to dental caries. *Aust. Dent. J.* **22,** 58–93 and 120–57.

Steinle, C. J., Madonia, J. V., and Bahn, A. N. (1967). Relationship of lactobacilli to the carious lesion. *J. Dent. Res.* **46,** 191–196.

Taylor, A. G. (1966). Dental conditions among the inhabitants of Easter Island. *J. Can. Dent. Ass.* **32,** 286–290.

Winter, G. B. (1968). Sucrose and cariogenesis: A review. *Brit. Dent. J.* **124,** 407–411.

Zipkin, I. (1969). Biology of the oral environment. *In* "Dental Science Handbook" (L. W. Morrey and F. J. Nelsen, eds.), pp. 37–51. American Dental Association and National Institute of Dental Research, Washington, D.C.

CHAPTER 37

Sugars and the Formation of Dental Plaque

KAUKO K. MÄKINEN

The general increase in the world sugar consumption per capita has created several health problems such as obesity and possibly some other internal diseases, as well as diseases of the denture and its supporting tissues. The severe effects of the overindulgence of sugars on human health have been known for a long time, but the effects on oral tissues may have received too little attention, although it is these tissues which are the first to react to the intake of sugars. It is not reasonable for the different fields of medicine to compete in regard to their importance in the life and death of man because any disease in any human organ may, at least indirectly, be fatal. Thus, for example, an untreated carious lesion in man, which almost certainly is related to the intake of certain carbohydrates, and neglect of oral hygiene may lead to severe inflammation in the dental pulp. These two reasons may also lead to gingival inflammations of various degrees and to unnecessary tooth detachment. In some patients dental plaque, a gelatinous material on the teeth formed during sugar intake, may become calcified with the formation of dental calculus which is another oral disease. Dentists are well aware of the severity of all these diseases; for example, through the oral inflammatory lesions an open channel exists into the internal systems of man, which can be used by different invasive pathogenic microorganisms and other foreign particles or compounds. The extreme result of these processes may be lethal to the tissue or man.

The above deliberate, but not exaggerated, statements involved a specific mediator which causes the chemical reactions ultimately responsible for the symptoms described: it is the dental plaque, the formation of which is decisively dependent on certain low-molecular-weight saccharides, particularly of sucrose. Dental plaque has been shown to be so intimately associated with dental caries (and other oral pathological conditions) that it is no exaggeration to state that there is no caries without dental plaque. This review will deal with the formation of dental plaque when affected by dietary sugars.

A. Terminology Related to Dental Plaque Formation

With dental plaque one generally means the white, grayish, or yellowish and adherent gelatinous material covering the teeth as a result of food (and especially sugar) consumption and neglected oral hygiene. Dental plaque consists mainly of bacteria and to a lesser extent of leukocytes and other cells all embedded in an organic matrix of a known nature. It is formed on recently cleaned teeth within 24 hr in the absence of oral hygiene procedures. It may not be easily removable by rinsing with water but can be detached with no difficulty with dental instruments by scraping and, to a certain extent, from accessible surfaces by tooth brushing. Some authors consider that plaque continues over the gingival margin to cover the gingival epithelium and other oral surfaces. The composition and metabolism of plaque may be different in different parts of the denture. A plaque sample from a deep fissura differs from that of open surfaces. The plaque in the former locations may contain only a few viable bacterial cells, whereas the other tooth surfaces may provide more active bacterial metabolism.

The *acquired pellicle* is a bacteria-free film of a protein nature, of about 0.001 mm in thickness, which is derived from salivary glycoproteins precipitating and adsorbing onto the surface of the *hydroxylapatite crystals* of enamel. The phenomenon is most likely a normal physiological consequence of the secretary action of salivary glands and it may provide some protection against tooth degrading agents. The formation of the acquired pellicle is said to be independent of the action of oral microorganisms whereas that of dental plaque is dependent. However, some authors consider that the acids formed by microbial metabolism could enhance the precipitation of the glycoproteins on enamel. Consequently, the formation of the acquired pellicle is most likely indirectly dependent on oral bacteria and sugar intake.

The plaque may gradually calcify to form *calculus*. In effect, dental plaque is often considered as early calculus once definite crystals of calcium phosphate have been deposited in the plaque (Dawes, 1968). Schroeder and de Boever (1970) stated that the salivary exogenous dental cuticles are not a part of the dental plaque, but they also admitted that these cuticles might be regarded as a structural part of rather old microbial deposits. Schroeder and de Boever (1970) described the mediators of the plaque–tooth interface in more detail as follows: (1) remnants of the epithelial attachment lamina, (2) a salivary cuticle, or (3) microbial extracellular polysaccharides.

Whatever the dental exogenous coverings are called, they are formed on the hydroxylapatite crystals of enamel. Human dental enamel contains as much as 93–99.5% inorganic material (as a type of hydroxylapatite), depending on the assay method, and, consequently, only little organic material. The latter comprises peptides, proteins, citric acid, and further organic components of which there is growing information (cf. Schumacher and Schmidt, 1972).

Dental plaque can be divided into two compartments: the *cellular compartment,* comprising primarily the plaque bacterial cells, and the *extracellular matrix,* containing organic material derived from salivary glycoproteins and dietary sugars. These sugars are converted by the bacterial cells to one particular component of the matrix: *the extracellular polysaccharides* of which levan and dextran types are important. Furthermore, the dietary sugars may in part be converted to *intracellular* glycogenlike *storage polysaccharide* which can be utilized by the cells when the endogenous sugar source is depleted. The dietary sugars discussed in this review are primarily the disaccharides, sucrose, lactose, and maltose, and the monosaccharides, glucose, fructose, sorbitol, mannitol, xylitol, and related sugars. *Cariogenic microorganisms* utilize particularly sucrose, but also some other saccharides for growth.

B. Acquired Pellicle

1. General Characteristics

The acquired pellicle is not generally included with dental plaque material because of its diverse mechanism of formation when compared to that of plaque. However, in the evolution of dental plaque the pellicle formation is very often a necessary factor. The acquired pellicle underlines plaque, and in some areas of the tooth old plaque may become almost totally calcified, in which case the pellicle may become an integral part of the hardened plaque. Therefore, it is necessary to refer to the known events in the formation of the acquired pellicle. The role of dietary sugars in the growth of plaque can be more easily understood if the earlier steps in the evolution of plaque are first considered.

Before the eruption of teeth, the surfaces which later will be exposed to the oral cavity are covered by enamel integuments of embryological origin. After eruption the wearing out of the integuments takes place. However, the enamel surface will still be covered by a thin organic membrane, the acquired pellicle, which is usually about 0.001 mm or less in thickness. The formation of the acquired pellicle is easy to demonstrate if a tooth is abraded with a polishing agent (such as pumice) so that the hydroxylapatite crystals are exposed. It is the first structure which appears after various prophylactic measures (Frank and Houver, 1970). The acquired pellicle is deposited on the surface of the enamel within a few minutes, but at least within 24 hr, *in vivo*. The acquired pellicle may also develop extensions into defects of the enamel surface (Meckel, 1965). It is evident that in certain conditions bacteria-free and stained pellicles are formed which may differ in thickness and composition from that pellicle material which has usually been termed "acquired pellicle"; for example, Manly (1943) and Meckel (1965) have described such a pellicle of 0.001–0.010 mm thickness on enamel surfaces in some people. A separate "denture pellicle," a structureless and brownish deposit, has been described (Millin and Smith, 1961; Smith, 1964), which may be similar to the acquired pellicle.

2. Composition of the Acquired Pellicle

All pellicle material so far analyzed seems to have originated from salivary glycoproteins of which there are more than twenty different types. Saliva is rich in these glycoproteins whose carbohydrate components include, for example, sialic acid, fucose, galactose, glucose, mannose, and hexosamines. Pellicle material does not contain hydroxylysin or hydroxyprolin, indicating that pellicle proteins are not of a collagenous nature. Nor does the pellicle material contain sulfur-containing amino acids. This indicates that the pellicle is not of a keratinous nature. Because the above-mentioned denture pellicle contains traces of cystine and methionine, it is possible that the denture pellicle and the acquired pellicle are not of identical structure (Dawes, 1968).

Although the acquired pellicle is a rather homogeneous structure, it still contains some bacterial material and proteins of the gingival fluid and other compounds which are always present in the oral cavity. The nature of the pellicle is such that it is rather resistant to the action of proteolytic enzymes. The only effective means in the oral cavity which

can partially remove it, is mechanical abrasion. New material is, however, continuously formed on the enamel surface. The resistance to the
action of proteolytic enzymes is related to the fact that pellicle material
can be isolated from the tooth surface by treating the enamel in diluted
(e.g., 2%) hydrochloric acid. This treatment will dissolve the enamel underlying the pellicle. The organic film is loosened from the enamel surface. The acquired pellicle may undergo a type of maturation because
it may become completely insoluble in 2% hydrochloric acid after a
period of about 12 days (Turner, 1958).

3. Mechanism of Formation of the Acquired Pellicle

The mechanism of the formation of the acquired pellicle was elucidated
by Dawes (1964), Hay (1966, 1967), Turner (1958), McDougall (1963a),
Leach (1963), McGaughey and Stowell (1966, 1967), Rølla and Mathiessen (1970), Ericson (1968), Pearce and Bibby (1966), Mayhall (1970,
1971), and others. In a progressive sequence of studies it was found that
certain salivary proteins were selectively precipitated or adsorbed onto
tooth surfaces. Salivary Cl⁻ ions can have either an enhancing or an inhibiting effect on the adsorption, depending on conditions not so far determined (McGaughey and Stowell, 1967, 1971). Lowering of the pH of
saliva by the addition of hydrochloric acid is accompanied by a substantial increase in the adsorption of the proteins. A probable cause of plaque
matrix deposition is the selective precipitation of certain salivary glycoproteins by calcium ions. Evidence in favor of this concept is that early
plaque contains a high concentration of calcium. Histochemical studies
by McDougall (1963a,b) and Meckel (1965) have also suggested that
plaque matrix resembles salivary mucoid.

The adsorption of proteins to hydroxylapatite (through the affinity between calcium of the solid phase of the apatite and the free carboxylate
groups of the proteins of the soluble phase) has been shown to be responsible for the formation of the acquired pellicle. Rølla and Mathiessen
(1970) claimed that certain commercial dextrans were adsorbed more
strongly to protein-coated hydroxylapatite than to untreated (not
"coated" with salivary proteins) hydroxylapatite. It is therefore possible
that small amounts of bacterial dextrans may be incorporated into the
acquired pellicle. This may in part cause the adhesion of bacteria. The
importance of dietary sugars in the reactions taking place in the interface
of the pellicle and at the merging points of pellicle and plaque becomes
apparent in this way, because the dextrans are formed particularly from
dietary sucrose.

The precipitating material contains relatively large amounts of acidic amino acids, an indication that acidic proteins are selectively adsorbed. The adsorbed glycoprotein also contains sialic acid (Rølla *et al.*, 1969; Mayhall, 1971). Since sialic acid is in a terminal position in the glycoprotein and since it possesses an ionizable carboxylate group, it could react with the calcium ions of the apatite. Also, phytate (a hexaphosphate ester of myoinositol), which is a natural constituent of unrefined sugars and cereals, is adsorbed onto enamel crystals (Kaufman and Kleinberg, 1970; Jenkins *et al.*, 1959a,b; McClure, 1963; Dawes and Shaw, 1965). This binding is not peculiar to phytate, but it may be characteristic of certain phosphate compounds in general (e.g. metaphosphate and pyrophosphate) which show inhibition of dental caries.

It has been shown in several papers that the mixed microbial flora of human oral fluid is able to form several extracellular glycosidases capable of releasing the carbohydrate components from the salivary glycoproteins that make up salivary mucin. The same carbohydrates are not to be found in more than trace amounts in the dental plaque (Dawes and Jenkins, 1963; Leach, 1963; Middleton, 1964; Leach and Critchley, 1966; Rølla, 1966; Hay, 1967; Harrap, 1968). In the mixed flora the majority of microorganisms are able to degrade, to some degree, the glycoproteins of salivary mucin. The elaboration of these enzymes is not uncommon (Leach and Melville, 1970).

4. Adherence of Bacteria onto Pellicle

Jenkins (1968) and Critchley *et al.* (1968), who have suggested the importance of the deposition of salivary glycoproteins as an initiating force in the development of plaque, also suggested that the influence exerted by extracellular polysaccharides on plaque development is directed more toward subsequent development of plaque rather than toward its initiation. This view seems to be generally accepted. The coating with salivary glycoproteins (and some components of bacterial origin) prepares the teeth for bacterial colonization. However, it should be emphasized that the enamel pellicle is not present over the whole of the enamel surface since the cells of plaque can be in direct contact with the enamel (Frank and Brendel, 1966).

Hay *et al.* (1971) recently described a high-molecular-weight glycoprotein in human whole saliva which may enhance the aggregation of oral microorganisms onto the enamel surface. This factor contained almost 3% sialic acid. There are, however, only low amounts of sialic acid in plaque. The glycoprotein of Hay *et al.* (1971) was either a completely developed mucin (i.e., containing sialic acid which would be hydrolyzed by bacterial enzymes during or after the aggregation of the cells) or this particular glycoprotein is one of those very specific compounds accounting for the low content of plaque sialic acid.

Leach (1963) showed that incubation of saliva with neuraminidase (EC 3.2.1.18, *N*-acetylneuraminate glycohydrolase) caused the hydrolysis of sialic acid and the precipitation of the residual proteins. The bacterial neuraminidase may affect plaque formation. There is a low concentration of sialic acid in mixed plaque, and bacteria rapidly metabolize added sialic acid. This mechanism of protein precipitation may assist plaque growth once bacteria have become established. It is uncertain whether or not it could account for the initial deposition of a bacteria-free matrix. The mechanism could not, however, act in the deposition of the matrix in germfree animals since neuraminidase seems to be of bacterial rather than of salivary origin. It is not known which dietary sugars specifically lead to the growth of neuraminidase-producing microorganisms, but at least high sucrose intake and high neuraminidase activity in plaque are correlated.

The adsorption of bacteria on the pellicle depends on the nature of the capsular structure of the cells and on the type of sugar consumed. Not all types of cells are fastened onto the glycoprotein layer; for example, salivary proteins may inhibit the adsorption of cariogenic streptococci by hydroxylapatite, because of the shielding of the hydroxylapatite surface from the bacteria by the protein (McGaughey *et al.*, 1971). It has been shown with cariogenic *Streptococcus mutans* that salivary proteins can either enhance (Gibbons and Spinell, 1970) or inhibit (McGaughey and Field, 1969) the binding of the bacteria by the teeth or hydroxylapatite, depending on the presence or absence of an extracellular capsule.

A schematic summary of the role of pellicle formation in the evolution of dental plaque is presented in Fig. 1.

5. Significance of the Acquired Pellicle

Although certain controversy has existed as to whether or not the acquired pellicle acts as a protective coating of the enamel against caries, it may perhaps be concluded on the basis of several studies that the natural and physiological action of the pellicle is to protect (cf. Pinter *et al.*, 1969; Dawes, 1968). It should be emphasized that the glycoproteins of the salivary glands have been most likely continuously precipitating onto the enamel surface during the evolution of man. At least the salivary glycoproteins of today are continuously present in the oral cavity with enzymes capable of splitting them. This does not disagree with suggestions that certain outer acellular or cellular layers of enamel would also

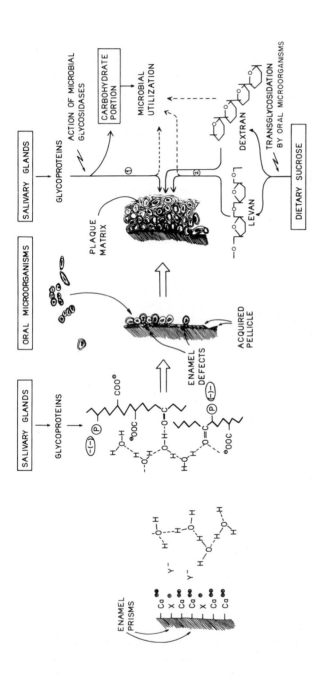

(a)

(b)

(c)

Fig. 1 Schematic presentation of the evolution of dental plaque, (a) The enamel surface is shown to have been abraded by a polishing agent. The hydroxylapatite crystals and other structural enamel compounds are exposed to the bathing influence of oral fluid. The enamel is shown to bear Ca^{2+} and other ions and groups with unknown structure (X). The salivary glycoprotein molecules and the enamel surface are covered by an icy water layer. The pellicle is formed by selective adsorption of salivary glycoproteins onto the crystal surface within a few minutes or hours. (b) Some oral microorganisms are adhered onto the surface of the pellicle (bacteria can also be found in direct contact with the enamel). The pellicle can either enhance or inhibit the adsorption of bacteria, depending on whether the cells are capsulated or not. The pellicle may extend into defects in the enamel surface. (c) The microbial cells of the plaque are multiplying. They elaborate glycosidases which release the carbohydrate portion of the secretory glycoproteins. If proper substrate (above all sucrose) of glycosyltransferases of cariogenic microorganisms is available, adhesive and hydrophilic polysaccharides are formed which together with the protein moiety of the gland product, form the plaque matrix. In the matrix, the most densely packed cells exhibit thicker cell walls and lower rate of division than in the outer layers. Plaque formation in fissures, interdental areas, and in the gingival crevices may differ considerably from that shown. Reactions (1) and (2) refer to processes necessary for the formation of the plaque matrix (the formation of the protein and carbohydrate constituents, respectively). The formation of the carbohydrate component evidently takes place

be important in creating resistance against caries. Pellicle formation has been discussed in reviews, for example, by Dawes (1968) and Jenkins (1969). It has become evident from most studies that in the growth of plaque, the pellicle formation is often an essential phenomenon which should be taken into account when considering the effect of dietary sugars on oral health.

C. Dental Plaque

1. General Composition and the Content of Microorganisms

The soft dental plaque is formed from bacterial cells, bacteria-free organic material, leukocytes, epithelial cell remnants, some inorganic constituents, and water. In spite of the presence of leukocytes and other material in plaque, its metabolism seems to be almost entirely bacterial. The earlier considerations differed from the present view; for example, Tonzetich and Friedman (1965) suggested that the main source of metabolic activity in saliva is the epithelial cell fraction or the leukocytes (Eichel and Lisanti, 1964). This view was challenged, and Molan and Hartles (1967) claimed that the glycolytic activity of saliva was mainly the result of its bacterial population. Kleinberg has come to the same conclusion in many papers. Also, Tatevossian and Jenkins (1969) showed that the activity is predominantly the result of the bacterial content; for example, epithelial cells did not show measurable oxygen uptake. In conclusion, the metabolism of dietary sugars in plaque is above all bacterial.

The total bacterial counts average about 2.5×10^{11} cells/gm of plaque. There are about 4.5×10^{10} anaerobic cells and about 2.5×10^{10} aerobic cells in 1 gm of plaque. Excluding the surface layers plaque can be considered essentially anaerobic (cf. Ritz, 1970). Strålfors (1956) has calculated that the enamel side of the plaque would be anaerobic at thicknesses at least >0.45 mm. It is evident that the absolute number of bacteria in plaque is somewhat higher than the amount of cultivable cells.

Tables I and II provide information on the occurrence of microorganisms in dental plaque. Theilade and Theilade (1970) estimated the amount of microorganisms in the dental plaque of students (Table II). Fusobacteria and filaments were found in low numbers. No *Spirilla* or spirochetes were found. Early plaque was dominated by streptococci, but the role of other organisms should also be investigated. The test material used represented a selective group of people but provided a rough idea

TABLE I

The Cultivable Microorganisms of Human Dental Plaque[a]

Microorganism	%	Microorganism	%
Streptococci (facultative)	27	*Veillonella*	6
Diphtheroids (facultative)	23	*Bacteroides*	4
Diphtheroids (anaerobic)	18	Fusobacteria	4
		Neisseria	3
Peptostreptococci	13	*Vibrios*	2

[a] From Gibbons *et al.* (1964).

of the occurrence of microbial cells in plaque. In old plaque lysis of bacterial cells is a normal phenomenon. Electron micrographs indicating this have been presented, for example, by Theilade and Theilade (1970). In the plaque studied by Saxton and Critchley (1970) most organisms were coccoidal in shape and filamentous organisms were observed only occasionally.

In 7- to 14-day-old human plaque the microorganisms occupy 70% and the extracellular space 30% of the area of plaque sections (Schroeder and de Boever, 1970). In older plaque there is an increasing number of unviable cells. In plaque younger than 14 days, cell division continues, however, at the base of the dental plaque. Although forming a minor portion in older plaque, the initial level of organisms such as *Neisseria* may decisively affect the rate of plaque formation over a 5-day period (Ritz,

TABLE II

Percentage Distribution of Microorganisms in
Gram-Stained Smears of 1- and 3-Day-Old Plaque
of 6 Persons[a]

Microorganisms	Day 1	Day 3
Gram-positive cocci	76–90	64–79
Gram-negative cocci	3–14	4–16
Gram-positive rods	4–8	3–7
Gram-negative rods	1–3	7–11
Filaments	0–1	0–5
Fusobacteria	0–3	1–6

[a] From Theilade and Theilade (1970).

1970). According to Bramstedt and Trautner (1971) aerobic *Neisseria* occurs in the surface layers of plaque, followed by facultative streptococci in the next deeper layer. In the innermost regions the strictly anaerobic *Veillonella* appear.

The relative portion of plaque microorganisms is altered when dietary circumstances are changed; for example, de Stoppelaar *et al.* (1970) observed during a carbohydrate-free period of 17 days that the percentage that *Streptococcus mutans* constituted in dental plaque of the total cultivable flora decreased to a very low or undetectable level, while simultaneously the percentage of *Streptococcus sanguis* increased. The percentage of bacteria producing intracellular polysaccharides was also lower during this period. Reinstitution of the normal diet (containing glucose and sucrose) led to almost normal original proportions. An inverse relationship between the occurrence of *S. mutans* and *S. sanguis* was observed.

2. The Ability of Microorganisms to Produce Plaque

Rosen (1969) described an experiment on gnotobiotic rats fed on either sucrose or glucose. The experiment showed that under certain circumstances less plaque-producing organisms may also be harmful on some areas of the teeth. Complete replacement of sucrose by glucose resulted in a higher incidence of dental caries in the test animal inoculated with *Lactobacillus casei, Streptococcus salivarius, Streptococcus faecalis,* and *Streptococcus* FA-1. Rosen found no observable accumulation of dental plaque. Caries was found only in the occlusal sulci. Rosen stated that these strains lacked plaque-forming potential under the conditions of the experiments. One could not expect them to initiate smooth surface caries, nor sucrose to be more cariogenic. It may be concluded that fissura caries was more easily induced by glucose than by sucrose.

Plaque was not produced by any of the streptococci studied by Sidaway (1970) unless carbohydrate was included in the medium. The only sugar to produce a significant increase in the rate of production of plaque, and the amount formed, was sucrose. The only difficulty in the interpretation of the results of such a comprehensive study as this is that the experiments were performed *in vitro* using the "artificial mouth." However, the simulation was evidently of high degree. All carbohydrates used (sucrose, glucose, and fructose at 5% concentration) were fermented and demineralization of teeth was usually evident within 3 days. The type of sugar fermented did not make a significant difference to the extent of the mineral lost.

Sidaway found that the lactobacilli and *Streptococcus mitis* used did not produce plaque in any of the mediums used (the basal medium was a solution of Oxoid Nutrient Broth No. 2, supplemented with 0.1% yeast extract). *Candida albicans, Leptotrichia dentium,* and *Actinomyces odontolyticus* did not produce any plaque, whereas *Nocardia* and *Coryne-bacterium* produced plaque in all media, as well as the true cocci. The presence of carbohydrates increased the amount of plaque formed. Coryne-bacteria in particular produced gross amounts of plaque in the presence of sucrose. In spite of the plaque-forming property of several oral micro-organisms, Sidaway stated that in the production of plaque in areas where the demineralization of enamel is usually found, the streptococci must play an important role.

The mutual criticism between authorized dental researchers indicates, for example, that the importance of extracellular dextrans in plaque growth is not known for sure. Indeed, it is at present difficult to estimate if the contribution of dextran-forming streptococci to the rate of plaque growth and to caries has been exaggerated. This view, however, is given some support from the results of Jensen *et al.* (1968), according to which subjects formed just as much plaque during a period of 22 days while using a vancomycin mouth rinse as they did while using a water mouth rinse. This suggests that dextran-forming streptococci (inhibited by vancomycin) had little or no role in affecting the rate of plaque forma-tion. Verifying experiments should be carried out. Although results from animal experiments are difficult to apply to humans, it is worth noting that *Streptococcus mutans* (strain OMZ 176E) produced in rats aggrega-tions of extracellular polysaccharides which contributed only little or nothing to the fissure content (Kalberer *et al.*, 1971).

Grenby (1971) draws attention to the result of his study that baboons fed for 6 months on a diet containing various sugars produced less plaque on sucrose diet than on a glucose or fructose diet. This indicates that the quality, and not necessarily the quantity alone, may determine the cariogeneity of plaque.

Polyols usually lead to lower bacterial growth in plaque (cf. Fig. 2). Although oral microorganisms may become adapted to grow in the pres-ence of polyols (cf. Mäkinen, 1972), Gehring (1971) has observed that in the mixed oral flora of human plaque and saliva it was not possible to demonstrate the existence of germs breaking down xylitol. This view does not necessarily differ from that of Mäkinen, because the results ob-tained in the author's laboratory (Mäkinen, 1971; Mäkinen and Scheinin, 1972) only maintain that the cells of *S. mutans* become adapted to *toler-ate* the presence of xylitol, whereas an adaptation to *metabolize* this sugar alcohol was only a possibility suggested.

The significance of cariogenic streptococci in the etiology of dental caries should not be overestimated. Other microorganisms may be equally

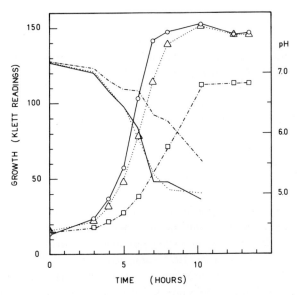

Fig. 2 Adaptation of the cells of a cariogenic streptococcus (*S. mutans,* strain Ingbritt) to utilize polyols (Mäkinen and Scheinin, 1972). The growth was followed in different culture media: (○) Growth in normal Trypticase Soy Broth, normal cells (stored in normal solid Trypticase soy broth); (△) growth of "xylitol cells" in xylitol-Trypticase Soy Broth (5% xylitol); and (□) growth of normal cells in xylitol-Trypticase Soy Broth (5% xylitol, before adaptation). Normal cells denote cells stored in solid Trypticase Soy Broth with monthly transfers into the same solid media. Xylitol cells refer to cells stored 9 months in Trypticase Soy Broth to which was added 5% xylitol. Descending curves: pH of the growth media; ascending curves; growth. Adaptation here indicates either an adaptation to tolerate the presence of xylitol or an adaptation to metabolize it.

important at different stages of caries (Larmas and Mäkinen, 1971), although it has been shown that there is a significant relationship between the plaque concentration of streptococcus and the subsequent dental caries experience. Larmas and Mäkinen (1971) found from *in vitro* studies that streptococci *Candida albicans* and two strains of *Lactobacillus casei* were able to soften human dentine rapidly.

The nonspecific, non-plaque-forming organisms may be active in the sulci only and the plaque-forming *S. mutans* may be active in other areas. McKendrick (1970), the chairman of a recent dental plaque symposium, mentioned that whereas streptococci can produce smooth-surface caries, lactobacilli may perhaps be the main causative organism in fissure caries. The mere presence of streptococci in large numbers does not necessarily make these bacteria the main causative factor in dental caries. McKendrick suggested that caution should be exercised in any future studies on the etiology of caries; for example, the predominant organism in the throat during diphtheria is said to be streptococcal, whereas the causative agent, *Corynebacterium diphteriae,* is present in relatively small numbers (McKendrick, 1970).

3. The Nongrowing Nature of Plaque Cultures

Most of the plaque mass on the surface of the teeth functions, at least most of the time, more like a nongrowing bacterial culture than a rapidly growing one (Tanzer *et al.,* 1969a,b). This may in part result from the amino acid deficiency in the deeper plaque layers. This suggestion was strengthened by the observations of Saxton and Critchley (1970) that only 4% of the coccoidal organisms in plaque exhibited signs of cell division in contrast to 35% of the organisms in exponential growth cultures. Also, Van Houte and Saxton (1971) and other authors have obtained results which support the assumption of slow streptococcal multiplication in plaque *in vivo.* Adhesive plaques *in vivo* do not increase in mass manyfold per day, in spite of the presence of a complex bacterial growth medium.

D. Effects of Sugars

The nutritional circumstances in the deeper layers of plaque may alter the morphology of the bacterial cells. The morphology of human dental plaque was studied by van Houte and Saxton (1971) with an electron microscope. In the inner part of the plaque, many microorganisms exhibited unusually thick cell walls and contained large amounts of intracellular polysaccharides. The magnitude of both phenomena diminished toward the saliva–plaque interface. Similar findings were earlier obtained by Critchley and Bowen (1970) with monkeys (*Macaca irus*). Sugar was found to be vital for the normal metabolism of the bacteria in the monkey plaque. The organisms starved of sugar had thin walls and less plaque was formed. In this study glucose facilitated the recovery of the bacteria more rapidly than sucrose, but both sugars yielded extracellular polysaccharides. The resulting compounds were found to be morphologically and chemically different.

The reduction in cell wall thickness and glycogen content occurred *in vitro* when certain amino acids were added to an otherwise adequate medium (Critchley and Bowen, 1970). This is in accordance with the results of Tanzer *et al.* (1969a,b). In plaque stained for polysaccharide material,

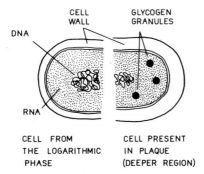

Fig. 3 Diagrammatic representation of some of the differences in morphology between microorganisms in the logarithmic phase of the growth and those present in plaque. The plaque cells (particularly in deeper layers of older plaque) exhibit less division and more stored intracellular polysaccharides. Adapted and slightly modified from Saxton and Critchley (1970).

the cytoplasm of the majority of organisms seemed to contain densely stained granules of intracellular polysaccharides. Although the peripheral organisms also contained intracellular polysaccharides, it was shown that in some plaque less polysaccharide was present in organisms at the periphery (Saxton and Critchley, 1970).

In the above-mentioned study of Saxton and Critchley (1970) the incubation of plaque in certain nutrient media was followed by incubation in Ringer's solution, which contained 2% sucrose. Transferring the plaque to this solution resulted in a rapid increase in both cell and wall thickness (from 22.5 to 29 nm, shown schematically in Fig. 3) in 30 min and synthesis of extracellular polysaccharides. Amino acids were claimed to be present only in very small amounts or totally absent from dental plaque, thus keeping the bacteria from growing rapidly in plaque. It is likely that relatively little protein in saliva can penetrate into the plaque. This means that for organisms comprising the greater part of the plaque, the plaque matrix probably functions as the major source of nutrient. However, compounds such as sucrose should have relatively free passage even into the deeper layers. This may be facilitated by the nonionic character of sucrose in contrast to amino acids, which in the pH range of the oral cavity are ionized and can be adsorbed to a variety of compounds already at the periphery of the plaque or at sites of formation. This view is supported by the suggestions of Hotz *et al.* (1972) that the high content of fermentable carbohydrates (5.6% of the plaque dry weight can be low molecular water-soluble carbohydrates) in dental plaque would allow normal microbial activity.

1. Composition of Dental Plaque as Affected by Dietary Sugars

According to the present knowledge the most interesting changes in the chemical composition of plaque affected by the consumption of sugars are those in the amount and nature of acids and polysaccharides. The gel-like plaque polysaccharides have been adopted by numerous laboratories as a main research subject. In this connection it is necessary to emphasize that it is the gel state which is most typical for these polysaccharides (Rees, 1969). These substances in plaque are able to retain water. Two- or three-day plaque is about 80% water. The organic and inorganic content depends upon the location of the plaque as well as upon the age (Dawes and Jenkins, 1962; Schroeder, 1963). The carbohydrate portion of the plaque extracellular matrix is largely derived from dietary sugars. Eastoe and Bowen (1971) concluded that salivary proteins rather than newly ingested food are the main immediate source of the entire protein content of dental plaque.

Naylor *et al.* (1970) showed that sucrose, glucose, fructose, glucose syrup, and a glucose–fructose mixture all supported plaque growth. This can be expected in view of the known nutritional requirements of plaque bacteria. Although these authors used one single type of microorganism (*Streptococcus mutans*, strain Ingbritt) in an "artificial mouth" system (Pigman *et al.*, 1952; Sidaway *et al.*, 1964), it may be of interest to emphasize that the plaque grown in the presence of sucrose (2%) contained more hexoses than the other plaques. Otherwise the plaques were similar (Table III). It was also found that much of the hexose in plaque was in the form of extracellular polysaccharides. Several authors have stressed

TABLE III

Mean Values for Plaque Produced by Various Sugars over 10 Days by an Experimental Plaque and Caries System[a]

Sugar tested	Total dry weight (mg)	% Dry weight	
		Total hexose	Total protein
Sucrose	28.37	23.22	26.73
Glucose	23.74	13.20	20.71
Fructose	25.67	8.74	30.93
Glucose syrup	22.43	16.73	25.86
Fructose-glucose	27.93	11.11	27.26

[a] From Naylor *et al.* (1970).

the lability of plaque levan. This lability of the plaque extracellular fructans is a known chemical fact and it can also explain the rather low figure for total hexose in the fructose plaque. The lability of plaque levan in relation to the development of caries will be dealt with later. McDougall (1964), Manly *et al.* (1966), and Wood and Critchley (1967) were among the first to show the presence of levan and dextran in plaque. The water-insoluble extracellular levans and dextrans are one of the structural components which occupy the intermicrobial space in dental plaques. There is evidence that suggests that the extent to which these components are found might be directly dependent on the intake of sucrose (Critchley *et al.*, 1967; Carlsson and Sundström, 1968). These polysaccharides are more prominent in the condensed microbial layer than at the surface of dental plaques. However, an excess of extracellular polysaccharide seems to be present at the surface of mature plaque based on electron micrographs.

No critical evaluation of the chemical data on plaque carbohydrates so far published is able to combine all results to a uniform presentation. This is because of (1) the great diversity of plaque samples studied, (2) the use of different isolation techniques for plaque fractions, and (3) the use of different chemical methods. The results have been expressed either per dry or fresh weight of whole plaque, per water-soluble or insoluble fraction, per dry weight of total carbohydrates and so on. The following examination includes a selection of papers.

Guggenheim's group (Hotz *et al.*, 1972) studied the plaque obtained from 3500 schoolchildren: 29.6% of plaque dry weight, containing 6.9% carbohydrates, 1.2% nitrogen, and more than 4% proteins, was water-soluble. The water-insoluble portion (67.1%), contained 11.3% carbohydrates and 7.4% nitrogen; in addition, 30.7% constituents insoluble in alkali (1 N KOH) were found. However, all subfractions prepared by ethanol precipitations contained substantial amounts of material other than carbohydrates. The sugar composition of acid hydrolysates revealed glucose as the main sugar constituent, but small amounts of pentoses, other hexoses, and disaccharides were also found. The water-insoluble plaque matrix polysaccharides (containing predominantly α-1,3 linkages) were calculated to account for 1.35% of the plaque dry weight; 5.6% of the plaque dry weight consisted of low molecular water-soluble carbohydrates. Glucose and oligosaccharides forming the bulk of this fraction were claimed to represent intermediates resulting from the enzymatic breakdown of α-1,6-linked dextran.

Higuchi *et al.* (1970) showed that dental plaque contained both dextran and levan which together accounted for an average of 4.2% by weight of dry plaque, levan comprising about one-eighth of the total polysaccharides. The results indicate that the activity of levan synthesis is higher than that of dextran in plaque, whereas the accumulation is slower, as discussed by Wood (1967), Wood and Critchley (1967), and McDougall (1964), who all reported that the amount of levan in dental plaque was about 5% of the total polysaccharides. However, the present investigation (Higuchi *et al.*, 1970) showed that the accumulation of levan in dental plaque may be somewhat greater than this. This may be correct because the other authors have used more vigorous conditions to hydrolyze levan than Higuchi *et al.* In the hydrolysis of levan it would be of the utmost importance to perform the hydrolysis so carefully that no levulinic acid or hydroxymethylfurfural is formed. The actual amounts of levan may in all cases be higher than reported.

Wood (1967, 1969) showed that 24-hr human dental plaque total hexoses content was almost 20% of the dry weight of the plaque. Approximately half of this hexose was extractable with water and alkali. After incubation at 37°C for 24 hr only 10% was estimated as hexose and again approximately 50% represented extractable hexoses. This loss was accompanied by the production of acid which, if not neutralized, partially inhibited the loss of hexose. These results showed the important fact that dietary sugar can be retained by human plaque as intra- and extracellular polysaccharides and that approximately half of this polysaccharide is metabolized in 24 hr. In the study of Wood (1969) monomeric glucose in plaque (percent dry weight of aqueous extract of plaque) ranged from 0.2 to 12.9.

In the study of Krembel *et al.* (1969) attention was paid to the analysis of plaque cellular and acellular fractions (Table IV). The values obtained for carbohydrates agree fairly well with those already given.

All the above-mentioned results can be compared to the results of Gib-

TABLE IV

Composition of Human Dental Plaque (Grown Approximately One Night)[a]

		% Dry weight			
Sample	Ash	Free lipids	Protein	Carbohydrates	UV-absorbing substances
Total plaque	10	10–14	40–50	13–17	10–15
Acellular fraction	15	26–30	6.7–7	31–41	2–6
Cellular fraction	11	1.3–5	40–70	7–14	10–15

[a]From Krembel *et al.* (1969).

bons and Socransky (1962). Plaque, incubated anaerobically for a 1-hr period in a buffer containing 0.5% glucose, accumulated approximately 250 μg of carbohydrate per milligram of plaque N. Figures 4 and 5 show results of two recent studies which indicated that xylitol diet reduced the amount of both plaque and carbohydrates of the matrix. The evolution of the "dextran era" in dentistry has been briefly discussed elsewhere (Mäkinen, 1972).

E. Plaque Polysaccharides

1. Heterogeneity

There are several points which can be made concerning the study of plaque carbohydrates. Above all it should be emphasized that the synthesized polysaccharides are not completely homogeneous products. Halhoul and Kleinberg (1970) showed that the type and heterogeneity

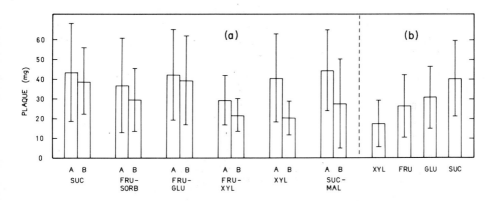

Fig. 4 Results from two recent studies showing the effect of xylitol and fructose to reduce the amount of dental plaque. (a) Scheinin and Mäkinen (1972). Five-day diet. The means and standard deviations of plaque fresh weight in different sugar groups determined at the end of period A and period B. Comparison between various sugar groups at period B: between sucrose and xylitol: **$p < 0.01$; between fructose-sorbitol and xylitol: °$p < 0.1$; between sucrose and fructose-sorbitol; **$p < 0.01$; between sucrose and sucrose-maltose: *$p < 0.05$; between sucrose and fructose-xylitol: **$p < 0.01$. Comparison between periods A and B: in fructose-xylitol group: **$p < 0.01$; in xylitol group: **$p < 0.01$. (b) Scheinin and Mäkinen (1971). Four-day diet. No analysis at the normal diet period was made. The means and standard deviations of plaque weight in different sugar groups. Comparison between fructose and sucrose groups: *$p < 0.05$; between xylitol and sucrose groups: ***$p < 0.001$.

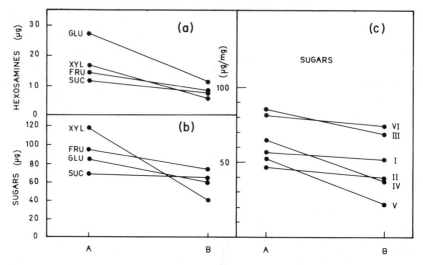

Fig. 5   Results from two recent plaque studies showing the relatively low amount of total sugars or hexosamines in the extracellular plaque material of test persons who were on a mild xylitol diet (and totally neglected their oral hygiene). (a,b) Changes of the amount of total sugars (determined as D-glucose with the Anthrone reagents; not with orcinol as erroneously mentioned earlier) and hexosamines (determined as D-glucosamine) in partially purified, dried, and pooled polysaccharide material, expressed as micrograms of sugars obtained from each dry polysaccharide pool. The pools represent the partially purified polysaccharide fraction of plaque extracellular material of four groups of test persons who were on the sugar diets indicated for 4 days. A, normal diet; B, the particular sugar diet (Scheinin and Mäkinen, 1971). (c) Change of the amount of total sugars (Anthrone) in pooled plaque water extracts between periods A and B. The plaque material in each sugar groups was pooled and suspended in cold water containing 0.9% NaCl. The suspensions were centrifuged and the supernatants were analyzed. For details, see Scheinin and Mäkinen (1972). The sugar groups in (c) (5-days diet) were: I, sucrose; II, fructose-sorbitol, 1:1; III, fructose-glucose, 1:1; IV, fructose-xylitol, 1:1; V, xylitol; VI, sucrose-maltose, 9:1.

of polyglucose and polyfructose is determined by the concentration of substrate (sucrose) and probably pH. Furthermore, the presence of sucrose greatly affects the stability of at least the levan component of plaque. Halhoul and Kleinberg (1972) showed than an acidic pH and continual presence of sucrose would favor accumulation of extracellular glucose and fructose polymers by reducing their rates of degradation.

Monkey plaque polysaccharides are also heterogeneous. The polysaccharide formed in the monkey plaque on glucose feeding did not resemble dextran in the electron microscope (Critchley and Bowen, 1970). The compound was termed a heteroglucan composed of glucose, galactose, hexosamine, and other sugars (Critchley, 1969). The extracellular polysacsaccharide preparation of two human caries-inducing streptococci (Ing-britt and Mats) (Dahlqvist *et al.*, 1967) contained 79–80% glucose and 17–18% fructose (the preparation being a mixture of glucan and fructan), but no hexosamines or uronic acids. The polysaccharides synthesized from sucrose by glycosyltransferases of oral streptococci were more homogeneous: 2–4% moisture, 85–98% total carbohydrate, 78–98% glucose, 0.9–2.6% fructose, and 0.5% nitrogen (Newbrun, 1972). The structures of the dextrans and levans are also controlled by the type of enzyme and initial receptor molecules involved or by the addition of branched low-molecular-weight polysaccharides which produce products of low molecular weight. Some dextrans may have D-fructose end groups as a result of sucrose acting as a receptor to initiate dextran synthesis. In conclusion, it should be emphasized that the plaque dextrans or levans do not represent any single compound of identical and constant structure. Their molecular weight and content of different glycosidic linkages vary considerably.

The purified polysaccharide preparations may contain impurities. However, Bramstedt and Lusty (1968) reported phosphate in a component of purified polysaccharide from plaque streptococci. The extracellular water-soluble polysaccharides have often been neglected. The presence of extracellular water-soluble polysaccharides in the extracellular space has been shown biochemically by several authors (Gibbons and Banghart, 1967; Dahlqvist *et al.*, 1967; Wood, 1967). Although the water-insoluble extracellular polysaccharides have gained much attention in dental research, other components of the extracellular space may be of equal importance.

The formation of plaque polysaccharides as a whole is a function of the changes in the environment of the cells and may vary greatly; for example, the production of extracellular polysaccharides by cariogenic streptococci may be subject to considerable variation. Such a variation observed in *Streptococcus mitis* (strain S3) raises the possibility that individuals ingesting small amounts of dietary sucrose may harbor potentially cariogenic bacteria in an avirulent noncapsulated phase. A dietary alteration resulting in increased sucrose ingestion could then favor the selection of extracellular polysaccharides producing cells, which would lead to gelatinous plaque formation (Gibbons and Banghart, 1967).

In his thorough review Kleinberg (1970) referred to a few papers on polysaccharides in plaque. An important conclusion from these was that levan is the minor component of the soluble polysaccharides in plaque matrix. Levan usually accounted for 1–2% of the total plaque. However, as stated above, the lability of levans and the presence or absence of sucrose at the moment of isolation may affect the results.

2. Formation of Plaque Polysaccharides

The plaque microbiota cannot only ferment single dietary sugars to acids. There are at least two separate mechanisms whereby these sugars are retained in the plaque in the form of polysaccharides which are broken down at a later time. The breakdown of these storage polysaccharides appears to be inhibited by their own end products. The ability of low-molecular-weight dextrans to reduce the gelatinous plaque formation has also been shown. The storage compounds are extra- and intracellular polysaccharides, the formation of which is a part of the growth of the dental plaque. Both types of compounds have been suggested to play an essential role in the etiology of caries. At least the extracellular ones may cause inflammatory reactions in gingiva and periodontium as well. The polysaccharides formed in the plaque are specific to the organism producing them. The microbial colonies appear to produce their own environment in the plaque by synthesizing their own particular types of polysaccharides.

a. Intracellular Polysaccharides

Unlike mammalian tissues and yeast, the synthesis of intracellular glycogen-type polysaccharides takes place with ADP-D-glucose as the donor of the D-glucosyl residues (Nikaido and Hassid, 1971). Under the normal physiological conditions some glycogen molecules or molecular fragments, or at least maltose, are probably always present, serving as a primer in the reaction catalyzed by glycogen synthetase. Intracellular polysaccharides in plaque cells have been demonstrated by several authors, for example, Saxton and Critchley (1970), who consider them at least partly as energy source.

Hammond (1971) has described intracellular polysaccharide production by human oral strains of *Lactobacillus casei*. The polysaccharides were formed when growing the cells in the presence of glucose, maltose, lactose, and fructose. They were found to be polymers of glucose linked by α-1,4-glycosidic bonds. The role of these polysaccharides as energy storage compounds in older and innermost plaque cells has been considered earlier.

b. Extracellular Polysaccharides

It is well known that on diets containing sucrose, a gelatinous and voluminous plaque develops on the tooth surface (Carlsson and Egelberg, 1965). Although evidently all plaque bacteria can also multiply on the natural components of the oral fluid, the plaque formation is, however, greatly increased when including higher amounts of carbohydrate in the diet (Keyes and Likins, 1946; van Houte, 1964). Some of these dietary carbohydrates lead to the formation of extracellular levans and dextrans in plaque. It has become clear that all the cariogenic streptococci so far isolated have the capacity to form dextran from sucrose. Although this property seems to be associated with cariogenicity, some microorganisms with this capacity are, however, incapable of producing caries in animals. Particularly *Streptococcus mutans* is known to produce extracellular dextran of plaque matrix (Edwardsson, 1968). Adhesive, high-molecular-weight dextrans of limited solubility are in general required for the formation of streptococcal plaque.

The available evidence seems to indicate that most of the known cariogenic microorganisms produce abundant amounts of extracellular polysaccharides (notably dextrans) when grown in the presence of sucrose, but that virtually none is produced from other sugars (Fitzgerald and Jordan, 1968). Even though it has been shown that certain caries-producing streptococci can ferment glucose, fructose, lactose, and maltose *in vitro* (Guggenheim *et al.*, 1965), in some systems sucrose is essential for plaque formation *in vivo* and *in vitro*. Robrish *et al.* (1972) demonstrated with three strains of *Streptococcus mutans* that the polysaccharide formation from sucrose was almost exclusively extracellular.

Almost all cariogenic hamster and rat strains uniformly (van Houte *et al.*, 1970) and consistently produced large amounts of iodophilic polysaccharides (IPS). However, noncariogenic hamster strains were less stable and produced IPS-positive and IPS-negative colonies. Almost all unidentified IPS-producing strains of streptococci from human carious lesions were found to be stable IPS producers, while most strains obtained from plaque of caries inactive persons were IPS producers. *Streptococcus mutans*, in contrast to other streptococcal species, was found characteristically to synthesize large amounts of IPS in a stable manner. This ability may significantly enhance acid production by these organisms and then contribute to its high cariogenic potential.

Jordan and Keyes (1966) reported that in an artificial mouth streptococci, found to be cariogenic in rats and hamsters, developed colonies in plaque, while noncariogenic strains did not. Other requirements for colony development were the presence of salivary glycoproteins and sucrose. Other carbohydrates were not so effective. Although these results were obtained in an artificial system, they still demonstrate that the salivary glycoproteins (involved in pellicle formation) are needed in plaque development.

After intake of sucrose a significant amount of the intermicrobial material in dental plaque is made up of levan (McDougall, 1964; Manly *et al.*, 1966). This polysaccharide may subsequently be utilized by the plaque flora, resulting in prolonged acid production periods in dental plaque (cf. Gibbons and Banghart, 1967; Wood, 1967; Carlsson and Sundström, 1968; daCosta and Gibbons, 1968). Both soluble and insoluble levan may be produced with only the former being degraded by oral levan-using organisms in the oral cavity. The levan may provide an additional substrate for other potentially cariogenic members of the oral flora (Gaffar *et al.*, 1970).

Streptococcus salivarius also produces levan but, since this organism is largely found on the tongue and is normally present at very low concentrations in dental plaque (Gibbons *et al.*, 1964), it may not contribute very much to the levans of plaque.

Carlsson and Egelberg (1965) confirmed that plaque will also develop in humans on a carbohydrate-free diet but that considerably larger amounts of plaque were formed when sucrose, rather than glucose, was the major carbohydrate in the diet. In a number of subjects on the sucrose diet the initial "plaque colonies" that developed resembled colonies of *S. salivarius*, in which much of the colony could produce an extracellular levan.

Most lactobacilli, particularly the heterofermentative strains, are likely to produce extracellular homopolysaccharides which resemble dextrans (Pedersen and Albury, 1955; Dunican and Seeley, 1965). Also, Gibbons and Banghart (1967) have shown that a cariogenic strain of *Lactobacillus acidophilus* produced an extracellular polysaccharide which was antigenically similar to the dextrans of the known cariogenic bacteria.

Hammond (1969) found with a strain of *Lactobacillus casei*, isolated from human gingival crevice, that the cells synthesized an adherent, dextranlike extracellular polysaccharide in which glucose units were the only components. The linkages found were of the type of α-1,6 and α-1,2. The synthesizing enzyme (dextransucrase) was constitutive and was produced in growing cells from glucose as well as sucrose, although sucrose was the preferred donor. This dextran was distinct from the capsular heteropolysaccharide produced by some *L. casei* strains, but was claimed to be identical to the dextrans of several cariogenic streptococci. This comparison was based on immunodiffusion. The strain produced 324 μg of dextran per milliliter of a standardized cell suspension of 300 Klett units from sucrose. The corresponding values (in micrograms) for other sugars were: 272 for raffinose, 150 for maltose, 144 for lactose, 96 for glucose, 70 for fructose, and 46 for the Rogosa base alone.

A recent study of Fry and Grenby (1972) suggested that prolonged reduction in sucrose intake might diminish plaque formation and in turn this might possibly reduce the caries rate. Soluble carbohydrate and phosphorus content of plaque was raised above normal levels on the low sucrose diet, but there was no significant difference in total carbohydrate, soluble levan, or calcium content of the plaque produced on low and normal sucrose diet.

3. Stability of Matrix Polysaccharides

Dextran is relatively resistant to hydrolysis in plaque (Leach, 1970). The finding that only a small amount of the extractable plaque hexose is generally in the form of glucose supports the idea that dextran is not metabolized very rapidly in the plaque. Furthermore, dextrans produced by carious strains of streptococci seem to be more resistant to hydrolysis than dextrans produced by noncariogenic streptococci.

Leach (1970), in a review of the biochemistry of dental plaque, emphasized the fact that the formation of dental plaque is related to a multiplicity of enzymes of bacterial origin in the oral cavity. He obtained evidence that certain enzyme-controlled reactions are responsible for the formation of both the major protein and carbohydrate components of the plaque matrix (cf. Fig. 1). According to him, enzymes responsible for the protein components are significantly active outside the plaque, whereas those responsible for the carbohydrate components exert their activities within the plaque. Several oral microorganisms synthesize enzymes which are able to hydrolyze the bond between the protein and carbohydrate moieties of the salivary glycoproteins. The carbohydrate components of the plaque matrix were considered to have been derived from dietary sucrose or other low-molecular-weight saccharides. Sucrose is, however, not unique among the small-molecular-weight sugars in its conversion to extracellular polysaccharides (Leach, 1970). The majority of the commonly available sugars, when exposed over an extended period of time to the oral organisms, were also claimed to be converted to extracellular polysaccharides.

The arguments set out by Leach (1970) included the statement that the inert extracellular polysaccharide (most likely of the dextran type) produced from sucrose reduces rather than promotes dental caries and that the levan component, which is labile in the plaque, might be more significant in the carious process. The plaque levan was found to be labile in all instances when the excess sucrose had been removed. The proportion of levan that was formed was found to increase as the sucrose concentration was increased. This was supported by the determination of the higher value of K_m for levansucrase than for dextransucrase (Hehre, 1961). It was stated that sticky sweets in direct contact with the plaque should favor the formation of levan. The slightly increased cariogenecity of sucrose compared to the other sugars could be explained by the lability of levan.

Amyloglucosidases were found to be largely ineffective in hydrolyzing bacterial glucans (Newbrun, 1972). Dextranase hydrolyzed various glucans at significantly different rates. The rate and degree of cleavage appeared to be related to the antigenic grouping of the strain. The relative resistance of some glucans to degradation may explain the ineffectiveness of the dextranase mouthwash in human clinical trials (Newbrun, 1972).

F. The Role of Sucrases, Invertases, and Hydrolases

1. Sucrases

In the development of the extracellular matrix of plaque the following two reactions are important:

$$n\ C_{12}H_{22}O_{11} + HOR \rightarrow H(C_6H_{10}O_5)_nOR + n\ C_6H_{12}O_6 \tag{1}$$
$$\text{Sucrose}\quad\text{Acceptor}\quad\text{Dextran}\qquad\quad\text{Fructose}$$

$$n\ C_{12}H_{22}O_{11} + HOR \rightarrow H(C_6H_{10}O_5)_nOR + n\ C_6H_{12}O_6 \tag{2}$$
$$\text{Sucrose}\quad\text{Acceptor}\quad\text{Levan}\qquad\qquad\text{Glucose}$$

Sucrose is converted to dextrans and levans in the dental plaque by various species of *Streptococcus* and *Lactobacillus*. Phosphate has not been shown to play any part in the synthesis, and the overall reaction in plaque seems to be irreversible. The corresponding enzymes, dextransucrase (EC 2.4.1.5; sucrose 6-glucosyltransferase, or α-1,6-glucan:D-fructose 2-glucosyltransferase) and levansucrase, have been demonstrated in cariogenic and noncariogenic microorganisms. The presence of membrane-bound and soluble dextransucrases has been demonstrated in the cells of *Streptococcus mutans* (strain Ingbritt) (Guggenheim and Schroeder, 1967). Such enzymes were earlier found in the genera *Leuconostoc* (Neely, 1960).

Newbrun and Carlsson (1969) showed the dextransucrase-catalyzed glucosyl transfer from sucrose to a growing chain of polyglucose. The enzyme was highly specific with regard to glucosyl donor, and it could utilize no substrate other than sucrose, but the glucosyl group could be transferred to a variety of acceptor molecules (sugars). Dextrans themselves formed an efficient class of acceptors. Various dextrans as well as maltose significantly increased the reaction rate, while levan, raffinose, and two isomers of sucrose phosphate had no effect on the reaction.

Streptococcus sanguis constitutes a large proportion of the microorganisms of the teeth (Carlsson, 1965). Dextransucrase of this organism was partially purified by Carlsson *et al.* (1969). This enzyme synthesizes a water-insoluble polyglucose from sucrose with 50% α-1,6 linkages. The enzyme was found to bind nonspecifically to a variety of cross-linked polymers. When bound to cross-linked dextrans the enzyme retained its activity. The enzyme did not release glucose when incubated with sucrose, indicating an absence of levansucrase and β-fructofuranosidase in the preparation. The following points, according to Carlsson *et al.* (1969), may be important when considering the role of the enzyme in plaque formation:

1. The range of optimal pH and temperature of the enzyme coincided with that of the oral cavity. (The determination of these parameters took place *in vitro*. This does not mean that the same pH optimum and temperature optimum would apply in the oral cavity as for the isolated enzyme in a reaction mixture.)

2. The comparatively low K_m indicated that dextran can be synthesized at the maximum rate at relatively low concentrations of sucrose. (This may indeed be the case for the enzyme reaction described, although a low K_m *in vitro* does not necessarily mean that a low amount of sucrose would lead to maximum rate of transglucosidation by the cells in the oral cavity. A low K_m may primarily indicate considerably high affinity between the enzyme and the substrate and not necessarily that the breakdown of the enzyme–substrate complex into free enzyme and reaction products would be high.)

3. The enzyme is adsorbed to dextran and retains its activity, suggesting that it may be kept in active form in the dental plaque.

4. The enzyme is constitutive.

The levansucrase of *Streptococcus mutans* was partially purified by Carlsson (1970) from cells cultured in a glucose broth. The substrates included sucrose and raffinose but not lactose, maltose, glucose, or fructose. In addition to levan (molecular weight greater than 25×10^6) free fructose and oligosaccharides were formed from sucrose. A similar high molecular weight has been suggested for levan from *Streptococcus salivarius* (Newbrun and Baker, 1968).

Further investigation (Newbrun, 1971) of the extracellular glucosyltransferase activity of *Streptococcus sanguis* (strain 804) revealed several glucosyltransferase enzymes with isoelectric points at pH 7.9, 6.4, and less often at about 4.5. The principal enzyme fraction, Ip 7.9, was shown to be a pure homogeneous protein moving as a single active band by polyacrylamide gel electrophoresis. The equilibrium of the reaction catalyzed by the glucosyltransferase Ip 7.9 was strongly in favor of polysaccharide formation (based on the observed disappearance of sucrose and the appearance of reducing sugars). No reversibility of the reaction to form sucrose from fructose and dextran could be demonstrated.

Challacombe *et al.* (1972) suggested that antibodies to glucosyltransferase of cariogenic microorganisms might prove a valuable measure in preventing dental caries. This would be a promising means which, of course, has also received much opposing opinion.

2. Invertases

Intertase (β-fructofuranosidase) is one of the transfructosylases which are also termed "sucrases." Consequently, the above-mentioned dextransucrases (transglycosylases) and levansucrases (transfructosylases) belong to the same enzyme group as the invertases, which may be called transfructosylases acting on sucrose. The possible significance of invertase in the metabolism of plaque has been dealt with earlier (Mäkinen, 1972). Plaque and salivary invertase are induced on a sucrose diet. The specific activity of invertase in the oral cavity is lower in persons on other sugar diets (Figs. 6 and 7). High invertase (as well as dextranase) activity in plaque and saliva can be claimed to indicate the presence of a higher number of microorganisms capable of utilizing sucrose. Unpublished re-

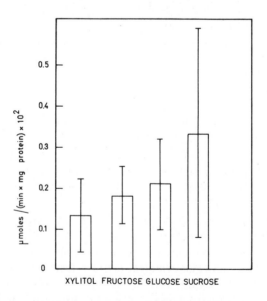

Fig. 6 The means and standard deviations of the activity of β-fructofuranosidase of centrifuged oral fluid (mixed saliva). Comparison between xylitol and sucrose: **p < 0.01; between fructose and sucrose: *p < 0.05 (Mäkinen and Scheinin, 1971).

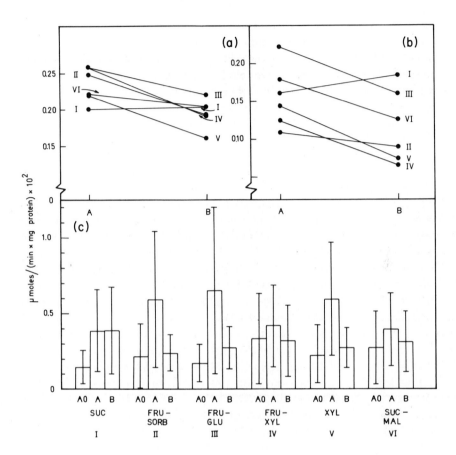

Fig. 7 Results from two recent plaque studies showing the induction of β-fructo-furanosidase and its high specific activity in some oral samples of persons who were on usual sucrose diet for 5 days (and neglected their oral hygiene) when compared to the effect observed with other diets which deliberately were made mild (Mäkinen and Scheinin, 1972). (a) The activity of β-fructofuranosidase in the supernatant fluid of plaque suspensions between periods A (normal diet) and B (the particular sugar diet). (b) Dextranase activity of the supernatant fluid of plaque suspension [as in (a)]. The neglect of oral hygiene had induced the formation of β-fructofuranosidase (the change from ΛO to A) and the specific activity was not reduced in the sucrose group when transferring from period A to period B. (c) The means and standard deviations of β-fructofuranosidase activities of centrifuged oral fluid (mixed saliva) determined at period AO (a control period with normal oral hygiene and dietary habits), period A (sucrose diet of 5 days and neglected oral hygiene), and B (the particular sugar diet and neglected oral hygiene). The sugar groups were the same as mentioned in Fig. 4. Comparison between periods A and B in xylitol group: **$p < 0.01$; in fructose-sorbitol group: *$p < 0.05$; in fructose-glucose group: *$p < 0.05$.

sults from the author's laboratory have shown that the induction of invertase (or rapid division of cells producing invertaselike activity) takes place within a few hours in the plaque of schoolchildren after sucrose intake but not after fructose intake. The recent results of Tanzer and Brown (1973) described an inducible invertase in the cells of *Streptococcus mutans*.

The animal experiments of Critchley and Bowen (1970) showed that if plaque were starved of sucrose (as a result of tube feeding) and then given this sugar after the starvation, the plaque organisms recovered very slowly. The authors correctly suggested that the difference in the response of the plaque bacteria to sucrose and glucose may be explained by the fact that glucose is considered as prime energy substrate for most bacteria. On the other hand, sucrose in plaque should undergo hydrolysis before transfer into the cells. The slow formation of polysaccharides observed in this case supports the idea that the activity of invertase is very low or absent in plaque formed in the absence of sugar (or better, sucrose). Critchley and Bowen referred to an analogous finding by Strange (1961) who discovered that starved cells of *Aerobacter aerogenes* failed to produce extracellular β-galactosidase.

Enzyme preparations obtained from sonically treated washed cells of *Streptococcus mutans* (strain GS5) have been found to contain sucrase activity which was distinct from that of the glucosyl- and fructosyltransferases known to be elaborated by this organism (Gibbons, 1972). This sucrase activity was formed constitutively and it could be separated from transferase activity by chromatography on agarose. The enzyme appeared to be very similar to invertase on the basis of its catalytic action, molecular size, and heat sensitivity. Its presence in cells of *S. mutans* (strain GS5) indicates that this organism was not solely dependent upon the glucosyl- and fructosyltransferases for the utilization of sucrose for growth. The total observable sucrase activity of glucose-grown cells of *S. mutans* (strains GS5 and 6715) was found to increase markedly when the organisms were treated with toluene. This may indicate that the cells were relatively impermeable to sucrose. Untreated sucrose-grown cells exhibited 10-fold higher total sucrase activity than did untreated glucose-grown cells. This suggests that the growth in sucrose broth resulted in the induction of a sucrose transport system.

3. Hydrolases

It may be that plaque which contains bacterial sucrases and polysaccharides also contains enzymes capable of hydrolyzing the polymers; for example, cariogenic streptococci, which synthesize extracellular dextrans also synthesize dextranase. The consumption of sucrose led to higher dextranase activity in plaque than that of other dietary common sugars (Fig. 7) (Mäkinen and Scheinin, 1972). For levanase there is at present no report available. The topical application of dextranase has received much attention as a possible clinical means in the removal of dental plaque. Some of the controversial opinions dealing with this problem have recently been discussed (Mäkinen, 1972).

A particular type of glucanohydrolase, mutanase, has been described by Guggenheim's group (cf. Guggenheim *et al.*, 1972). This enzyme (α-1,3-glucan 3-glucanohydrolase, free of 1,6 activity) inhibited fissura caries in rats while dextranase had no significant cariostatic effect.

A large proportion of the cultivable bacteria present in human dental plaque was found to hydrolyze levan of high molecular weight. All the levan-hydrolyzing organisms encountered proved to be streptococci which formed extracellular polysaccharides when grown in sucrose broth. Caries-inducing streptococci were also found to hydrolyze levan (daCosta and Gibbons, 1968). The fructan hydrolase of one strain was studied. The enzyme was inducible and it was formed in broths containing levan, inulin, or sucrose but not in broths containing glucose or fructose.

G. Metabolic Traits of Plaque as Affected by Dietary Sugars

Finally, the following metabolic characteristics of human dental plaque are briefly emphasized:

Plaque formed from sucrose generally has a greater acid-producing capacity than that formed from glucose and other sugars. However, under certain laboratory conditions acids are formed by plaque more rapidly from glucose or fructose or certain other monosaccharides than from sucrose. This most likely results from the time required for the action of invertase and related enzymes on sucrose to yield glucose and fructose (Skinner and Naylor, 1972).

Sorbitol is metabolized normally by plaque samples, although it may hinder the mixed oral flora. Plaque ferments mannitol and sorbitol mostly by glycolysis. Isolated plaque streptococci produce lactate, ethanol, and formate as the most important final products of this metabolism (Stegmeier *et al.*, 1971). Also, Brown and Patterson (1973) reported significant ethanol production with *S. mutans* grown in the presence of mannitol and sorbitol. The results of Stegmeier *et al.* (1971) and Brown (Brown and Wittenberger, 1972) agree also in other aspects: The cells fermented both polyols by a pathway which involved phosphorylation of the substrates prior to their oxidation. The hexitol phosphate dehydrogenases were inducible (Brown and Patterson, 1972).

Sandham and Kleinberg (1969) found that if formate is an end product of glucose catabolism in plaque, it would be rapidly converted to carbon dioxide through the action of formate dehydrogenase, hydrogenlyase, etc.

The only weakness in the use of the suspended salivary sediment system (SSS) as a good substitute for plaque is the possibility that the suspended material does not represent the plaque *in situ*, where separate plaque layers with different aerobicity and other properties can be encountered. The suspensions represent mixtures where the integrity of the plaque structure has been destroyed. To be honest, this is done, of course, in most studies, excluding perhaps the use of certain types of histochemical or electron microscopic techniques. Still, the studies carried out by Kleinberg and his co-workers have provided much valuable evidence on the plaque metabolism. Their statement "the metabolism of plaque and SSS is similar" is in many respects valid.

Using the SSS system Kleinberg showed that lactate was an intermediate not an end product of glucose metabolism. Lactate was not a substrate from which significant amounts of sediment carbohydrate could be formed (Sandham and Kleinberg, 1969).

According to the hetero acid idea of Kleinberg (Sandham and Kleinberg, 1970), acetate and propionate were more important in determining the minimum pH of plaque (cf. Fig. 8).

Under anaerobic conditions the resting cells of *S. mutans* (strain PK-1) metabolized glucose via the Embden-Meyerhof pathway. About two moles of lactate were formed for each mole of glucose consumed. Under aerobic conditions the greater part of glucose was also metabolized by the Embden-Meyerhof pathway to pyruvate, but the pyruvate was converted not only to lactate, but to acetoin, carbon dioxide, and acetate. Under aerobic conditions a part of the glucose consumed was considered to be metabolized via the hexose monophosphate shunt but none via the tricarboxylic acid cycle or the Entner-Doudoroff pathway (Yamada *et al.*, 1970).

It seems obvious that while the effect of dietary sugars on the metabolism of isolated plaque microorganisms can be easily studied, and is actually rather well known, the behavior of plaque as an entity is not a sum of these isolated metabolic traits; for example, it has been shown that acetate and propionate were end products of the glucose metabolism when *Veillonella* was mixed with *Streptococcus salivarius*, whereas in pure cultures the final acid found with the latter cells was lactate (un-

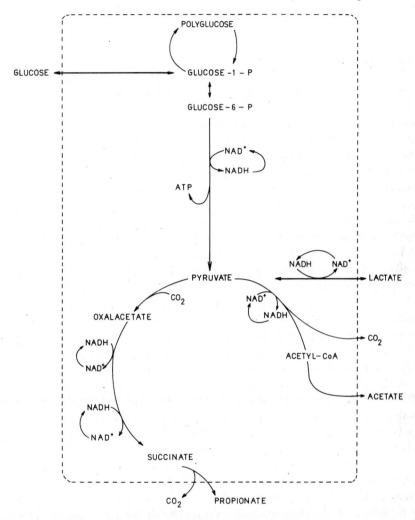

Fig. 8 Proposed scheme for the regulation of the formation of lactate, acetate, and propionate acids from glucose by a SSS system. The reactions requiring the $NAD^+/NADH + H^+$ system would include: (1) glyceraldehyde-3-P $+ NAD^+ + P_i \rightarrow$ 1,3-diphosphoglycerate $+ NADH + H^+$, (2) pyruvate $+ NADH + H^+ \rightarrow$ lactate $+ NAD^+$, (3) pyruvate $+ CoASH + NAD^+ \rightarrow$ acetyl-CoA $+ CO_2 + NADH + H^+$, (4) oxalacetate $+ NADH + H^+ \rightarrow$ L-maltate $+ NAD^+$, and (5) fumarate $+ NADH + H^+ \rightarrow$ succinate $+ NAD^+$. According to Standham and Kleinberg (1970).

published results of Ng and Hamilton, quoted by Sandham and Kleinberg, 1970). Sandham and Kleinberg referred also to a related finding of Hungate (1962). The composition of true plaque is very complex when compared to that of cultures of single bacterial species.

H. Research Needs

The study of the mechanism of the propagation of caries and periodontal disease has passed through numerous eras with regard to which chemical reactions or physical phenomena have been considered as the most important causative agents. Many have considered the dental decay as a simple acid dissolution of an inorganic matrix and others have also pointed out the inhibitory action of sucrose on a possible hormone, diffused from the pulp to the enamel, which should in part account for the mineralization. Although the above-mentioned oral diseases are now considered as infectious, it has not been possible to provide the final answer as to what the ultimate causative reactions are (cf. Mäkinen, 1972), even though the intake of sucrose can be regarded as one necessary exogenous factor.

Meanwhile, nutritional scientists have organized systematic searches for proper alternatives which could replace sucrose as a sweetener and energy source. Whatever the possibilities may be, they should all be *physiological*. Any sucrose substitute should normally occur in the metabolism of man and in various natural products. However, a mere sugar change may not be enough without proper oral hygiene measures and the use of fluorine prophylactics.

Although a certain type of plaque is formed in persons whose diet is deficient in sugars, it is nevertheless a fact that most dental plaque is formed when diets contain certain sugars, especially sucrose. Consequently, research in the field of nutritional dentistry (for want of a better expression) should be directed to elucidate the following points:

1. To study which of the naturally occurring sugars are least cariogenic and able to induce least periodontal diseases in man. This includes the study of the following problem:

2. Which of the natural sugars (normally occurring in human metabolism) produces least plaque and which is qualitatively least harmful to the teeth and its supporting tissues. This again requires the elucidation of the following problem.

3. What are the particular qualitative traits of dental plaque which make it caries-inducing? Is it the levan content and the lability of levan? Is it, simultaneously, the effect of acetate and propionate instead of lactate? Is it the content of hydrolytic enzymes in the plaque, in direct contact with the enamel organic matrix, as suggested on several occasions, for example, by Mäkinen (cf. Mäkinen, 1972), or is it the presence in plaque of bacterial toxins? But the following points should also be studied.

In what way (s) do dietary sugars contribute to the properties of plaque in different locations on the teeth? Are polyols able to produce, on a long-term diet, extracellular polysaccharides or other compounds harmful to the teeth? In what way does the consumption of polyols alter the oral microflora? As a direct consequence of the preliminary short-term dietary studies (Scheinin and Mäkinen, 1971, 1972), an extensive 2-year study with more than 130 test persons, divided into sucrose, fructose, and sucrose groups, is now being carried out in the authors' laboratory in order to provide in part an answer to these questions. The very preliminary results of this study (after a 7-week diet) have shown that the amount of plaque in the xylitol and fructose groups was reduced when compared to the sucrose group (Scheinin and Mäkinen, 1974).

Whatever the final mechanism of the carious process will be, the dietary sugars will be included in it. While awaiting the answer, we may feel satisfied when aware of the fact that the continuous acquisition of various organic material (lipids, polysaccharides, etc.) of saliva, food, or bacterial origin modifies the form of early enamel caries, contributing to the increased caries resistance and pigmentation of old teeth (Bibby, 1971).

Dental plaque is a gel-like adhesive structure on the teeth, which consists largely of oral microorganisms embedded in an extracellular matrix. Plaque is developed either directly onto the dental enamel surface, or, as is usually the case, onto a thin (approximately 0.001 mm) film, the acquired pellicle, which mainly consists of salivary glycoproteins (mucins) adsorbed and precipitated on the hydroxylapatite crystals of enamel. The plaque extracellular matrix contains proteins derived from the salivary mucins and adhesive gel-like polysaccharides (chiefly of the dextran and levan type) derived from dietary sugars (especially sucrose) through the action of microbial glycosyltransferases. The plaque microorganisms also act on the mucins with the aid of their glycosidases, liberating the protein moiety for the development of plaque extracellular matrix.

It has been almost unequivocally proved that sucrose leads in man to more voluminous mass of adhesive plaque and higher caries incidence than other sugars. Glucose may, under certain conditions, also cause considerable plaque growth, but other disaccharides, fructose, and polyols (sugar alcohols) generally lead to slow plaque growth. Of the polyols, xylitol in particular is able to inhibit the growth of cariogenic streptococci, as well as to reduce the amount of plaque extracellular polysaccharides, and the activity of certain sucrose-utilizing plaque enzymes. Oral cariogenic microorganisms may become adapted at least to tolerate the presence of xylitol and to form acids from other sugars simultaneously present. This type of adaptation is, however, not harmful, because the present information suggests that the type and amount of polysaccharides produced extracellularly are simultaneously altered to a more inoccuous form.

Plaque in different locations on the teeth may have different composition, mode of development, and contribution to caries. In addition to plaque mass, its qualitative properties are important for the propagation of caries. Plaque containing levan may be more harmful than plaque containing occasionally only small amounts of levan. The consumption of sucrose and its continuous presence in plaque leads to the retaining of levan in plaque. The mere acid production in plaque may not be the key process, but only when combined with the simultaneous formation of extracellular polysaccharides does the acid formation become important. The necessary industrial facilities for the production of such dietary sugars exist, and according to present dental views, these would be more healthy than sucrose and glucose.

References

Bibby, B. G. (1971). Organic enamel material and caries. *Caries Res.* **5**, 305–322.

Bramstedt, F., and Lusty, C. J. (1968). The nature of the intracellular polysaccharides synthesised by streptococci in the dental plaque. *Caries Res.* **2**, 201–213.

Bramstedt, F., and Trautner, K. (1971). Zuckeraustauschstoffe und Biochemie der Zahnplaques. *Deut. Zahnaerztl. Z.* **26**, 1135–1144.

Brown, A. T., and Patterson, C. E. (1973). Ethanol production and alcohol dehydrogenase activity in *Streptococcus mutans. Arch. Oral Biol.* **18**, 127–131.

Brown, A. T., and Patterson, C. E. (1972). Heterogeneity of *Streptococcus mutans* strains based on their mannitol-1-phosphate dehydrogenases: Criterion for rapid classification. *Infec. Immunity* **6**, 422–424.

Brown, A. T., and Wittenberger, C. L. (1972). Mannitol and sorbitol catabolism in *Streptococcus mutans. Arch. Oral Biol.* **17**, (in press).

Carlsson, J. (1965). Zooglea-forming streptococci resembling *Streptococcus sanguis* isolated from dental plaque in man. *Odontol. Revy* **16**, 348–358.

Carlsson, J. (1970). A levansucrase from *Streptococcus mutans. Caries Res.* **4**, 97–113.

Carlsson, J., and Egelberg, J. (1965). Effect of diet on early plaque formation in man. *Odontol. Revy* **16**, 112–125.

Carlsson, J., and Sundström, B. (1968). Variations in composition of early dental plaque following ingestion of sucrose and glucose. *Odontol. Revy* **19**, 161–169.

Carlsson, J., Newbrun, E., and Krasse, B. (1969). Purification and properties of dextransucrase from *Streptococcus sanguis. Arch. Oral Biol.* **14**, 469–478.

Challacombe, S. J., Lehner, T., and Guggenheim, B. (1972). Serum and salivary antibodies to glucosyltransferase in dental caries in man. *Nature (London)* **238**, 219.

Critchley, P. (1969). The formation of extracellular polysaccharides from glucose in monkey plaque *in vivo. Caries Res.* **3**, 205–206.

Critchley, P., and Bowen, W. H. (1970). Correlation of the biochemical composition of plaque with the diet. *In* "Dental Plaque" (W. D. McHugh, ed.), pp. 157–169. D. C. Thomson & Co., Ltd., Dundee, Scotland.

Critchley, P., Wood, J. M., Saxton, C. A., and Leach, S. A. (1967). The polymerisation of dietary sugars by dental plaque. *Caries Res.* **1**, 112–129.

Critchley, P., Saxton, C. A., and Kolendo, A. B. (1968). A histology and histochemistry of dental plaque. *Caries Res.* **2**, 115–129.

daCosta, T., and Gibbons, R. J. (1968). Hydrolysis of levan by human plaque streptococci. *Arch. Oral Biol.* **13**, 609–617.

Dahlqvist, A., Krasse, B., Olsson, I., and Gardell, S. (1967). Extracellular polysaccharides formed by caries-inducing streptococci. *Helv. Odontol. Acta* **11**, 15–21.

Dawes, C. (1964). Is acid-precipitation of salivary proteins a factor in plaque formation? *Arch. Oral Biol.* **9**, 375–376.

Dawes, C. (1968). The nature of dental plaque, films, and calcareous deposits. *Ann. N.Y. Acad. Sci.* **153**, 102–119.

Dawes, C., and Jenkins, G. N. (1962). Some inorganic constituents of dental plaque and their relationship to early calculus formation and caries. *Arch. Oral Biol.* **7**, 161–172.

Dawes, C., and Jenkins, G. N. (1963). Studies related to the formation of dental plaque. *IADR Program Abstr. Pap.* No 362.

Dawes, C., and Shaw, J. H. (1965). Dietary phosphate supplementation and its effects on dental caries and salivary and serum concentrations of calcium and inorganic phosphate in the rat. *Arch. Oral Biol.* **10**, 567–577.

de Stoppelaar, J. D., van Houte, J., and Backer Dirks, O. (1970). The effect of carbohydrate restriction on the presence of *Streptococcus mutans, Streptococcus sanguis* and iodophilic polysaccharide-producing bacteria in human dental plaque. *Caries Res.* **4**, 114–123.

Dunican, L., and Seeley, H. W. (1965). Extracellular polysaccharide synthesis by members of the genus *Lactobacillus:* Conditions for formation and accumulation. *J. Gen. Microbiol.* **40**, 297–308.

Eastoe, J. E., and Bowen, W. H. (1971). Effects of changes in feeding on the amino acid composition of protein in dental plaque from the monkey, *Macaca irus. Caries Res.* **5**, 101–110.

Edwardsson, S. (1968). Characteristics of caries-inducing streptococci resembling *Streptococcus mutans. Arch. Oral Biol.* **13**, 637–646.

Eichel, B., and Lisanti, V. F. (1964). Leucocyte metabolism in human saliva. *Arch. Oral Biol.* **9**, 299–314.

Ericson, T. (1968). Adsorption to hydroxylapatite of proteins and conjugated proteins from human saliva. *Caries Res.* **1**, 52–58.

Fitzgerald, R. J., and Jordan, H. V. (1968). Polysaccharide-producing bacteria and caries. *In* "The Art and Science of Dental Caries Research" (R. S. Harris, ed.), pp. 79–86. Academic Press, New York.

Frank, R. M., and Brendel, A. (1966). Ultrastructure of the approximal dental plaque and the underlying normal and carious enamel. *Arch. Oral Biol.* **11**, 883–912.

Frank, R. M., and Houver, G. (1970). An ultrastructural study of human supragingival dental plaque formation. *In* "Dental Plaque" (W. D. McHugh, ed.), pp. 85–108. D. C. Thomson & Co, Ltd., Dundee, Scotland.

Fry, A. J., and Grenby, T. H. (1972). The effects of reduced sucrose intake on the formation and composition of dental plaque in a group of men in the Antarctic. *Arch. Oral Biol.* **17**, 873–882.

Gaffar, A., Coleman, E. J., Marcussen, H., and Kestenbaum, R. C. (1970). Effects of inoculating a levan-forming cariogenic streptococcus on experimental caries in hamsters. *Arch. Oral Biol.* **15**, 1393–1396.

Gehring, F. (1971). Saccharose und Zuckeraustauschstoffe im mikrobiologischen Test. *Deut. Zahnäerztl. Z.* **26**, 1162–1171.

Gibbons, R. J. (1972). Presence of an invertase-like enzyme and a sucrose permeation system in strains of *Streptococcus mutans. Caries Res.* **6**, 122–131.

Gibbons, R. J., and Banghart, S. B. (1967). Synthesis of extracellular dextran by cariogenic bacteria and its presence in human dental plaque. *Arch. Oral Biol.* **12**, 11–24.

Gibbons, R. J., and Socransky, S. S. (1962). Intracellular polysaccharide storage by organisms in dental plaque. Its relation to dental caries and microbial ecology of the oral cavity. *Arch. Oral Biol.* **7**, 73–80.

Gibbons, R. J., and Spinell, D. M. (1970). Salivary induced agglutination of plaque bacteria. *IADR Program Abstr. Pap. No.* 157.

Gibbons, R. J., Socransky, S. S., deAraujo, W. C., and Van Houte, J. (1964). Studies of the predominant cultivable microbiota of dental plaque. *Arch. Oral Bio.* **9**, 365–370.

Grenby, T. H. (1971). Dental plaque studies on baboons fed on diets containing different carbohydrates. *Arch. Oral Biol.* **16**, 631–638.

Guggenheim, B., and Schroeder, H. E. (1967). Biochemical and morphological aspects of extracellular polysaccharides produced by cariogenic streptococci. *Helv. Odontol. Acta.* **11**, 131–152.

Guggenheim, B., König, K. G., and Mühlemann, H. R. (1965). Modifications of the oral bacterial flora and their influence on dental caries in the rat. I. The effects of inoculating 4 labelled strains of streptococci. *Helv. Odontol. Acta* **9**, 121–129.

Guggenheim, B., Regolati, B., and Mühlemann, H. R. (1972). Caries and plaque inhibition by mutanase in rats. *Caries Res.* **6**, 289–297.

Halhoul, M. N., and Kleinberg, I. (1970). Effect of sucrose concentration on polymer formation by salivary sediment. *IADR Abstr.* No. 587.

Halhoul, M. N., and Kleinberg, I. (1972). Effect of pH and sucrose on the degradation of extracellular glucose and fructose polymers by salivary sediment. *IADR Abstr.* No. 343.

Hammond, B. F. (1969). Dextran production by a human oral strain of *Lactobacillus casei. Arch. Oral Biol.* **14**, 879–890.

Hammond, B. F. (1971). Intracellular polysaccharide production by human oral strains of *Lactobacillus casei. Arch. Oral Biol.* **16**, 323–338.

Harrap, G. J. (1968). Glycoside hydrolases in whole human saliva. *J. Dent. Res.* **47**, 984.

Hay, D. I. (1966). The adsorption of saliva proteins onto apatite and enamel powder. *IADR Program Abstr. Pap. No.* 457.

Hay, D. I. (1967). The adsorption of salivary proteins by hydroxylapatite and enamel. *Arch. Oral Biol.* **12**, 937–946.

Hay, D. I., Gibbons, R. J., and Spinell, D. M. (1971). Characteristics of some high molecular weight constituents with bacterial aggregating activity from whole saliva and dental plaque. *Caries Res.* **5**, 111–123.

Hehre, E. J. (1961). Levansucrase. *In* "Biochemists' Handbook" (C. Long, ed.), pp. 514–515. Spon, London.

Higuchi, M., Iwami, Y., Yamada, T., and Araya, S. (1970). Levan synthesis and accumulation by human dental plaque. *Arch. Oral Biol.* **15**, 563–567.

Hotz, P., Guggenheim, B., and Schmid, R. (1972). Carbohydrates in poded dental plaque. *Caries Res.* **6**, 103–121.

Hungate, R. E. (1962). The physiology of growth. *In* "The Bacteria" (I. C. Gunsalus and R. Y. Stanier, eds.), Vol. 4, p. 95. Academic Press, New York.

Jenkins, G. N. (1968). The mode of formation of dental plaque. *Caries Res.* **2**, 130–138.

Jenkins, G. N. (1969). Composition and formation of dental plaque. *Parodontologie* **23**, 30–38.

Jenkins, G. N., Forster, M. G., Speirs, R. L., and Kleinberg, I. (1959a). The influence of the refinement of carbohydrates on their cariogenicity. *In vitro* experiments on white and brown flour. *Brit. Dent. J.* **106**, 195–208.

Jenkins, G. N., Forster, M. G., and Speirs, R. L. (1959b). The influence of the refinement of carbohydrates on their cariogenicity. *In vitro* studies on crude and refined sugars and animal experiments. *Brit. Dent. J.* **106**, 362–374.

Jensen, S. B., Löe, H., Schiött, C. R., and Theilade, E. (1968). Experimental gingivitis in man. IV. Vancomycin induced changes in bacterial plaque composition as related to development of gingival inflammation. *J. Periodontal Res.* **3**, 284–292.

Jordan, H. V., and Keyes, P. H. (1966). *In vitro* methods for the study of plaque formation and carious lesions. *Arch. Oral Biol.* **11**, 793–801.

Kalberer, P. U., Schroeder, H. E., Guggenheim, B., and Mühlemann, H. R. (1971). The microbial colonization in fissures. A morphological and morphometric study in rat molars. *Helv. Odontol. Acta* **15**, 1–14.

Kaufman, H. W., and Kleinberg, I. (1970). The effect of pH on the adsorption properties of the phytate molecule. *Arch. Oral Biol.* **15**, 917–934.

Keyes, P. H., and Likins, R. C. (1946). Plaque formation, periodontal disease and dental caries in Syrian hamsters. *J. Dent. Res.* **25**, 166.

Kleinberg, I. (1970). Biochemistry of the dental plaque. *Advan. Oral Biol.* **4**, 44–90.

Krembel, J., Frank, R. M., and Deluzarche, A. (1969). Fractionation of human dental plaques. *Arch. Oral Biol.* **14**, 563–565.

Larmas, M. A., and Mäkinen, K. K. (1971). The ability of various microorganisms to produce histochemically determinable enzyme activity in human dentine. *Acta Odontol. Scand.* **29**, 471–486.

Leach, S. A. (1963). Release and breakdown of sialic acid from human salivary mucin and its role in the formation of dental plaque. *Nature (London)* **199**, 486–487.

Leach, S. A. (1970). A review on the biochemistry of dental plaque. *In* "Dental Plaque" (W. D. McHugh, ed.) pp. 143–156. D. C. Thomson & Co, Ltd., Dundee, Scotland.

Leach, S. A., and Critchley, P. (1966). Bacterial degradation of glycoprotein sugars in human saliva. *Nature (London)* **209**, 506.

Leach, S. A., and Melville, T. H. (1970). Investigation of some human oral organisms capable of releasing the carbohydrates from salivary glycoproteins. *Arch. Oral Biol.* **15**, 87–88.

McClure, F. J. (1963). Further studies on the cariostatic effect of organic and inorganic phosphates. *J. Dent. Res.* **42**, 693–699.

McDougall, W. A. (1963a). Studies on the dental plaque. I. The histology of the dental plaque and its attachment. *Aust. Dent. J.* **8**, 261–273.

McDougall, W. A. (1963b). Studies on the dental plaque. II. The histology of the developing interproximal plaque. *Aust. Dent. J.* **8**, 398–407.

McDougall, W. A. (1963c). Studies on the dental plaque. III. The effect of saliva on salivary mucoids and its relationship to the regrowth of plaques. *Aust. Dent. J.* **8,** 463–467.

McDougall, W. A. (1964). Studies on dental plaque. IV. Levans and the dental plaque. *Aust. Dent. J.* **9,** 1–5.

McGaughey, C., and Field, B. D. (1969). Effect of salivary proteins on adsorption of cariogenic streptococci by hydroxylapatite. *IADR Program Abstr. Pap.* No. 169.

McGaughey, C., and Stowell, E. C. (1966). Plaque formation by purified salivary mucin *in vitro:* Effects of incubation, calcium and phosphate. *Nature (London)* **209,** 897–899.

McGaughey, C., and Stowell, E. C. (1967). The adsorption of human salivary protein and porcine submaxillary mucin by hydroxylapatite. *Arch. Oral Biol.* **12,** 815–828.

McGaughey, C., and Stowell, E. C. (1971). A specific effect of hydrogen ions on the adsorption of salivary proteins by hydroxylapatite. *Caries Res.* **5,** 373–377.

McGaughey, C., Field, B. D., and Stowell, E. C. (1971). Effects of salivary proteins on the adsorption of cariogenic streptococci by hydroxylapatite. *J. Dent. Res.* **50,** 917–922.

McKendrick, A. J. W. (1970). Prevention and control of dental plaque. *In* "Dental Plaque" (W. D. McHugh, ed.), pp. 297–298. D. C. Thomson & Co, Ltd., Dundee, Scotland.

Mäkinen, K. K. (1971). Enzyme dynamics of a cariogenic streptococcus: The effect of xylitol and sorbitol. *J. Dent. Res.* **51,** 403–408.

Mäkinen, K. K. (1972). The role of sucrose and other sugars in the development of dental caries. *Int. Dent. J.* **22,** 363–386.

Mäkinen, K. K., and Scheinin, A. (1971). The effect of the consumption of various sugars on the activity of plaque and salivary enzymes. *Int. Dent. J.* **21,** 331–339.

Mäkinen, K. K., and Scheinin, A. (1972). The effect of various sugars and sugar mixtures on the activity and formation of enzymes of dental plaque and oral fluid. *Acta. Odontol. Scand.* **30,** 259–275.

Manly, R. S. (1943). A structureless recurrent deposit on teeth. *J. Dent. Res.* **22,** 479–86.

Manly, R. S., Liberfarb, R., Freese, M., and O'Brien, A. (1966). Determination of levan content and formation rate in dental plaque. *IADR Program Abstr. Pap.* No. 201.

Mayhall, C. W. (1970). Concerning the composition and source of the acquired enamel pellicle of human teeth. *Arch. Oral Biol.* **15,** 1327–1341.

Mayhall, C. W. (1971). Amino acid composition of total and fractionated *in vitro* salivary pellicles. *IADR Program Abstr. Pap.* No. 848.

Meckel, A. H. (1965). The formation and properties of organic films on teeth. *Arch. Oral Biol.* **10,** 585–598.

Middleton, J. D. (1964). Some observations of methyl pentoses in plaque and saliva. *J. Dent. Res.* **43,** No. 5, Suppl., 955.

Millin, D. J., and Smith, M. H. (1961). Nature and composition of dental plaque. *Nature (London)* **189,** 664–665.

Molan, P. C., and Hartles, R. L. (1967). The source of glycolytic activity in human saliva. *Arch. Oral Biol.* **12,** 1593–1603.

Naylor, M. N., Wilson, R. F., and Melville, M. R. B. (1970). Mono- and disaccharide solutions and the formation of plaque *in vitro*. *In* "Dental Plaque" (W. D. McHugh, ed.), pp. 41–47. D. C. Thomson & Co., Ltd., Dundee, Scotland.

Neely, B. W. (1960). Dextran: Structure and synthesis. *Advan. Carbohyd. Chem.* **15**, 341–369.

Newbrun, E. (1971). Dextransucrase from *Streptococcus sanguis*. Further characterization. *Caries Res.* **5**, 124–134.

Newbrun, E. (1972). Extracellular polysaccharides synthesized by glycosyltransferases of oral streptococci. *Caries Res.* **6**, 132–147.

Newbrun, E., and Baker, S. (1968). Physico-chemical characteristics of the levan produced by *Streptococcus salivarius*. *Carbohyd. Res.* **6**, 165–170.

Newbrun, E., and Carlsson, J. (1969). Reaction rate of dextran-sucrose from *Streptococcus sanguis* in the presence of various compounds. *Arch. Oral Biol.* **14**, 461–468.

Nikaido, H., and Hassid, W. Z. (1971). Biosynthesis of saccharides from glycopyranosyl esters of nucleoside pyrophosphates ("sugar nucleotides"). *Advan. Carbohyd. Chem.* **26**, 351–483.

Pearce, E. I. F., and Bibby, B. G. (1966). Effects of time, surface area, pH and some ions on protein adsorption by bovine enamel. *Arch. Oral Biol.* **11**, 825–832.

Pedersen, C. S., and Albury, M. N. (1955). Variation among the heterofermentative lactobacilli. *J. Bacteriol.* **70**, 702–708.

Pigman, W., Elliott, H. C., and Laffre, R. O. (1952). An artificial mouth for caries research. *J. Dent. Res.* **31**, 627–633.

Pinter, J. K., Hayashi, J. A., and Bahn, A. N. (1969). Carbohydrate hydrolases of oral streptococci. *Arch. Oral Biol.* **14**, 735–744.

Rees, D. A. (1969). Structure, conformation, and mechanism in the formation of polysaccharide gels and networks. *Advan. Carbohyd. Chem.* **24**, 267–332.

Ritz, H. L. (1970). The role of aerobic *Neisseriae* in the initial formation of dental plaque. *In* "Dental Plaque" (W. D. McHugh, ed.), pp. 17–26. D. C. Thomson & Co, Ltd., Dundee, Scotland.

Robrish, S. A., Reid, W., and Kritchevsky, M. I. (1972). Distribution of enzymes forming polysaccharide from sucrose and the composition of extracellular polysaccharide synthesized by *Streptococcus mutans*. *Appl. Microbiol.* **24**, 184–190.

Rølla, G. (1966). Neuraminidase in human sputum. *Acta Odontol. Scand.* **24**, 431–442.

Rølla, G., and Mathiessen, P. (1970). The adsorption of salivary proteins and dextrans to hydroxylapatite. *In* "Dental Plaque" (W. D. McHugh, ed.), pp. 129–140. D. C. Thomson & Co, Ltd., Dundee, Scotland.

Rølla, G., Kornstad, L., Mathiessen, P., and Povatong, L. (1969). The adsorption by an acidic salivary glycoprotein to tooth surfaces *in vitro*. *J. Periodontal Res., Suppl.* **7**, 8–9.

Rosen, S. (1969). Comparison of sucrose and glucose in the causation of dental caries in gnotobiotic rats. *Arch. Oral Biol.* **14**, 445–450.

Sandham, H. J., and Kleinberg, I. (1969). Utilization of glucose and lactic acid by salivary sediment. *Arch. Oral Biol.* **14**, 597–602 and 603–618.

Sandham, H. J., and Kleinberg, I. (1970). Contribution of lactic and other acids to the pH of a human salivary sediment system during glucose catabolism. *Arch. Oral Biol.* **15**, 1263–1283.

Saxton, C. A., and Critchley, P. (1970). An electron microscope investigation of the effect of diminished protein synthesis on the morphology of the organisms in dental plaque *in vitro*. *In* "Dental Plaque" (W. D. McHugh, ed.), pp. 109–127. D. C. Thomson & Co, Ltd., Dundee, Scotland.

Scheinin, A., and Mäkinen, K. K. (1971). The effect of various sugars on the formation and chemical composition of dental plaque. *Int. Dent. J.* **21**, 302–321.

Scheinin, A., and Mäkinen, K. K. (1972). Effect of sugars and sugar mixtures on dental plaque. *Acta Odontol. Scand.* **30**, 235–257.

Scheinin, A., and Mäkinen, K. K. (1974). *Acta Odontol. Scand.* (To be published).

Schroeder, H. E. (1963). Inorganic content and histology of early dental calculus in man. *Helv. Odontol. Acta* **7**, 17–30.

Schroeder, H. E., and de Boever, J. (1970). The structure of microbial dental plaque. *In* "Dental Plaque" (W. D. McHugh, ed.), pp. 49–75. D. C. Thomson & Co, Ltd., Dundee, Scotland.

Schumacher, G.-H., and Schmidt, H. (1972). "Anatomie und Biochemic der Zähne," p. 264, Fischer, Stuttgart.

Sidaway, D. A. (1970). The bacterial composition of natural plaque and the *in vitro* production of artificial plaque. *In* "Dental Plaque" (W. D. McHugh, ed.), pp. 225–240. D. C. Thomson & Co, Ltd., Dundee, Scotland.

Sidaway, A. B., Marsland, E. A., Rowles, S. L., and MacGregor, A. B. (1964). The artificial mouth in caries research. *Proc. Roy. Soc. Med.* **57**, 1065–1069.

Skinner, A., and Naylor, M. N. (1972). Influence of sugar type on the pattern of acid production by *Streptococcus mutans*. *J. Dent. Res.* **51**, 1022–1024.

Smith, M. H. (1964). Amino acid analyses of denture pellicle and a glycoprotein-containing component of saliva. *J. Dent. Res.* **43**, 302.

Stegmeier, K., Dallmeier, E., Bestmann, H.-J., and Kröncke, A. (1971). Untersuchungen über den Sorbitabbau unter Verwendung von ^{14}C-markierten Substanzen und der Gaschromatographie. *Deut. Zahnaerztl. Z.* **26**, 1129–1134.

Strålfors, A. (1956). An investigation of the respiratory activities of oral bacteria. *Acta Odontol. Scand.* **14**, 153–186.

Strange, R. E. (1961). Induced enzyme synthesis in aqueous suspensions of starved stationary phase *Aerobacter aerogenes*. *Nature (London)* **191**, 1272.

Tanzer, J. M., and Brown, A. T. (1973). To be published.

Tanzer, J. M., Kritchevsky, M. I., and Keyes, P. H. (1969a). The metabolic fate of glucose catabolized by a washed stationary phase caries-conducive *Streptococcus*. *Caries Res.* **3**, 167–177.

Tanzer, J. M., Wood, W. I., and Krichevsky, M. I. (1969b). Growth kinetics of plaque-forming streptococci in the presence of sucrose. *IADR Program Abstr. Pap.* No. 643.

Tatevossian, A., and Jenkins, G. N. (1969). The source of metabolic activity in human saliva. *Arch. Oral Biol.* **14**, 1121–1123.

Theilade, E., and Theilade, J. (1970). Bacteriological and ultrastructural studies of developing dental plaque. *In* "Dental Plaque" (W. D. McHugh, ed.), pp. 27–40. D. C. Thomson & Co, Ltd., Dundee, Scotland.

Tonzetich, J., and Friedman, S. D. (1965). The regulation of metabolism by the cellular elements in saliva. *Ann. N.Y. Acad. Sci.* **131**, 815–829.

Turner, E. P. (1958). The integument of the enamel surface of the human tooth. *Dent. Practit.* **8**, 341–348 and 373–382.

van Houte, J. (1964). Relationship between carbohydrate intake and polysaccharide-storing microorganisms in dental plaque. *Arch. Oral Biol.* **9**, 91–93.

van Houte, J., and Saxton, C. A. (1971). Cell wall thickening and intracellular polysaccharide in microorganisms of the dental plaque. *Caries Res.* **5**, 30–43.

van Houte, J., deMoor, C. E., and Jansen, H. M. (1970). Synthesis of iodophilic polysaccharide by human oral streptococci. *Arch. Oral Biol.* **15**, 263–266.

Wood, J. M. (1967). The amount, distribution and metabolism of soluble polysaccharides in human dental plaque. *Arch. Oral Biol.* **12,** 849–858.

Wood, J. M. (1969). The state of hexose sugar in human dental plaque and its metabolism by the plaque bacteria. *Arch. Oral Biol.* **14,** 161–168.

Wood, J. M., and Critchley, P. (1967). The soluble carbohydrates of the plaque matrix. *J. Dent. Res.* **46,** Suppl., 129–130.

Yamada, T., Hojo, S., Kobayashi, K., Asano, Y., and Araya, S. (1970). Studies on the carbohydrate metabolism of cariogenic *Streptococcus mutans* strain PK-1. *Arch. Oral Biol.* **15,** 1205–1217.

CHAPTER 38

Carbohydrate Metabolism in Caries-Conducive, Oral Streptococci

ALBERT T. BROWN

Extensive evidence exists in the literature for the role of oral micro-organisms in various caries processes. Oral streptococci compose a large proportion of the cultivable microflora from saliva, surface of the tongue, oral epithelium, and plaque. However, the species of streptococci present in the oral environment vary from site to site; for example, *Streptococcus salivarius* constitutes a high percentage of the oral bacteria present on the tongue and in saliva but is present in relatively low numbers in dental plaque (Carlsson, 1965b, 1967; Krasse, 1953, 1954). In contrast to *S. salivarius*, *Streptococcus sanguis* comprises a high proportion of the facultative streptococci in dental plaque but is seldom found on the tongue, in saliva, or on the oral epithelium (Carlsson, 1965a, 1967). Because of their distribution and relatively high numbers in the oral environment, a great deal of attention has focused upon members of this group of organisms as specific etiological agents of dental caries. Recently, it has been established that the group of plaque-forming streptococci designated *Streptococcus mutans* (Edwardsson, 1968; Fitzgerald and Keyes, 1960;

689

Guggenheim, 1968) plays a role as infectious agents in multisurface caries in rodents (Fitzgerald and Keyes, 1960; Gibbons, 1964; Gibbons *et al.*, 1966; Zinner *et al.*, 1965) and humans (de Stoppelaar *et al.*, 1969; Ikeda and Sandham, 1971; Krasse *et al.*, 1968; Littleton *et al.*, 1970).

Originally isolated from a human carious lesion by Clarke in 1924, it was not until 1960 that Fitzgerald and Keyes (1960) were able to implant hamster strains of *S. mutans*-like organisms back into hamsters and successfully induce caries. Subsequently, *S. mutans* strains isolated from human sources were successfully implanted into gnotobiotic or nearly gnotobiotic rats and hamsters and were observed to cause caries in these rodents (Gibbons *et al.*, 1966; Krasse, 1966; Zinner *et al.*, 1965). Although caries-conducive in rodents, existing evidence does not definitely support involvement of *S. mutans* in human caries. However, *S. mutans* probably has the potential to cause caries in humans based upon the animal model experiments discussed above and the following studies involving human subjects. Littleton *et al.* (1970) observed that *S. mutans* were found in high numbers in lesions involving human coronal tooth surfaces, whereas this same organism occurred in low numbers on sound human tooth surfaces. In addition, de Stoppelaar *et al.* (1969) showed that a correlation existed between smooth surface caries activity and the presence of *S. mutans* in plaque samples taken from 13-year-old children. Krasse *et al.* (1968) also established a correlation between the presence of *S. mutans* and the frequency of smooth surface lesions in school children who showed high caries activity. Recently, Ikeda and Sandham (1971) reported that *S. mutans* constitute 39% of the total streptococci from the pits and fissures of Negro school children with a significant caries incidence.

The organisms which have been classified as *S. mutans* are an extremely heterogeneous group of bacteria based upon their GC content (Coykendall, 1970), DNA homology (Coykendall, 1971), cell wall antigens (Bratthall, 1970), and characteristics of several enzymes (Brown and Patterson, 1972), and recently, the classification of *S. mutans* into four separate varieties (*S. mutans* Var. clarke, *S. mutans* Var. rattus, *S. mutans* Var. cricetus, and *S. mutans* Var. SL) has been proposed by Coykendall (1972). However, certain biochemical and physiological characteristics related to the dissimilation of sucrose are possessed by all *S. mutans* strains examined and appear to contribute to the caries potential of these organisms. Of all the carbon compounds which are able to serve as a primary energy source for *S. mutans*, only sucrose will initiate the broad spectrum of biochemical events thought to be germane to the virulence of these organisms. Since the role of dietary sucrose in the frequency and incidence of dental caries is well established (Fitzgerald and Keyes,

1960; Keyes, 1968; Larson *et al.*, 1967; Mandel, 1970; Takeuchi, 1961), the metabolism of this disaccharide by *S. mutans* is probably a major factor in the initiation of various caries processes. Several aspects of *S. mutans*-mediated sucrose dissimilation thought to contribute to the caries potential of these organisms are summarized in Fig. 1. The role(s) of these various aspects of sucrose metabolism in *S. mutans*-dependent dental caries will now be discussed.

A. Lactic Acid Production by *S. mutans* via Glycolysis

Years ago, Miller proposed that destruction of teeth in dental caries resulted from the acid produced from carbohydrate metabolism by oral bacteria (Miller, 1890). The pH of *S. mutans* plaque is extremely acidic (Charlton *et al.*, 1971a,b) and the terminal pH of *S. mutans* strains grown in sucrose- or glucose-containing medium has been shown in a number of laboratories to be well below pH 4.5. All *S. mutans* strains studied have been shown to be typical homolactic fermenters which convert over 90% of their glucose carbon source to lactic acid (Drucker and Melville, 1968; Jordan, 1965; Tanzer *et al.*, 1969c). Essentially the same situation exists for sucrose-grown *S. mutans* strains. At low sucrose concentrations, both the fructosyl and glucosyl moieties of sucrose are ultimately converted to lactic acid which subsequently initiates the demineralization of teeth (Tanzer, 1972; Tanzer *et al.*, 1972).

The production of lactic acid from glucose and sucrose by *S. mutans* probably proceeds strictly by glycolysis (Brown and Wittenberger, 1971b). These organisms lack at least the oxidative portion of the hexose monophosphate shunt since no glucose-6-phosphate dehydrogenase or 6-

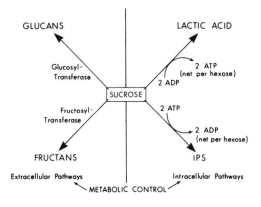

Fig. 1 Metabolic fate of sucrose in a variety of *S. mutans* strains.

TABLE I

Levels of Glucose-6-Phosphate Dehydrogenase
and 6-Phosphogluconate Dehydrogenase in Cell
Extracts from Various Streptococcal Species[a]

Organism	Enzymatic activities (units/mg protein)	
	G6PDH	6PGDH
S. faecalis MR	0.0659	0.0740
S. faecium N-55	0.0701	0.0653
S. sanguis ATCC 10558	0.0952	0.0917
S. mutans strains		
FA-1	0.0009	<0.0003
SL-1	0.0005	<0.0003
21-Typical	0.0004	<0.0003
6715–9	0.0006	<0.0003
3720	0.0008	<0.0003
NCTC 10449	0.0009	<0.0003

[a] Organisms were grown anaerobically in complex medium with glucose (0.5% w/v) as the primary energy source and cell extracts prepared as described previously (Brown and Wittenberger, 1971b). Glucose-6-phosphate dehydrogenase (G6PDH) and 6-phosphogluconate dehydrogenase (6PGDH) were assayed in crude extracts of the indicated organism by monitoring the enzyme-dependent reduction of NADP at 340 nm as also described previously (Brown and Wittenberger, 1971b). All streptococcal hexose monophosphate pathway dehydrogenases were specific for NADP and showed no activity with equimolar concentrations of NAD.

phosphogluconate dehydrogenase activity is detectable in cell-free extracts from *S. mutans* using assay conditions which show these enzymes to be present in other streptococci (Table I). Also, when *S. mutans* strains are incubated in the presence of glucose[^{14}C-1], less than 1% of the labeled carbon metabolized is converted to $^{14}CO_2$ which indicates that little, if any, glucose carbon is metabolized via the oxidative portion of the hexose monophosphate pathway (Table II). Utilization of glucose[^{14}C-1] by the hexose monophosphate pathway would result in the production of significant amounts of $^{14}CO_2$ from the intracellular conversion of 6-phosphogluconate[^{14}C-1] to $^{14}CO_2$ and ribulose-5-phosphate.

Although the Entner-Doudoroff pathway has been reported to be op-

TABLE II

$^{14}CO_2$ Production from Glucose [^{14}C-1] by Glucose-Adapted
Cell Suspensions of Various *S. mutans* Strains[a]

S. mutans strain	Glucose used (μmoles)	$^{14}CO_2$ evolved (μmoles)	μmoles of $^{14}CO_2$ evolved/ 100 μmoles of glucose used
FA–1	39.3	0.14	0.36
E–49	44.4	0.30	0.68
6715–9	41.9	0.21	0.50
NCTC 10449	48.6	0.37	0.76

[a] An assessment of hexose monophosphate pathway activity was made by determining the amount of $^{14}CO_2$ produced from glucose [C^{14}-1] by cell suspensions of various *S. mutans* strains. This procedure has been described previously (Brown and Wittenberger, 1971b). Glucose was routinely determined using the Worthington glucose oxidase method. In all instances, less than 1 % of the glucose [C^{14}-1] metabolized appeared as $^{14}CO_2$.

erative in gluconate-adapted strains of *Streptococcus faecalis*, this pathway is not operative when these organisms are grown with glucose as the sole energy source (Sokatch and Gunsalus, 1957). The existence of Entner-Doudoroff activity in *S. mutans* is extremely unlikely since these organisms cannot be adapted to grow on gluconate (A. T. Brown, unpublished observation). In addition, glucose-6-phosphate dehydrogenase, which catalyzes the initial step in the Entner-Doudoroff pathway, is not detectable in cell-free extracts from glucose-adapted *S. mutans* strains (Table I).

B. Extracellular and Cell Surface Glucan Synthesis by *S. mutans*

Plaque formation by *S. mutans in vivo* is dependent upon the availability of dietary sucrose (Fitzgerald and Keyes, 1960; Gibbons *et al.*, 1966; Krasse, 1966; Larson *et al.*, 1967; Zinner *et al.*, 1965), and likewise the ability of *S. mutans* to adhere to wires and form plaque *in vitro* is a phenomenon which is similarly dependent upon the presence of sucrose in the culture medium (Gibbons and Nygaard, 1968; Jordan and Keyes, 1966; McCabe *et al.*, 1953; Tanzer *et al.*, 1969b). Plaque formation by these organisms *in vitro* is a sucrose-specific process since other carbohydrates such as glucose do not impart to *S. mutans* the ability to adhere to smooth surfaces. *Streptococcus mutans* produces two classes of glucose polymers called glucans (dextrans) from sucrose. One of these classes of glucans is cell-associated and the other is extracellular and present in the culture supernatant. The glucans associated with the cell surface are generally water insoluble while those in the culture supernatant are always water soluble (Dahlqvist *et al.*, 1967; Donahue *et al.*, 1966; Fitz-gerald and Jordan, 1968; Gibbons and Banghart, 1967; Gibbons and Nygaard, 1968; Guggenheim, 1968; Wood and Critchley, 1966). These glucans are produced from sucrose by the action of two types of glucosyl-transferase enzymes, one associated with the cell surface and the other with the culture liquor (Dahlqvist *et al.*, 1967; Gibbons and Nygaard, 1968; Tanzer *et al.*, 1969b).

Since the extracellular glucosyltransferase has been relatively easy to obtain, this enzyme has been partially characterized from both *S. mutans* (Guggenheim and Newbrun, 1969; Newbrun, 1972) and *S. sanguis* (Carlsson *et al.*, 1969; Newbrun, 1971; Newbrun and Carlsson, 1969; Newbrun *et al.*, 1971). The extracellular glucosyltransferase from *S. mutans* (Gibbons and Nygaard, 1968; Guggenheim and Newbrun, 1969; Tanzer *et al.*, 1969b) is constitutive and catalyzes the polymerization of the glucose moeities of sucrose while generating free fructose. The enzyme functions as an α-1,6-glucan: D-fructose 2-glucosyltransferase (Gibbons and Nyga-ard, 1968; Guggenheim and Newbrun, 1969; Guggenheim and Schroeder, 1967). Recently, Guggenheim and Newbrun (1969) have found that mul-tiple glucosyltransferases exist in the culture supernatant of a strain of *S. mutans*. Other workers have shown that both the cell surface and extra-cellular glucans are heterogeneous in regard to both molecular weight and the nature of their cross-linkages (Dahlqvist *et al.*, 1967; Guggen-heim, 1970; Guggenheim and Haller, 1972; Guggenheim and Schroeder, 1967; Newbrun *et al.*, 1971; Sidebotham *et al.*, 1971; Tanzer, 1972; Tan-zer *et al.*, 1972). There is also evidence that some of the glucans are α-1,6 linked, α-1,3 linked, and that some might be branched by the occur-rence of both α-1,3 and α-1,6 linkages in these polymers. It is attractive to hypothesize that specific extracellular glucans are a reaction product of one of the multiple glucosyltransferases reported by Guggenheim and Newbrun.

The sucrose-dependent synthesis of glucans by *S. mutans* permits them to adhere to smooth surfaces and agglutinate. The cell-surface insoluble glucans appear to be involved in the adhesion of *S. mutans* to smooth tooth surfaces while the extracellular water-soluble glucans appear to be associated with the cohesion of these bacteria. The ability of microorga-nisms to adhere and agglutinate is essential if they are to play a significant role in plaque formation. Correlations between the presence of plaque and the incidence of dental caries are well known.

The evidence that insoluble glucans play a central role in adhesion is as follows. Dextranase preparations from *Penicillium funiculosum* are effective in removing sucrose-dependent streptococcal plaque deposits *in vitro* (Fitzgerald *et al.*, 1968a) and in experimental animals (Fitzgerald *et al.*, 1968b). Glucosyl acceptors such as low-molecular-weight soluble dextrans which simulate the production of water-soluble glucans inhibit the production of insoluble glucans and prohibit sucrose-dependent plaque formation to take place *in vivo* (Gibbons and Nygaard, 1968). Also, the addition of water-soluble glucans to *S. mutans* suspensions in the absence of sucrose did not promote the production of plaque deposits *in vitro* (Gibbons and Nygaard, 1968). In addition, mutants of *S. mutans* which are defective in glucan synthesis from sucrose are unable to adhere to smooth surfaces *in vitro* (de Stoppelaar *et al.*, 1971; Freedman and Tanzer, 1972). Thus, it appears that the production of glucans, specifically cell-associated insoluble glucans, impart to these organisms the ability to adhere to smooth surfaces.

Glucans also contribute to the ability of *S. mutans* to agglutinate. The addition of exogenous water-soluble glucans to a suspension of *S. mutans* results in their cohesion as will the addition of sucrose (Gibbons and Fitzgerald, 1969). Mutants which are defective in extracellular glucan synthesis and unable to adhere to smooth surfaces do not lose their ability to agglutinate (Freedman and Tanzer, 1972). Hence, adhesion and cohesion may be distinct processes which are dependent upon the production of different classes of extracellular glucans.

C. Extracellular Fructan Synthesis by *S. mutans*

Extracellular fructans, or levans, have been found in cultures of *S. mutans* as well as in human dental plaque (Carlsson, 1970; Dahlqvist *et al.*, 1967; Gibbons and Nygaard, 1968; Guggenheim and Schroeder, 1967; Wood and Critchley, 1966). These polymers are produced only from sucrose by the action of a fructosyltransferase which is in the cell-free supernatant fluid of an *S. mutans* culture. The enzyme is constitutive and functions as a β-2,6-fructan:D-glucose-6-fructosyltransferase, the reaction products being a fructose polymer and free glucose (Carlsson, 1970). Unlike extracellular glucans, fructans play no role in the adhesion or agglutination of *S. mutans*-type organisms. Oral microorganisms which produce copious amounts of extracellular levans from sucrose but not cell-associated insoluble glucans are unable to adhere to smooth surfaces or form plaque *in vitro*. This is not to say, however, that the production of extracellular fructans by *S. mutans* and other oral bacteria (Carlsson and Sundstrom, 1968; Higuchi *et al.*, 1970; Wood, 1964, 1967) have no role in plaque formation and hence do not contribute to the virulence of these organisms. The constitutive nature of the fructosyltransferase means that this enzyme is present in plaque regardless of the composition of the diet. Exposure to sucrose will then result in the synthesis of fructans by plaque microorganisms including *S. mutans*. A variety of oral microorganisms are able to degrade high molecular weight fructans to free fructose (DaCosta and Gibbons, 1968; Manly and Richardson, 1968; Parker and Creamer, 1971; van Houte, 1968). Since *S. mutans* is able to utilize free fructose for growth (Edwardsson, 1968), fructans may be considered as extracellular storage polysaccharides which can be mobilized for energy production when the supply of exogenous carbohydrates is depleted. Consequently, the long-term degradations of fructans may contribute to the survival potential of these organisms and enable them to produce lactic acid via glycolysis over extended periods of time. The long-term production of lactic acid by oral microorganisms has long been thought to be responsible for the demineralization of teeth in various caries processes.

D. Synthesis of Intracellular Polysaccharides by *S. mutans*

Streptococcus mutans, like a variety of other microorganisms, are able to convert exogenous carbohydrates into intracellular polysaccharides (IPS) (Berman and Gibbons, 1966; Loesche and Henry, 1967; van Houte, 1967; van Houte *et al.*, 1969). In *S. mutans*, these IPS are of the amylopectin type (Berman and Gibbons, 1966) and appear to compose a large fraction of the intracellular composition of *S. mutans* strains when they are grown under optimal nutritional conditions (Gibbons, 1964). The staining of plaque by the use of iodophilic stains specific for amylopectin-type IPS has been used as a means of screening samples for the presence of *S. mutans* (Loesche and Henry, 1967; van Houte *et al.*, 1969). Nothing is known, however, of the biochemical mechanisms responsible for the synthesis and mobilization of the IPS in *S. mutans*. Regardless of the biochemical mechanisms involved, it is necessary that these processes be under stringent cellular control since the incorporation of a mole of hexose into IPS in these organisms probably requires the expenditure of 2 moles of ATP. The contribution of IPS to the disease potential of *S. mutans* is twofold. The mobilization of IPS during periods when exogenous carbohydrates are unavailable provides these organisms with an energy source during adverse nutritional conditions thus permitting them to survive and remain viable for extended periods of time in the oral environment. Also, the metabolism of IPS via glycolysis over extended periods of time results in the long-term production of lactic acid which is essential for the demineralization of teeth and yields an acidic environment which favors the growth of aciduric microorganisms such as *S. mutans* over that of acid-sensitive microorganisms in plaque (Loesche and Henry, 1967; van Houte, 1967).

E. Regulation of Metabolic Processes in *S. mutans* and Other Oral Streptococci

The author's recent interests have focused upon the regulation of carbohydrate metabolism in *S. mutans*. Specifically, he has been interested in those mechanisms responsible for regulating carbon flow within the glycolytic pathway and between this pathway and other carbohydrate dissimulative pathways in these organisms. Since glycolysis is the only known pathway by which *S. mutans* is able to generate ATP for biosynthesis and growth, the dissimilation of carbohydrates by pathways other than glycolysis must be done at the expense of ATP production. Since only 2 moles of ATP are generated per mole of hexose catabolized via glycolysis, metabolic control over utilization of carbon by other diverse carbohydrate dissimulative pathways in *S. mutans* is essential in order for these organisms to survive in an environment where their biosynthetic potential must be continually expressed.

Initial studies have shown that carbohydrate metabolism in *S. mutans* and other oral streptococci is under stringent cellular control both from the standpoints of regulation of enzyme synthesis (e.g., induction and repression) and regulation of enzymatic activity (e.g., allosteric modulation). The following sections will be devoted to a summary of our work with cellular regulatory mechanisms in oral streptococci including *S. mutans*.

1. Regulation of Mannitol and Sorbitol Metabolism in *S. mutans*

A nutritional characteristic which serves to distinguish *S. mutans* from other homolactic, oral streptococci is their ability to utilize either mannitol or sorbitol as a primary energy source (Edwardsson, 1968; Fitzgerald and Keyes, 1960; Guggenheim, 1968). The ability to grow on mannitol and sorbitol does not contribute to the caries-conducive nature of these organisms since spontaneous mutants of *S. mutans* unable to grow on mannitol or sorbitol do not lose their ability to cause sucrose-dependent caries in animal models (Edwardsson, 1970). However, fermentation of these hexitols may contribute to the survival potential of *S. mutans* in the oral environment. It has been found that extracts from mannitol- and sorbitol-adapted *S. mutans* strains possess elevated levels of NAD-dependent mannitol-1-phosphate dehydrogenase and NAD-dependent sorbitol-6-phosphate dehydrogenase activity (Brown and Wittenberger, 1973). The level of the former enzyme is higher in mannitol-adapted cells while the level of the latter enzyme is higher in sorbitol-adapted cells. The synthesis of both hexitol phosphate dehydrogenases is repressed by glucose (Table III). The mannitol-1-phosphate and sorbitol-6-phosphate dehydrogenase activity from hexitol-adapted *S. mutans* result from separate hexitol phosphate dehydrogenases and not from a single enzyme with a specificity for both mannitol-1-phosphate and sorbitol-6-phosphate since these activities may be resolved by gel filtration on a Sephadex G-200 column (Brown and Wittenberger, 1973) and poly-acrylamide disc gel electrophoresis (Brown and Wittenberger, 1973).

TABLE III

Effect of Growth Substrate on the Specific Activity of Mannitol-1-Phosphate Dehydrogenase and Sorbitol-6-Phosphate Dehydrogenase in Cell-Free Extracts of *S. mutans* NCTC 10449[a]

Carbon source	Mannitol-1-phosphate dehydrogenase	Sorbitol-6-phosphate dehydrogenase
Mannitol	0.490	0.070
Sorbitol	0.090	0.480
Glucose	0.015	0.001
Mannitol plus glucose	0.030	0.003
Sorbitol plus glucose	0.010	0.003
Sucrose	0.015	0.001
Fructose	0.020	0.002
Raffinose	0.025	0.001

[a] *Streptococcus mutans* NCTC 10449 was grown anaerobically in complex medium supplemented with 0.5% of the appropriate carbon source(s) and cell-free extracts prepared as previously described (Brown and Wittenberger, 1973). Mannitol-1-phosphate dehydrogenase and sorbitol-6-phosphate dehydrogenase activities were assayed by monitoring the enzyme-dependent reduction of NAD as also described previously (Brown and Wittenberger, 1973).

The elution profile from the Sephadex G-200 column indicated that the molecular weight of the sorbitol-6-phosphate dehydrogenase was significantly greater than that of the mannitol-1-phosphate dehydrogenase (Fig. 2) (Brown and Wittenberger, 1973). The product of the reaction catalyzed by both hexitol phosphate dehydrogenases appeared to be the glycolytic intermediated fructose-6-phosphate since two activity peaks from the G-200 column, which corresponded exactly with the mannitol-1-phosphate and sorbitol-6-phosphate dehydrogenase activity peaks catalyzed the fructose-6-phosphate-dependent oxidation of NADH (Fig. 2). Also, "fructose-6-phosphate reductase" activity, like the hexitol phosphate dehydrogenase activities, was elevated in extracts from mannitol- and sorbitol-adapted cells (Brown and Wittenberger, 1973). No activity was observed when either mannitol or sorbitol was substituted for the phosphorylated hexitols in assays using either crude, cell-free extracts or Sephadex G-200 fractions. It thus appears that mannitol and sorbitol metabolism in *S. mutans* proceeds exclusively by the phosphorylated pathway for the catabolism of polyols. In addition, the mannitol-1-phos-

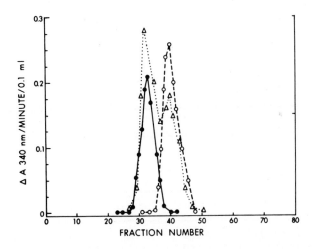

Fig. 2 Elution profiles of (○) mannitol-1-phosphate dehydrogenase, (●) sorbitol-6-phosphate dehydrogenase, and (△) fructose-6-phosphate reductase from a Sephadex G-200 column (Brown and Wittenberger, 1973). Equal volumes of extracts from mannitol- and sorbitol-adapted *S. mutans* NCTC 10449 were combined and 1.5 ml of the mixture was then applied to a Sephadex G-200 column and the sample eluted as described previously (Brown and Wittenberger, 1973). Mannitol-1-phosphate dehydrogenase and sorbitol-6-phosphate dehydrogenase activity was assayed by monitoring the hexitol phosphate-dependent reduction of NAD at 340 nm (Brown and Wittenberger, 1973), whereas fructose-6-phosphate reductase activity was assayed by monitoring the fructose-6-phosphate-dependent oxidation of NADH (Brown and Wittenberger, 1973).

phate dehydrogenase from members of the four *S. mutans* groups show different mobilities upon polyacrylamide disc gel electrophoresis (Brown and Patterson, 1972). The electrophoretic characterization of this enzyme from these different varieties of *S. mutans* may have a clinical value in providing a method for rapidly identifying which *S. mutans* types are present in an oral isolate.

2. Regulation of Alcohol Dehydrogenase Activity in S. mutans

Mannitol- and sorbitol-adapted *S. mutans* strains possess elevated levels of alcohol dehydrogenase activity in addition to induced levels of mannitol-1-phosphate dehydrogenase and sorbitol-6-phosphate dehydrogenase activity (Brown and Patterson, 1973) (Table IV). The fermentation of mannitol and sorbitol by *S. mutans* results in a shift from the homolactic fermentation observed when these organisms are grown on glucose or sucrose toward the production of high levels of ethanol (Brown and Patterson, 1973). The anaerobic catabolism of mannitol and sorbitol by *S. mutans*, in contrast to glucose or sucrose metabolism by these organisms (Drucker and Melville, 1968; Jordan, 1965; Tanzer *et al.*, 1969c)

TABLE IV

Production of Alcohol Dehydrogenase Activity in a Variety of *S. mutans* Strains[a]

S. mutans strain	Alcohol dehydrogenase activity in cells grown on specified carbon source (units/mg protein)			
	Glucose	Sucrose	Mannitol	Sorbitol
NCTC 10449	0.009	0.008	0.168	0.166
FA-1	0.020	0.012	0.227	0.210
E-49	0.015	0.010	0.198	0.187
SL-1	0.017	0.014	0.159	0.162

[a] All *S. mutans* strains were grown anaerobically in complex medium supplemented with the appropriate carbon source (0.5% w/v) and cell-free extracts prepared as described previously (Brown and Patterson, 1973). Alcohol dehydrogenase activity was assayed in cell-free extracts by monitoring the ethanol-dependent reduction of NAD at 340 nm as also described previously (Brown and Patterson, 1973). The alcohol dehydrogenase from all *S. mutans* strains examined was specific for NAD.

TABLE V

Production of Ethanol from Mannitol and Sorbitol by a
Variety of *S. mutans* Strains[a]

| *S. mutans* strain | Ethanol produced from indicated carbon source (μmoles/ml of medium) | | | |
	Sucrose	Glucose	Mannitol	Sorbitol
NCTC 10449	0.27	0.31	9.1	9.1
FA-1	0.20	0.17	8.7	8.9
E-49	0.35	0.22	9.5	9.4
SL-1	0.19	0.28	9.7	9.3

[a] All *S. mutans* were grown anaerobically in complex medium supplemented with the indicated carbon source (10 μmoles/ml) as previously described (Brown and Patterson, 1973). The culture supernatant was acidified after removing the cells by centrifugation and 5 μl of the acidified supernatant were assayed for ethanol content by the gas chromatography method of Rogosa and Love (1968).

results in an additional mole of NADH being generated at the level of the inducible hexitol phosphate dehydrogenases. This additional mole of NADH must be reoxidized by reaction with a suitable electron acceptor, and this is accomplished in *S. mutans* by means of the NAD-linked alcohol dehydrogenase.

The alcohol dehydrogenase present in cell-free extracts from mannitol- or sorbitol-adapted *S. mutans* E-49 possesses several properties which distinguish it from other bacterial alcohol dehydrogenases. The *S. mutans* enzyme has a specificity for *n*-butanol, a pH optimum of 10.5–11, and a molecular weight greater than 200,000 (A. T. Brown, unpublished observation). Multiple alcohol dehydrogenase activity bands may be resolved after cell-free extracts from mannitol-adapted cells have been exposed to polyacrylamide gel electrophoresis (Fig. 3). The occurrence of multiple alcohol dehydrogenase isozymes has been reported previously for the enzyme from horse liver (Pietruszko *et al.*, 1966, 1969; Theorell *et al.*, 1966). These alcohol dehydrogenase isozymes have different physiological functions since some of them catalyze the oxidation of primary alcohol while others specifically catalyze the oxidation of hydroxylated steroids (Pietruszko *et al.*, 1966, 1969; Theorell *et al.*, 1966). Differences in physiological functions, biochemical characteristics, or regulatory properties for the alcohol dehydrogenase isozymes from *S. mutans* have not yet been found. However, studies to determine differences in the properties of the alcohol dehydrogenase isozymes from *S. mutans* are currently in progress.

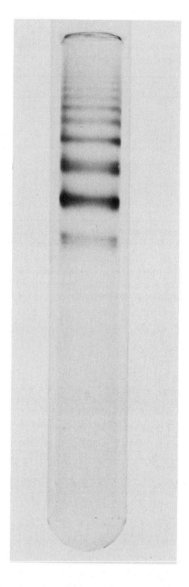

Fig. 3 Resolution of alcohol dehydrogenase activity from *S. mutans* E-49 into multiple isozymes. A cell-free extract from mannitol-adapted *S. mutans* E-49 was applied to the top of polyacrylamide gels (pH 9.3) which contained 5.0% acrylamide and 0.2% bis-acrylamide. After electrophoresis was completed (approximately 3 hr), the gels were stained for alcohol dehydrogenase activity with the following system: Tris-HCl buffer, pH 10.0, 100 μmoles/ml; NAD, 2.0 μmoles/ml; n-butanol, 50 μmoles/ml; nitrobluetetrazolium chloride, 0.8 μmoles/ml; phenazine methosulfate, 0.15 μmoles/ml. After staining the gels were stored in 7% acetic acid.

3. Regulation of Invertase Activity in S. mutans

Glucosyl- and fructosyltransferase systems which catalyze the conversion of sucrose to extracellular polymers and free glucose or fructose could be the only systems operative in *S. mutans* responsible for the production of hexoses to be utilized for energy production via glycolysis. However, at low extracellular sucrose concentrations almost all of the carbon catabolized is converted to lactic acid and little appears as extracellular glucans and fructans (Tanzer, 1972; Tanzer *et al.*, 1972). But as the extracellular sucrose concentration is raised, increasing proportions of the hexose moieties are incorporated into extracellular glucans and fructans (Tanzer, 1972; Tanzer *et al.*, 1972). These data indicate that a pathway for sucrose catabolism exists in *S. mutans* in addition to the glycosyl- and fructosyltransferase systems. We have recently found that sucrose-adapted *S. mutans* strains possess significant levels of invertase activity (Tanzer *et al.*, 1971). This enzyme functions as a β-D-fructofuranoside fructohydrolase which hydrolyzes sucrose to free glucose and fructose, has not been partially purified and characterized, and appears to be subject to cellular control since it is activated by inorganic phosphate (A. T. Brown and J. M. Tanzer, unpublished observation). Phosphate accumulation is coupled to acid production from sucrose in cell suspensions of *S. mutans* (Luoma, 1964; Tanzer *et al.*, 1969a). Thus, *S. mutans* on the tooth surface can produce acid and accumulate phosphate at a high rate. The activation of invertase by inorganic phosphate is probably only one of many mechanisms by which phosphate regulates and controls sucrose dissimilation in these organisms.

4. Occurrence of Multiple Glyceraldehyde-3-Phosphate Dehydrogenases in S. mutans

A novel metabolic situation exists within the glycolytic pathway of all *S. mutans* strains examined. These organisms possess two separable glyceraldehyde-3-phosphate dehydrogenases which differ in their coenzyme specificity and molecular weights (Brown and Wittenberger, 1971a). One of these enzymes is NAD-dependent while the other exhibits specificity for NADP. These activities result from distinct enzymes since they can be separated by Sephadex G-200 gel filtration and DEAE cellulose ion exchange, column chromatography (Brown and Wittenberger, 1971a). A possible role for the NADP-specific enzyme in the metabolism of *S. mutans* is the generation of NADPH for reductive biosynthesis since these organisms lack both the oxidative portion of the hexose monophosphate pathway as well as transhydrogenase activity. Other oral streptococci which have the oxidative portion of the hexose monophosphate pathway lack detectable levels of NADP-dependent glyceraldehyde-3-phosphate dehydrogenase activity (Table VI) (Brown and Wittenberger, 1971a). All organisms examined which had no hexose monophosphate dehydrogenases contained significant levels of NADP-dependent glyceraldehyde-3-phosphate dehydrogenase activity. Hence, an absolute correlation exists in all streptococcal species studied between the presence of NADP-dependent glyceraldehyde-3-phosphate dehydrogenase activity and the absence of NADP-linked hexose monophosphate pathway dehydrogenases.

The existence of both an NAD- and an NADP-linked glyceraldehyde-3-phosphate dehydrogenase within the glycolytic pathway of *S. mutans* raises important questions concerning metabolic control. Under anaerobic growth conditions these organisms are homofermenters and convert about

TABLE VI

Glyceraldehyde-3-Phosphate Dehydrogenase Activity with NAD or NADP in Cell Extracts from Various Streptococcal Strains[a,b]

Organism	GA3PDH (NADP)	GA3PDH (NAD)	G6PDH (NADP)	6PGDH (NADP)
S. faecalis 10CI	<0.001	0.455	0.085	0.072
S. faecium N-55	<0.001	0.375	0.070	0.065
S. sanguis ATCC 10558	<0.001	0.370	0.095	0.091
S. lactis	<0.001	0.410	0.082	0.075
S. mutans strains				
6715-9	0.128	0.332	<0.001	<0.001
FA-1	0.120	0.342	<0.001	<0.001
NCTC 10449	0.100	0.390	<0.001	<0.001
E-49	0.108	0.280	<0.001	<0.001
S. salivarius SS 2	0.140	0.374	<0.001	<0.001

[a] From Brown and Wittenberger (1971a).

[b] All organisms were grown anaerobically in complex medium supplemented with glucose (0.5 % w/v) and cell-free extracts prepared as described previously (Brown and Wittenberger, 1971b). Glyceraldehyde-3-phosphate dehydrogenase (GA3PDH) activity was assayed in cell extracts by monitoring the glyceraldehyde-3-phosphate (GA3P)-dependent reduction of NAD or NADP at 340 nm as also described previously (Brown and Wittenberger, 1971a). Glucose-6-phosphate dehydrogenase (G6PDH) and 6-phosphogluconate dehydrogenase (6PGDH) activities were assayed in cell extracts by monitoring the enzyme-dependent reduction of NADP at 340 nm (Brown and Wittenberger, 1971b).

90% of their glucose carbon to lactic acid. Since the lactate dehydrogenase in these organisms is NAD specific (Brown and Wittenberger, 1972b), and no transhydrogenase activity is present in these cells, the percentage of glucose carbon which is ultimately oxidized by each of these two glyceraldehyde-3-phosphate dehydrogenases must be under stringent cellular control. Studies are now in progress to determine how the distribution of carbon flow between these two glyceraldehyde-3-phosphate dehydrogenases is regulated in *S. mutans*.

5. Regulation of Lactate Dehydrogenase Activity in *S. mutans*

Lactate dehydrogenase, the enzyme which catalyzes the terminal step in glycolysis, is also a focal point for metabolic control in *S. mutans* (Brown and Wittenberger, 1972b). This enzyme from *S. mutans*, like the lactate dehydrogenase from other streptococcal sources (Wolin, 1964), has an absolute dependence upon the glycolytic intermediate, fructose-1,6-diphosphate, for catalytic activity (Fig. 4A) (Brown and Wittenberger, 1972b). Enzymatic activity is a sigmoidal function of the fructose-1,6-diphosphate concentration and Hill plots of the fructose-1,6-diphosphate saturation data indicate that there are at least two distinct fructose-1,6-diphosphate binding sites on the enzyme with some degree of cooperative interaction between them (Fig. 4B) (Brown and Wittenberger, 1972b). As also shown in Fig. 4A the $(M)_{0.5 \, v}$ value for fructose-1,6-diphosphate is approximately 5 mM. This is 100 times higher than the value reported for the lactate dehydrogenases from other streptococcal sources (Wittenberger and Angelo, 1970; Wolin, 1964) and may indicate that high intracellular pool levels of phosphorylated glycolytic intermediates exist in *S. mutans*.

The lactate dehydrogenase from *S. mutans* 10449 also exhibits homotropic interactions with its substrate, pyruvate (Brown and Wittenberger, 1972b). Enzymatic activity is a sigmoidal function of the pyruvate concentration, and Hill plots of the pyruvate saturation data indicate at least two pyruvate binding sites on the enzyme with cooperative interactions between them (Brown and Wittenberger, 1972b). The kinetic response of the *S. mutans* lactate dehydrogenase to certain pyruvate analogs indicate that this enzyme possesses two types of substrate binding sites: one site being a catalytic site and the other an effector site (Brown and Wittenberger, 1972b). α-Ketobutyrate appeared to have a preferential affinity for the effector site. This analog caused a shift in the pyruvate saturation curve from sigmoidal to hyperbolic and decreased the Hill co-

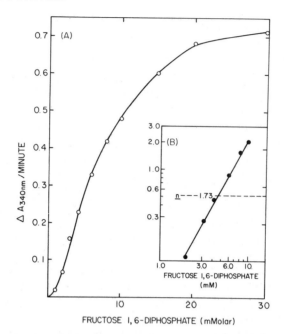

FRUCTOSE I, 6-DIPHOSPHATE (mMolar)

Fig. 4(A)   Effects of increasing concentrations of fructose-1,6-diphosphate on the activity of lactate dehydrogenase from *S. mutans* NCTC 10449 (Brown and Wittenberger, 1972b). The enzyme was purified from mannitol-adapted cells as previously described (Brown and Wittenberger, 1972b) and lactate dehydrogenase activity assayed in the presence of increasing levels of fructose-1,6-diphosphate by monitoring the pyruvate-dependent oxidation of NADH (Brown and Wittenberger, 1972b). (B) Hill plot of the fructose-1,6-diphosphate saturation data for the lactate dehydrogenase from *S. mutans* NCTC 10449 (Brown and Wittenberger, 1972b). The fructose-1,6-diphosphate saturation data used for determining the Hill coefficients were the same as shown in Fig. 4A (Brown and Wittenberger, 1972b).

efficients for the pyruvate saturation data from 1.7 to 1.0 without inhibiting enzyme activity or effecting V_{max}. Oxalate, on the other hand, appeared to be preferentially bound at the catalytic site. This pyruvate analog inhibited enzymatic activity at all pyruvate concentrations without shifting the sigmoidal nature of the pyruvate saturation curve or altering the Hill coefficient for the substrate.

The lactate dehydrogenase from *S. mutans* is subject to negative regulation by ATP (Brown and Wittenberger, 1972b). Since the primary function of the glycolytic pathway in *S. mutans* is the generation of ATP for biosynthesis and growth, negative control of the lactate dehydrogenase in *S. mutans* by this nucleoside triphosphate may provide this organism with a means of conserving substrate carbon when the intracellular pool levels of ATP are high.

6. Regulation of Glucose Carbon Flow between Diverse Catabolic Pathways in S. faecalis

Early studies have indicated that the sole product of the glucose fermentation in *S. faecalis* is lactic acid and that the lactic acid is derived exclusively by the Embden-Meyerhof pathway (Platt and Foster, 1957). It has been shown, however, that glucose-6-phosphate dehydrogenase and 6-phosphogluconate dehydrogenase, enzymes responsible for the conversion of the glycolytic intermediate glucose-6-phosphate to CO_2 and ribulose-5-phosphate, are present constitutively in *S. faecalis* (Table VII) (Brown and Wittenberger, 1971b). This raised the question as to how *S. faecalis* is able to utilize glucose carbon preferentially by the glycolytic pathway when at least the oxidative portion of the hexose monophosphate pathway is present constitutively in the cell. One possible mechanism for directing glucose carbon flow through the glycolytic pathway would be the inhibition of one or both of the hexose monophosphate

TABLE VII

Effect of Various Growth Substrates on the Levels of Glucose-6-Phosphate Dehydrogenase and 6-Phosphogluconate Dehydrogenase in Cell Extracts of *S. faecalis* MR[a]

Growth substrate	Enzymatic activity (units/mg protein)	
	G6PDH	6PGDH
Sodium D-gluconate	0.0691	0.0740
D-Glucose	0.0788	0.0675
D-Fructose	0.0482	0.0643
D-Ribose	0.0772	0.0723
D-Mannitol	0.0675	0.0691
D-Sorbitol	0.0675	0.0659
D,L-Serine	0.0370	0.0498

[a] The cells were grown anaerobically in complex medium supplemented with the appropriate carbon source (0.5% w/v) and extracts prepared as described previously (Brown and Wittenberger, 1971b). All extracts were assayed for glucose-6-phosphate dehydrogenase (G6PDH) and 6-phosphogluconate dehydrogenase (6PGDH) activities by monitoring the enzyme-dependent reduction of NADP at 340 nm as also described previously (Brown and Wittenberger, 1971b).

pathway dehydrogenases by one or more glycolytic intermediates. Of a variety of glycolytic intermediates tested singly and in combination for their effect on enzymatic activity only fructose-1,6-diphosphate was found to be an effective inhibitor of hexose monophosphate dehydrogenase activity (Table VIII) (Brown and Wittenberger, 1971b). In addition, the fructose-1,6-diphosphate inhibition was specific for 6-phosphogluconate dehydrogenase, having no effect on glucose-6-phosphate dehydrogenase activity. The concentrations at which fructose-1,6-diphosphate is an effective inhibitor of 6-phosphogluconate dehydrogenase activity are within

TABLE VIII

Effect of Various Glycolytic Intermediates on Glucose-6-Phosphate Dehydrogenase and 6-Phosphogluconate Dehydrogenase Activities from *S. faecalis* MR[a]

| | % Inhibition at 1 mM inhibitor concentration | |
Compounds	G6PDH	6PGDH
Glucose-6-phosphate (G-6-P)	—	0
Glucose-1-phosphate (G-1-P)	0	<5
Fructose-6-phosphate (F-6-P)	0	12
Fructose-1-phosphate (F-1-P)	<5	<5
Fructose-1,6-diphosphate (FDP)	0[b]	71[b]
Glyceraldehyde-3-phosphate	0	0
3-Phosphoglycerate	<5	0
2-Phosphoglycerate	10	0
Phosphoenolpyruvate	<5	<5
Pyruvate	0	0
Lactate	0	0
ATP	10	0
G-1-P + G-6-P	—	<5
F-1-P + F-6-P	<5[b]	<5[b]
G-1-P + F-1-P	<5	0
G-1-P + F-6-P	0	5
F-1-P + G-6-P	—	<0
G-6-P + F-6-P	—	<5

[a] Assays for glucose-6-phosphate dehydrogenase (G6PDH) and 6-phosphogluconate dehydrogenase (6PGDH) were performed by monitoring the enzyme-dependent reduction of NADP at 340 nm in the presence and absence of various inhibitors listed above as described previously (Brown and Wittenberger, 1971b).

[b] Here underlining indicates metabolites with potential regulatory activity.

the same range as those concentrations of fructose-1,6-diphosphate which
activate the lactate dehydrogenase from *S. faecalis* (Wittenberger and
Angelo, 1970). It is interesting to note that the 6-phosphogluconate dehydrogenase from *S. faecalis* could not be inhibited 100% by fructose-1,6-
diphosphate (Fig. 5) (Brown and Wittenberger, 1972a). This may be significant in that even at high intracellular fructose-1,6-diphosphate levels
enough glucose carbon may be metabolized via the hexose monophosphate
pathway to supply the ribose-5-phosphate necessary for nucleic acid biosynthesis. Thus, fructose-1,6-diphosphate serves to direct glucose carbon
flow preferentially down the glycolytic pathway in *S. faecalis* by serving
as a specific activator of the terminal step in glycolysis, lactate dehydrogenase (Wittenberger and Angelo, 1970; Wolin, 1964), and a specific inhibitor of the second step in the oxidative portion of the hexose mono-

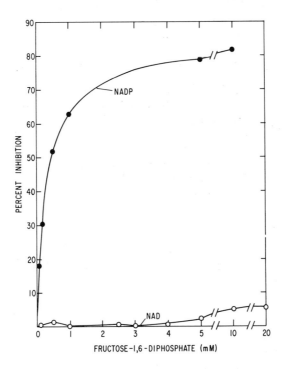

Fig. 5 Effect of fructose-1,6-diphosphate on the NAD- and NADP-dependent
6-phosphogluconate dehydrogenases from gluconate-adapted *S. faecalis* (Brown and
Wittenberger, 1972a). Assays for the (○) NAD or (●) NADP-specific 6-phospho-
gluconate dehydrogenases were carried out by monitoring the 6-phosphogluconate-
dependent reduction of the respective pyridine nucleotide at 340 nm in the presence
of increasing levels of fructose-1,6-diphosphate as previously described (Brown
and Wittenberger, 1972a).

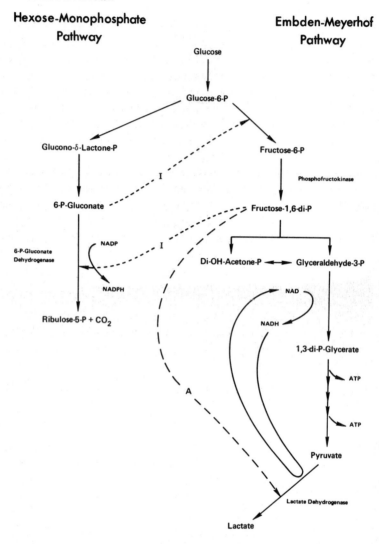

Hexose-Monophosphate
Pathway

Embden-Meyerhof
Pathway

Glucose

Glucose-6-P

Glucono-δ-Lactone-P

Fructose-6-P

Phosphofructokinase

I

6-P-Gluconate

Fructose-1,6-di-P

6-P-Gluconate
Dehydrogenase

NADP

I

Di-OH-Acetone-P ⟶ Glyceraldehyde-3-P

NADPH

NAD

Ribulose-5-P + CO₂

NADH

1,3-di-P-Glycerate

ATP

A

ATP

Pyruvate

Lactate Dehydrogenase

Lactate

Fig. 6 Schematic representation of the relationship between the hexose monophosphate and Embden-Meyerhof pathways in *S. faecalis* (Brown and Wittenberger, 1971b). Regulatory sites indicated by I (inhibition) and A (activation).

phosphate pathway, 6-phosphogluconate dehydrogenase (Fig. 6) (Brown and Wittenberger, 1971b).

Earlier reports, however, have indicated that *S. faecalis* metabolizes sodium gluconate by a combination of the Entner-Doudoroff and hexose monophosphate pathways (Sokatch and Gunsalus, 1957). Since these organisms produce lactic acid from gluconate (Sokatch and Gunsalus,

1957), and the lactate dehydrogenase from glucose-adapted *S. faecalis* exhibits an absolute dependence upon fructose-1,6-diphosphate for catalytic activity (Wittenberger and Angelo, 1970; Wolin, 1964) intracellular pool levels of fructose-1,6-diphosphate sufficiently high enough to activate the lactate dehydrogenase should effectively inhibit the NADP-linked 6-phosphogluconate dehydrogenase activity and thus prohibit gluconate metabolism via the hexose monophosphate pathway. It was initially postulated that two possibilities might exist which would enable gluconate metabolism to proceed by the hexose monophosphate pathway with the subsequent production of lactic acid. One possibility was that gluconate-adapted cells possess a lactate dehydrogenase which was not dependent upon the presence of fructose-1,6-diphosphate for catalytic activity, and the other was that gluconate-adapted cells possess an NADP-dependent 6-phosphogluconate dehydrogenase that was insensitive to fructose-1,6-diphosphate inhibition. As shown in Table IX (Brown and Wittenberger, 1972a) neither of these possibilities occurred. Gluconate-adapted *S. faecalis* strains, however, do possess induced levels of an NAD-dependent 6-phosphogluconate dehydrogenase in addition

TABLE IX

Effect of Fructose-1,6-Diphosphate upon Lactate Dehydrogenase and 6-Phosphogluconate Dehydrogenase Activity from *S. faecalis* MR Grown on Various Carbon Sources[a,b]

	Enzymatic activity (units/mg protein)			
	Lactate dehydrogenase		6-Phosphogluconate dehydrogenase	
Carbon source	−FDP	+FDP	−FDP	+FDP
Glucose	0.020	1.43	0.072	0.018
Gluconate	0.015	1.10	0.075	0.016
Sucrose	0.020	1.84	0.070	0.015
Mannitol	0.010	1.52	0.069	0.017
Ribose	0.010	1.30	0.064	0.013

[a] From Brown and Wittenberger (1972a).

[b] Cells were grown anaerobically in complex medium and extracts prepared as described previously (Brown and Wittenberger, 1971b). Enzyme assays were performed in the presence and absence of 1 mM FDP. Lactate dehydrogenase activity was assayed by monitoring the pyruvate-dependent oxidation of NADH at 340 nm (Wittenberger and Angelo, 1970) and 6-phosphogluconate dehydrogenase activity was assayed by monitoring the 6-phosphogluconate-dependent reduction of NADP (Brown and Wittenberger, 1971b).

to the NADP-specific enzyme found in cells grown on glucose and sucrose (Table X) (Brown and Wittenberger, 1972a). The NAD-specific
6-phosphogluconate dehydrogenase unlike the NADP-linked enzyme is
insensitive to inhibition by fructose-1,6-diphosphate (Fig. 5) (Brown and
Wittenberger, 1972a). Consequently, even at intracellular fructose-1,6-
diphosphate levels high enough to activate the lactate dehydrogenase,
gluconate carbon flow may proceed through the NAD-specific 6-phos-
phogluconate dehydrogenase which is insensitive to inhibition by fructose-
1,6-diphosphate. The NAD-dependent enzyme, however, is subject to
negative control since it is inhibited by ATP (Brown and Wittenberger,
1972a). This is in contrast to the NADP-specific 6-phosphogluconate dehydrogenase which is insensitive to ATP inhibition (Table XI). Inhibition of the NAD-linked enzyme by ATP may provide *S. faecalis* with
a mechanism for conserving gluconate carbon when the intracellular
levels of ATP are high.

TABLE X

6-Phosphogluconate Dehydrogenase Activity
with NADP or NAD in Extracts of *S. faecalis*
MR Grown on a Variety of Energy Sources[a,b]

Growth substrate	6PGDH activity (units/mg protein)	
	NADP	NAD
Glucose	0.072	0.004
Gluconate	0.068	0.068
Glucose plus gluconate	0.080	0.008
Ribose	0.078	0.004
Sucrose	0.062	0.003
Mannitol	0.068	0.002
Fructose	0.070	0.005

[a] From Brown and Wittenberger (1972a).

[b] Cells were grown anaerobically in complex
medium supplemented with the appropriate
carbon source(s) and extracts prepared as described previously (Brown and Wittenberger,
1971b). 6-Phosphogluconate dehydrogenase
(6PGDH) activity was assayed by monitoring
the 6-phosphogluconate-dependent reduction of
NADP or NAD at 340 nm as also described previously (Brown and Wittenberger, 1971b).

TABLE XI

Differential Effects of Fructose-1,6-Diphosphate and ATP
on the NADP and NAD-Specific 6-Phosphogluconate
Dehydrogenase from *S. faecalis* MR[a]

Ligand (10 mM)	% Inhibition of designated 6-phosphogluconate dehydrogenase	
	NADP-specific 6PGDH	NAD-specific 6PGDH
Fructose-1,6-diphosphate	85	5
ATP	0	68

[a] Cell extracts were prepared from gluconate-adapted *S. faecalis* grown anaerobically in complex medium as described previously (Brown and Wittenberger, 1971b). 6-Phosphogluconate dehydrogenase (6PGDH) activity was assayed by monitoring the 6-phosphogluconate-dependent reduction of NAD or NADP at 340 nm in the presence and absence of the inhibitors listed above as also described previously (Brown and Wittenberger, 1972a).

F. Research Needs

1. Control of Sucrose Dissimilation in S. mutans

Since *S. mutans*-mediated caries processes are dependent upon the dissimilation of sucrose, studies designed to obtain information as to how sucrose metabolism is controlled at the cellular level are necessary in order to understand the biochemical bases for the pathogenicity of these organisms. An understanding of the basic biochemical processes responsible for the virulence of these organisms cannot but lead to therapeutic measures which would decrease the frequency and incidence of dental caries. Specifically, studies should be initiated which will determine those mechanisms by which *S. mutans* regulates the distribution of sucrose carbon flow between glycolysis, intracellular polysaccharide production, and extracellular glucan and fructan synthesis. Termination of one or more of these processes will serve to either eliminate *S. mutans* completely from the oral environment or markedly decrease their virulence. Those processes which regulate sucrose carbon flow between the two intracellular pathways for sucrose catabolism, glycolysis, and intracellular polysaccharide synthesis (Fig. 1) are currently being studied. However, these studies should be expanded to include the regulation of sucrose carbon flow between these intracellular pathways and extracellular glucan and fructan synthesis.

2. Studies of Microbial Interactions in Dental Plaque

Plaque is composed of a variety of microorganisms other than oral streptococci. Studies should be initiated to determine the carbohydrate-dependent microbial interactions operative within plaque which serve to make it a balanced ecological system and allow specific oral bacteria to metabolize sucrose and remain viable. An in-depth knowledge of the microbial interactions within dental plaque will contribute to an understanding of the dynamics of microbial-mediated oral disease. Work in the author's laboratory and in other laboratories is now being done to determine the ability of fermentation products and extracellular polymers from certain plaque bacteria to support the growth of other components of the plaque microflora, particularly oral streptococci.

3. Selective Elimination of Sucrose-Dependent Glucan Production

As stated previously, the production of extracellular, insoluble glucans impart to *S. mutans* the ability to adhere to smooth tooth surfaces. *S. mutans* which could not adhere to smooth tooth surfaces would be unable to play a role in smooth surface caries. The production of insoluble glucans might be stopped by exposure of *S. mutans* to chemical agents such as sucrose analogs which might serve as specific inhibitors of glucosyltransferase activity. Also, immunization with purified glucosyltransferases might inhibit plaque formation *in vivo* through the production of salivary antibodies. The induction of dextranase synthesis by other plaque microorganisms might result in the production of an enzyme which would specifically degrade *S. mutans*-associated glucans. The production of dextranase by an indigenous organism would probably not upset the delicate ecological balance which exists in the oral environment. A search for useful antiplaque agents which interfere specifically with sucross-dependent extracellular glucan synthesis may provide information essential for the control of *S. mutans*-induced dental caries.

Acknowledgments

The author would like to thank Mrs. Diane Hallo for her help in the preparation of this manuscript.

References

Berman, K. S., and Gibbons, R. J. (1966). Iodophilic polysaccharide synthesis by human and rodent oral bacteria and its relation to cariogenicity. *Arch. Oral Biol.* **11**, 533.

Bratthall, D. (1970). Demonstration of five serological groups of streptococcal strains resembling *Streptococcus mutans*. *Odontol. Revy* **21**, 143.

Brown, A. T., and Patterson, C. E. (1972). Heterogeneity of *Streptococcus mutans* strains based upon their mannitol-1-phosphate dehydrogenases. A criterion for rapid classification. *Infect. Immunity* **6**, 422.

Brown, A. T., and Patterson, C. E. (1973). Ethanol production and alcohol dehydrogenase activity in *Streptococcus mutans*. *Arch. Oral Biol.* **18**, 127.

Brown, A. T., and Wittenberger, C. L. (1971a). The occurrence of multiple glyceraldehyde-3-phosphate dehydrogenases in cariogenic streptococci. *Biochem. Biophys. Res. Commun.* **43**, 217.

Brown, A. T., and Wittenberger, C. L. (1971b). Mechanism for regulating the distribution of glucose carbon between the Enbden-Meyerhof and hexose monophosphate pathways in *Streptococcus faecalis*. *J. Bacteriol.* **106**, 456.

Brown, A. T., and Wittenberger, (1972a). Induction and regulation of a nicotinamide adenine-specific 6-phosphogluconate dehydrogenase in *Streptococcus faecalis*. *J. Bacteriol.* **109**, 106.

Brown, A. T., and Wittenberger, C. L. (1972b). Fructose-1,6-diphosphate dependent lactate dehydrogenase from a cariogenic *Streptococcus:* Purification and regulatory properties. *J. Bacteriol.* **110**, 604.

Brown, A. T., and Wittenberger, C. L. (1973). Mannitol and sorbitol catabolism in *Streptococcus mutans*. *Arch. Oral Biol.* **18**, 117.

Carlsson, J. (1965a). Zooglea-forming streptococci resembling *Streptococcus sanguis* isolated from dental plaque in man. *Odontol. Revy* **16**, 55.

Carlsson, J. (1965b). Effect of diet on the presence of *Streptococcus salivarius* in dental plaque and saliva. *Odontol. Revy* **16**, 336.

Carlsson, J. (1967). Presence of various types of non-haemolytic streptococci in dental plaque and in other sites of the oral cavity in man. *Odontol. Revy* **18**, 55.

Carlsson, J. (1970). A levansucrase from *Streptococcus mutans*. *Caries Res.* **4**, 97.

Carlsson, J., and Sundström, B. (1968). Variations in composition of early dental plaque following ingestion of glucose and sucrose. *Odontol. Revy* **19**, 161.

Carlsson, J., Newbrun, E., and Krasse, B. (1969). Purification and properties of dextran sucrase from *Streptococcus sanguis*. *Arch. Oral Biol.* **14**, 469.

Charlton, G., Fitzgerald, R. J., and Keyes, P. H. (1971a). Determination of saliva and dental plaque pH in hamsters with glass microelectrodes. *Arch. Oral Biol.* **16**, 649.

Charlton, G., Fitzgerald, D. B., and Keyes P. H. (1971b). Hydrogen ion activity in dental plaques of hamsters during metabolism of sucrose, glucose, and fructose. *Arch. Oral Biol.* **16**, 655.

Clarke, J. K. (1924). On the bacterial factor in the etiology of dental caries. *Brit. J. Exp. Pathol.* **5**, 141.

Coykendall, A. (1970). Base composition of deoxyribonucleic acid isolated from cariogenic streptococci. *Arch. Oral Biol.* **15**, 365.

Coykendall, A. (1971). Genetic heterogeneity in *Streptococcus mutans*. *J. Bacteriol.* **106**, 192.

716   *Albert T. Brown*

Coykendall, A. (1972). A proposal to divide *Streptococcus mutans* into four varieties. *Int. Ass. Dent. Res.* Abstr. No. 66.

daCosta, T., and Gibbons, R. J. (1968). Hydrolysis of levan by human plaque streptococci. *Arch. Oral Biol.* **13**, 609.

Dahlqvist, A., Krasse, B., Olsson, I., and Gardell, S. (1967). Extracellular polysaccharides formed by caries inducing streptococci. *Helv. Odontol. Acta* **11**, 15.

de Stoppelaar, J. D., van Houte, J., and Backer Dirks, O. (1969). The relationship between extracellular polysaccharide-producing streptococci and smooth surface caries in 13-year-old children. *Caries Res.* **3**, 190.

de Stoppelaar, J. D., Konig, K. G., Plasschaert, A. J. M., and van der Hoeven, J. S. (1971). Decreased cariogenicity of a mutant of *Streptococcus mutans*. *Arch. Oral Biol.* **16**, 971.

Donahue, J. J., Kestenbaum, R. C., and King, W. J. (1966). The utilization of sugar by selected strains of oral streptococci. *Int. Ass. Dent. Res. Abstr.* No. 58.

Drucker, D. B., and Melville, T. H. (1968). Fermentation end-products of cariogenic and non-cariogenic streptococci. *Arch. Oral Biol.* **13**, 563.

Edwardsson, S. (1968). Characteristics of caries inducing streptococci resembling *Streptococcus mutans*. *Arch. Oral Biol.* **13**, 637.

Edwardsson, S. (1970). The caries-inducing property of variants of *Streptococcus mutans*. *Odontol. Revy* **21**, 153.

Fitzgerald, R. J., and Jordan, H. V. (1968). Polysaccharide producing bacteria and caries. *In* "The Art and Science of Dental Caries Research" (R. S. Harris, ed.), p. 167. Academic Press, New York.

Fitzgerald, R. J., and Keyes, P. H. (1960). Demonstration of the etiologic role of streptococci in experimental caries in the hamster. *J. Amer. Dent. Ass.* **61**, 9.

Fitzgerald, R. J., Spinell, D. M., and Stoudt, T. H. (1968a). Enzymatic removal of artificial plaques. *Arch. Oral Biol.* **13**, 125.

Fitzgerald, R. J., Keyes, P. H., Stoudt, T. H., and Spinell, D. M. (1968b). The effects of a dextranase preparation on plaque and caries in hamsters, a preliminary report. *J. Amer. Dent. Ass.* **76**, 301.

Freedman, M. L., and Tanzer, J. M. (1972). Dissociation of plaque formation from glucan-induced agglutination in mutans of *Streptococcus mutans*. *Abstr. Amer. Soc. Microbiol.*, p. 287.

Gibbons, R. J. (1964). Metabolism of intracellular polysaccharide by *Streptococcus mitis* and its relation to inducible enzyme formation. *J. Bacteriol.* **87**, 1512.

Gibbons, R. J., and Banghart, S. B. (1967). Synthesis of extracellular dextran by cariogenic bacteria: Its presence in human dental plaque. *Arch. Oral Biol.* **12**, 11.

Gibbons, R. J., Berman, K. S., Knoettner, P., and Kapsimalis, B. (1966). Dental caries and alveolar bone loss in gnotobiotic rats infected with capsule forming streptococci of human origin. *Arch. Oral Biol.* **11**, 549.

Gibbons, R. J., and Fitzgerald, R. J. (1969). Dextran induced agglutination of *Streptococcus mutans* and its potential role in the formation of microbial dental plaque. *J. Bacteriol.* **98**, 341.

Gibbons, R. J., and Nygaard, M. (1968). Synthesis of insoluble dextran and its significance in the formation of gelatinous deposits by plaque-forming streptococci. *Arch. Oral Biol.* **13**, 1249.

Guggenheim, B. (1968). Streptococci of dental plaques. *Caries Res.* **2**, 147.

Guggenheim, B. (1970). Enzymatic hydrolysis and structure of water-insoluble glucan produced by glucosyl transferases from a strain of *Streptococcus mutans*. *Helv. Odontol. Acta* **14**, 89.

Guggenheim, B., and Haller, R. (1972). Purification and properties of an α-(1 → 3) glucan hydrolase from *Trichoderma harzianum. J. Dent. Res.* **51**, 394.

Guggenheim, B., and Newbrun, E. (1969). Extracellular glucosyl transferase activity of an HS strain of *Streptococcus mutans. Helv. Odontol. Acta* **13**, 84.

Guggenheim, B., and Schroeder, H. E. (1967). Biochemical and morphological aspects of extracellular polysaccharides produced by cariogenic streptococci. *Helv. Odontol. Acta* **11**, 131.

Higuchi, M., Iwami, Y., Yamada, T., and Araya, S. (1970). Levan synthesis and accumulation by human dental plaque. *Arch. Oral Biol.* **15**, 563.

Ikeda, T., and Sandham, H. J. (1971). Prevalence of *Streptococcus mutans* on various tooth surfaces in Negro children. *Arch. Oral Biol.* **16**, 1237.

Jordan, H. V. (1965). Bacteriological aspects of experimental dental caries. *Ann. N.Y. Acad. Sci.* **131**, 905.

Jordan, H. V., and Keyes, P. H. (1966). *In vitro* methods for the study of plaque formation and carious lesions. *Arch. Oral Biol.* **11**, 793.

Keyes, P. H. (1968). Research in dental caries. *J. Amer. Dent. Ass.* **76**, 1357.

Krasse, B. (1953). The proportional distribution of different types of streptococci in saliva and plaque material. *Odontol. Revy* **4**, 304.

Krasse, B. (1954). The proportional distribution of *Streptococcus salivarius* and other streptococci in various parts of the mouth. *Odontol. Revy* **5**, 203.

Krasse, B. (1966). Human streptococci and experimental caries in hamsters. *Arch. Oral Biol.* **11**, 429.

Krasse, B., Jordan, H. V., Edwardsson, S., Svenssen, I., and Trell, L. (1968). The occurrence of certain "caries-inducing" streptococci in human dental plaque material. *Arch. Oral Biol.* **13**, 911.

Larson, R. H., Theilade, E., and Fitzgerald, R. J. (1967). The interaction of diet and microflora in experimental caries in the rat. *Arch. Oral Biol.* **12**, 663.

Littleton, N. W., Kakehashi, S., and Fitzgerald, R. J. (1970). Recovery of specific "caries-inducing" streptococci from various lesions in the teeth of children. *Arch. Oral Biol.* **15**, 461.

Loesche, W. J., and Henry, C. A. (1967). Intracellular microbial polysaccharide production and dental caries in a Guatemalan Indian village. *Arch. Oral Biol.* **12**, 189.

Luoma, H. (1964). Lability of inorganic phosphate in dental plaque and saliva. *Acta Odontol. Scand.* **22**, 7.

McCabe, P. M., Keyes, P. H., and Howell, A., Jr. (1953). An *in vitro* method for assessing the plaque forming ability of oral bacteria. *Arch. Oral Biol.* **12**, 1963.

Mandel, I. D. (1970). Effects of dietary modifications on caries in humans. *J. Dent. Res.* **49**, 1201.

Manly, R. S., and Richardson, D. T. (1968). Metabolism of levan by oral samples. *J. Dent. Res.* **47**, 1080.

Miller, W. D. (1890). "The Micro-organisms of the Human Mouth." S. S. White Dental Manufacturing Company, Philadelphia, Pennsylvania.

Newbrun, E. (1971). Dextransucrase from *Streptococcus sanguis* further classification. *Caries Res.* **5**, 124.

Newbrun, E. (1972). Extracellular polysaccharides synthesized by glucosyl transferases of oral streptococci. *Caries Res.* **6**, 132.

Newbrun, E., and Carlsson, J. (1969). Reaction rate of dextransucrase from *Streptococcus sanguis* in the presence of various compounds. *Arch. Oral Biol.* **14**, 461.

Newbrun, E., Lacy, R., and Christie, T. (1971). The morphology and size of the extracellular polysaccharides from oral streptococci. *Arch. Oral Biol.* **16**, 863.

Parker, R. B., and Creamer, H. R. (1971). Contribution of plaque polysaccharides to growth of cariogenic microorganisms. *Arch. Oral Biol.* **16**, 855.

Pietruszko, R., Clark, A., Graves, J. M. H., and Ringold, H. J. (1966). The steroid activity and multiplicity of crystalline alcohol dehydrogenase. *Biochem. Biophys. Res. Commun.* **4**, 526.

Pietruszko, R., Ringold, H. J., Li, T. K., Vallee, B. L., Akeson, A., Theorell, H., (1969). Structure and function relationships in isozymes of horse liver alcohol dehydrogenase. *Nature (London)* **221**, 440.

Platt, T. B., and Foster, E. M. (1957). Products of glucose metabolism by homofermentative streptococci under anaerobic conditions. *J. Bacteriol.* **75**, 453–459.

Rogosa, M., and Love, L. L. (1968). Direct quantitative gas chromatographic separation of C_2–C_6 fatty acids, methanol, and ethyl alcohol in aqueous microbial fermentation media. *Appl. Microbiol.* **16**, 285.

Sidebotham, R. L., Weigel, H., and Bowen, W. H. (1971). Studies on dextrans and dextranases. *Carbohyd. Res.* **19**, 151.

Sokatch, J. T., and Gunsalus, I. C. (1957). Aldonic acid metabolism. I. Pathway of carbon in an inducible gluconate fermentation by *Steptococcus faecalis*. *J. Bacteriol.* **73**, 452.

Takeuchi, M. (1961). Epidemiological study on dental caries in Japanese children before, during, and after World War II. *Int. Dent. J.* **11**, 443.

Tanzer, J. M. (1972). Studies on the fate of the glucosyl moiety of sucrose metabolized by *Streptococcus mutans*. *J. Dent. Res.* **51**, 415.

Tanzer, J. M. (1973). The role of microorganisms in dental caries. *Proc. Int. Cong. Microbiol. 10th, 1970* (In press).

Tanzer, J. M., Krichevsky, M. I., and Keyes, P. H. (1969a). The coupling of phosphate accumulation to acid production by a non-growing streptococcus. *J. Gen. Microbiol.* **55**, 351.

Tanzer, J. M., Wood, W. I., and Krichevsky, M. I. (1969b). Linear growth kinetics of plaque-forming streptococci in the presence of sucrose. *J. Gen. Microbiol.* **58**, 125.

Tanzer, J. M., Krichevsky, M. I., and Keyes, P. H. (1969c). The metabolic fate of glucose catabolized by a washed stationary phase caries-conducive streptococcus. *Caries Res.* **3**, 167.

Tanzer, J. M., Brown, A. T., and Meyers, K. I. (1972). Sucrose dissimilation by *Streptococcus mutans* independently of glucosyl and fructosyl transferase. *Int. Ass. Dent. Res.* Abstr. No. 232.

Tanzer, J. M., Chassy, B. M., and Krichevsky, M. I. (1972). Sucrose metabolism by *Streptococcus mutans* SL-1. *Biochem. Biophys. Acta* **261**, 379.

Theorell, H., Taniquchi, S., Akeson, A., and Skursky, L. (1966). Crystallization of a separate steroid-active liver alcohol dehydrogenase. *Biochem. Biophys. Res. Commun.* **24**, 603.

van Houte, J. H. (1967). Iodophilic polysaccharide in bacteria from the dental plaque. Ph.D. Dissertation, University of Utrecht.

van Houte, J. H. (1968). Levan degradation by streptococci isolated from human dental plaque. *Arch. Oral Biol.* **13**, 827.

van Houte, J. H., Backer Dirks, O., de Stoppelaar, J. D., and Jansen, H. M. (1969). Iodophilic polysaccharide-producing bacteria and dental caries in children consuming flouridated and non-flouridated drinking water. *Caries Res.* **3**, 178.

Wittenberger, C. L., and Angelo, N. (1970). Purification and properties of a fructose-1,6-diphosphate activated lactic dehydrogenase from *Streptococcus faecalis. J. Bacteriol.* **101,** 717.

Wolin, M. J. (1964). Fructose-1,6-diphosphate requirement of streptococcal lactic dehydrogenases. *Science* **146,** 775.

Wood, J. M. (1964). Polysaccharide synthesis and utilization of dental plaque. *J. Dent. Res.* **43,** 955.

Wood, J. M. (1967). The amount, distribution and metabolism of soluble polysaccharides in human dental plaque. *Arch. Oral Biol.* **12,** 849.

Wood, J. M., and Critchley, P. (1966). The extracellular polysaccharide produced from sucrose by a cariogenic streptococcus. *Arch. Oral Biol.* **11,** 1039.

Zinner, D. D., Jablon, J. M., Aran, A. P., and Saslaw, M. S. (1965). Experimental caries induced in animals by streptococci of human origin. *Proc. Soc. Exp. Biol. Med.* **188,** 766.

Author Index

Numbers in italics refer to the pages on which the complete references are listed.

A

Abei, T., 607, *612*
Abell, L. L., 407, 408, *432*
Abrahamsson, L., 330, *336*
Acree, T. E., 20, *36*, 69, 70, *79*, *80*
Acton, E. M., 27, *34*
Adachi, F., 363, *370*
Adachi, K., 235, *237*, 617, *629*, *630*
Adachi, R. A., 322, *334*
Adams, L. C., 477, *480*, 494, *509*
Adams, M., 405, *436*
Adcock, L. D., 549, *563*
Adelman, R. C., 358, *368*
Adnitt, P. I., 457, *465*
Åkerblom, H. K., 445, *446*
Åkesson, B., 189, *210*
Agarwal, L. P., 496, 497, *507*
Agarwal, P. K., 496, 497, *507*
Aguire, L. M., *605*
Ahlfeld, H., 82, 84, *92*
Ahmad, Z., 155, *168*
Ahrens, E. H., 307, *311*, 408, 409, 420, 421, *429*
Ahuja, M. M. S., 457, 459, *465*
Ainsworth, S. K., 577, *588*
Akeson, A., 701, *718*
Akinyanju, P. A., 411, 427, *429*
Alakija, W., 181, *185*
Albrecht, G. J., 294, *298*
Albury, M. N., 668, *685*
Aldor, T. A. M., 182, *184*
Alhough, I., 268, *279*
Al-Khalidi, U., 414, *436*
Allen, F. H., Jr., 291, *300*
Allen, R. J. L., 309, *311*, 405, *429*
Allerton, R., 27, *36*
Allison, F., 577, *588*
Almen, M. C., 548, *566*
Alpers, D. H., 150, *165*

Altschul, R., 405, *435*
Altshuler, J. H., 160, *168*
Alvarado, F., 155, 161, *165*, *166*
Alvarez, W. C., 314, *333*
Amador, F., 599, 600, 601, *605*
Amherdt, M., 498, 503, *510*, *511*
Ammann, R., 188, *212*
Amundson, C. H., 282, *301*
Andersen, D. H., 341, *353*
Anderson, C. M., 196, *212*, 306, *312*
Anderson, E. P., 289, *300*, 338, *355*, 377, *386*
Anderson, G. E., 541, 543, 544, 545, 546, 547, 548, 559, *563*, *565*
Anderson, J., 321, 330, 331, *335*
Anderson, J. T., 409, 410, 413, 414, 415, 416, 417, 418, 419, 420, *429*, *431*, *432*, *433*
Anderson, J. W., 163, *166*
Anderson, L., 235, *238*
Anderson, P. R., 182, *185*, 200, *212*, 283, *300*
Anderson, S. C., 403, 404, 405, 414, *437*
Anderson, T. A., 404, 415, *429*, *431*
Andersson, B., 28, *34*
Andersson, G., 578, *588*
Andersson, H., 195, *209*
Andreae, U., 353, *356*
Andreev, D., 494, *507*
Andrus, S. B., 403, 405, *431*, *435*
Angelo, N., 705, 709, 711, *719*
Anitschkow, N., 402, *430*
Anno, K., 73, *80*
Antar, M. A., 409, 418, 423, *430*, *434*, 440, 442, *446*, *447*
Anthony, S., 471, 473, *482*
Antonini, E., 282, *301*
Antonis, A., 408, 416, 417, *430*
Anzai, M., 567, *590*
Applegarth, D. A., 231, *238*

721

G

Gabbay, K. H., 228, *299*, 384, 385, *386,* 469, 470, 471, 472, 473, 475, 476, 477, 478, 480, *481, 482*
Gabrielson, I. W., 160, *168,* 306, 309, *311*
Gaffar, A., 668, *682*
Gallagher, T. F., Jr., 163, *166*
Galloway, B., 298, *301*
Gander, J. E., 285, *299*
Garces, L. Y., 454, *465*
Gardell, S., 665, *681,* 694, 695, *716*
Gardiner, B. L., 320, *335*
Garfinkel, A. S., 406, *430*
Garnett, E. S., 398, *399*
Gatti, E., 426, *431*
Gauhe, A., 221, *225*
Geddes, R., 147, *168*
Gehring, F., 656, *682*
Gelman-Malachi, E., 182, *184*
Gentili, B., 27, *35*
Gentz, J. C. H., 452, 453, *466*
Genuth, S. M., 501, *509*
Gericke, C., 277, *279*
Gerritsen, G. C., 501, 503, *507, 508*
Geser, C. A., 266, 271, *280*
Geyer, R. P., 528, 558, *564*
Gibbons, R. J., 636, 637, *642, 643,* 650, 651, 654, 662, 665, 668, 674, 675, *681, 682,* 690, 693, 694, 695, 696, *715, 716*
Gibbons, T., 407, *431*
Gibbs, G. E., 494, *509*
Gifford, H., Jr., 494, *509*
Gilat, T., 182, *184,* 283, *299*
Gilles, K. A., 316, *334*
Gilligan, J. E., 237, *239,* 269, *280,* 539, 559, *566,* 568, 570, 583, 585, *589, 590,* 591, 592, *606,* 625, *631*
Ginsberg, J. L., 245, *257,* 358, *369*
Ginsburg, V., 158, *168,* 407, *431*
Gitzelmann, R., 288, 295, *299,* 317, *334,* 338, 339, *354*
Glaser, L., *299*
Gleason, R. E., 456, *465*
Gluck, C. M., 395, *400*
Glynn, L. E., 586, *589,* 592, *605*
Goatcher, W. D., 26, 32, *35*
Gobble, W. G., Jr., 531, *565*
Goddard, J. W., 494, *509*
Goebell, H., 363, *368*

Goetz, F. C., 600, 601, 602, *605*
Gold, E. M., 394, *399*
Goldblatt, P. J., 363, *369*
Goldfinger, S., 580, *589*
Goldkamp, A. H., 27, *36*
Goldman, B. M., 640, *643*
Goldman, M. H., 538, *563*
Goldman, R. F., 395, *400*
Goldner, A. M., 155, *168*
Goldrick, R. B., 429, *431*
Goldsworthy, N. E., 639, *644*
Gomori, G., 339, *354*
Gomyo, T., 316, *334*
Gonet, A. E., 495, 503, 505, *509*
Gonon, W. F., 149, 150, *169*
Gonzalez, N. S., 358, *371*
Goodale, F., 417, *435*
Gorbachow, S. W., 48, *63*
Gordon, H. A., 637, *644*
Gordon, R. S., 406, *430*
Gorlin, R., 578, *588*
Gorman, R. E., 30, *34, 36*
Gossain, V. V., 457, 459, *465*
Goto, Y., 567, *590*
Gottenlos, J. J., 402, *436*
Gotterer, G. S., 154, *170*
Gracey, M., 206, *212*
Graff, J., 415, *437*
Graham, G. G., 206, *212*
Grande, F., 404, 405, 406, 409, 410, 412, 413, 414, 415, 416, 417, 419, 420, *429, 431, 432, 433, 435*
Grant, R. T., 484, *507*
Grant, W. C., 404, 405, *432*
Graves, J. M. H., 701, *718*
Gray, C. H., 549, *563*
Gray, G. M., 195, *212,* 284, *299*
Grebin, B., 538, *564*
Green, J. W., 70, *80*
Green, L. F., 306, 309, *311*
Greenawalt, J. W., 150, *166*
Greenberg, D. M., 360, *369*
Greene, H. L., Jr., 158, 160, 164, *168, 170,* 183, *184, 186*
Greengard, P., 30, *35*
Greenwood, C. T., 147, *168*
Greeves, A., *63*
Gregersen, G., 477, *481*
Greisbach, W., 548, *564*

Mintz, D., 454, *465*
Mitchell, H. S., 375, 376, *386*
Mitra, R. C., 155, *169*
Mittle, R., 495, 501, *507*
Miyakawa, T., 615, *630*
Moertel, C. G., 181, *186*
Molan, P. C., 653, *684*
Moloney, P. J., 484, *509*
Mommaerts, W. F. H. M., 343, *355*
Mondal, A., 155, *170*
Monod, J., 163, *169*, 178, *185*
Moonsammy, G. I., 231, 238, 470, 471, 472, 475, 479, *482*
Moore, F. D., 538, *563*, 607, 608, 609, *611, 612*
Moore, P., Jr., 181, *184*
Moore, S., 426, *431*
Moorhouse, J. A., 444, *448*
Morgan, A. G., 549, 561, *564*, 610, *612*
Morgan, C. R., *465*, 505, *509*
Morita, T. N., 607, 608, *612*
Morland, J., 425, *437*
Moro, E., 216, *226*
Morris, J. A., 23, 25, 27, *36*, 121, *125*
Morris, R. C., Jr., 360, *370*
Morrison, A. D., 230, 232, *239*, 472, 479, *481, 482*
Morrison, E. S., 417, *435*
Morrison, J. F., 286, *300*
Mosbach, K., *211*
Mosettig, E., 27, *36*
Mosinger, M., 594, 596, 597, *606*
Moskowitz, H. R., 40, 48, 51, 52, 53, 54, 55, 58, 60, *63*
Mossel, D. A. A., 188, 204, *214*
Mothon, S., 503, *507*
Motomura, Y., 75, *79*
Mottu, F., 324, *334*
Motulsky, A. G., 624, *631*
Mougin, J., 503, *509*
Moyer, J. C., 315, *335*
Mühlelann, H. R., 656, 667, 675, *682, 683*
Muehrcke, R. C., 611, *612*
Mürset, G., 188, 195, 196, 204, *209, 213*
Mukherjee, T. M., 570, *588*
Muleahy, R., 426, *431*
Mumford, P., 395, 397, *400*
Munday, K. A., 151, 154, *166*
Muntz, J. A., 549, *565*

Murata, A., 320, 323, *335*
Murisasco, A., 549, *565*
Murphy, E. L., 330, 332, *334, 335*
Murray, E. G., 320, *334*
Murray, R. L., 179, *185*
Murthy, D. Y. N., 461, 462, 463, *466*
Mushett, C. W., 402, *436*
Myers, C. S., 39, *63*
Myers, M. L., 409, 410, 414, 416, *432, 434*

N

Nadkarni, M. V., 484, *510*
Nagarajan, R., 75, *79*
Nagy, K. P., 181, *184*
Nahas, G. G., 575, 583, *590*
Naismith, D. J., 406, *434*
Naito, C., 405, *431*
Nakada, H. I., 358, *370*
Nakajima, Y., 625, *630*
Nakamura, N., 316, *334*
Nakashima, K., 490, *510*
Nasrallah, S. M., 231, *238*, 607, 608, *612*
Nath, N., 406, *435*
Naylor, M. N., 660, 675, *684, 686*
Neely, B. W., 670, *685*
Nelsestuen, G. L., 293, *300*
Nelson, D., 405, *435*
Nes, W. R., 27, *36*
Nestel, P. J., 419, 420, 423, 428, 429, *435, 440, 447*
Neufeld, A. H., 30, *34*
Neufeld, E. F., 352, *353, 355, 356*
Newbrun, E., 665, 670, 671, *680, 685*, 694, *715, 717, 718*
Newburgh, L. H., 397, *400*
Newman, E. V., 610, *612*
Ng, W. G., 289, *298*
Nichoalds, G. E., 614, *630*
Nichols, E. L., 41, *63*
Nicholson, J. F., 538, *564*
Nicolai, von H., 222, *226*
Nielsen, K., 494, *510*
Nikaido, H., 666, *685*
Nikkilä, E. A., 406, 407, 428, *435*, 439, 442, 444, *448*
Ninomiya, H., 617, 621, *630*
Nishika, M., 345, *356*
Noack, R., 318, *336*
Noel, P., 309, *311*

Subject Index

A

Absorption, *see also* Malabsorption, Metabolism
and digestion of carbohydrates, physiology of, 189–193
and hydrolysis of carbohydrates, 145–172

Acetone, breath analyses in diabetics, 522–524

N-Acetylglucosamine, and alkyl derivatives in active bifidus factor, 221, 222

N-Acetylhexosamine, in active bifidus factor, 221

Acquired pellicle, *see* Pellicle

Acrodermatitis enteropathica, effect of human milk on, 224

Adenyl cyclase
activity of, 364
and metabolic pathway in small intestine, 146
monitoring membrane events by, 30–32

Adipose tissue
effect of maltose on, 308
fructose, sorbitol and xylitol metabolism by, 243–247

Alanine
as ligand in sweetness studies, 27
relative sweetness of D- and DL-, 45

Alcohol dehydrogenase
activity regulation in *Streptococcus mutans*, 700–702
in metabolism of fructose and D-glyceraldehyde, 359, 360

Aldehyde dehydrogenase, in metabolism of fructose and D-glyceraldehyde, 359

Aldohexoses, configuration and conformation of naturally occurring, 68, 69

Aldolase, liver, in fructose metabolism, 359, 362

Aldopentoses, configuration and conformation of naturally occurring, 68, 69

Aldose reductase
in cataractogenesis, and inhibitors, 383–385
in diabetes, inhibitors of, 15, 477
in metabolism of fructose and D-glyceraldehyde, 359, 360
in peripheral and spinal nerves, 473
in polyol formation and deposition, 231
in sorbitol pathway, 470–472

1-Allyloxy-2-amino-4-nitrobenzene, relative sweetness of, 44

Amino acids
in cataract development, 377
catfish barbel binding of, 26
effect on dental plaque composition, 658
in galactose-exposed lens, 379, 380
as ligands in sweetness studies, 27

2-Amino-4-nitrophenol, relative sweetness of, 44

Ammonium N-cyclohexylsulfamate, relative sweetness of, 44

α-Amylase
assay of activity, 196
carbohydrate hydrolysis by, 190
specificity of, 193
starch hydrolysis by, 117, 147, 149, 150

γ-Amylase, in intestinal mucosa, 190

n-Amylchloromalonamide, relative sweetness of, 44

Amylo-1,6-glucosidase, deficiency, 344

Amylopectin, structure of, 147

Amylose, structure of, 147

Amylum, *see* Starch

Anemia, occurrence in Scandinavian countries, 89

749

L

α-Lactalbumin, in lactose synthetase, 285, 286
Lactase
  activity and deficiency, 201–203
  deficiency, in adults, 188
    congenital, 201, 203
    and lactose feeding, 181, 182
    and lactose intolerance, 188
    and milk consumption, 282–284
    and protein deficiency, 206
  effect of glucose and sucrose diets on jejunal activities of, 173–177
  localization in small intestine, 193
  in mammalian digestion, 282
  in small intestine mucosa, 194
Lactate dehydrogenase, activity regulation in *Streptococcus mutans*, 705, 706
Lactic acid
  from glucose fermentation in *Streptococcus faecalis*, 707
  production by *Streptococcus mutans* by glycolysis, 691
Lactitol
  as sweetener, 138
  patents, 142
Lactobacillus
  acidophilus, in dental caries, in animals, 636
    effect on intestinal flora, 216
    in oral biology, 14
  bifidus, flora, specific role of, 224
    growth factor, 218–222
    in intestinal flora of breast-fed and bottle-fed infants, 216–218
    var. *Pennsylvanicus,* 218, 222, 223
  bulgaricus, effect on intestinal flora, 216
  parabifidus, 217
Lactoferrin, in human milk, 225
Lacto-*N*-fructopentose, in active bifidus factor, 221
Lactose
  absorption by small intestine, 160
  biosynthesis, 285, 286
  catabolism, 281–285
  cataractogenic effects of, 375, 376
  consumption of, 190

in diet in United States, changes in, 98
  digestion by mammals, 282–284
  and flatus formation, 323
  intolerance, 204, 283, 284
    and lactase deficiency, 188
  lactase deficiency and feeding of, 181, 182
  malabsorption, in Eskimos, 207
  metabolism and research needs, 124
  naturally occurring in foods, 71
  relative sweetness of, 46
  tolerance test, 197
Lactose synthetase, composition of, 285, 286
Lactosuria, occurrence of, 160
Legumes
  role in history, 3
  sugars in, 73, 78
Leguminous seeds
  evolution in, 315
  and flatus formation, 313, 321, 323
  heat treatment in presence of water, effect on oligosaccharide content, 331
  oligosaccharides in, 316
Leloir pathway, and galactosemia, 289, 295
Levans
  and dental caries, 636, 669
  in dental plaque, 661, 670
Levoglucosan, reversion product from glucose, 73, 75, 79
Levulose, *see* Fructose
Levulose-containing syrups, production of, 119, 120
Lipids
  dietary, effect on cardiovascular disease, 13
  effect of dietary carbohydrates on serum, of hyperlipemic patients, 420–423
    of normolipemic subjects, 408–420
  effect of sugars and complex carbohydrates on serum, in experimental animals, 404–407
    in man, 407–408
Lipogenesis, hepatic enzymes in, 487–490
Liver
  effect of glucose syrup and maltose polysaccharides on, 307